This volume presents the results of a major international and interdisciplinary study of the mechanism and consequences of surface water acidification, which is a growing environmental problem. This detailed, long-term study provides an authoritative overview of the chemical process of acidification and of its biological impact on freshwater life. The surface waters acidification programme (SWAP) has drawn together for the first time the many facets of this complex phenomenon: the volume includes sections on hydrochemical studies in catchments, catchment process studies, catchment manipulation experiments, chemical processes, palaeolimnological studies, biological effects on freshwater life, and modelling studies. The volume finishes with an assessment of the SWAP and its conclusions. Though the results are presented in a predominantly European context, the findings and conclusions are applicable to all those areas suffering the consequences of acidification.

The volume will appeal to all those with an interest in acid rain and acidification: water-research scientists, environmental scientists and planners, foresters, hydrologists, freshwater ecologists and biologists. The papers presented here are the proceedings of the final SWAP conference held at the Royal Society in March 1990.

The surface waters acidification programme

The surface waters acidification programme

EDITED BY B. J. MASON

Director, UK–Scandinavian Surface Waters Acidification Programme, Centre for Environmental Technology, Imperial College of Science, Technology and Medicine, London

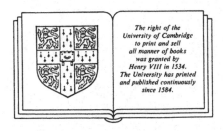

The right of the University of Cambridge to print and sell all manner of books was granted by Henry VIII in 1534. The University has printed and published continuously since 1584.

CAMBRIDGE UNIVERSITY PRESS

Cambridge
New York   Port Chester
Melbourne   Sydney

CAMBRIDGE UNIVERSITY PRESS
Cambridge, New York, Melbourne, Madrid, Cape Town, Singapore, São Paulo

Cambridge University Press
The Edinburgh Building, Cambridge CB2 2RU, UK

Published in the United States of America by Cambridge University Press, New York

www.cambridge.org
Information on this title: www.cambridge.org/9780521395335

First published 1990
This digitally printed first paperback version 2006

A catalogue record for this publication is available from the British Library

ISBN-13 978-0-521-39533-5 hardback
ISBN-10 0-521-39533-X hardback

ISBN-13 978-0-521-03167-7 paperback
ISBN-10 0-521-03167-2 paperback

CONTENTS

The rationale, design and management of the Surface Waters Acidification Programme

By Sir John Mason.

Centre for Environmental Technology, Imperial College of Science, Technology and Medicine, 48 Prince's Gardens, London, SW7 2PE, U.K.

1. Origin of the programme

The initial concept of the Surface Waters Acidification Programme (SWAP) was proposed by Sir Water Marshall (now Lord Marshall), Chairman of the Central Electricity Generating Board, in the Spring of 1983. At that time, the CEGB was being strongly criticized, especially in Norway and Sweden, on the grounds that the emissions of SO_2 and NO_x from its power stations were making a major contribution to the acidification of surface waters in southern Scandinavia. Following the researches done in Norway during 1972–80, the results of which were published in 1980 in a report entitled, *Acid precipitation – effects on forest and fish*, the CEGB had set in hand a major research programme of their own. Although the CEGB scientists became understandably rather defensive in the face of exaggerated statements that appeared in the media they, nevertheless, did a good deal of high-quality research and supported additional work in the universities. Even so, Marshall realized that however sound the work of his own organization might be, the results would be regarded with suspicion by its critics and he foresaw the need for a completely independent investigation, which he was willing to fund.

In March 1983 Marshall consulted me, as Treasurer of the Royal Society, as to whether the Society might be willing to coordinate and supervise a substantial five-year programme over which it would have full control, with no interference from the sponsor, once the objectives, scope and scale had been agreed. The Royal Society had never managed an operational research programme of this type and size, but I saw no reason, in principle, why it should not. Furthermore, we both agreed on the desirability of involving both the Norwegian Academy of Science and Letters and the Royal Swedish Academy of Sciences, as this would greatly facilitate much of the field work being conducted in these two countries, where concern over acid rain was most acute and where investigations and measurements had been made on a national basis for at least a decade.

I was personally strongly attracted to the idea of the Royal Society becoming actively involved in a major research project in close collaboration with foreign Academies, and had little difficulty in persuading my fellow Officers that the Society should accept the challenge and responsibility subject, of course, to agreement on the essential requirements of a completely independent investigation, including full publication of its results, 'irrespective of whether they would bring comfort or dismay to the energy consumer in this country.' Sir Walter actually volunteered these conditions and, furthermore, gave a commitment to take remedial action in light of the findings. Because such action might entail great costs, they had to be based on sound scientific knowledge and advice.

[1]

The objectives of the programme were subsequently drawn up and agreed, subject to endorsement by the Scandinavian Academies, in an exchange of two short letters between the CEGB and the Royal Society. The objectives were, and have remained, to find answers to the following questions.

1. In the affected areas of Norway and Sweden, what factors, in addition to pH, in practice, determine the fishery status of lakes?

2. What are the biological, chemical and hydrogeological characteristics of catchments that determine whether the composition of surface waters falls within a range acceptable to fish?

3. To what extent are these characteristics being adversely affected by the acid deposition itself?

4. What changes would be brought about in water chemistry and fishery status by given levels of reduction of man-made sulphur deposition?

At this stage, the Royal Society sought the participation of the Norwegian and Swedish Academies, while Marshall enquired of the National Coal Board, the main suppliers of the fuel largely responsible for the acidic emissions, whether they would wish to contribute to the funding of the Programme. Despite the financial problems of the Coal Board, Sir Ian MacGregor, the newly appointed Chairman, agreed to contribute half the total cost, with the memorable remark, 'Unlike Sir Walter, I'm committing money that I haven't got!'. Accordingly, it was agreed that a sum of £5 million, to be spent over a period of about five years, should be administered by the Royal Society who, in concert with the Scandinavian Academies, would have complete discretion as to how this would best be spent to fulfil the agreed objectives.

2. Management of the programme

The next step was to appoint a Programme Director. I was asked, and readily agreed, to accept the post on a half-time basis on my retirement from the Meteorological Office on 30 September 1983. An immediate start was made possible when Imperial College kindly offered me office accommodation in the Centre for Environmental Technology.

Meanwhile, the three Academies had decided to form a steering group to oversee the Programme, approve major expenditures and ensure that the planned research was in line with the agreed objectives. The first meeting, held at the Royal Society on 23 November 1983, confirmed the proposed objectives, adopted the title, Surface Waters Acidification Programme with the acronym SWAP, and established a formal Management Group consisting of four members appointed by the Royal Society and two from each of the Scandinavian Academies, the Director, and an Observer from each of the sponsoring bodies, with Sir Richard Southwood, F.R.S., as Chairman. It was expected to meet about twice a year. It was decided that commissioning of research, the judging of applications and the signing and monitoring of contracts should be done as expeditiously as possible, without elaborate reviewing procedures and with a minimum of bureaucracy. It was agreed that about 60% of the funds would be spent in Scandinavia and about 40% in the United Kingdom.

The Director was given authority to commission research and place contracts for projects approved by the Management Group. The Norwegian and Swedish Academies formed small consultative committees of senior scientists to screen and assess research proposals before submission to the Management Group, their recommendations being coordinated by Professor Hans-Martin Seip, who was appointed part-time consultant to the Director responsible for liaison with the Scandinavian research groups.

The first task of the Director was to assess the current state of knowledge, to identify the important gaps, and to outline a coherent, viable programme that would build upon on-going national programmes but have an identity of its own. This resulted in the publication of a review article by Mason & Seip (1985). These early planning stages involved visits to most of the laboratories and research groups in the three countries that had either participated in related research or had the potential to do so. One was thereby able, quite quickly, to identify those centres and senior scientists best equipped to undertake the required research and to invite them to submit proposals within the guidelines described below. The response was most encouraging, with many scientific groups expressing interest and enthusiasm for participating in a novel international experiment to tackle an important environmental problem that would bring together many different disciplines and organizations, and that promised to be unbureaucratic, if not entirely democratic!

The first contracts were placed in June 1984 and work started very soon thereafter. Research groups were requested to submit only a short annual report and reprints of any publications. A summary of these was prepared by the Director and submitted with his comments in an annual report to the Management Group. Once the programme was established, the Management Group found it necessary to meet only once a year.

3. The research strategy

The main goal of the research programme was to improve our knowledge and understanding of the physical, chemical and biological processes involved in the acidification of the streams and lakes and thereby make better predictions of future trends. The general strategy was to carry out detailed extensive investigations on a limited number of sites in the three countries that would allow comparisons between highly acidified, pristine and transitional catchments, and between forested and unforested sites. It was already known that broad correlation exists between the distribution and loading of acid deposition and the occurrence of acid lakes and streams. Moreover, strongly acidified waters with impoverished fish populations tend to occur in areas with hard granite rocks and thin soils having low cation-exchange capacity, whereas comparable levels of acid deposition cause little acidification if the soil is deep, and rich in calcium. There was also evidence to suggest that acidic deposition on a catchment can be appreciably influenced by trees and other vegetation, whereas the chemistry of the run-off into lakes and streams may be affected by changes in land use.

SWAP would therefore involve studies of catchment hydrogeology and vegetation, the chemical composition of the soils and underlying rocks, changes in the chemistry of rain/snow water as it percolates through the different soil layers, and measurements of the water flow through the soil to determine the residence times available for the chemical and biological processes to act. Changes in the chemistry of stream waters would be related to the stream flow, with particular attention to the high flows that follow heavy rains and melting snow and are accompanied by pulses of high acidity and aluminium that are especially toxic to fish. The last link in the chain would concern the effects of changes in water chemistry on fishes at various stages of their life history and on other aquatic animals in the food chain.

A comprehensive field programme, involving several different disciplines and

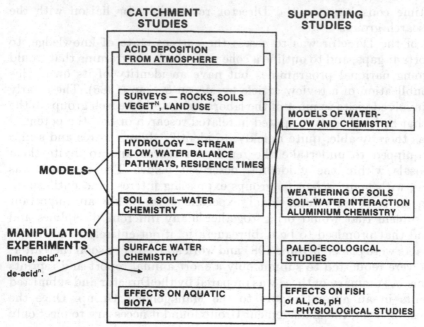

Figure 1. The integrated research programme.

research groups, and structured as shown in figure 1, would be established at as many of the sites as possible. This was largely achieved at the highly acidified sites of Birkenes in Norway, Loch Ard in Scotland, and Lake Gårdsjøn in Sweden, and also at the 'intermediate' Allt a'Mharcaidh catchment in the Cairngorms and at the pristine Høylandet site in northern Norway. Rather less comprehensive instrumentation and measurements were carried out at the other sites (shown in figure 1 of Harriman & Ferrier (this symposium) and figure 1 of Seip *et al.* (this symposium). Most of the key sites were inspected by members of the Management Group in the summer of 1984.

The most comprehensively instrumented site was the Allt a'Mharcaidh, the following description of which may perhaps serve to illustrate the experiment design, organization and implementation of a major field experiment. A more detailed account is given by Ferrier & Harriman in the subsequent paper.

4. The Allt a'Mharcaidh experiment

The catchment, located in the Cairngorm Mountains, is a hanging valley that drains an area of about 12 km², dropping from a height of about 1000 m to 300 m at its outlet into the River Feshie, a tributary of the River Spey. The upland plateau lies above 750 m and rises gently to about 1000 m and is largely snow-covered in winter. The valley floor, rising from 300 m to 700 m, carries the main Allt a'Mharcaidh stream and is covered by a thick layer of boulder clay and extensive peat deposits and moraine. The underlying rocks are entirely granite with little variation in structure and composition. Detailed surveys of the geology, soil and vegetation have been done by the Macaulay Institute.

One of the aims of the survey was to locate areas of uniform soils that could be used as experimental plots. These had to be near a stream and representative of the dominant topography and soil in that part of the catchment. The chosen experimental plots are marked P1, P2, P3 in figure 2 of the paper by Ferrier & Harriman (this symposium).

The Allt a'Mharcaidh stream contains salmon and trout but is low in calcium and inorganic acids, which makes it sensitive to changes in pH, especially during the early stages of snow-melt in the spring.

The main activities at this site, all of which were closely coordinated, were assigned as follows.

Measurements of acid deposition including tree throughfall, stem flow, interception by vegetation, etc.	MLURI
Stream flow and water-quality monitoring	IOH
Weather and rainfall observations	IOH, FFL
Snow surveys and snow-melt chemistry	IOH
Stream chemistry	FFL
Stream biology	FFL
Soil and soil-water chemistry	MLURI
Soil-water flow and pathways on hill slopes	ICL
Palaeolimnological studies	UCL

(IOH, Institute of Hydrology, Wallingford; FFL, Freshwater Fisheries Laboratory, Pitlochry; MLURI, Macaulay Land Use Research Institute, Aberdeen; ICL, Imperial College, London; UCL, University College, London (where Dr Battarbee coordinates the work of several university groups)).

Readings of steam-flow, pH, conductivity and temperature, were taken at 20 minute intervals at the outflow of the Mharcaidh stream, and on two incoming streams from the upper part of the catchment. Hourly rainfall totals were measured at six sites, and comprehensive hourly meteorological data are provided at two sites. Automatic snow-melt samplers were installed at three sites, and snow-depth surveys were made weekly and more frequently during snow-melt episodes. Chemical analyses of bulk precipitation, snow and melt-water for pH, alkalinity, TOC and all the main ionic species were done on a weekly basis, but much more frequently during storms and melting-snow episodes. On such occasions, automatic samplers were used to obtain 1–3 hour measurements of streamwater chemistry.

A series of borehole piezometers, installed along three profiles perpendicular to the main stream and extending some 200 m upslope, were monitored to determine the height of the water table during storm events, and hence the contribution of the shallow groundwater to the stream.

The structure and chemistry of the soils and the chemistry and flow of the percolating soil water were determined at the three experimental sites on peaty podzol, alpine podzol and peat, all close to a stream and under upland heath vegetation.

The soils were characterized by particle size and shape, packing, porosity and crystalline structure. Soils from different horizons were analysed for their major inorganic and organic constituents, for potentially toxic heavy metals, humic substances, organic acids, C/N ratios, etc. Soil moisture was measured at, and soil water extracted from, an array of points in each of three pits dug in each of the three sites. Soil-water potential (pressure) was measured by an array of some 20 porous cup tensiometers, which provide a three-dimensional representation of soil-water pressure from which flow directions can be deduced. They also help to identify areas of saturated, unsaturated, channel and macropore flow. The soil water was also extracted by simple lysimeters at three different horizon depths in each pit. Water flow from the lysimeters was monitored by using tilting-bucket collectors, and samples were collected for chemical analysis of all the major ionic species, pH, conductivity, alkalinity and TOC, every two weeks. Concentrations of water-soluble aluminium species were determined by the Driscoll method. Hourly sampling during episodes is provided by an automated system in one pit in each site. Probes to measure the concentration of CO_2 in the soils were installed on the two podzol sites.

The amounts of soluble and adsorbed sulphate in the soils have been determined. Sulphate adsorption appears to be an important parameter in determining stream-water chemistry. In organic soils, mineralization of organic sulphur can lead to substantial production of sulphate and enhance the store of man-made sulphate available for leaching.

The Imperial College group used the data from their own close network of tensiometers and from the IOH network of piezometers to study the water-flow paths on the plot and hill-slope scales, and to incorporate this understanding into both plot- and catchment-scale models.

The populations of trout and salmon were determined from electro-fishing surveys made over 30–50 m stretches of the stream at three sites in the upper, middle and lower part of the stream. The effect of changes in stream chemistry, especially during highly acidic episodes, on the hatching of salmon eggs, has been studied by Harriman *et al.* (this symposium). Sampling of invertebrates was also undertaken at regular intervals to determine seasonal and annual changes in numbers and diversity, and to relate these to changes in stream and water chemistry.

Sediment cores were taken from Lochnagar, which is some considerable distance from, and at greater altitude than, the Allt a'Mharcaidh stream, but it lies on similar granitic bedrock and is probably fairly representative of conditions at the head of the Mharcaidh catchment. This lake having low Ca concentration and low buffering capacity, has undergone progressive acidification, the pH having fallen from 5.7 to 4.8 since 1850.

Similar integrated investigations were done on the acidified Loch Ard sites and in the Norwegian and Swedish experimental catchments. These are described in later papers in this symposium.

5. Supporting studies

There was clear recognition that the field experiments would need to be supplemented by laboratory studies of key processes such as the chemical weathering of rocks and soils, the speciation and toxicity of aluminium, the role of organic acids in surface-water chemistry, and the physiological reaction of fishes to water quality. The results of such investigations are reported in several papers in this volume.

In view of the lack of a long time series of direct measurements, it was also decided to mount a major investigation into the history of lake acidification in the three countries, by using the palaeolimnological techniques developed notably by Battarbee in the U.K., Birks in Norway and Renberg in Sweden. Evidence for the progressive acidification of lakes since the industrial revolution comes from the analysis of acid-sensitive species of diatoms from radioactively dated lake sediments. The layers of sediments, laid down over hundreds of years, are dated by using radioactive lead isotopes and the diatoms found in the layers are correlated with similar populations found in the uppermost layers of lakes of known pH. It is thus possible to reconstruct the pH–age profile of the sediments. The technique has been employed extensively in SWAP, involving some 20 sites and 15 research groups. The results were presented at a conference held in the Royal Society in August 1989, and have been published recently in The Transactions of the Royal Society and in book form under the title, *Palaeolimnology and lake acidification* (Battarbee *et al.* 1990). The main results and conclusions are summarized by Battarbee, Birks and Renberg in the three separate contributions to this symposium.

Several groups in SWAP have exploited and further developed a mathematical model, the Norwegian Birkenes model, to simulate short-period changes in stream flow and chemistry, and to account for observed episodic and seasonal variations. We have also used the MAGIC model, developed in the U.S.A., to simulate long-term changes in surface water chemistry at several experimental sites, and to infer what changes may occur in the future in response to various assumptions about future levels of acid deposition. There was also an intention to use the output of these and other models to guide the experimental design and to help integrate and interpret the data. The main results of these modelling studies are described by Christophersen, Seip, Whitehead and Beck *et al.* (this symposium).

In addition, we accepted an invitation to participate in two Scandinavian manipulation experiments. One of these, the RAIN Project in Norway (see Wright, this symposium) seeks to study the effects of stream chemistry by either excluding acid deposition from a small catchment or by deliberately acidifying a small pristine catchment by spraying it with sulphuric and nitric acid. In the Lake Gårdsjøn catchment, one of the main national field sites in southwest Sweden, we have funded an experiment to study the changes in water chemistry produced by treating two forested catchments, one with elemental sulphur and the other with soluble sodium sulphate. The results will be reported by Hultberg *et al.* (this symposium).

6. Standardization and quality control of measurements

With so many institutions and research groups involved in SWAP, it was clear that it would be necessary to take special measures to ensure compatibility and direct comparability between the measurements, especially the chemical analyses, made by the different groups. Difficulties are particularly acute in the measurement of the pH of weakly buffered solutions and of aluminium speciation at very low concentrations.

Special workshops were therefore formed to establish standardized protocols and arrange for intercomparisons of methods, intercalibration of instruments, uniform methods of data analysis and quality control to be adopted by all participants. The special experimental and statistical procedures adopted by contributors to the analysis of diatoms in lake sediments are described in Battarbee *et al.* (1990).

Workshops were held to plan and coordinate the studies of chemical weathering of rocks and mineral soils and the measurement of aluminium species, and seminars were organized, at roughly yearly intervals, to coordinate and discuss the various modelling studies, often with the invited participation of American colleagues.

A 'mid-term' conference, attended by all the main participants, was held in Bergen in June 1987, during which the results to date were presented, discussed and assessed in relation to the original programme objectives. Even before this, the results of the work were sufficiently clear that Walter Marshall felt able to keep his promise and, in 1986, he recommended a programme of flue-gas desulphurization to the British Government. These actions and the conference results had an important bearing on the Government's response to the Brussels Large Plant Directive and its decision, in the Autumn of 1987, to authorize a larger flue-gas desulphurization programme for existing and all new power stations.

The SWAP has therefore been a unique research programme with important policy consequences. It may perhaps serve as a paradigm for future international collaborative projects requiring strong coordination and integration with a minimum of bureaucracy and administrative costs.

References

Battarbee, R. W., Mason, B. J., Renberg, I. & Talling, J. F. (eds) 1990 *Palaeolimnology and lake acidification* (*Phil. Trans. R. Soc. Lond.* B 327), 219 pp. London: The Royal Society.

Mason, B. J. & Seip, H. M. 1985 *Ambio* **14**, 45–51.

Pristine, transitional and acidified catchment studies in Scotland

COMPILED BY ROBERT C. FERRIER† AND RON HARRIMAN‡

1. Introduction

As part of the U.K.–Scandinavian Surface Waters Acidification Programme, catchment basins were selected to study the hydrological and hydrochemical consequences of surface water acidification. The purpose of this paper is to indicate why different catchments have been selected, their geographical location in relation to anthropogenic deposition loadings, and to describe in detail the nature of the collaborative research done and the equipment used.

Four main institutes were involved in the planning and execution of the research programme: they were; the Macaulay Land Use Research Institute (MLURI), the Freshwater Fisheries Laboratory (FFL), the Institute of Hydrology (IH) and Imperial College (IC). Other collaborating institutes were Stirling University (SU) and the Forth River Purification Board (FRPB). The identifiers shown in parentheses will be used throughout the text to identify the specific institute involved in any particular area of field research. The field research programme was initiated in late 1985 and continued until mid-1989 in all the Scottish sites.

Instrumentation and equipment for routine and experimental studies were designed to fulfil the four main research objectives.

1. What are the factors, in addition to pH, that determine the fisheries status of surface waters?

2. What are the biological, chemical and hydrological characteristics of catchments that determine whether the composition of surface waters falls within a range acceptable for fish?

3. To what extent are these characteristics being affected by acid deposition itself?

4. What changes would be brought about in water chemistry and fisheries status by given levels of reduction of man-made sulphur deposition?

Within the project timescale and resources it was impossible to address all these objectives in each catchment, therefore the following sections not only provide a summary of the catchment attributes at each site, but identify the key locations where process studies, leading to the development and improvement of long- and short-term predictive models, were completed.

2. Selection of study catchments

The research brief from the Management Group required a three tier separation of sites based on acidification status and pollution loading, i.e. pristine, transitional and acidified. The geographical location of the study catchments as classified under this regime is shown in figure 1.

Because the cleanest site in Scotland could only be classed as semi-pristine (excess

On behalf of: †Macaulay Land Use Research Institute, Aberdeen AB9 2QJ, Scotland, U.K. and ‡Freshwater Fisheries Laboratory, Pitlochry PH16 5LB, Scotland, U.K.; Institute of Hydrology, Wallingford, Oxon., OX10 8BB, U.K.; Imperial College of Science & Technology, London SW7 2BU, U.K.

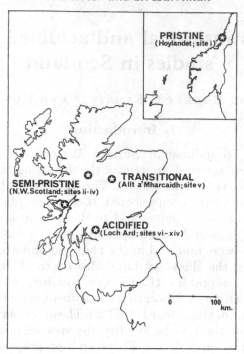

Figure 1. The geographical location of the SWAP study catchments.

Table 1. *Mean pH and excess sulphate* (µeq l⁻¹) *in bulk deposition during the study period (1986–88).*
The site classification on this basis is also shown

sites	bulk deposition pH	bulk deposition SO_4 (µeq l⁻¹)	project site classification
(i) Hoylandet	5.1	8	pristine
(ii) Allt na Graide	5.0	16 ⎫	
(iii) Coire nan Con	4.9	21 ⎬	semi-pristine
(iv) Choire Duibh	4.9	21 ⎭	
(v) Allt a'Mharcaidh	4.7	32	transitional
(vi) Corrie	4.5	43 ⎫	
(vii) Burn 6	4.5	43	
(viii) Burn 5	4.5	43	
(ix) Burn 2	4.5	43	
(x) Burn 14 (Chon)	4.5	43 ⎬	acidified
(xi) Burn 7	4.5	43	
(xii) Burn 10	4.5	43	
(xiii) Burn 11	4.5	43	
(xiv) Burn 9 (Kelty)	4.5	43 ⎭	

sulphate exceeding 15 µeq l⁻¹ in bulk deposition) a site at Hoylandet in mid-Norway was included to represent a truly pristine site (Seip, this symposium). Site classification on this basis is presented in table 1, along with the mean pH and excess sulphate levels during the study period (1986–88). At this stage it should be emphasized that these Scottish sites have been subject to a changing pollution climate during the past decade and the implications of this scenario are discussed in several of the following papers.

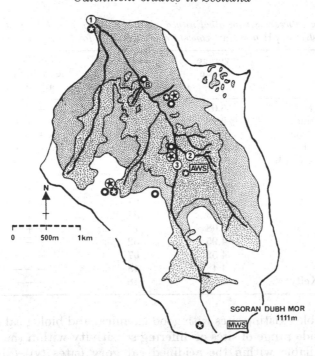

Figure 2. The nature and location of all the field instrumentation in the Allt a'Mharcaidh catchment, Cairngorms, N.E. Scotland; (◑), detailed plot studies; (①), gauging sites; (◐), input gauges; (✪), tensiometer networks; (Ⓑ), boreholes; (▢), alpine soils; (▦), peaty podzols; (▥), peats; MWS, mountain weather station; AWS, automatic weather station.

Table 2. *Summary of the edaphic and land-use characteristics of the study catchments*

sites	dominant geology	dominant soils	dominant vegetation	percentage forest
(i) Hoylandet	granitic	podzols, peats	F[b]	natural
(ii) Allt na Graide	schists	podzols, peats, gleys	F	95
(iii) Coire nan Con	schists	podzols, peats	F/M[c]	45
(iv) Choire Duibh	schists	podzols, peats	M	—
(v) Allt a'Mharcaidh	granitic	podzols, peats	M	—
(vi) Corrie	boundary fault[a]	peats	F	90
(vii) Burn 6	schists	podzols, peats	F/M	45
(viii) Burn 5	schists	podzols, peats	F	55
(ix) Burn 2	schists	podzols, peats	M	—
(x) Burn 14 (Chon)	schists	podzols, gleys	F	80
(xi) Burn 7	schists	podzols, peats	F	60
(xii) Burn 10	schists	gleys, peats	F	95
(xiii) Burn 11	schists	gleys, peats	F	90
(xiv) Burn 9 (Kelty)	schists	gleys, peats	F	95

[a] Predominantly schists with doloritic intrusions.
[b] Forest.
[c] Moorland.

Table 3. *The relation between stream alkalinity during low, mean and high flow and fish survival.*
Average stream pH *and* Ca^{2+} *concentration* (μeq l^{-1}) *for the study catchments*

sites	average stream pH	average stream Ca^{2+} (μeq l^{-1})	low flow/ mean/high flow alkalinity	fish present
(i) Hoylandet	6.01	40	+ + +	yes
(ii) Allt na Graide	6.16	113	+ + +	yes
(iii) Coire nan Con	5.78	52	+ + +	yes
(iv) Choire Duibh	6.24	70	+ + +	yes
(v) Allt a'Mharcaidh	6.49	40	+ + −	yes
(vi) Corrie	6.57	179	+ + +	yes
(vii) Burn 6	6.08	77	+ + −	yes
(viii) Burn 5	5.47	67	+ + −	yes
(ix) Burn 2	5.43	54	+ + −	yes
(x) Burn 14 (Chon)	5.08	72	+ + −	yes
(xi) Burn 7	4.95	52	+ − −	no
(xii) Burn 10	4.52	57	− − −	no
(xiii) Burn 11	4.48	48	− − −	no
(xiv) Burn 9 (Kelty)	4.47	48	− − −	no

Wherever possible existing sites with good chemical and biological databases were selected, thus a wide range of sites of differing sensitivity within the same pollution climate were available within the acidified category (sites (vi)–(xiv), tables 1–3) (Harriman & Morrison 1982). Only one transitional site was studied but this was the most extensively investigated (site (v)). The nature and location of all the field instrumentation at this site is shown in figure 2. The four remaining sites (sites (i)–(iv)) covered a range of sensitivities within the pristine/semi-pristine category. A summary of the edaphic and land-use characteristics of the study catchments is shown in table 2, and reveals almost uniform geology and soil types but with a range of vegetation characteristics.

Although the term transitional was originally envisaged to indicate a catchment with intermediate sulphate loadings along a particular sulphate gradient it was also used to describe the chemical properties of each catchment in terms of alkalinity status during periods of high-, mean- and low-flow conditions. Therefore, catchments (v), (vii), (viii), (ix) and (x) were defined as transitional based on their positive alkalinity status during mean flow conditions and negative alkalinity during high-flow conditions. These were identified as key sites in determining the relation between water quality and fish survival (table 3). Apart from sites (ii) and (vi) the catchment sensitivity range is quite small as reflected in mean stream calcium levels (40–80 μeq l^{-1}).

Of the four most intensively studied catchments (table 4), the pristine Hoylandet catchment in Norway has been described in detail elsewhere (Seip, this symposium) and the following site descriptions of the study catchments will be confined to the Allt a'Mharcaidh in the Cairngorms and two stream systems in the Loch Ard area of West Central Scotland.

3. Allt a'Mharcaidh (Cairngorms, N.E. Scotland)

The study site lies on the western edge of the Cairngorm Mountains, draining an area of 10 km into the River Feshie, which is a tributary of the River Spey. The mean

Table 4. *Summary of equipment installations at the study catchments*
(X, measured; E, estimated)

sites	inputs	vegetation through-puts	soil through-puts	out-puts	event sampling	weather-station data	soil moisture measurement	borehole sampling	soil/vegetation survey	laboratory-based experimentation	fish/invertebrate survey
(i) Hoylandet	X	X	X	E	X	—	—	—	X	X	X
(ii) Allt na Graide	X	—	—	X	—	—	—	—	—	—	X
(iii) Coire nan Con	X	—	—	X	—	—	—	—	—	—	X
(iv) Choire Duibh	X	—	—	X	—	—	—	—	—	—	X
(v) Allt a'Mharcaidh	E	X	X	E	X	X	X	X	X	X	X
(vi) Corrie	E	—	—	X	—	—	—	—	—	—	X
(vii) Burn 6	E	—	—	E	—	—	—	—	—	—	X
(viii) Burn 5	E	—	—	E	—	—	—	—	—	—	X
(ix) Burn 2	X	X	X	X	X	—	—	—	—	—	X
(x) Burn 14 (Chon)	E	—	—	E	—	—	X	—	X	X	X
(xi) Burn 7	X	—	—	X	—	—	—	—	—	X	X
(xii) Burn 10	X	—	—	X	—	—	—	—	—	—	X
(xiii) Burn 11	X	—	—	X	—	—	—	—	—	—	X
(xiv) Burn 9 (Kelty)	X	X	X	X	X	—	X	—	X	X	X

R. C. Ferrier and R. Harriman

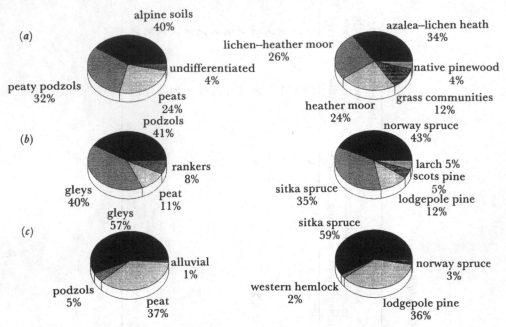

Figure 3. The component soils and dominant vegetation of the (a) Allt a'Mharcaidh
(b) Chon and (c) Kelty catchments.

annual rainfall is approximately 1000 mm. The catchment consists of three broad
geomorphological regions; (i) the valley floor, (ii) the valley sides and (iii) the upland
plateau. The main stream (Allt a'Mharcaidh) leaves the catchment at a height of
320 m and the highest point in the catchment (1111 m) is the summit of Sgoran Dubh
Mor (NH 905002). The valley floor extends from 320 m to around 550 m and the
valley sides to approximately 700 m, followed by a gentle rising broad plateau.

The catchment is underlain by intrusive biotite–granite of Lower Red Sandstone
age associated with the late stages of the Caledonian Orogeny. Thick deposits of
boulder clay, derived from local rock, cover much of the valley floor. The catchment
soils were investigated in 1985 and classified according to the method used by the Soil
Survey of Scotland (1984). Once the various soil types on the catchment were
identified, samples were taken from selected soil profiles for mechanical and chemical
characterization. Fifteen soil map units were initially identified, these falling into
three broad categories; alpine soils (including alpine podzols, alpine rankers and
alpine gleyed soils); peaty podzolic soils (peaty podzols, humus–iron podzols and
peaty rankers); peats (shallow peats to deep blanket peat (1 m)) (Nolan et al. 1985).
The vegetation of the Allt a'Mharcaidh can also be described in relation to the three
major landform units. At the lower altitudes and on the gently sloping topography,
the dominant community is northern blanket bog with a range of associated flush
communities. On the catchment walls the vegetation is that of lichen-rich boreal
heather moor, with the upper plateau colonized by alpine azalea–lichen heath and
fescue-wooly fringe-moss heath. The component soils and dominant vegetation of the
Allt a'Mharcaidh are summarized in figure 3a.

4. Loch Ard forest (West Central Scotland)

Two stream systems were selected for plot-based studies in the Loch Ard area, which receives a mean annual rainfall of approximately 2000 mm. The first forms part of the Kelty Water (NS 455960), approximately 35 km west of Stirling, the second lies 12 km northwest of the Kelty Water and flows into Loch Chon (NN 423055). These two systems were selected because both have broadly similar vegetation and geology (table 2) yet differ in their capacity to support fish populations (table 3). The component soils and dominant vegetation types of the Chon and Kelty catchments are summarized in figure 3*b*, *c*, respectively.

(i) *Kelty Water* (Burn 9)

The upper reaches of the catchment were selected for study, draining an area of approximately 1.32 km². The altitude ranges from 175 m in the east to 505 m in the west of the catchment. The soils comprise 37% deep and shallow peats, 57% peaty gleys, and 5% podzols (total of eight map units identified), and the underlying geology is primarily mica-schist from the Dalradian series. Maturing Sitka spruce (30–35 years old) in the lower reaches cover 25% and newly planted Sitka spruce (3 years) in the upper reaches account for a further 35% of the vegetation. About 5% of the catchment area is unforested (Hudson & Hipkin 1985).

(ii) *Chon* (Burn 14)

The catchment is roughly circular, the highest point being 392 m and the downstream limit of the catchment is at NN 413073 at an altitude of 122 m. Thirteen soil map units were initially identified and the dominant soils of this catchment are humus–iron podzols and brown forest soils, which account for 41% of the total area. Gleyed soils cover a further 40%, with some peats (11%), and rankers (8%) cover the higher level and steeply sloping areas. The main plantation crops are Norway spruce on the podzolic soils and Sitka spruce in areas where gleyed soils have developed. Approximately 20% of the catchment is unforested. Again the underlying geology is composed of mica-schists from the Dalradian series (Henderson & Campbell 1985).

5. Instrumentation and equipment

Input water quality and quantity were monitored at the study sites by using two types of gauges. Bulk deposition was measured by using Nipher-shielded rain gauges (MLURI), which consist of a collector housed within an aerodynamic screen to minimize wind turbulence, in addition to measurements from standard Rimco 12″ ground-level collectors (IH). An assessment of the enhancement of deposition through the interception of cloud or mist water utilized filter gauges comprising a double cylinder of polythene-coated wire mesh mounted vertically above collecting funnels (MLURI) (Miller *et al.* 1990). Hydrological data on varying time-bases were provided by logging systems connected to tipping-bucket gauges. Snow hydro-chemistry (where appropriate) was monitored by using remotely controlled *in situ* daily samplers (IH, MLURI) and regular snow depth survey and profile analysis (SU, IH). Rime formation and chemistry were regularly sampled by using polypropylene deposition plates (MLURI, IH).

A variety of vegetation types was studied, requiring different collectors depending on the type, size and distribution of the vegetation. Discrete throughfall and stemflow were collected from below forest stands and combined throughfall plus stemflow from below ground and shrub vegetation (MLURI). Surface flow, litter and subsurface flow from selected soil horizons were collected by installing permanent soil throughflow collectors into pit faces on suitable hillslopes. Supplementary information on unsaturated soil water utilized tension lysimeters that were installed adjacent to the throughflow pits. As well as collecting water for chemical analysis, water flux through the vegetation and soil was monitored by using tipping-bucket gauges connected to digital logging recorders (MLURI) (Miller *et al.* 1990). Detailed soil-water hydrological data for the hillslope based studies were obtained from automatic three-dimensional tensiometer array systems. In the experimental plots an array of ceramic-cup tensiometers was connected to a Scanivalve wafer switch, which was stepped every 2.5 minutes to connect in turn the tensiometers and reference heads to a single pressure transducer. Transducer output was recorded onto a data logger (IC) (Wheater *et al.* 1987).

Stream pH, temperature and electrical conductivity were monitored continuously by using pHOX 100 DPM monitoring systems; pH was automatically compensated for stream temperature. Stream level was monitored by using two pressure transducers installed in a stilling well. These were accurate to ±1 mm, and are unaffected by changes in stream temperature. Stream level was related to discharge through a rating equation constructed by dilution gauging techniques. All continuously monitored variables were logged at 20 minute intervals (IH, FRPB) (Jenkins 1989). Automatic liquid samplers (QS1500 Quality Environment Ltd) were used to collect composite water samples for chemical analysis on a flow-related basis (FFL, MLURI). A discussion of the collecting characteristics of these samplers as compared with dip samples is given by Harriman *et al.* (1987).

During storm or snow-melt events, water samples of inputs (bulk deposition and interception), vegetation and soil throughflow, and streamwater were collected (IH, FFL, MLURI) usually on an hourly or three-hourly interval with input, soil and streamwater discharge being continuously recorded (IH, MLURI, IC). At one site (Allt a'Mharcaidh), the monitoring network was expanded to include a series of borehole piezometer profiles running perpendicular to the mainstream and extending upslope. These were manually monitored for water-table height during storm events and also pumped out for chemical analysis routinely and during events (IH).

Meteorological data were collected at 5 minute intervals at an automatic weather station, monitoring wind run and direction, incoming and net radiation, wet and dry bulb temperature and rainfall. The mountain weather station monitored humidity, incoming and reflected radiation, temperature, wind run and direction (IH).

Many of the study sites underwent extensive soil and vegetation surveys to identify type, cover and distribution (MLURI) (Nolan *et al.* 1985; Hudson & Hipkin 1985; Henderson & Campbell 1985) and all the stream systems studied were assessed for salmonid distribution and abundance and invertebrate classification (FFL). A summary of all equipment installations at the various study catchments is presented in table 4.

6. Chemical and biological methodology

Samples returned to the laboratory were filtered through 0.45 μm filters. Analyses were done in duplicate along with control standards used for quality control schemes with expected standard deviation of less than 5%. Ammonium-N and nitrate-N were determined colorimetrically by salicylate-dichloroisocyanurate nitroprusside and reduction to nitrite followed by diazotization, respectively. Potassium and sodium were analysed by flame emission and calcium and magnesium by atomic absorption spectrophotometry. Chloride and sulphate were measured by ion chromatography and pH by using a remote KCl electrode system. Dissolved organic carbon was measured by ultraviolet persulphate digestion. Measurement of total monomeric-aluminium was based on the pyrocatechol violet method and labile and non-labile fractions were determined after ion-exchange separation. Other chemical analyses pertinent to specific research objectives will be considered in detail elsewhere.

After air-drying and sieving to less than 2 mm, or in the case of organic soils, milling to less than 1 mm, soils were analysed by standard methods (Soil Conservation Service 1972; Sheldrick 1984) to give routine soil analytical data, and by non-standard techniques, as noted elsewhere (Walker *et al.*, this symposium).

The fish populations were estimated by repeated electrofishing and removal of fish from within an area of stream (normally 100 m) enclosed by stop-nets. This removal method allows an estimate of the population density to be calculated by using a formula devised by Zippin (1958). Fork length and weight of the fish were recorded and a sample of scales for age determination was taken from each fish before returning it to the stream.

7. Summary

The experimental catchment selection and the nature of the data collected have fulfilled all of the original programme objectives. The geographical location of the various catchments in areas subject to different pollution loadings allows for regional assessment of the effects of anthropogenic deposition. The original site selection to include catchments with existing biological and chemical data bases has elucidated time trends within the collected data and allowed an evaluation of hydrochemical response under a changing pollution environment. The identification of processes within the catchments has improved the application of long- and short-term models to specific sites and process studies have also played a key role in model improvement and validation allowing for application to other sites, outside the scope of the programme.

References

Harriman, R. & Morrison, B. R. S. 1982 *Hydrobiologia* **88**, 251–263.

Harriman, R., Wells, D. E., Gillespie, E., King, D., Taylor, E. & Watt, A. W. 1987 In *SWAP Mid-term Conference, Bergen, Norway.*

Henderson, D. J. & Campbell, C. G. B. 1985 *The Surface Water Acidification Programme; soil survey of the North Loch Chon catchment.* Aberdeen: Macaulay Land Use Research Institute. (Restricted circulation.)

Hudson, G. & Hipkin, J. A. 1985 *The Surface Water Acidification Programme: soil survey of the Kelty Water catchment.* Aberdeen: Macaulay Land Use Research Institute. (Restricted circulation.)

Jenkins, A. 1989 *Hydrol. Sci. J.* **34**, 393–404.

Miller, J. D., Stuart, A. W. & Gaskin, G. J. 1990 *Br. geomorph. Res. Bull.* **29**. (In the press.)

Nolan, A., Lilly, A. & Robertson, J. S. 1985 *The Surface Water Acidification Programme: the soils of the Allt a'Mharcaidh catchment.* Aberdeen: Macaulay Land Use Research Institute. (Restricted circulation.)

Sheldrick, B. H. (ed.) 1984 *Analytical methods manual, 1984.* Ottawa, Ontario: Land Resource Research Institute.

Soil Conservation Service 1972 *Soil survey laboratory methods and procedures for collecting soil samples.* Washington, DC.: U.S.D.A.

Soil Survey of Scotland 1984 Organisation and methods: soil and land capability for agriculture. 1–250000 soil survey. Aberdeen: Macaulay Land Use Research Institute.

Wheater, M. S., Langan, S. J., Miller, J. D. & Ferrier, R. C. 1987 In *Forest hydrology and watershed management*. IAHS-AISH publn no. 167.

Zippin, C. 1958 *Wildl. Mgmnt J.* **22**, 82–90.

Hydrochemical studies in Scandinavian catchments

By Hans M. Seip[1], Inggard A. Blakar[2], Nils Christophersen[3], Harald Grip[4] and Rolf D. Vogt[1]

[1] Department of Chemistry, University of Oslo, P.O. Box 1033 Blindern, 0315 Oslo 3, Norway
[2] Norwegian Institute for Nature Research, Tungasletta 2, N-7004 Trondheim, Norway
[3] Center for Industrial Research, P.O. Box 124 Blindern, 0314 Oslo 3, Norway
[4] Department of Forest Site Research, Swedish University of Agricultural Sciences, S-90183 Umeå, Sweden

The research activities at the three Norwegian and one Swedish SWAP field sites are described: Birkenes (southernmost Norway), Atna (central Norway), Høylandet (central Norway), and Svartberget (northeastern Sweden). The sites are to various degrees exposed to anthropogenic pollutants. Birkenes has the highest S-deposition (about $1.7\,\mathrm{g\,S\,m^{-2}\,a^{-1}}$) and the most acid streamwater; pH reaches as low as 4.2 and inorganic, monomeric aluminium (Al_i) may exceed 20 µM during highflow periods. The lowest pH values and highest Al_i values (in micromoles) in streamwater at the other sites are approximately: 4.7 and 5 (Atna); 4.8 and 3 (Høylandet); 4.2 and 2 (Svartberget).

Acidification studies started in Birkenes in 1972. Long-term streamwater data of major ions indicate that soil acidification has occurred. A decline in sulphate concentrations has been observed in precipitation and streamwater since 1985. The other sites have shorter data records and concentration trends cannot be detected. The high sulphate deposition at Birkenes is, most likely, the major cause of the acid, aluminium rich streamwater in the catchment. Organic acids play an important role for the acid episodes at Svartberget. Studies of precipitation, stream and soil water at these sites have provided important information about the mechanisms which determine water chemistry. Several other studies, e.g. on fish populations, acidification history, land-use changes, flow paths, and soil comparisons between catchments, have also been done.

1. Introduction

The main goal of the Surface Waters Acidification Programme (SWAP) field studies was to improve our understanding of acidification mechanisms and thereby make better predictions of future trends possible. Because long data series are imperative in this connection, sites with ongoing studies were preferred. Three field sites were selected in Norway: Birkenes, Atna, and Høylandet and one in Sweden (Svartberget) (figure 1). All the sites are susceptible to surface water acidification. The Norwegian catchments were selected to enable comparison of sites along a gradient with respect to air-borne pollutants. Birkenes is in the area of maximum sulphur deposition in Norway (about $1.7\,\mathrm{g\,S\,m^{-2}\,a^{-1}}$, including dry deposition). The deposition at Atna is much less (about $0.25\,\mathrm{g\,S\,m^{-2}\,a^{-1}}$) because of lower sulphate concentration in precipitation, less precipitation and less dry deposition. At Høylandet the deposition

Figure 1. SWAP field sites in Scandinavia. The studies discussed in this paper include the sites Birkenes (two catchments, Birkenes I and Birkenes II), Atna (the Storbekken catchment), Høylandet (the Ingabekken catchment), and Svartberget.

Table 1. *Characterization of Scandinavian SWAP field sites*

	Birkenes I	Atna (Storbekken)	Høylandet (Ingabekken)	Svartberget
area of catchment studied/km²	0.41	5.80	0.21	0.50
altitude/m	205–300	780–1537	280–370	235–310
bedrock	granite	quartzite	granite	gneiss, schist
precipitation/mm	1500	550	1300	710
excess sulphate in precipitate/(μeq l^{-1})	55	29	9	40
approximate pH range stream	4.2–5.3	4.7–5.5	4.8–6.6	4.2–6.8
characterization	acidified	transitional	pristine	transitional

is about $0.35\,\mathrm{g\,S\,m^{-2}\,a^{-1}}$. However, the sulphate concentration in precipitation is quite low and more than half of the sulphate is from sea-spray. The site can therefore be regarded as nearly unaffected by air-borne pollutants. Svartberget has also a fairly low S-deposition (about $0.6\,\mathrm{g\,S\,m^{-2}\,a^{-1}}$). However, the surface waters at Svartberget are periodically quite acid, mainly due to organic acids. Some key characteristics for the sites are given in table 1. The Birkenes site comprises two catchments, Birkenes I, where most of the studies have been done, and another nearby catchment (Birkenes II) with somewhat higher stream pH.

Table 2. *Ionic concentrations* (μeq l^{-1}) *in precipitation at Scandinavian SWAP sites*

	Birkenes[a]	Atna[b]	Høylandet[c]	Svartberget[d]
pH	4.28	4.7	4.99	4.44
Ca^{2+}	9	2	7	9
Mg^{2+}	13	1	16	2
Na^+	49	3	66	5
NH_4^+	41	12	17	18
Cl^-	57	3	76	10
NO_3^-	38	11	4	19
SO_4^{2-}	61	30	17	52
SO_4^{2-} (excess)	55	29	9	51

[a] Volume weighted averages for 1981–83 + 1985–88 (SFT 1989).
[b] Mean values for five stations, 1986–88.
[c] NH_4^+ and NO_3^- have been measured close to the Ingabekken catchment from September 1986 to December 1987. The other data are for samples collected at Høylandet about 5 km from Ingabekken for the period 1987–88.
[d] From Grip & Bishop (this symposium).

Several other papers in this volume deal with studies in these catchments: Andersen *et al.*, Blakar *et al.*, Bishop & Grip, Christophersen *et al.*, Grip & Bishop, Stone & Seip, Stone *et al.*, Taugbøl *et al.*, and Vogt *et al.*, all this symposium.

2. Birkenes I

Studies in the Birkenes I catchment were initiated as a part of the Norwegian SNSF Project (Acid precipitation – effects on forest and fish) in 1972 (Overrein *et al.* 1980). Recent papers include: Christophersen *et al.* 1990*a,b*, this symposium; Mulder *et al.* 1990; Seip *et al.* 1989; Stone & Seip, 1989; Sullivan *et al.* 1986; Vogt *et al.* 1990, this symposium. The site, which is covered by spruce forest and has shallow soils, is part of the Norwegian monitoring programme for acidification of precipitation, water and soils. In this programme, streamwater is sampled weekly by the Norwegian Institute for Water Research (NIVA). Precipitation chemistry and air concentrations of SO_2 and SO_4^{2-} are measured by the Norwegian Institute for Air Research (NILU). Birkenes is also included in a forest monitoring programme conducted by the Norwegian Institute for Forest Research (NISK). Throughfall and litterfall are collected weekly.

Volume-weighted concentrations in precipitation are given in table 2 and concentrations in streamwater for the main brook (BIE01) in table 3. The input and output of sulphur are roughly in balance. The ratio [Na^+]/[Cl^-] in streamwater is 0.92, close to that in seawater (0.86). Several tributaries have also been sampled periodically. Lysimeters have been installed for soil water sampling (Seip *et al.* 1989; Vogt *et al.* 1990, this symposium).

Concentrations of major ions observed in streamwater at Birkenes during the period 1972–88 have been analysed statistically by Christophersen *et al.* (1990*a*), cf. figure 2. They found a decreasing trend in [Ca^{2+}] and [Mg^{2+}]. Sulphate concentrations have declined only since 1985. The change in Ca^{2+} and Mg^{2+}, at least up to 1985, was ascribed to soil acidification. Decreasing sulphate concentrations in precipitation and surface waters have also been observed at other sites in southern Norway (SFT 1989).

Table 3. *Concentrations (mean or approximate range) for streamwater at the SWAP sites.*

(The unit is μeq l^{-1} except for pH, TOC (mg l^{-1}), Al$_i$ (μM) and Al$_o$ (μM). Note that averages for Birkenes II are biased towards the highflow composition due to over-representation of episodic samples).

	Birkenes I[a] (BIE01)	Birkenes II[b]	Atna[c] (Storbekken)	Høylandet[d] (Ingabekken)	Svartberget[e]
pH	4.56	5.0	5.2	5.1	4.7
Ca^{2+}	51	56	11	23	100
Mg^{2+}	31	35	6	34	54
K$^+$	5	8	7	6	10
Na$^+$	109	78	9	118	53
Cl$^-$	118	81	71	152	21
NO$_3^-$	9	13	2	3	1
SO$_4^{2-}$	122	85	30	29	105
Al$_i$	4–24	1–13	0.3–5	< 0.2–3	0.5–2
Al$_o$	2–9	1–10	—	< 0.2–5	1–33
TOC	3–11	1–7	< 1–4	2–13	5–30

[a] Volume-weighted averages for (1981–83)+(1985–88) (SFT 1989). Al$_i$ Al$_o$ and TOC are from episode studies 1984–88.

[b] Volume-weighted averages for four locations along the main brook.

[c] Averages for 1986–88.

[d] Volume-weighted averages for 1986–88 (Mainly from Chrisophersen *et al.* 1990*b*).

[e] Volume-weighted averages for June 1985–June 1989 (Grip & Bishop, this symposium). Al$_i$ from episodic studies 1988–89 (Bishop & Grip, this symposium).

Episodic studies at Birkenes have played an important role within SWAP. One summer, six autumn and three spring seasons have been followed closely by frequent sampling of stream and soil waters with particular emphasis on H$^+$ and aluminium species. The concentrations of these species in water from soil lysimeters show considerable spatial variations, but surprisingly small temporal variations except in the upper horizon despite sampling under widely varying hydrological conditions (Sullivan *et al.* 1986; Seip *et al.* 1989; Vogt *et al.* 1990). The studies have clearly shown that short-term variations in [H$^+$] and aluminium do not correspond to gibbsite equilibrium (i.e. the equation [H$^+$]$^{-3}$[Al^{3+}] = constant, is not satisfied). This is illustrated in figure 3, showing saturation indices for some minerals. Patterns in concentrations of H$^+$ and Al-species in three tributaries are remarkably similar to that found in the main brook, but the actual levels differ to some extent. This seems to indicate that similar processes are important despite considerable differences between the sub-catchments.

To obtain information on water residence times and flowpaths, samples of precipitation, meltwater, stream and soil water have been analysed for ^{18}O/^{16}O ratios, and piezometers and tensiometers have been installed to measure groundwater table and soil-water potential (Sullivan *et al.* 1987; Christophersen *et al.*, this symposium). The studies strongly indicate that short-term stream chemistry is primarily determined by changes in water pathways (cf. Rosenqvist 1978). At high flow, more water flows through upper, acid soil horizons, which apparently contain less easily mobilized aluminium. Further discussions of the results are given in Sullivan *et al.* 1986; Seip *et al.* 1989; Mulder *et al.* 1990; Christophersen *et al.*, this symposium; Vogt *et al.* 1990, this symposium).

Palaeolimnological studies have been carried out on sediments from Lake

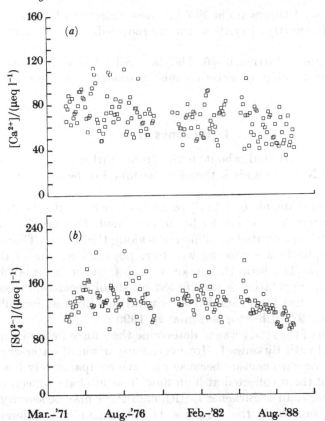

Figure 2. Monthly averages of (a) calcium and (b) sulphate concentrations in streamwater at Birkenes for the period July 1972–November 1988 (data from NIVA).

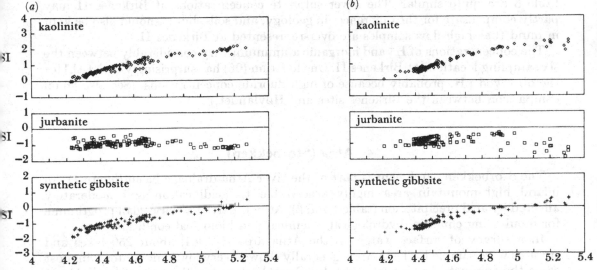

Figure 3. Saturation indices (SI) for Birkenes I streamwater samples; SI is the logarithm of the ratio of the ion product to the solubility constant, i.e. for gibbsite $SI = 3pH - pAl^{3+} - \log K$. (a) Autumn; (b) spring.

Verevatn (Berge *et al.* 1990) near the SWAP sites. Inferred pH was close to 5.0 for many centuries until shortly after 1900. During recent decades the lake has acidified to about 4.4.

Soil data are discussed further in §6. The data collected at Birkenes I have been used extensively for developing water-chemistry models (see Christophersen *et al.*, this symposium).

3. Birkenes II

This catchment is situated about 6 km from Birkenes I. The atmospheric deposition of S and N compounds is therefore assumed to be nearly the same as for Birkenes I.

The Birkenes II catchment (0.30 km^2) consists of two steep subcatchments that merge in a valley where a second-order brook is formed. This brook is disturbed by beaver ponds, which have created small marshes along the stream. The stream enters a lake that still supports a sparse brown-trout population; one of the very few remaining in the area. The bedrock is mainly gneiss, richer in amphibole minerals than Birkenes I. The vegetation is dominated by small deciduous trees of various species. Streamwater has been sampled at several locations at irregular intervals since the autumn of 1985 (Christophersen *et al.* 1990*b*).

The main purpose of the study was to determine the causes for differences in water chemistry compared with Birkenes I. However, comparison of water-chemistry data (table 3) must be done with caution because we have comparatively few samples for Birkenes II, most of them collected at high flow. The sulphate concentration seems to be somewhat lower than at Birkenes I. Although there may be less dry deposition, there being fewer conifers in the Birkenes II catchment, the difference in total S-deposition is likely to be small. It is therefore possible that some sulphate reduction occurs in the catchment. Considering the differences in geology one might expect [Ca^{2+}] to be higher at Birkenes II than at Birkenes I. The average values given in table 3 are quite similar. The lower sulphate concentrations at Birkenes II may partly compensate for the differences in geology, and soils, but it should also be kept in mind that high-flow samples are over-represented at Birkenes II.

The concentrations of H$^+$ and inorganic aluminium vary considerably between the six sampling locations at Birkenes II. One location (06) has surprisingly high [Al$_i$] for the observed pH, probably because of high fluoride concentrations. (See also §6 on comparison between the Birkenes sites and Høylandet.)

4. Atna (Storbekken)

The Storbekken catchment is part of the river Atna drainage basin (figure 1), an inland high-mountain area highly susceptible to acidification and moderately affected by acid precipitation (table 1 and 2). Atna is part of a Norwegian programme for monitoring climatic, hydrological, chemical and biological conditions.

In a survey of surface waters in the Atna area in 1984, about 250 lakes and streams were sampled. The samples generally showed very low ionic content, in most cases the conductivity was less than 13 µS cm^{-1}, often less than 10 µS cm^{-1}. About 37 % of the samples showed pH ⩽ 5.5. Survey of the main river (Atna) with tributaries and small headwater lakes did not indicate significant loss of fish populations (Hesthagen *et al.* 1989; J. Skurdal, personal communication). A preliminary study of diatoms in sediment cores from two lakes does not show recent acidification. Because few diatom species were represented in the study, results may not be very accurate (F. Berge, unpublished results).

Dahl studied soils from this mountain area during the period 1942–49. In 1984 he relocated many of the sample plots and compared pH of the various soil layers between the two periods. Statistically significant acidification was found; pH decreases ranged from 0.31 in the A0 layer to 0.88 in the C layer (Dahl 1988). He concluded that the only factor that can reasonably explain the observed acidification is the leaching of nutrients due to acid deposition.

A tributary of Atna (Storbekken) was chosen for intensive studies (Blakar *et al.*, this symposium). The Storbekken catchment has a sparse pine forest in its lower regions; a zone with birch then follows, but about 90% of the catchment is alpine. The catchment is mainly covered by glacial till of variable thickness with very thin topsoil. Annual mean precipitation (at 730 m a.s.l. (above sea level)) is about 550 mm.

Water samples from Storbekken have been analysed since 1983. More frequent sampling started in spring 1986. Approximately 250 streamwater samples have been collected annually during the period 1986–88. Snow and rainwater have also been collected. All samples have been analysed for chemical components (tables 2 and 3). Discharge and water temperatures have been measured.

In comparing streamwater chemistry in Storbekken with the other sites, we must again bear in mind differences in sampling periods and the number of samples. The very low ionic content in Storbekken is, however, evident (conductivity 5–13 μS cm^{-1}). As found in the other areas, pH drops during high flow (see fig. 2 in Blakar *et al.*, this symposium). The pH at peak discharge is similar to that observed at Høylandet and considerably higher than at Birkenes (table 1). With the very low sum of mobile anions (Cl^-, NO_3^-, SO_4^{2-}), it seems reasonable from simple charge balance considerations that pH does not drop much below 5 (cf. Seip 1980). Considerable contributions of organic anions can be excluded as total organic carbon (TOC) is low (table 3). The average sulphate concentrations in precipitation and stream are equal. Because annual discharge is considerably smaller than annual precipitation, and dry deposition of S compounds increases the input slightly, there must be some sulphate reduction or adsorption in the catchment.

5. Høylandet (Ingabekken)

The Høylandet area (figure 1, table 1) has been selected as a pristine reference site suitable for comparison with acidified areas such as Birkenes. A regional survey of water chemistry in lakes and streams (about 200 locations) has been done by Muniz and co-workers (Muniz 1987). They have also studied fish, zooplankton, and benthic fauna in several streams and lakes. Lower parts of the area are forested (mainly spruce) and may be compared with Birkenes. A site close to the SWAP research area is included in the Norwegian forest monitoring program.

Palaeolimnological studies of sediments from a lake near the SWAP site (Røyrtjørna) indicated that pH has been remarkably stable (5.6–5.9) since about A.D. 650 (Berge *et al.* 1990).

The subcatchment chosen for intensive stream waterstudies lies in the alpine, upper part of the Høylandet site, making comparison with Birkenes somewhat

Table 4. *Mean values and standard deviations for cation exchange capacity* (CEC, *in* meq kg^{-1}), *base saturation* (BS (%)) *and exchangeable aluminum* (*in* mmol kg^{-1}) *for soils in the Birkenes I, Birkenes II and Ingabekken catchments.*

(Soil samples were extracted with 0.1 M BaCl$_2$ (Hendershot & Duquette 1986). Soil analyses at Svartberget have been carried out with a different method making the comparison more uncertain. However, it appears that exchangeable aluminium is lower at Svartberget than at the Norwegian catchments included in the table).

| horizon | podzol/brown earth | | | peat |
	O/H	A/E	B	
Birkenes I				
CEC	206 (54)	24 (10)	36 (13)	208 (55)
BS	57 (20)	15 (5)	7 (2)	30 (22)
Al	13.0 (14.1)	4.9 (1.8)	10.1 (3.6)	35.5 (19.3)
n	47	6	9	42
Birkenes II				
CEC	272 (40)	44 (14)	30 (12)	240 (90)
BS	72 (3)	36 (11)	15 (5)	56 (12)
Al	4.5 (0.4)	7.0 (3.1)	7.9 (3.3)	25.6 (12.2)
n	2	8	10	6
Ingabekken				
CEC	265 (104)	28 (8)	43 (18)	188 (91)
BS	74 (13)	29 (17)	15 (5)	47 (23)
Al	5.5 (4.1)	4.8 (2.9)	11.3 (4.7)	21.3 (12.4)
n	5	6	5	12

† 1 ha = 10^4 m^2.

uncertain. Streamwater sampling started in autumn 1986. Although pH decreases during high-flow periods, streamwater is much less acid than at Birkenes and the concentration of inorganic aluminium very low (table 3). [Ca^{2+}] is much lower than in Birkenes whereas concentrations of chloride and sodium are slightly higher. The ratio [Na$^+$]/[Cl$^-$] is 0.78, slightly lower than in seawater. Total organic carbon seems to be close to that found at Birkenes.

Cation exchange capacity, base saturation and exchangeable Al for some soil samples from the Ingabekken catchment are given in table 4. These are discussed further in §6.

6. Comparison of Birkenes I, Birkenes II and Høylandet

The two Birkenes catchments and the Ingabekken catchment have been compared by Christophersen *et al.* (1990 b). The differences in streamwater chemistry for the sites Birkenes I, Birkenes II and Ingabekken can be seen from table 3. For the important species H$^+$ and Al$_i$ the differences are illustrated in more detail in figure 4 (Christophersen *et al.* 1990 b). The observations fall in three clusters with very low Al$_i$ values at Ingabekken, intermediate at Birkenes II, and high values at Birkenes I.

The work by Christophersen *et al.* (1990 b) was hampered by lack of comparable soil data. Some recent results obtained by using the same methods in the three areas are given in table 4 (J. Mulder, M. Pijpers, P. H. Fjeldal and R. D. Vogt, unpublished results). The base saturation (BS) is seen to be generally higher in the Ingabekken catchment than in Birkenes I, but about equal to or lower than the BS in Birkenes II. The aluminium content in the upper (O/H) horizon and in the peat is apparently highest at Birkenes I. The trends in table 4 agree with analyses of soils from these

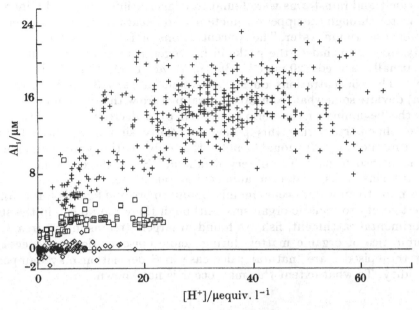

Figure 4. Streamwater $[Al_i]$ against $[H^+]$ for Birkenes I ($+$), Birkenes II ($\square$), and Ingabekken ($\diamond$) (from Christophersen *et al.* (1990*b*)).

catchments done at NISK (J. Esser, personal communication). Dahl and co-workers have also compared the soils in the Birkenes and Høylandet area (Aune *et al.* 1989). They found slightly lower pH and base saturation at Høylandet in all soil layers (down to 1 m). The differences in soil chemistry do not seem sufficient to explain the differences in water chemistry. The higher sulphate concentrations at Birkenes apparently affect the acidity and concentration of inorganic aluminium.

7. Svartberget

The Svartberget site is situated some 70 km inland on the east coast of Sweden. The catchment is forested with mature Scots pine on higher ground and Norway spruce in lower lying areas, except for 8 ha† at the head of the catchment which is an open mire with deep peat deposits (Bishop *et al.* 1990; Grip & Bishop, this symposium; Taugbøl *et al.*, this symposium). Mean concentrations of major anions in precipitation and runoff are given in tables 2 and 3, respectively. Sulphate deposition is approximately equal to sulphate output ($\approx 0.6 \, \mathrm{g \, S \, m^{-2} \, a^{-1}}$). Dry deposition accounts for about 3% of the deposition.

Several zero-tension lysimeters, groundwater tubes and tensiometers were located in the stream and at various distances from the stream. Measurements of $^{18}O/^{16}O$ in precipitation, stream-, soil- and groundwater have been done.

Streamwater pH generally drops at high discharge, sometimes to values below 4.4 (see figure 3 in Grip & Bishop, this symposium), i.e. about the same as found at Birkenes (cf. table 3). The decreases cannot be explained simply by increased sulphate concentrations or decreased cation concentrations. Increases in flow accompanying

† 1 ha $= 10^4 \, \mathrm{m}^2$.

spring snowmelt and rainstorms were found to trigger acidic episodes by increasing the flow of water through the upper decimeters of soil near the stream channel where organic acids enter runoff water. The concentrations of TOC showed a pattern that was broadly similar to that of the peaks in hydrogen concentrations. During peak flow TOC usually exceeds 20 mg l^{-1} (Bishop *et al.* 1990; Grip & Bishop, this symposium). The concentrations during snowmelt in 1988 (Grip & Bishop, this symposium) deviate somewhat from this general picture with only a modest increase in TOC in the beginning of the period followed by a decrease. The variations in streamwater chemistry during this period are fairly similar to those found at Birkenes during snowmelt. It should be noted that only relatively weak organic acids have been identified to date. The observed falls in pH are therefore not presently explicable in terms of dissociated organic acid anions alone.

Aluminium in streamwater occurs mainly bound to organic ligands (table 3). This reduces the toxicity to aquatic organisms. Although there are no fish in the stream in the experimental catchment, fish are found in larger streams in the area.

The prominence of organic matter during acidic episodes does not necessarily imply that the episodes are 'natural'; decrease in S deposition may dampen the peaks in acidity. To what extent this may occur is not known.

8. Conclusions

The Scandinavian SWAP sites span from the highly acidified Birkenes I catchment, which is representative of the most acidified areas in southern Scandinavia, to the pristine Høylandet area. The differences in deposition, in particular of S compounds, offer possibilities for quantifying the role of acid precipitation and other potential causes for freshwater acidification. Detailed studies, particularly at Birkenes and Svartberget, show that water pathways are extremely important for water chemistry. The pH in streamwater tends to decrease with increasing flow at all sites, but the minimum values are higher during these periods at the least polluted sites (Ingabekken and Storbekken) than at Birkenes or Svartberget. At Ingabekken this may partly be due to the presence of more easily weatherable minerals than in the Birkenes I catchment. The importance of the soil and bedrock composition is also clear from a comparison of Birkenes I and Birkenes II. The latter has higher base saturation in soils and 'better' water quality. The high sulphate concentrations at Birkenes I tend to increase the concentrations of H^+ and inorganic aluminium species. The streamwater acidity at Svartberget is mainly due to organic acids occurring naturally in the catchment. The concentrations of monomeric, inorganic aluminium are therefore low at Svartberget and the water quality much less toxic to aquatic biota than at Birkenes. The differences in water chemistry at Svartberget and Birkenes illustrate that natural organic acids are not the main cause of the acidity at Birkenes. The aluminium concentrations in streamwater do not correspond to a simple equilibrium with gibbsite. In the upper soils aluminium bound to organic matter may control the concentration in soil water.

The Birkenes data have been extremely valuable in developing mathematical modes for water acidification. Work is also in progress to model the other catchments although some of the records are rather short for this purpose.

References

Aune, E. I., Dahl, E. & Løes, A. K. 1989 *Medd. Nor. inst. skogforsk.* **42** (1), 135–147.

Berge, R., Brodin, Y.-W., Cronberg, G., El-Daoushy, F., Høeg, H. I. Nilssen, J. P., Renberg, I., Rippey, B., Sandøy, S., Timberlid, A. & Wik, M. 1990 In *Palaeolimnology and lake acidification* (ed. R. W. Battarbee, Sir John Mason, I. Renberg & J. F. Talling) (*Phil. Trans. R. Soc. Lond.* B **327**), pp. 159–163. London: The Royal Society.

Bishop, K. H., Grip, H. & O'Neill, A. 1990 *J. Hydrol.* (In the press.)

Christophersen, N., Robson, A., Neal, C., Whitehead, P., Vigerust, B. & Henriksen, A. 1990*a J. Hydrol.* (In the press.)

Christophersen, N., Neal, C., Vogt, R., Esser, J. & Andersen, A. 1990*b Sci. tot. Envir.* **96**. (In the press.)

Dahl, E. 1988 *Økoforsk report* 1988:1, Økoforsk, P.O. Box 64, 1432 Ås-NLH, Norway.

Hendershot, W. H. & Duquette, M. 1986 *Soil Sci. Soc. Am. J.* **50**, 605–608.

Hesthagen, T., Hegge, O., Dervo, B. K. & Skurdal, J. 1989 *MVU-report* B60, NINA, Tungasletta 2, 7004 Trondheim, Norway.

Mulder, J., Christophersen, N., Hauhs, M., Vogt, R. D., Andersen, S. & Andersen, D. O. 1990 *Wat. Resour. Res.* **26**, 611–622.

Muniz, I. P. 1987 *SWAP mid-term review conference, Lindås, Norway*, pp. 259–265. London: The Royal Society.

Overrein, L. N., Seip, H. M. & Tollan, A. 1980 *Research Report 19/80*, SNSF-Project, 1432 Ås-NLH, Norway.

Rosenqvist, I. T. 1978 *Sci. tot. Envir.* **10**, 39–49.

Seip, H. M. 1980 In *Ecological impact of acid precipitation* (ed. D. Drabløs & A. Tollan), pp. 358–366. SNSF-Project, 1432 Ås-NLH, Norway.

Seip, H. M., Andersen, D. O., Christophersen, N., Sullivan, T. J. & Vogt, R. D. 1989 *J. Hydrol.* **108**, 387–405.

SFT, 1989 *Statlig program for forurensningsovervåking*, report no. 375/89, Oslo: Statens forurensningstilsyn.

Stone, A. & Seip, H. M. 1989 *Ambio* **18**, 192–199.

Sullivan, T. J., Christophersen, N., Muniz, I. P., Seip, H. M. & Sullivan, P. D. 1986 *Nature, Lond.* **323**, 324–327.

Sullivan, T. J., Christophersen, N., Hooper, R. P., Seip, H. M., Muniz, I. P., Sullivan, P. D. & Vogt, R. D. 1987 *International symposium on acidification and water pathways, Bolkesjø, Norway, 4–8 May, 1987*.

Vogt, R. D., Seip, H. M., Christophersen, N. & Andersen, S. 1990 *Sci. tot. Envir.* **96**. (In the press.)

Long- and short-term hydrochemical budgets in Scottish catchments

BY R. HARRIMAN[1], R. C. FERRIER[2], A. JENKINS[3] AND J. D. MILLER[2]

[1] *Freshwater Fisheries Laboratory, Pitlochry, Perthshire PH16 5LB, Scotland, U.K.*
[2] *Macaulay Land Use Research Institute, Craigiebuckler, Aberdeen AB9 2QJ, Scotland, U.K.*
[3] *Institute of Hydrology, Wallingford, OX10 8BB, Oxfordshire, U.K.*

Although the key role of the mobile sulphate anion in surface water acidification is now generally recognized, the transfer function between sulphate inputs and catchment output is still difficult to predict. Within-catchment process studies have been used extensively to elucidate the complex vegetation and soil interactions that determine sulphate pathways; however, input–output budgets collected on a short- and long-term basis from a wide range of catchments, can provide useful insights into the present sulphate status and timescale of catchment response.

Estimates of retention and release of sulphate and other major ions were made at 13 SWAP sites in Scotland. The results have highlighted the requirement for accurate input data and emphasized the need for caution when interpreting annual budgets, especially for atmospherically derived constituents such as sulphate and chloride. Bearing in mind the likely errors involved in these estimates there appears to be a 10–30 % net retention of sulphate at most of the acidified sites whereas the pristine sites are in approximate balance over the three-year study period.

The transient response of sulphate on an hourly basis was also investigated at transitional site 4. Net alkalinity production was shown to be modified by organic contributions at high flow, which resulted in a stable alkalinity output. In contrast, sulphate contributions increased directly with increasing flow suggesting a source of soluble sulphate in the upper soil horizons.

Long-term sulphate data from the 'acidified' SWAP sites indicate little delay in stream sulphate response to the reduction in sulphate inputs over the past decade.

1. Introduction

The acidification of terrestrial and freshwater ecosystems and its associated biological consequences is now recognized as a major environmental problem in Europe and North America.

While recognizing the key role of the mobile sulphate anion (and to a lesser extent nitrate) (Seip 1980) the transfer function between sulphate inputs and outputs is not easy to predict. Even when catchments of similar sensitivity, based on soil and geological characteristics, are compared, the response in terms of stream acidity can vary considerably. Ryan *et al.* (1989) noted that surface waters in the southeastern United States had lower sulphate concentrations than those in northeast regions and southeastern Canada despite the similar magnitude of atmospheric sulphur depositions. This variability, which has been attributed to differences in sulphate adsorption capacity of soils, (Johnson *et al.* 1979; Singh 1984) causes sulphate to be removed from solution at different rates thus modifying base cation release.

Changes in land-use may also affect the sulphur dynamics in sensitive catchments both in terms of increasing sulphate inputs (Harriman & Morrison 1982) and by increasing the quantities of sulphur cycled between vegetation and soil (Miller *et al.* 1990).

Quantification of these factors is extremely important because many predictive models depend on accurate sulphate input and soil adsorption data to estimate accurately pH changes in surface waters (Cosby *et al.* 1985). The mode of sulphate adsorption and desorption is complex, particularly in acid, forest soils (Johnson & Todd 1983); therefore the timescale to reach equilibrium after increased or decreased sulphate inputs is difficult to predict. As part of the Surface Waters Acidification Programme (SWAP) input–output budgets were constructed for sulphate and other major ions for catchments of differing sensitivities along the sulphate deposition gradient in Scotland. Both long-term (three year) and short-term (36 h) budgets were used to investigate the present-day acidification status of these catchments by using different input scenarios. The results are then discussed in relation to the four major SWAP objectives (Mason & Seip 1985) and compared with the long-term changes in inputs and stream concentrations at these sites.

2. Site description

The SWAP sites in the U.K. are located in Scotland and were selected to cover a range of catchment sensitivity, sulphate inputs and land use. Where possible, existing sites with historical chemical and fishery databases were selected if they fitted the SWAP criteria. Catchments were classified as acidified, transitional and pristine based on sulphate input data. Most of the 'acidified' catchments (numbered 5–13, table 1) were those described by Harriman & Morrison (1982) but with an extra site (12) to provide a stream with an adequate trout population. The transitional site (4) was a 'new' site as were the three semi-pristine sites in northwest Scotland (1–3). Physical attributes relevant to the budget studies are given in table 1 and a summary of the more general catchment attributes, locations, instrumentation and measurement techniques are described by Ferrier & Harriman (this symposium). Detailed plot, catchment, vegetation and input studies were limited to catchments 4, 9 and 13.

3. Data collection and processing

(a) Inputs

Precipitation volume was measured within or adjacent to all catchments to varying degrees of accuracy. At sites 4, 9 and 13 the catchment input was derived from bulk collectors (Rimco, ground level) and Nipher-shielded rain gauges at 3 m height. The volumes collected in all polypropylene 'chemistry' collectors provided an extra check on the standard rain collectors. Gauges were located along the elevation gradient to provide average inputs. Standard meteorological rain collectors were deployed at the remaining sites. In addition, a series of filter gauges were installed alongside the Nipher gauges at sites 4, 9 and 13 to obtain estimates of enhancement deposition. The average 'enhancement' (EH) values were then added to bulk inputs

Table 1. *Catchment attributes of SWAP sites (1986–88)*

site	name	G.R.[a]	catchment area/ha[b]	elevation range/m	% forest cover/age[c]	rainfall/ mm	output/ mm
1	Coire nan Con	NM793688	790	74–750	45/0	2813	2352
2	Choire Duibh	NM824652	144	150–500	0	2609	2166
3	Allt na Graidhe	NH346073	275	90–500	95/0	1361	951
4	Allt a'Mharcaidh	NH882045	987	330–1110	5/0	923	855
5	Burn 2	NN388043	440	150–974	2/Y	2900	2405
6	Burn 5	NS438992	170	125–573	55/Y	2850	2353
7	Burn 6	NS438989	113	125–573	45/Y	2850	2592
8	Burn 7	NS451985	135	100–425	60/0	2730	1703
9	Burn 9	NS466967	225	160–507	65/0	2800	1694
10	Burn 10	NS469988	90	100–220	95/0	2575	1785
11	Burn 11	NS470988	145	100–220	90/0	2575	1781
12	Corrie	NS485758	255	100–460	90/0	2750	1692
13	Burn 14	NN414069	300	100–392	80/0	2725	1701

[a] G.R., National Grid Reference Number.
[b] 1 ha = 10^4 m².
[c] 0, greater than 15 years; Y, less than 15 years.

for all the sites. A more detailed assessment of this technique is given by Ferrier *et al.* (this symposium); however, it should be noted that this method is purely operational and does not necessarily include all possible contributions to total deposition or attribute inputs from particular sources. A second method of estimating inputs was based on the widely used assumption that chloride is conservative over a reasonable timescale (more than a year). The inputs from this method (CL) are based on a factor derived from the Cl output divided by Cl input (bulk deposition), which is then applicable to other ions on the basis of their ratio to chloride in sea-salts. The consequences of using this factor to calculate budgets and the validity of the assumed chloride balance are discussed later. A third estimate of inputs was made by using best present estimates (BPE) of the influence of vegetation, mist and orographic cloud on input quantity and quality. At elevations greater than 750 m at site 4, Ferrier *et al.* 1990*a* showed that enhancement inputs were four times greater than those at lower elevations. Recent studies of the composition differences in orographic cloud compared with bulk collector composition (Fowler *et al.* 1988) indicate that this deposition mechanism is significant for catchments with maximum elevation more than 500–600 m.

Because of the large differences between elevation, vegetation type and dry deposition inputs at the SWAP sites, a wide range of BPE factors were produced. The highest values for sulphate are those at high rainfall, high elevation sites with a large percentage forest cover. Calculations were therefore based on a dry-deposition component (Williams *et al.* 1989), a vegetation-interception component and an elevation component (Fowler *et al.* 1989).

Standard chemical methods were used for rainwater and snow samples. Major anions were determined by ion chromatography and cations by flame emission or atomic absorption spectroscopy. Volume-weighted concentrations for all ions were used to calculate bulk-deposition inputs.

(b) *Outputs*

Accurate outputs are technically much easier to measure and tend to be more accurate as only one site, for measurement of flow and chemistry is required for each catchment. Stream-level recording gauges were operative at sites 1, 3, 4, 7, 9, 10 and 11 during the study period and any short-term gaps were filled in by using flow relations with the nearest stream. Flows for other sites were estimated from rainfall/flow relations at the nearest sites. Mean weekly chemistry was obtained by using stage-related composite samplers fixed on the stream bed and operated automatically by the rising stream level. Sample quantity increased as a function of stream height ensuring that increments of all flow types were continually sampled. Full chemical analysis was usually completed within 7–10 days. Chemical methods were identical to those for precipitation and details of pH, alkalinity, aluminium and total organic carbon (TOC) methodology are given by Harriman *et al.* (1990) and Ferrier *et al.* (1990*b*).

Charge-balance errors for precipitation and stream-water chemistry were invariably less than 10 % therefore it was assumed that estimates of organic anion contribution from charge balance estimates for streamwater would be reasonably accurate.

4. Results

(a) *General synthesis*

Mean bulk precipitation and catchment-output chemistry for the period 1986–88 are given in table 2. Within this timescale there was no significant variation in bulk precipitation composition between the acidified sites (5–13) (Miller *et al.* 1990) therefore the same composition values were used for all these sites. Removal of sea-salt components (based on chloride relations) eliminated all Na^+ and Mg^{2+} leaving a residual base cation $(Ca^{2+} + K^+)$ input of less than 10 µeq l^{-1} at all sites. Excess SO_4^{2-} and pH reflect the distance from sources of atmospheric pollution giving a bottom of the gradient range of 16–21 µeq l^{-1} at the semi-pristine sites to 30 µeq l^{-1} at the transitional site and rising to a top of the gradient value of 42 µeq l^{-1}. NO_3^{2-} and NH_4^+ were present in approximately equivalent amounts with the highest values measured at the acidified sites.

Apart from sites 3 and 12 catchment output concentrations reflected excess sulphate and sea-salt inputs within a relatively narrow sensitivity range (40–80 µeq l^{-1} Ca^{2+}). The highest excess SO_4 level (126 µeq. l^{-1}) was recorded in the alkaline, afforested acidified site (12) and the lowest (31 µeq l^{-1}) in the moorland, semi-pristine site (2). This low excess SO_4^{2-} value is about 10 µeq l^{-1} above the pre-acidification level of 22 µeq l^{-1} calculated for this site (CLTAP 1989). This pre-acidification value is probably applicable to the other sites apart from sites 3 and 12, which would have higher values because of their higher base cation levels. Associated with this SO_4^{2-} gradient from sites 1 up to 13 is a corresponding decline in pH for sites with similar base cation levels, cf. sites 2 and 6; sites 1 and 10, etc. Only site 4 is atypical in its base cation composition where significant excess sodium levels are associated with high silicate weathering rates. This catchment derived sodium source contributes about 50 % of the base cations associated with alkalinity at base flow. Labile Al levels were less than 10 µeq l^{-1} at sites 1–4 whereas values up to 119 µeq l^{-1} were found at site 8, which, surprisingly, was the least acid of the sites within the pH 4–5 range.

Table 2. *Mean precipitation and stream chemistry for SWAP sites (1986–88)*

(Units as μeq l^{-1} except SiO$_2$ and Al species (μg l^{-1}) and TOC (mgC l^{-1}).)

site	rain/mm	pH	Alk	Na$^+$	K$^+$	Ca^{2+}	Mg^{2+}	NH$_4^+$	Cl$^-$	SO$_4^{2-}$	SO$_4^{2-a}$	NO$_3^{2-}$
						(a) bulk precipitation						
1	2813	4.94	−12	133	6	12	30	12	157	37	21	8
2	2609	4.94	−12	133	6	12	30	12	157	37	21	8
3	1361	4.98	−11	113	4	12	27	3	136	30	16	5
4	923	4.64	−23	61	4	10	16	8	74	37	29	13
5	2900	4.54	−29	91	4	11	21	21	106	53	42	19
6	2850	4.54	−29	91	4	11	21	21	106	53	42	19
7	2850	4.54	−29	91	4	11	21	21	106	53	42	19
8	2730	4.54	−29	91	4	11	21	21	106	53	42	19
9	2800	4.54	−29	91	4	11	21	21	106	53	42	19
10	2757	4.54	−29	91	4	11	21	21	106	53	42	19
11	2575	4.54	−29	91	4	11	21	21	106	53	42	19
12	2750	4.54	−29	91	4	11	21	21	106	53	42	19
13	2725	4.54	−29	91	4	11	21	21	106	53	42	19

(b) stream chemistry (volume-weighted averages)

site	pH	Alk	Cond	Na$^+$	K$^+$	Ca^{2+}	Mg^{2+}	Cl$^-$	SO$_4^{2-}$	SO$_4^{2-a}$	NO$_3^{2-}$	SiO$_2$	Al$_t$	Al$_{nl}$	Al$_l$	TOC	A^b	CD^c
1	5.69	30	42	232	8	56	61	256	66	39	3	3455	50	44	8	3	7	2.3
2	6.24	50	37	202	8	70	53	222	54	31	1	2092	28	22	6	2.8	6	2.14
3	6.16	70	43	218	8	113	63	229	66	42	8	2832	73	65	8	6.6	27	4.1
4	6.49	43	22	124	6	40	26	91	52	43	1	5671	30	26	4	1.7	9	5.3
5	5.43	−3	27	110	7	54	39	118	73	61	7	691	54	33	21	3	15	5
6	5.24	5	28	118	5	70	41	124	74	61	8	888	73	58	15	4	19	4.2
7	6.08	30	28	118	5	77	52	123	78	65	9	1201	43	35	8	2.8	12	4.3
8	4.95	−13	38	155	5	52	55	173	105	87	12	2003	194	77	119	3	4	1.3
9	4.47	−34	44	141	10	48	48	166	93	76	5	1690	140	78	62	8.2	37	4.5
10	4.52	−30	45	165	5	57	59	167	133	116	19	1948	217	106	112	6	11	1.8
11	4.48	−33	43	159	3	48	43	159	103	86	4	1894	155	93	61	7.9	28	3.5
12	6.57	137	54	183	7	179	147	205	147	126	24	2800	51	42	9	4.4	4	0.91
13	5.08	−8	39	177	5	72	44	186	94	75	10	1639	129	82	47	4.8	25	5.2

[a] Excluding sea-salt contribution.
[b] A, Organic anions (from charge balance).
[c] CD, Charge density (μeq per milligram TOC), where CD = A/TOC.

(b) Long-term budgets

Input/output budgets, constructed by using the three methods previously described, are presented for excess SO$_4^{2-}$, excess base cations and Cl$^-$ in figure 1. The BPE method invariably produced the highest input estimates although the EH method occasionally gave higher values for the non-forested sites. On a quantitative basis the differences between the three input estimates were relatively small for most ions apart from SO$_4^{2-}$, NO$_3^{2-}$ and NH$_4^+$ for which the BPE was significantly higher. Because base cation inputs were relatively low at all sites the visual comparison of base cation outputs gives an adequate indication of base cation weathering. For chloride the general impression is one of an input–output balance at all sites for the three-year period. However, a closer analysis shows a small net retention of chloride at most sites. As noted previously the BPE method for sulphate produces significantly higher input estimates than the other methods, which, for some sites, means the difference between SO$_4^{2-}$ retention or SO$_4^{2-}$ release. Despite this prognosis we felt justified in taking the BPE estimate at face value and using it as the best present

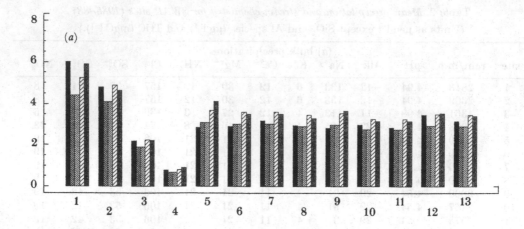

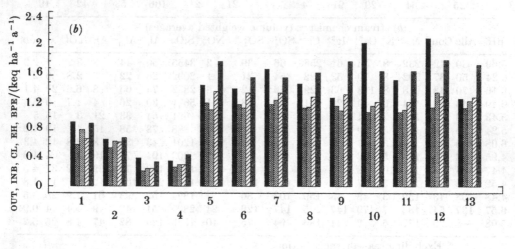

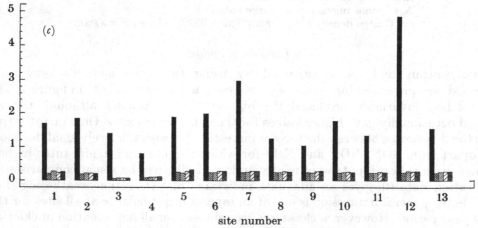

Figure 1. Estimate of (i) OUT (stream output); INB (bulk precipitation); (iii) CL (chloride factor); (iv) EH (enhancement deposition); (v) BPE (best present estimate) for (*a*) chloride, (*b*) excess sulphate and (*c*) excess base cations at SWAP sites. Units as keq ha^{-1} a^{-1} (see text for description of methodology); (■), OUT; (■), INB; (▨), CL; (▨), EH (▨), BPE.

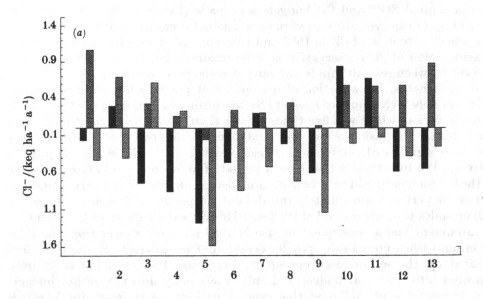

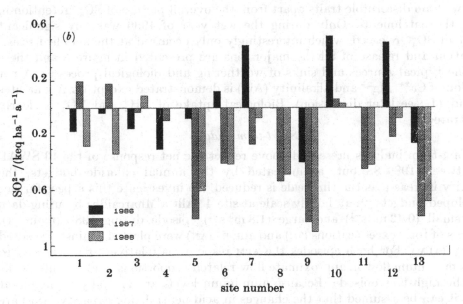

Figure 2. Annual variation in catchment yield for (*a*) chloride and (*b*) sulphate using BPE input estimates (negative values indicate retention). (■), 1986; (▨), 1987; (▨), 1988.

estimate of inputs for the following analysis. Thus in figure 1, by using BPE inputs, most of the sites are retaining SO_4^{2-} in the catchment to varying degrees. For example, sites 4, 5, 9 and 13 retain between 21 and 31 % of SO_4^{2-} input; sites 6 and 8, 14 % of the input and apart from site 10 the rest can be considered to be in approximate balance. The 12 % release of SO_4^{2-} at site 10 is probably due to clear-felling operations that commenced during the study period.

Because rainfall and inputs of SO_4^{2-} and Cl^- vary from year to year, a more detailed analysis of annual SO_4^{2-} and Cl^- budgets was made (figure 2). For Cl^- there was evidence of significant year-to-year variations and between site variations with most sites showing a retention of Cl^- in 1986 and 1988 compared with Cl^- release in 1987. If the assumption of Cl^- conservation is to be retained then there must be a delay mechanism between sea-salt inputs and output response that could, of course, vary between catchments. Examination of sea-salt input patterns is quite revealing in that an extremely high input occurred at all sites during the middle of the first month of the study period, which was five-times greater than the second highest input that occurred during the last week of the study period. Average Cl^- concentrations in rain were significantly different between years being 116, 58 and 77 µeq l^{-1}, respectively. We hypothesize that over a period of about one year after a sea-salt event, the Cl^- budget should be in balance provided no further episodes occur towards the end of the period. Unfortunately this did not happen in 1986 because of a series of small episodes towards the end of the year. This caused an apparent Cl^- retention in the catchment and a consequent release in 1987 (as a carry-over from the 1986 winter inputs) when no sea-salt episodes occurred. The apparent retention in 1988 was caused by the second high episode in December 1988, which was entirely incorporated into the input budget and only fractionally into the output budget. Most of the catchments followed this general pattern apart from site 5, which indicated a retention in all years. The between year SO_4^{2-} pattern is more difficult to explain with no discernible traits apart from the overall picture of SO_4^{2-} retention in most of the catchments. Only during the wet year of 1986 was any significant quantity of SO_4^{2-} released, which interestingly only occurred at the acidified sites.

Retention and release of all the major ions are presented in figure 3 and these reflect the typical sources and sinks of weathering and biological processes. A net generation of Ca^{2+}, Mg^{2+} and alkalinity (Alk) is demonstrated except for Alk at sites 9, 10 and 11 (see later discussion). Biological uptake of NH_4^+ and NO_3^- is clearly demonstrated.

(c) Short-term budgets

The long-term budgets presented above reflect the net response of the 13 SWAP sites between 1986–88 but, as indicated by the annual chloride budgets, the complexity increases as the timescale is reduced. To investigate this aspect further we developed budgets on an hourly scale at site 4 (Allt a'Mharcaidh) by using data from a small ($0.42 \text{ m}^3 \text{ s}^{-1}$) and large ($1.8 \text{ m}^3 \text{ s}^{-1}$) episode during 1988 (figure 4). Quantities of total excess cations (TC) and anions (AN) were plotted against measured alkalinity (ALM). For both episodes the ALM response is relatively flat whereas TC peaks at maximum flow in approximate flow-related proportions, and AN only peaks during the high-flow episode. Because aluminium levels are virtually negligible at this site it can be assumed that the changes in acid neutralizing capacity (ANC) are equivalent to the ALT values. Thus on a charge balance basis:

$$\text{ANC} = \Sigma\text{TC} - \Sigma\text{AN} = \text{ALT}.$$

Where TC = total excess base cations; AN = total excess anions; ALT = theoretical alkalinity from charge balance.

When this value (ALT) is calculated for both episodes distinct peaks of alkalinity are generated compared with the measured values. If the composition of the stream water during hydrological events is viewed as a mixture of groundwater and soil throughflow water, then:

$$\text{ALT} = \text{ALM} + \text{AO} = \text{ALMO}.$$

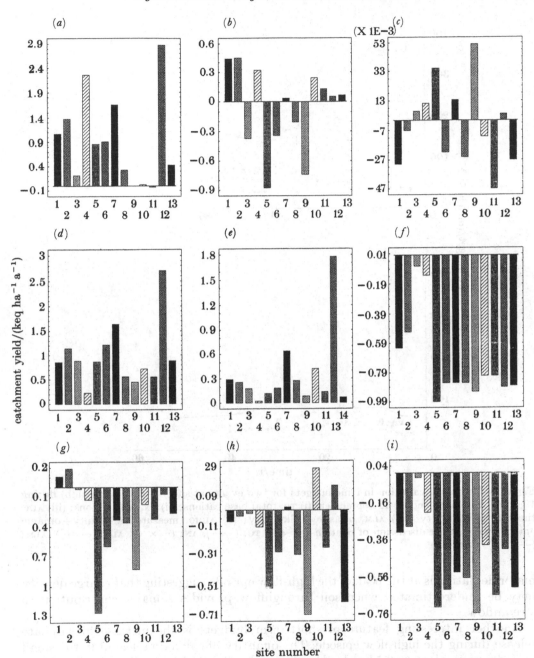

Figure 3. Catchment yield for nine major cations and anions using the BPE input estimates (1986–88); (*a*) Alk; (*b*) Na$^+$; (*c*) K$^+$; (*d*) Ca^{2+}; (*e*) Mg^{2+}; (*f*) NH$_4^+$; (*g*) Cl$^-$; (*h*) SO$_4^{2-}$; (*i*) NO$_3^{2-}$.

Where, AO = organic acid contribution; ALMO = measured + equivalent AO alkalinity contribution. The AO contribution was determined from TOC values in each sample multiplied by mean charge density in table 2.

A comparison of the revised alkalinity generation (ALMO) with the theoretical value (ALT) shows a much closer match, particularly in the low-flow episode. The

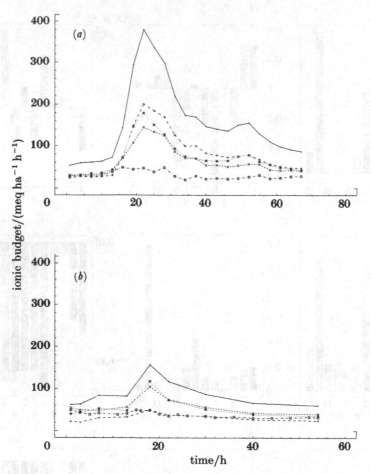

Figure 4. Temporal variation in ionic budgets for two events at site 4 (Allt a'Mharcaidh) during 1988; (*a*) 18 March and (*b*) 12 September; (i) TC, total base cations; (ii) AN, total anions; (iii) ALT, theoretical alkalinity; (iv) ALM, measured alkalinity; (v) ALMO, measured alkalinity + organic anions (see text for discussion of relations); (–●–), TC; (–+–), AN; (···×···), ALT; (–·☐·–), ALM; (––×––), ALMO.

major deviation is at the peak of the high-flow episode suggesting that charge density may be underestimated when soil throughflow provides a major contribution to streamflow.

Another interesting feature of short-term budgets is the increase in sulphate release during the high-flow episode. The quantity of sulphate released in the small episode, equivalent to $0.5\ \mathrm{keq\ ha^{-1}}$† $\mathrm{a^{-1}}$, is only slightly higher than the three year average of $0.44\ \mathrm{keq\ ha^{-1}\ a^{-1}}$ whereas the output of $2.0\ \mathrm{keq\ ha^{-1}\ a^{-1}}$ during the high-flow episode is nearly five-times the three-year average. This indicates the presence of a highly reactive, soluble store of sulphate in those soil horizons that provide the major routing pathways in saturated, high flow conditions.

Short-term budgets can also be used to investigate the relation between base

† $1\ \mathrm{ha} = 10^4\ \mathrm{m^2}$.

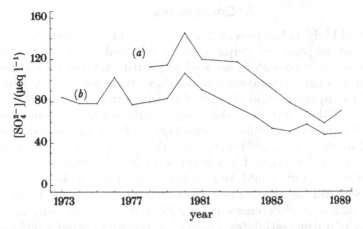

Figure 5. Long-term trends in stream and bulk precipitation sulphate concentrations for site 5 (Burn 2) between 1973 and 1989. (a) Stream site 5, (b) rain site 5.

cation weathering and exchangeable base cations in the context of critical load assessment. At the beginning of the larger episode, when streamflow was low, 60% of the excess base cations are associated with alkalinity compared with 38% at peak flow. If this difference is taken as an approximate measure of the exchangeable fraction of the base cation component (i.e. zero exchangeable cation at base flow) then about 40% of the base cation production at peak flow could be allocated as an exchangeable form. When extrapolated to the three year average chemistry data the mean exchangeable base cation component contributes about 10% of the total base cation content.

(d) Historical perspective

In this final section we attempt to set the current status of the SWAP sites into a historical context.

Scotland is unusual in that SO_4^{2-} concentrations in precipitation show both a steep gradient from southeast to northwest and an overall decline during the past decade (Harriman 1989). This change has been reflected as a reduction in loch acidity at sensitive acidified sites in S.W. Scotland (Harriman & Wells 1987) and at the acidified SWAP sites. At site 5, an unmanaged, unafforested acidified site, this decline can be seen both in terms of precipitation and stream SO_4^{2-} decline (figure 5). Linear regression analysis of the 1981–89 mean data indicates an approximate 5 µeq l^{-1} a^{-1} decline in precipitation and 8 µeq l^{-1} a^{-1} change in stream 5. During the SWAP programme, stream and precipitation SO_4^{2-} levels appear to have stabilized within the same timescale thus implying a short delay between precipitation and stream response. A major concern in interpretation of the data was that output budgets would be influenced not by present-day inputs but by the release of pollutants accumulated in the acidified catchments over the past decades. The fact that most catchments appear to be still retaining up to 25% of the BPE input of SO_4^{2-} suggests that these fears may be somewhat unfounded, however two of the acidified catchments did show evidence of SO_4^{2-} release, which again highlights the problems of catchment comparisons.

5. Conclusions

The use of chemical budgets has provided a useful insight into the relations between inputs and the net response of major ions as reflected in stream output. This technique is not sensitive enough to be used to identify the complex chemical and biological processes within vegetation and soils and these aspects are discussed by Ferrier et al. (this symposium) and Walker et al. (this symposium) in the context of plot-based studies. Although most of the budget analysis was based on best present estimates (BPE) of inputs the cumulative errors associated with the budget estimates are uncertain. Rochelle et al. (1987) recognized this problem when comparing SO_4^{2-} budgets of 40 extensively studied catchments in North America. A systematic underestimate of SO_4^{2-} inputs could well explain why many of these sites appeared to have a large SO_4^{2-} release. Errors in output estimates are probably smaller than input estimates although differences in sampling design and data analysis can produce a variety of output estimates. Dann et al. (1986) reported a 30 % difference between seven methods of estimating SO_4^{2-} export from two catchments in Virginia, U.S.A. and derived a best possible method from these data. If a 10 % error band is applied to the SO_4^{2-} and Cl^- budgets (being the most conservative input ions) then 8 of the 13 sites would be in balance for Cl^- compared with 5 of the 13 for SO_4^{2-}. Because SO_4^{2-} retention values are generally higher than those for Cl^-, and the inter-annual Cl^- variations (apart from site 5) switch from retention to release we are reasonably confident about drawing the following general conclusions from the data.

1. The sulphate and acidification status at the SWAP sites reflects the present sulphate deposition gradient in Scotland.

2. Inter-annual variations in SO_4^{2-} and Cl^- budgets indicate that rainfall patterns and timing of sea-salt inputs may distort annual budget estimates.

3. For most acidified and transitional sites there is evidence of between 10–30 % retention of SO_4^{2-} inputs within the catchments whereas the pristine sites appear to be in balance.

4. Data for short-term episodes at site 4 reveal a source of highly mobile, soluble SO_4^{2-} in the surface soil horizons that generate storm-flow water. This SO_4^{2-} source, which also generates an equivalent (exchangeable) cation component, maintains stream SO_4^{2-} concentration at a relatively constant level.

5. Long-term data from the acidified SWAP sites point to a fairly rapid response in stream SO_4^{2-} levels to changes in SO_4^{2-} inputs. This finding should therefore assist in the debate over critical loads and their implementation.

The authors express their appreciation to those who assisted in any way with this project over the past three years. In particular the contributions of E. Taylor, E. Gillespie, D. King, M. Edwards, T. Edwards and J. Porter are recognized. The financial support of the Surface Waters Acidification Programme is gratefully acknowledged.

References

Cosby, B. J., Hornberger, G. M., Galloway, J. N. & Wright, R. F. 1985 *Environ. Sci. Technol.* **19**, 1144–1149.

CLTAP (Convention on long-range transboundary air pollution) 1989 *Draft manual on: Mapping critical levels/loads.* Bad Harzburg, F.R.G.

Dann, M. S., Lynch, J. A. & Corbett, E. S. 1986 *J. Environ. Qual.* **15** (2), 140–145.

Ferrier, R. C., Jenkins, A., Miller, J. D., Walker, T. A. B. & Anderson, H. A. 1990a *J. Hydrol.* **113**, 285–296.

Ferrier, R. C., Walker, T. A. B., Harriman, R., Miller, J. D. & Anderson, H. A. 1990*b* *J. Hydrol.* **115**. (In the press.)

Fowler, D., Cape, J. N., Leith, I. D., Charleston, T. W., Gray, M. J. & Jones, A. 1988 *Atmos. Environ.* **22**, 1355–1362.

Fowler, D., Cape, J. N. & Unsworth, M. H. 1989 *Phil. Trans. R. Soc. Lond.* B **324**, 247–265.

Harriman, R. & Morrison, B. R. S. 1982 *Hydrobiologia* **88**, 251–263.

Harriman, R. 1989 In *Acidification in Scotland, symposium proceedings, Scottish Development Department*, pp. 72–79. Edinburgh.

Harriman, R., Gillespie, E., King, D., Watt, A. W., Christie, A. E. G., Cowan, A. A. & Edwards, T. 1990 *J. Hydrol.* **116**, 267–285.

Harriman, R. & Wells, D. E. 1987 In *Acid rain: scientific and technical advances* (ed. R. Perry, R. Harrison, J. N. B. Bell & J. N. Lester), pp. 287–292. London: Selper Ltd.

Johnson, D. W., Cole, D. W. & Gessel, S. P. 1979 *Biotropica* **11**, 38–42.

Johnson, D. W. & Todd, D. E. 1983 *Soil Sci. Soc. Am. J.* **47**, 792–800.

Mason, J. & Seip, H. M. 1985 *Ambio* **14** (1), 45–51.

Miller, J. D., Anderson, H. A., Ferrier, R. C. & Walker, T. A. B. 1990 *Forestry* **163** (3). (In the press.)

Rochelle, B. P., Church, M. R. & David, M. B. 1987 *Wat. Air Soil Pollut.* **33**, 78–83.

Ryan, P. F., Hornberger, G. M., Cosby, B. J., Galloway, J. N., Webb, J. R. & Rostetter, E. B. 1989 *Wat. Resour. Res.* **25** (10), 2091–2099.

Seip, H. M. 1980 In *Proceedings of an international conference on the ecological impact of acid precipitation* (ed. D. Drabløs & A. Tollan), pp. 358–366. Oslo:SNSF project.

Singh, B. R. 1984 *Soil Sci.* **138** (3), 189–197.

Williams, M. L., Atkins, D. H. F., Bower, J. S., Campbell, G. W., Irwin, J. G. & Simpson, D. 1989 Warren Spring laboratory report LR723(AP) Stevenage, U.K.

Discussion

R. A. SKEFFINGTON (*National Power Technology and Environmental Centre, Kelvin Avenue, Leatherhead, U.K.*). The budgets presented by Mr Harriman indicated that some of the Scottish sites are currently retaining sulphate to a significant extent. Presumably when sulphur deposition was higher a few years ago they were retaining at least as much. This implies a very considerable store of sulphur in these catchments, and yet we also have the statement that streamwater sulphate responds rapidly and proportionately to changes in sulphur deposition. Is this true of the sulphur-absorbing catchments and if so why does the large store of sulphur not buffer the streamwater response? The formulation of sulphur retention in (for instance) the MAGIC model implies that there ought to be considerable delays in response to reductions of sulphate inputs in this situation. Have the soil scientists attempted to track down the nature and amount of stored sulphur in these catchments?

R. HARRIMAN. By using the best present estimates (BPE) of sulphate inputs to the SWAP catchments, a total of 5 of the 13 catchments were estimated to be in steady state (inputs = outputs) based on a 10% error band. By using a 20% error band only three catchments were retaining sulphate and one was releasing sulphate. Clearly the conclusions regarding sulphate retention depend on the sensitivity analysis performed and not on the errors of the input/output budgets, which are not easy to quantify. The main objective of the budget estimates was to determine whether the present outputs of sulphate (which are measured reasonably accurately) were closer

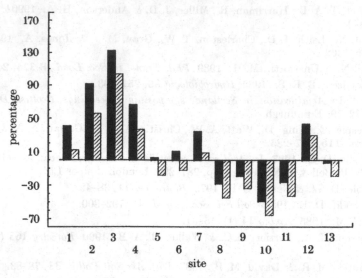

Figure D1. For explanation see text.

to present-day inputs or to inputs 10 years ago (which were approximately twice the present-day values). On balance the results supported the former case implying a relatively short delay in catchment response to reduced sulphate inputs.

This conclusion was supported by similar responses in the sulphate concentrations in precipitation and in streamwater at acidified site 5 (figure 5). Although there is a significant downward trend in sulphate levels it is more difficult to interpret these changes in terms of sulphur-storage processes in catchments, and indeed the SWAP programme was not designed to quantify this aspect.

Although it is interesting to speculate about the past rates of sulphate retention we are still uncertain as to the types of sulphur presently stored in different soil types/horizons.

From the detailed episodic soil and stream studies we can hypothesize a two-phase sulphate response to reduce inputs by dividing the soil types into those with weak adsorption characteristics (e.g. peats, peaty gleys) and these with strong adsorption characteristics (e.g. iron podzols, humus podzols). The specific hydrological routing within catchments may preferentially remove the weakly adsorbed sulphate giving a quick response in the stream to reduced inputs. This form of sulphate may only represent a minor fraction of the total sulphur store. If the present sulphate status at the SWAP sites reflects a steady/state after the complete depletion of the weakly held sulphate then further reductions in sulphate input may not produce a similar rate of decline in stream sulphate because the remaining sulphate would be more strongly adsorbed. However, if steady state has not been reached then the pool of weakly adsorbed sulphate may continue to be depleted and sulphate levels in the SWAP streams may continue to fall even if inputs remain at their present level. Both scenarios are possible; the important factor being the fraction of sulphate input which was incorporated into the soil as strongly adsorbed sulphate.

Any formulation of sulphur retention in models such as MAGIC depends on a lumped value that attempts to describe an overall catchment response. This value may bear no direct relation to any individual response in either weakly or strongly adsorbing soil types. We are certainly in agreement with your view that the mechanisms of sulphur storage and release should be further investigated, particularly if it results in realistic estimates of the hysteresis of sulphate reversibility and a better estimate of the adsorption potential used in models.

Additional comment from R. Harriman

The question of critical loads and emission reductions has been applied to the U.K., SWAP sites by using the water-chemistry approach. This assumes a steady state between sulphate deposition and sulphate in run-off water. Two scenarios have been used that define the critical alkalinity for a stream or lake as zero or 20 µequiv. l^{-1} (equivalent to an approximate pH of 5.5 and 6.0, respectively). These calculations (see figure D1; (■), ANC = 0; (▨), ANC = 20) indicate that for the pristine and transitional sites (1–4) an increase in sulphate inputs of between 40 and 130% would be required before the critical load was exceeded. For the acidified sites (5–13) the most acid ones (8–11) require a further reduction in sulphate ranging from around 30% at site 9 to nearly 70% at site 10.

Integrated hydrochemical responses on the catchment scale

By Alan Jenkins[1], R. Harriman[2] and S. J. Tuck[1]

[1] Institute of Hydrology, Wallingford, Oxfordshire OX10 8BB, U.K.
[2] Freshwater Fisheries Laboratory, Pitlochry, Perthshire PH16 5LB, Scotland, U.K.

At the Allt a'Mharcaidh, a transitional site in N.E. Scotland, mean streamwater pH lies around 6.0 yet the system is subjected to short-term acute acidic shocks. These acid episodes are always associated with high-flow events generated by rainfall and snowmelt. Three modes of hydrological response are identified from the flow record; (i) a relatively constant baseflow augmented by (ii) slow drainage of catchment soils to produce a 'whaleback' response and (iii) with short-term hydrograph 'spikes' superimposed. These three responses are linked to the hydrological characteristics of the major soil types in the catchment. Chemical observations indicate a similar three phase response that can be directly related to the flow pathways. During low flow, water chemistry is dominated by alkaline baseflow water. As flow increases, water reaching the stream via preferred pathways predominates and the chemistry rapidly changes to reach a fairly constant level reflecting the chemical composition of the upper, acidic soil layers. As flow recesses, the chemistry slowly returns to pre-storm levels fed by contribution from the slowly draining mineral soils. The data emphasize the importance of flow pathways and soil acidification status in determining streamwater chemistry.

1. Introduction

Long-term build up of chronic acidity in surface waters has been widely reported in Europe, Scandinavia and North America. There now exists a considerable weight of evidence that these long-term changes in mean pH of lakes and streams are caused by atmospheric pollution associated with emissions of sulphur and nitrogen (Battarbee 1990; Reuss et al. 1987). The degree of acidification is determined by total acidifying input and the timing of pH change reflects the physio-chemical characteristics of the catchment, notably the base status and weatherability of the bed rock, the base status of the soils and the ability of the soils to adsorb sulphur.

Not all surface waters in an area receiving high sulphur deposition are chronically acidified. The surface water in many catchments demonstrates a mean pH around 6.0 and only becomes acid during short lived, high-flow events. In these transitional catchments the dynamic hydrological and chemical response is determined mainly by water flowpaths, storages and transit times within the catchment and the way in which these change as the storm progresses. Characteristically, these low-pH episodes occur with irregular frequency and yet may adversely affect fish populations. In particular, declining fish stocks have been attributed to increased mortalities at the hatching and emergence stages resulting in the failure of juvenile recruitment (Howells 1983). Studies to date indicate that pH (< 5.5), calcium (< 50 μeq l^{-1})

[47]

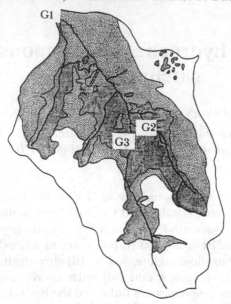

Figure 1. The Allt a'Mharcaidh catchment; (▨), peaty podzols;
(▦), blanket peat; (□), alpine podzols.

and ionic forms of aluminium (> 50 µg l^{-1}) are the key chemical parameters that
influence fish survival (Brown 1983; Harriman *et al.* 1987). The determination of the
timing, frequency and magnitude of episodes with chemistry similar to these critical
levels is essential if the relation between water quality and fish survival is to be
quantified.

This paper examines the short-term hydrochemical response at the Allt
a'Mharcaidh, a transitional site in Scotland (Ferrier & Harriman, this symposium),
to determine the changes in chemical composition of the incident precipitation as it
travels through the catchment and to assess the importance of the factors influencing
runoff chemistry.

2. Sampling and methodology

Flow and pH were recorded continuously at three sites within the Allt
a'Mharcaidh; at the basin outflow (G1) and at the two major sub-catchment outflows
(G2 and G3). The position of these sites and the distribution of the three major soil
types is given in figure 1. Detailed descriptions of the catchment, recording
equipment and soils are given elsewhere (Ferrier & Harriman, this symposium;
Jenkins 1988; Ferrier *et al.* 1990). In addition, water samples were collected for
chemical analysis twice weekly and more frequently during high-flow episodes. Full
laboratory procedures and mean chemistry at the three sites are described by
Harriman *et al.* (1990).

3. Hydrological response

Table 1 shows the wide range of flows recorded at each of the sites. The mean flows
are towards the low end of the ranges indicating that for the majority of the time the
stream is close to baseflow. The flow duration curves for G1, G2 and G3 (figure 2a)
support this observation indicating that for 60 % of the time, flows are less than

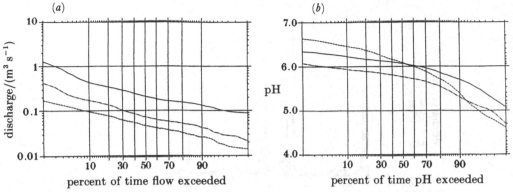

Figure 2. (*a*) Flow and (*b*) pH duration curves for G1 (——), G2 (----) and G3 (·—·—·).

Table 1. *Annual mean, maximum and minimum instantaneous discharges* ($m^3 \, s^{-1}$) *at the three monitoring sites*

year	G1 min.	G1 mean	G1 max.	G2 min.	G2 mean	G2 max.	G3 min.	G3 mean	G3 max.
1986	0.08	0.3	5.74	—	—	—	—	—	—
1987	0.07	0.24	3.38	0.02	0.06	0.26	0.03	0.11	0.58
1988	0.05	0.28	12.9	0.01	0.06	0.34	0.02	0.1	1.13

approximately 0.25, 0.06, and 0.09 $m^3 \, s^{-1}$ at G1, G2 and G3, respectively. These flows represent only three times the minimum levels recorded (table 1). The slopes of all three duration curves are almost constant although this belies the flashy nature of the catchment in response to precipitation input. The hydrological significance of the high flow lies in the observation that 8.1% of the total flow volume at G1, in 1988, occurred over less than 1% of the time.

High-flow episodes are generated by three distinct mechanisms: snowmelt, rain on snow and rainfall (Jenkins 1989). Snowmelt produces a noticeably pyramidal hydrograph shape because of the gradual increase in flow as the melt intensifies, compared with the sharp rise in response to rainfall. Snowmelt and rain-on-snow events tend to be characterized by greater water yields although this is predominantly a function of the antecedent wetness of the catchment. The effect of storage dynamics is demonstrated by the hydrograph record in which three types of response are evident (figure 3). A sharp peak (spike) occurs in response to rainfall and during periods of prolonged rainfall (and so high antecedent wetness) this is superimposed on an underlying 'whaleback' that is in turn superimposed on a relatively constant baseflow. The implication is that the underlying baseflow is fed from groundwater whereas the whaleback represents the slow wetting of soils (especially the podzolic soils) in response to rainfall input and the slower draining of these soils following wet periods. The sharp hydrograph peaks are a result of preferential pathways operating during the storm events and the relative magnitude of the peak depends, in part, on its timing in relation to the position on the whaleback. By employing an arbitrary (but objective) hydrograph separation technique, and by using July 1988 as an example (figure 3), the contribution of each of the three modes of response has been calculated (table 3). The whaleback response is smallest at G2 and flows are dominated by baseflow and quickflow spikes reflecting

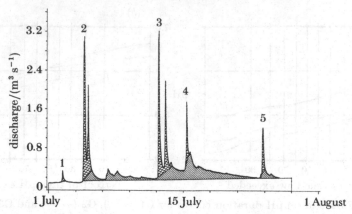

Figure 3. An example of the three modes of flow response at G1 during July 1988. Cross hatching delineates the spike, hatching represents the whaleback and the baseflow is unshaded.

Table 2. *Rainfall, peak flow (qpeak) and runoff yield at the three monitoring sites for a series of storms in 1988*

	rain/ mm	G1		G2		G3	
		qpeak/ $m^3 s^{-1}$	yield (%)	qpeak/ $m^3 s^{-1}$	yield (%)	qpeak/ $m^3 s^{-1}$	yield (%)
storm one 1–3/7/88	15.0	0.35	3.6	0.08	5.8	0.15	5.2
storm two 4–6/7/88	38.3	3.09	27.0	0.29	22.1	0.56	18.0
storm three 13–16/7/88	22.4	4.35	22.4	0.34	19.2	0.82	21.4
storm four 16–18/7/88	15.6	1.76	20.6	0.23	15.5	0.49	22.8
storm five 23–26/7/88	24.5	1.22	11.89	0.22	14.5	0.32	12.5
storm six 31/7–1/8/88	16.0	2.6	28.9	—	—	0.53	21.2

the relative absence of alpine and peaty podzol soils in the subcatchment (Ferrier *et al.* 1990). On the other hand, the G3 catchment has a high proportion of podzolic soils and 27% of the flow is contributed in the whaleback.

The influence of catchment wetness on storm yield is emphasized in table 2. During storm one, at the beginning of July 1988 (figure 3) when the catchment was dry, a total rainfall input of 15 mm produced a streamflow peak of 0.3 $m^3 s^{-1}$ at G1 and the quickflow yield (i.e. in the spike) was only 3%. The bulk of the rainfall was absorbed in making up soil moisture deficit and other storage deficits within the catchment. On the other hand, storm four, on 16–18 July, coincides with the peak of the whaleback response when the catchment is wetted up and, consequently, a flow peak of 1.7 $m^3 s^{-1}$ and a yield of 21% is achieved in response to a similar total rainfall input. Rainfall intensity can also affect runoff response, however, storm one (mean intensity 3.2 mm h^{-1}) was more intense than storm four (mean intensity 1.7 mm h^{-1}) and so the differences in flow peak and yield are directly attributable to catchment wetness.

These inferences regarding processes on the catchment scale can only be regarded

Table 3. *The contribution of the three modes of response to total streamflow at the three monitoring sites during July 1988*

		spike		whaleback		baseflow	
	total flow/mm	mm	(%)	mm	(%)	mm	(%)
G1	78	16	(21)	12	(15)	50	(64)
G2	90	10	(11)	8	(9)	72	(80)
G3	95	10	(11)	26	(27)	59	(62)

as generalizations of the complex hydrological interactions involved in the rainfall–runoff mechanism. Representative plot studies on the three main soil types in the catchment by using distributed tensiometer arrays and throughflow lysimeters have enabled an evaluation of the hydrological responses of the soils and these may be extrapolated to explain catchment hydrological response (Wheater *et al.* 1990). The picture that emerges is of deep drainage of the alpine soils and mid-slope generation of rapid throughflow in the peaty podzols, overlying a deeper damped water-table response. A locally perched water-table may feed preferred flowpaths in the peaty podzols that are then observed at the base of the podzol slopes. This highly transient response is rapidly transmitted under the peat to the streams providing a major contribution to the quickflow spike. The deeper, damped water-table response in the peaty podzol drains slowly downslope and the observed tensiometer response mirrors the whaleback response observed in the stream. The underlying baseflow in the stream is consistent between the sub-catchments and is maintained by slow drainage of deeper groundwater sources. This is entirely consistent with the catchment scale data.

4. Chemical response

Chemical response of the catchment and its sub-basins is inferred from continuously monitored pH data and from spot samples. Long-term chemical characteristics, based on annual average concentrations, are reported in Harriman *et al.* (this symposium). The percentage of time for which continuously monitored pH exceeded given values in 1988 is shown in figure 2*b* as pH duration curves. The curves for all three sites exhibit a similar mean pH (G1 = 6.1, G2 = 6.1 and G3 = 5.75) although the response to high flows is different. At G1 a pH of less than 5.0 is not recorded during 1988 whereas at G2 and G3, pH lower than 5.0 is observed for 5% and 2% of the time, respectively. The pH duration curves for all sites are not as flashy as the corresponding flow duration curves because the pH is slower to return to pre-storm conditions. G2 shows the most dynamic pH response of the three sites, characterized by the steepest slope.

The observed relations between flow and alkalinity, total organic carbon, calcium and sulphate are shown in figure 4; more detail is given by Harriman *et al.* (1990). In summary, calcium demonstrates a negative correlation with flow indicating the catchment derived nature of the calcium load as it dilutes with high flows. Sulphate is the most stable ion showing the least variation with flow and indicates a steady contribution of sulphate to the stream is maintained even at high flows. TOC increases consistently with increasing discharge and as a catchment-derived solute this is conventionally attributed to a greater proportion of the higher flow taking near surface pathways through organic soils.

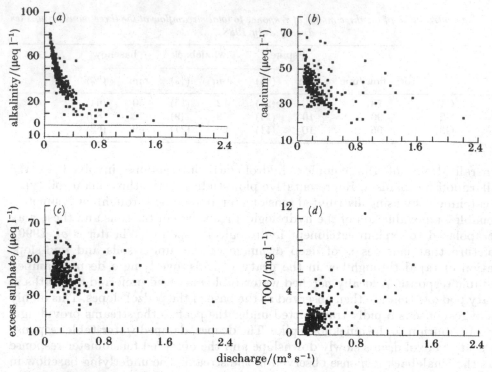

Figure 4. The relations of flow with (a) alkalinity, (b) calcium, (c) seasalt excess sulphate
and (d) total organic carbon, at G1.

Alkalinity exhibits the best correlation with flow and three distinct response areas
can be identified. In the flow range 0–0.2 m³ s⁻¹ alkalinity decreases rapidly and this
decline continues in the flow range 0.2–0.8 m³ s⁻¹ although the decline is somewhat
slower and the relation is less clear cut. This results from sampling on the rising and
recession limbs of the hydrograph that are characterized by different chemistries for
similar flow levels. Above flows of 0.8 m³ s⁻¹ alkalinity remains essentially constant
and no change is observed even up to flows greater than 10 m³ s⁻¹.

5. Integration

The crude separation of the hydrological response into three components
corresponds well with the fairly crude categorization of the alkalinity response to
flow. This provides a good starting point for integrating the observed responses by
incorporating a mixing approach. During low flow, when the soil is dry, a rain event
rapidly dilutes the high alkalinity baseflow water without any contribution from
upper soil horizons. As flow increases through the 0.2–0.25 m³ s⁻¹ level a more
complex mixing process occurs whereby the relative contributions from the different
soil horizons gradually changes as the catchment wets up. Any rainfall event
producing peak flows in the 0.2–0.8 m³ s⁻¹ range will have different chemical starting
points depending on the antecedent wetness of the catchment, which is reflected in
the whaleback hydrological response. The effect of antecedent wetness, therefore,
will be to introduce variability into the chemical composition of the whaleback
component. Above a flow of about 0.8 m³ s⁻¹ the output alkalinity remains stable,

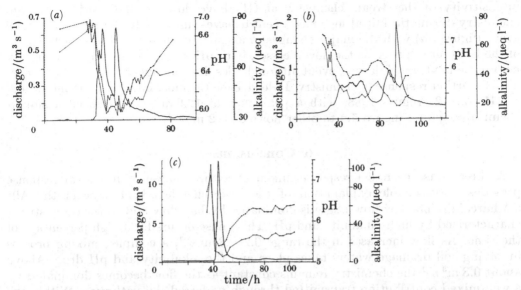

Figure 5. Alkalinity (. . . .), pH (–––––) and flow (——) for three storm events at G1; (*a*) 12–15 June 1989, (*b*) 26 February–3 March 1987, and (*c*) 12–15 August 1988.

indicating that either preferred pathways have become established, which contribute water of constant chemistry, or that efficient mixing of water from more than one source or pathway has become established, and contribute at a constant level. This general link between the hydrological and chemical responses is supported by the continuously monitored pH data (figure 1*b*). Although the general pattern of response between the sites is broadly similar, G2 has a higher baseflow pH and yet drops to the lowest pH during storm events (figure 2*b*). This reflects the dominance of peaty soils in the G2 sub-catchment and so high flows are dominated by contributions from soil pipes and other preferential pathways, in particular, beneath the peat at the interface with the mineral layer. As there is less slow drainage of podzols, than in G1 and G3, the pH returns to pre-storm levels more rapidly than at the other sites. The relative proportions of the three flow contributions at each site (table 3) support this hypothesis, the whaleback contributing only 8% of total flow at G2 compared with 12% and 26% at G1 and G3, respectively.

Given that water contributions from distinctly different flowpaths have been identified, the chemical changes in the stream can be interpreted as the result of mixing these flow components. Figure 5*a* shows a rainfall event on 13 June 1989 that is typical of events with peak flows up to about 0.8 m³ s⁻¹. On the rising limb of the hydrograph, both alkalinity and pH drop rapidly as baseflow water is neutralized by contributions from acidic soil water transmitted rapidly along preferred pathways. During the flow recession, when the contribution from these pathways has declined, continued slow drainage of acidic soil water ensures that little recovery occurs between the two flow peaks. Chemistry only slowly returns to pre-storm levels as the proportional contribution of baseflow water increases with respect to the slow draining soil water. This change in the relative contributions from baseflow and soil drainage is responsible for observed hysteresis in the chemical response.

The event on 26 February 1987 (figure 5*b*) comprises several peaks dominated by a snowmelt episode followed quickly by a rain-on-snow episode and demonstrates the insensitivity of the stream chemistry at G1 above flows of *ca.* $0.8 \, \text{m}^3 \, \text{s}^{-1}$. Stream chemistry during the initial snowmelt event behaves similarly to the summer rainfall event (figure 5*a*) with the major changes in chemistry occurring in the early stages of the hydrograph rise. As the flow recesses, both pH and alkalinity recover slightly before the next, much larger, event although, despite the higher flow reached, there is little further response in chemistry. Indeed, even the most intense storm monitored at G1, on 12 August 1988, with a peak flow of $12.9 \, \text{m}^3 \, \text{s}^{-1}$, produced a similar chemical response (figure 5*c*) to other flows of 1–$2 \, \text{m}^3 \, \text{s}^{-1}$.

6. Conclusions

A three-phase chemical response linked to a three-phase hydrological response provides a reasonable explanation of the hydrochemical behaviour of the Allt a'Mharcaidh. The low flow phase is dominated by baseflow and water chemistry is characterized by high alkalinity and pH. This phase occurs for a high percentage of the time. As flow increases in the range 0.2–$0.8 \, \text{m}^3 \, \text{s}^{-1}$, a complex mixing process involving soil drainage waters takes place and the alkalinity and pH drop. Above about $0.8 \, \text{m}^3 \, \text{s}^{-1}$ the chemistry remains constant as the flow becomes dominated by a well-mixed contribution transmitted through preferred soil pathways. Within the intermediate phase the slow drainage of soil waters of variable composition suppresses the recovery of alkalinity, pH and base cations whilst maintaining levels of organic carbon and non-labile aluminium. Only when this contribution begins to decline does stream chemistry slowly return towards pre-event levels.

This general model for hydrochemical response is grossly simplified. In reality, a myriad of hydrological pathways exist in the catchment that contribute to the whaleback and hydrograph spike responses and so process identification is difficult. The same is true of chemical response, in particular the second-phase mixing where different soil profiles and types are contributing water of varying quantity and quality. Despite the limitations of the mixing approach, the observed hydrological and chemical responses emphasise the importance of flowpaths and soil type in controlling stream chemistry.

The future acidification status of the Allt a'Mharcaidh depends upon the long-term response of the soils; in particular, their ability to adsorb sulphate and to exchange calcium and sodium. If the base saturation of the peaty podzol soil decreases, perhaps as a result of continued high sulphate loading, storm events will cause greater drops in pH, which will recover more slowly and only contribution from the well buffered baseflow will prevent chronic acidification. On the other hand, a decrease in deposition might increase the base status of the podzols and pH may recover more rapidly after storm events.

We are indebted to Jo Porter for his efforts in sample collection, data transfer and site maintenance. Our colleagues in the Macaulay Land Use Research Institute and Imperial College provided invaluable advice.

References

Battarbee, R. W. 1990 In *Palaeolimnology and lake acidification* (ed. R. W. Battarbee, Sir John Mason, I. Renberg & J. F. Talling) (*Phil. Trans. R. Soc. Lond.* B **327**), pp. 113–121. London: The Royal Society.

Brown, D. J. A. 1983 *Bull. Envir. Contam. Toxicol.* **30**, 582–587.

Ferrier, R. C., Walker, T. A. B., Harriman, R., Miller, J. D. & Anderson, H. A. 1990 *J. Hydrol.* (In the press.)

Harriman, R., Morrison, B. R. S., Caines, L. A., Collen, P. & Watt, A. W. 1987 *Wat. Air Soil Pollut.* **32**, 89–112.

Harriman, R., Gillespie, E., King, D., Watt, A. W., Christie, A. E. G. & Edwards, T. 1990 *J. Hydrol.* (In the press.)

Howells, G. D. 1983 *Adv. appl. Biol.* **9**, 143–155.

Jenkins, A. 1989 *Hydrol. Sci. J.* **34**, 393–404.

Reuss, J. O., Cosby, B. J. & Wright, R. F. 1987 *Nature, Lond.* **329**, 27–32.

Wheater, H. S., Tuck, S., Ferrier, R. C., Jenkins, A., Kleissen, F. M. & Walker, T. A. B. 1990 (In preparation.)

Discussion

M. CRESSER (*Department of Plant and Soil Science, Aberdeen University, Old Aberdeen, U.K.*). Is Dr Jenkins familiar with the papers of a research student of mine, Murray Reid, published in 1979 and the early 1980s, on the relations between water chemistry, soils and the hydrological pathway, in which the mixing model was used to explain solute composition changes during episodes for the River Dye in northeast Scotland? Even then, the approach had been used even earlier by Webb and Walling. In my experience of long-term studies in 22 catchments in the Scottish uplands, soil chemistry and hydrological pathway can always be used to explain river water chemistry and its changes during episodes, using this simple model. We have also found that, except in the case of peat-dominated catchments, outgassing of carbon dioxide explains why baseflow river-water pH exceeds the pH of lower mineral soil, pH by up to 2 pH units. We have never needed the ground-water from an unexplained source, advocated by Jenkins and Harriman. We have made long-term measurements of soil atmosphere carbon dioxide concentrations at two series of moorland sites, and found it to lie in the partial pressure range 0.01–0.03. I thought this concept was now widely recognized and accepted. Are Jenkins and Harriman now saying that this mechanism is not important?

A. JENKINS. We are familiar with the work cited and have not claimed that the mixing concept is new. We would, however, like to point out that soil chemistry is not as important as soil *water* chemistry in determining mixing relations. Furthermore, we would hope that, as the question rightly states, soil water chemistry and hydrological pathway *can* always be used to explain river water chemistry and its change through episodes, as there are no other significant parameters or processes involved. We have hypothesized a mixing scheme for the Allt a'Mharcaidh that utilizes several different water sources of different chemistry, driven by hydrological processes that are determined by catchment wetness.

As regards degassing of soil-water carbon dioxide, we have never suggested that this mechanism is unimportant. It should be remembered, however, that soil solutions with pH less than about 5.5 at the partial pressure stated in the question, will not degas to pH 6.8 (i.e. Allt a'Mharcaidh baseflow pH). The BC horizon soil water pH at Allt a'Mharcaidh is 4.7 (Ferrier *et al.*, this symposium) and so there must be some contribution of high pH water at baseflow. Stream baseflow pH cannot be explained solely by degassing of carbon dioxide from soil water. In the same respect, alkalinity concentrations at baseflow are higher than in soil water and so a source of high alkalinity water is needed to match observations as alkalinity is conserved on degassing.

Hydrochemical changes associated with vegetation and soils

By Robert C. Ferrier, John D. Miller, T. A. Bruce Walker
and Hamish A. Anderson

*Macaulay Land Use Research Institute, Craigiebuckler, Aberdeen AB9 2QJ,
Scotland, U.K.*

The total wet deposition loading of a catchment will comprise the product of the amount and chemical content of wet deposition as snow and rain (bulk deposition) and an enhancement contribution from the interception of mist and fog waters (enhancement deposition). Studies in the Allt a'Mharcaidh catchment (Cairngorms, Scotland) have indicated that bulk deposition is constant throughout the whole catchment; however, enhancement deposition was greatest at higher altitude. In the Høylandet, Loch Chon and Kelty Water forested systems, interception or scavenging of air-borne material alters the quality and quantity of hydrochemical inputs. The importance of these additional loadings in the assessment of total catchment inputs at the range of field sites studied, is discussed. The ability of vegetation to modify input chemistries will be illustrated by using selected data from a variety of forest and ground vegetation ecosystems in Scotland and Norway. Soil-water flow in different soil horizons is dependent upon the nature of the vegetation cover, soil type and hill-slope characteristics. The relative contribution of different soil waters to stream output chemistry is discussed in relation to potential sources of sulphate release, and to buffering processes that may alleviate deleterious soil-derived leaching.

---

1. Introduction

A broad correlation exists between the distribution of rainfall acidity and the occurrence of acid streams and lakes. However, the interaction of acid rain with rock, soil and vegetation cover modifies the effect on surface water acidity in two principal ways. Rock, soil and vegetation can act to neutralize rainwater acidity or to exacerbate its effect by the release of other acidifying components such as aluminium or organic acidity. The rates and pathways of water movement through, and the chemical reactions within, vegetation and soils are of critical importance, as are management regimes and practices that may enhance the total loading of inputs to systems.

Both losses and gains of certain elements result from the transfer to the soil surface of materials captured, washed and leached from different kinds of vegetation. The response of any particular vegetation type is dependent upon the prevailing anthropogenic climate and the nature of deposition mechanism (Fowler & Cape 1982; Coenan *et al.* 1987), altitude and exposure (Fowler *et al.* 1989), seasonality (Roberts *et al.* 1984), and the age and nutrient status of the vegetation (Matzner *et al.* 1983; Miller *et al.* 1987). The dominant soil processes in altering the composition of percolating waters are: the routing pathways and residence times of water in different horizons (Ferrier *et al.* 1990 a; Miller *et al.* 1990 a), sulphate adsorption/desorption dynamics (Walker *et al.* 1990 a), the nature and extent of the cation exchange complex, and the production of organic acidity (Walker *et al.*, this symposium).

This paper reports research undertaken at the primary SWAP sites to elucidate the dominant processes and factors controlling the modification of incoming anthropogenic pollution by vegetation and soils. The main aims can be summarized as follows: (i) to monitor the chemistry of anthropogenic deposition in areas known to experience different pollution loadings; (ii) to measure changes in the chemistry of water passing through the canopies of moorland and forest plant communities; (iii) to determine the spatial and temporal variations in soil hydrochemistry in the dominant soil types of the study catchments.

2. Sampling and methodology

Hydrological and hydrochemical measurement of inputs, vegetation and soil throughputs and stream outflow was done at the three major study sites, as described by Miller *et al.* (1989), Gaskin *et al.* (1989) and Ferrier & Harriman (this symposium). Samples were collected initially at two-week intervals, subsequently sampling intensity was increased to daily and finally hourly to identify specific process-related responses under different conditions.

3. Results and discussion

(a) Inputs

A network of collectors was installed at various altitudes and degrees of exposure in the Allt a'Mharcaidh catchment, northeast Scotland, in an attempt to obtain an accurate assessment of wet deposition loading (Ferrier *et al.* 1990 b). Results indicate that the quantity and quality of bulk deposition is constant over the whole catchment, during the period of the investigation. 'Enhancement deposition' as measured by filter gauge interception collectors indicated that there was approximately four-times more deposition at altitudes greater than 750 m. The concentrations of all elements in enhancement deposition were greater than those of bulk deposition (except for hydrogen) at the higher altitudes; at lower altitude, enrichment was only appreciable for sodium and chloride. Input–output chloride budgets were used to assess catchment evapotranspiration rates and the relative proportions of enhancement deposition within different altitudinal ranges. The percentage increase in total input to the catchment when enhancement deposition is included is shown in table 1. The increase is greatest for sea-derived salts and nitrogen compounds but negligible for hydrogen. The altitudinal range within the catchment is characterized by a change in the dominant vegetation communities (Nolan *et al.* 1985). Because of altitudinal differences in the amount and quality of enhancement deposition it is clear that different vegetation communities will experience different input loadings. Ionic enrichment above 750 m will affect predominantly the alpine azalea/lichen heath vegetation, whereas below 750 m the vegetation is mainly *Calluna* and *Nardus* with correspondingly different inputs.

Detailed process studies have concentrated on the effect of snowmelt on streamwater output at the SWAP sites in Scotland (Jenkins 1989) and in Norway

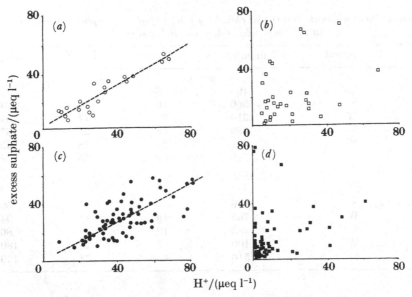

Figure 1. Ionic relations of wet deposition inputs at the Allt a'Mharcaidh catchment, Cairngorms, N.E. Scotland. Winter inputs ((a) rime, (b) snow) from winter season of 1987–88; Summer inputs ((c) rain, (d) interception) from 1986–88.

Table 1. *Total loadings* (kg) *of elements in bulk deposition and enhancement deposition* (*June–December 1986*)

(Allt a'Mharcaidh catchment, Cairngorms, N.E. Scotland. (From Ferrier *et al.* 1990*b*.))

element	measured bulk deposition	enhancement > 750 m	enhancement < 750 m	total input	percentage increase
K	630	57	18	705	12
Ca	900	57	22	979	9
Mg	730	93	36	859	18
Na	5930	651	265	6846	15
NH_4-N	540	48	10	598	11
NO_3-N	470	51	8	529	12
Cl	11284	1321	493	13098	16
SO_4-S	3060	180	51	3291	7
H	110	2.2	0.3	112.5	2

(Ferrier *et al.* 1989). Up to 20 % of the total volumetric input at the Allt a'Mharcaidh catchment comes as winter inputs, and the ionic relations of winter inputs are very different from those during the summer. Figure 1 shows the relation between hydrogen and excess sulphate in snow, rime, rain and interception (water collected by the filter gauges). The strong correlation of hydrogen to sulphate (5 % level) in rainwater implies that the incoming acidity is of anthropogenic origin. Water intercepted by filter gauges does not show a relation between H^+ and excess SO_4^{2-}, and indeed this input is predominantly less acidic than rain. Snow does not show any direct ionic relation because of the filtering of non-acidic particulates from the atmosphere, whereas rime (the deposition of supercooled water droplets) chemistry indicates the presence of an anthropogenic pollution climate during the winter months.

Table 2. *Seasonality in amounts* (mm) *and chemistry* (μeq l^{-1}) *of bulk deposition and throughfall at the Høylandet, Chon and Kelty sites*

site	period	amount/mm	H$^+$	NH$_4$-N	SO$_4$-S
		bulk deposition			
Høylandet	W	495	8	4	17
	S	366	10	16	34
Chon	W	1310	20	14	54
	S	680	28	24	57
Kelty	W	1930	25	18	45
	S	710	40	45	78
		throughfall			
Høylandet	W	354	3	4	33
(below understory)	S	146	1	11	50
Chon	W	765	19	18	94
	S	485	12	18	86
Kelty	W	1100	62	28	180
	S	516	45	31	130

The forested Surface Waters Acidification Programme (SWAP) catchments also show seasonality in the amount and chemistry of bulk deposition inputs (table 2). Winter volumes (excluding snow) always exceed those of the summer period, however higher concentrations of H$^+$, NH$_4$-N, and SO$_4$-S in rainfall are found in summer rainfall. The concentration difference between the periods is proportionally greatest at Hoylandet, where there were increases of four-times and twice in the concentration of ammonium and sulphate in bulk deposition, respectively. This large increase in ammonium concentration presumably reflects the effect of organic farming practices prevalent in this region of Norway (Ferrier *et al.* 1990c).

(b) Vegetation effects

The fate of the seasonal input of ammonium as it passes through the vegetation canopy is shown in table 2. As water passes through the vegetation, evaporation processes cause an increase in concentration (assuming no interaction with the canopy). There was such a concentration increase in winter throughfall values, but decreases in summer indicate the seasonal uptake of ammonium. Uptake of ammonium by vegetation canopies can result in an exchange for hydrogen with a resultant decrease in the pH of throughfall water, as seen for the Sitka spruce at Kelty (Miller 1984). Clearly this is not the case at Chon and Høylandet (both *Picea abies*) as there is a reduction in the hydrogen ion concentration of throughfall, implying base cation release to balance NH$_4$-N uptake. Sulphate concentrations in throughfall water show an increase throughout the year that is proportionally greatest at the Kelty site during the winter period. The calculated fluxes of SO$_4^{2-}$, K$^+$ and H$^+$ ions above and below the vegetation canopies are shown in table 3. Input water represents the product of bulk deposition loading plus an additional contribution of enhancement deposition (as calculated above) and water collected below the vegetation is throughfall plus stemflow for all sites. In the forested systems there is an increase in the amount of potassium in water as it passes through the canopy, which is proportionally greatest under the Norway spruce at Chon. This is also true for the two dominant ground vegetation types in the Mharcaidh catchment

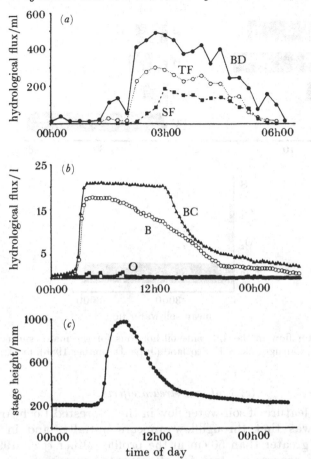

Figure 2. Hydrological fluxes through vegetation and soil and the subsequent stream response, Chon catchment, Loch Ard area, west central Scotland. (*a*) Vegetation; (*b*) soil; (*c*) stream; BD, bulk deposition; TF, throughfall; SF, stemflow. (From Miller *et al.* 1990*b*.)

(Ferrier *et al.* 1990*a*). Acidity is reduced under all vegetation types except for Sitka spruce in the Kelty catchment, and under this species there is an increase in the amount of sulphate-S present in throughfall and stemflow (Miller *et al.* 1990*a*, *b*). In the afforested systems (Chon and Kelty) water from below the canopy arrives at the soil surface, whereas at Høylandet there is further modification and neutralization as water passes through the *Vaccinium* understory.

Laboratory analysis of tree foliage has shown that Sitka spruce needles from Kelty have higher sulphur concentrations than Norway spruce at Chon, that is 0.2% to 0.1%, although needle-litter sulphur is similar at both sites. Much more of this sulphur is water soluble and exchangeable in the Sitka compared with the Norway spruce (40% at Kelty compared with 15% at Chon). Water collected below accumulating litterfall is also enriched in sulphate at Kelty compared with Chon so it therefore seems that the difference in sulphate in throughfall and stemflow are true species effects due to the Sitka spruce.

R. C. Ferrier and others

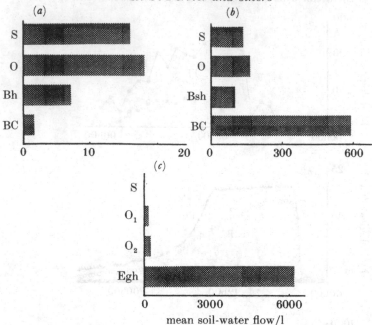

Figure 3. Mean soil-water flow in the different soil horizons for the major soil types in the Allt a'Mharcaidh catchment, Cairngorms, N.E. Scotland (June–December 1986). (*a*) Alpine podzol; (*b*) peaty podzol; (*c*) peat.

(*c*) *Soil and stream effects*

The most striking feature of soil-water flow in the afforested Loch Ard study sites (Chon and Kelty) was that throughflow appears initially, and in the greatest amounts, at depths greater than 50 cm in the profile (Miller *et al.* 1990*a*). During rain/snow events, a progressive transfer of saturated water up the profile then follows (figure 2). A more detailed evaluation of soil hydrological processes is presented by Wheater *et al.* (this symposium), however basal horizon flow is the most important hydrochemical contribution to stream flow.

Soil-water movement on the hill slopes of the Allt a'Mharcaidh is shown in figure 3. Lateral water flow in the alpine podzol was predominantly on the surface and in the organic horizon, water penetrating the upper horizons was subsequently channelled vertically through the highly permeable lower mineral horizons to a depth below that of the deepest lysimeter installation (80 cm). In the lower peaty podzol soils the predominant flow appears at depth in the BC horizon. This represents water flow at depth from upslope and infiltration of input water to the peaty podzol hillslopes. The volume of flow in the organic and surface horizons exceeds that of Bsh flow as a result of the peaty podzols on these mid-altitude transit slopes having developed iron pan and Bx (indurated) horizons, which although discontinuous, will restrict water movement down the profile. The western side of the catchment is characterized by an area of blanket peat. Water movement at this site was mainly below the peat at the junction of the organic/mineral interface being supplied from upslope and at depth. The peat is also characterized by the presence of pipes and incised eroded channels to the level of the mineral Egh horizon (Wheater *et al.*, this symposium). Clearly, again it is water movement at depth that is the major contributor to stream flow.

Table 3. *Elemental loadings (including enhancement deposition) above and below vegetation canopies* ((All data as kg ha^{-1} a^{-1}). Gains/losses greater than 10% are indicated at $+/-$.)

site/species	K above	K below	SO$_4$-S above	SO$_4$-S below	H$^+$ above	H$^+$ below
		forest vegetation				
Høylandet	2.18	10.2	4.8	4.8	0.20	0.04
(*Picea abies*)		+				−
Chon	6.7	32.6	22.0	23.0	0.48	0.24
(*Picea abies*)		+ +				−
Kelty	5.6	18.8	23.3	38.6	0.72	0.80
(*Picea sitchensis*)		+		+ +		+
		ground vegetation				
Mharcaidh	1.6	3.9	7.3	4.0	0.24	0.12
(alpine heath)		+		−		−
Mharcaidh	1.3	4.1	6.3	3.7	0.22	0.10
(*Calluna*)		+		−		−
Høylandet	10.2	11.9	4.8	3.0	0.04	0.01
(*Vaccinium* spp.)		+		−		−

Table 4. *Selected soil-water leachate chemistry for the organic and BC horizons of the forested Chon and Kelty catchments*

(All concentrations as mg l^{-1}, except hydrogen, as µeq l^{-1}.)

site	H$^+$	K	SO$_4$-S	DOC
	organic horizon leachate			
Chon	68	1.95	1.96	21
Kelty	104	0.35	1.84	12
	BC horizon leachate			
Chon	37	0.27	1.33	3
Kelty	79	0.35	1.17	13

Water chemistry in the organic and BC horizons of the forested Chon and Kelty catchments is shown in table 4. Organic horizon leachate is consistently more acidic (due to organic acidity) than BC flow at both sites. The importance of organic acidity on soil processes will be considered in detail elsewhere (Walker *et al.*, this symposium), however it is clear that there is a reduction in DOC concentration as water passes from the upper horizons to the horizons in the humus–iron podzol at Chon. The peaty gley soils of the Kelty catchment do not show a concentration reduction and the total organic contribution from soil-water output (basal flow is the major contributor to streamflow) is considerably greater at this site. Potassium concentrations are reduced in the BC Chon flow indicating tight cycling of this element, as there is a significant increase in the amount of potassium in throughfall (table 3). Sulphate concentrations in organic horizon leachate at both sites are greater than that of BC flow, reflecting mineralization of organic-S (Walker *et al.*, this symposium), however the contribution to total soil flow of organic horizon leachate is much smaller than that from the basal horizons.

Differences in stream chemistry between the Chon and Kelty sites reflect a combination of vegetation and soil processes in terms of both chemistry and hydrology. The changes in concentrations of calcium and sulphate from inputs to

Table 5. *Calcium and sulphate concentrations* (μeq l^{-1}) *at the Chon and Kelty catchments, Loch Ard Area, west central Scotland*

	calcium		sulphate	
	Chon	Kelty	Chon	Kelty
bulk deposition	19	17	56	60
throughfall	50	45	92	164
stemflow	64	62	136	170
BC soil water	19	14	89	73
streamwater	43	19	93	93

Table 6. *Range of calcium concentration and mean pH of basal soil horizon drainage water, stream-water at catchment outflow, and shallow groundwater from the riparian zone*

(Allt a'Mharcaidh, Cairngorms, N.E. Scotland (all concentrations as μeq l^{-1}.)

	calcium concentration (range)	mean pH
Egh horizon (peat)	10–20	4.4
BC horizon (peaty podzol)	10–20	4.7
streamwater (catchment outflow)	35–45	6.5
shallow groundwater	40–60	7.0

outputs illustrate some of these mechanisms (table 5). The calcium concentration in the Chon stream was higher than values from the soil horizons which indicates an additional source. This has been traced to a small doleritic intrusion (Bain *et al.*, this symposium) present in the upper catchment. This has a greater influence on stream chemistry, which is an integral from a large area, compared with the more localized plot chemistry, and is the major contributor in maintaining water quality at a standard viable for fish survival (Ferrier & Harriman, this symposium). Analysis of soil-water throughflow from pits close to this more base-rich material contained much more calcium (greater than 100 μeq l^{-1} Ca^{2+}) compared to the study plots (19.5 μeq l^{-1}) (Miller *et al.* 1990*b*). The upper stream also had higher levels of calcium (99.4 μeq l^{-1}) compared with the lower sampled stream. Battarbee (1988) analysed Loch Chon water as part of a sediment survey, and this data confirms these enhanced calcium levels (79.4 μeq l^{-1}). This demonstrates the importance of any small variation in catchment geology leading to altered weathering inputs. This mechanism leading to changes in stream chemistry may mask any vegetational effects.

Sulphur dynamics also differ between Chon and Kelty, so that although the stream sulphate concentrations are similar, these are the product of different processes. As previously mentioned, there is a species effect of Sitka spruce at the Kelty site, with elevated levels of sulphur being found below the vegetation. At Chon, throughfall sulphate concentrations (92 μeq l^{-1}), are similar to soil-water (89 μeq l^{-1}), and stream outflow (93 μeq l^{-1}). This implies that root uptake and sulphur retention (i.e. adsorption, immobilization, or ion pairing) by the soils is in balance with sulphur release by the soils (either as mineralization, desorption, or ion exchange processes). At Kelty, the large reduction in sulphate concentration from vegetation throughfall (164 μeq l^{-1}) to soil-water (78 μeq l^{-1}) implies a greater uptake of sulphate by the Sitka spruce and a retention of sulphate by the soil. Indeed, mineralization experiments have indicated that there is 20% more total mineralizable sulphur

Table 7. *Summary of vegetation, soil and stream responses to incoming anthropogenic deposition at the various study catchments*

site	inputs	vegetation response	soil effects	outputs
Høylandet	forest interception	neutralization	S retention	buffered by groundwater
Mharcaidh	altitude enhancement	neutralization	S retention	buffered by long residence-time water
Chon	forest interception	neutralization	output = input	buffered by doloritic dyke
Kelty	forest interception	acidification	S retention	no buffering

present in the soils at Kelty (220 kg ha^{-1}†) as compared with Chon (183 kg ha^{-1}). The nature of sulphur dynamics in the soils of the study catchments is discussed in full in Walker *et al.* (this symposium). Streamwater concentrations at Kelty are elevated (93 µeq l^{-1}) with respect to soilwater because of organic horizon leachate contributing directly to the stream in areas where drainage ditches intercept the main stream channel.

Streamwater quality at the Allt a'Mharcaidh should reflect a mixing of soil waters from the Egh horizon of the peat soils and the BC horizon of the peaty podzols where the dominant flow occurs. Table 4 shows the range of calcium concentrations and average pH of these two soil horizons, and of main stream outflow. It is not possible to reconcile soil-water chemistry with that of the stream, so evidently another source of calcium exists within the catchment. Detailed on site survey and mineralogical analysis (Bain *et al.* this symposium), has revealed no geologially distinctive base-rich material. Bore-hole excavation within the riparian zone and steep transit slopes on the east of the catchment have subsequently indicated the presence of shallow (2–3 m) water with a long residence time. Base flow conditions are primarily influenced by this water, maintaining high alkalinity, with associated stream buffering (Harriman *et al.* 1987, 1990; Jenkins *et al.* this symposium).

4. Conclusions

The plot-based studies have elucidated some of the major processes governing stream-flow hydrochemical response. The prevailing anthropogenic climate, the type and response of the vegetation, and the dominant catchment soils all contribute to the observed changes in percolating waters. The major conclusions of the research are summarized for each site in table 7, and as follows.

1. Høylandet. The dominant processes at this site include interception and neutralization of low levels of anthropogenic deposition by the vegetation (Ferrier *et al.* 1990*c*), and the retention of sulphate by the soils (Walker *et al.* 1990). A high alkalinity baseflow component (Christophersen *et al.* 1990) contributes substantially to streamflow buffering at this site.

2. Mharcaidh. Increased deposition loadings, due to the interception of cloud/mist water are a direct result of the altitudinal range in the catchment (Ferrier *et al.* 1990*b*). Different vegetation communities receive different deposition loadings, however there is a neutralization of inputs as water passes through the different

† 1 ha = 10^4 m^2.

vegetation canopies (Ferrier *et al.* 1990*a*). Sulphate is retained within the catchment (Jenkins *et al.* 1989), and the presence of long residence-time water results in stream buffering (Harriman *et al.* 1990).

3. Chon. Although in an area subject to a high anthropogenic loading, incoming deposition is partly neutralized by the vegetation. S release from the soil must be approximately balanced by vegetation uptake and immobilization as sulphate concentrations in throughfall, soil and streamwater are similar. The major factor governing streamwater quality at this site is the presence of a doloritic dyke in the upper region of the catchment (Miller *et al.* 1990*b*; Bain *et al.*, this symposium).

4. Kelty. Sulphate inputs to this site are increased under the Sitka spruce and sulphur cycling by the vegetation is a dominant process in controlling soil-water S concentrations (Miller *et al.* 1990*b*). The vegetation causes a net acidification of input water. There is retention of S by the soil but the lack of a further buffering results in stream water quality that cannot support a fish population (Ferrier & Harriman, this symposium).

We are indebted to those who assisted in all aspects of this research work, especially, Jo Porter for field assistance, and our colleagues at IH, FFL, FC, IC, University of Trondheim, and NIVA. The support of members of the Analytical Division of MLURI is greatly appreciated.

References

Battarbee, R. W. 1988 Lake acidification in the United Kingdom. DOE Report.

Christophersen, N., Vogt, R. D., Anderson, H. A., Ferrier, R. C., Miller, J. D., Neal, C. & Seip, H.-M. 1990 *Wat. Resour. Res.* **26**, 59–67.

Coenan, B., Ronneau, C. & Cara, J. 1987 *Wat. Air Soil Pollut.* **36**, 231–237.

Ferrier, R. C., Anderson, J. S., Miller, J. D. & Christophersen, N. 1989 *Wat. Air Soil Pollut.* **44**, 321–337.

Ferrier, R. C., Walker, T. A. B., Harriman, R., Miller, J. D. & Anderson, H. A. 1990*a* *J. Hydrol.* **115**. (In the press.)

Ferrier, R. C., Jenkins, A., Miller, J. D., Walker, T. A. B. & Anderson, H. A. 1990*b* *J. Hydrol.* **113**, 285–296.

Ferrier, R. C., Miller, J. D., Walker, T. A. B., Anderson, H. A. & Christophersen, N. 1990*c* *Wat. Air Soil Pollut.* (Submitted.)

Fowler, D. & Cape, J. N. 1982 In *Precipitation scavenging* (ed. H. R. Pruppacher, R. G. Semonin & W. G. N. Slinn), pp. 763–773. London: Elsevier.

Fowler, D., Cape, N. J. & Unsworth, M. H. 1989 *Phil. Trans. R. Soc. Lond.* B **324**, 247–265.

Gaskin, G. J., Miller, J. D. & Stuart, A. N. 1989 *Brit. Geomorph. Res. Group Tech. Bull.* **37**. (In the press.)

Harriman, R., Gillespie, E., King, D., Watt, A. W., Christie, A. E. G., Cowan, A. A. & Edwards, T. 1990 *J. Hydrol.* **115**. (In the press.)

Harriman, R., Wells, D. E., Gillespie, E., King, D., Taylor, E. & Watt, A. W. 1987 *SWAP mid-term conference, Bergen, 1987*.

Jenkins, A. 1989 *Hydrol. Sci.* **34**, 393–404.

Jenkins, A., Ferrier, R. C., Walker, T. A. B. & Whitehead, P. G. 1988 *Wat. Air Soil Pollut.* **40**, 275–291.

Matzner, E., Khanna, P. K., Meiwes, K. J. & Ulrich, B. 1983 *Pl. Soil* **74**, 343–358.

Miller, J. D., Stuart, A. W. & Gaskin, G. J. 1989 *Brit. Geomorph. Res. Group Tech. Bull.* (In the press.)

Miller, J. D., Anderson, H. A., Ferrier, R. C. & Walker, T. A. B. 1990*a* *Forestry.* **63** (3).

Miller, J. D., Anderson, H. A., Ferrier, R. C. & Walker, T. A. B. 1990*b* *Forestry.* **63** (4). (In the press.)

Miller, H. G. 1984 *Phil. Trans. R. Soc. Lond.* B **305**, 339–352.

Miller, H., Miller, J. D. & Cooper, J. M. 1987 In *Pollutant transport & fate in ecosystems* (ed. P. J. Goughtrey, M. H. Martin & M. H. Unsworth. *British Ecological Society*, publication no. 6. Oxford: Blackwell Scientific.

Nolan, A., Lilly, A. & Robertson, J. S. 1985 Surface Waters Acidification Programme: the soils of the Allt a'Mharcaidh catchment. Aberdeen: Macauley Land Use Research Institute. (Restricted circulation.)

Roberts, G., Hudson, J. A. & Blackie, J. R. 1984 *Agric. Wat. Mgmnt* **9**, 177–191.

Walker, T. A. B., McMahon, R. G., Hepburn, A. & Ferrier, R. C. 1990 *Sci. tot. Envir.* **92**, 235–247.

Precipitation and streamwater chemistry at an alpine catchment in central Norway

By Inggard A. Blakar[1], Ivan Digernes[1] and Hans M. Seip[2]

[1] Norwegian Institute for Nature Research, Tungasletta 2, N-7004 Trondheim, Norway
[2] Department of Chemistry, University of Oslo, P.O. Box 1033 Blindern, 0315 Oslo 3, Norway

The catchment studied (Storbekken) is part of the river Atna drainage basin, an inland high-mountain area highly susceptible to acidification and moderately affected by acid precipitation. About 250 streamwater samples have been collected annually during the period 1986–1988. Snow and rain have also been sampled. All samples have been analysed for chemical components. Discharge and water temperatures have been measured.

The pH in precipitation normally varied between 4.3 and 5.4 (mean 4.7), while the mean concentration of sulphate was about 30 µequiv. l^{-1}. Ionic concentrations in streamwater were very low; the conductivity was 5–13 µS/cm. Concentrations of most ions increased during winter and decreased during snowmelt and summer periods with high discharge. However, turbidity and colour (which may be regarded as a measure of total organic carbon) increased during highflow. At highflow pH reached as low as 4.7–4.8 and increased to about 5.5 at baseflow. Inorganic, monomeric aluminium (Al_i) increased at high flow to levels that may stress fish (4–5 µM). The mean streamwater sulphate concentration was the same as in precipitation showing that sulphate retention occurs in the catchment.

1. Introduction

The Storbekken drainage basin, which is susceptible to acidification and moderately affected by acid precipitation, was chosen as one of three SWAP catchments in Norway (Seip *et al.*, this symposium) for intensive study of water chemistry to obtain a better understanding of acidification processes and for use in validating the Birkenes model of water acidification (Stone & Seip 1989; Christophersen *et al.*, this symposium). The location and main characteristics of the catchment are described by Seip *et al.* (this symposium).

2. Sample collection and analyses

During the period 1986–88, approximately 250 streamwater samples have been collected annually. Snow samples were collected from profiles and with a snow-melt gauge. Rain was collected with three samplers at 750, 1000 and 1200 m a.s.l. and one pluviograph at 1000 m a.s.l. functioning as a bulk sampler. All samples were analysed for chemical components (table 1). A dam was constructed with a thin plate V-notch weir with a notch angle of 120° to monitor discharge. The head and water level at a distance above the dam were recorded manually. The discharge was also

Table 1. *Analytical methods*

parameter	method
pH	radiometer PHM 62 pH-meter
alkalinity	conductivimetric titration
Ca, Mg, Na, K	atomic absorption, Perkin-Elmer 603
NH_4-N	Shimadzu UV-160, indophenol blue
SSAA	autoanalyser, FIA Star 5020 ion exchange
	and conductivimetry
Cl	autoanalyser, Fe-thiocyanate
NO_3-N	autoanalyser, nitrite (azo-red)
SO_4	$= $ SSAA $- $ Cl $- NO_3$
Si	autoanalyser, molybdate blue
conductivity	Pt-electrode, Philips PW 9501
colour	OD-410, Shimadzu UV-160
turbidity	Nefelometric, HACH 2100
Al_a	autoanalyser, pyrochatechol violet
Al_o	autoanalyser, ion exchange
Al_i	$= Al_a - Al_o$

monitored by one limnograph in the dam and one in a natural profile downstream. The water temperature was measured with a thermistor.

3. Results

The three-year mean rain and stream concentrations for a variety of chemical species are shown in table 2. Discharge, and some streamwater chemical parameters; pH, calcium, sum of strong acid anions (SSAA), nitrate, and acid reactive aluminium (Al_a) are presented in figure 2. The mean annual precipitation for the period 1986–88 was approximately 550 mm. The corresponding estimate of runoff from the catchment was about 300 mm. This gives an estimated evaporation of about 40% of the precipitation. This estimate is not adjusted for the effect of increased precipitation with height, so the real evaporation is probably higher.

4. Discussion

The snow-free period is only 5–6 months in the lower part of Storbekken at about 730 m a.s.l. and even less higher in the catchment. Atna is located in the rain-shadow from west, and most of the rain falls as convective precipitation during July and August.

Most of the ion concentrations increased during autumn and winter and decreased during snowmelt and quickflows in the summer, whereas turbidity and colour (humic concentration) increased. Snowmelt and spring flood started in April and culminated in May–June. The water temperature was at its highest (10–13 °C) in July–August. The average value of turbidity was 0.2 FTU and of the colour 7 mg Pt l⁻¹. The colour may be regarded as a measure of total organic carbon (TOC) and the given value corresponds to TOC less than 2 mg C l⁻¹. Organic acids therefore do not contribute greatly to the acidity, except probably in some episodes when TOC reached 5–6 mg C l⁻¹.

The ion content of the streamwater is extremely low with a conductivity in the range 5–13 µS cm⁻¹. The highest conductivity values are due to a high H⁺

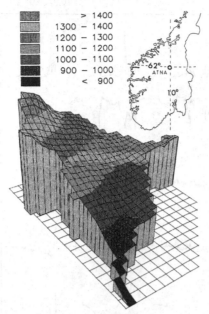

> 1400
1300 – 1400
1200 – 1300
1100 – 1200
1000 – 1100
900 – 1000
< 900

Figure 1. Topographic view of the Storbekken catchment (150 × 150 m² grid). The catchment is 5.8 km² and ranges from 780 to 1537 m a.s.l.

Table 2. *Chemical composition of precipitation and streamwater at Storbekken (Atna). Mean of three years and standard deviation*

	precipitation $n \approx 230$		streamwater $n \approx 750$		
pH	4.7	(0.3)	5.2	(0.2)	
alkalinity	0		1	(1)	μequiv. l⁻¹
Ca	2	(1)	11	(2)	μequiv. l⁻¹
Mg	1	(2)	6	(2)	μequiv. l⁻¹
Na	3	(3)	9	(1)	μequiv. l⁻¹
K	1	(1)	7	(2)	μequiv. l⁻¹
NH_4	12	(15)	1	(1)	μequiv. l⁻¹
SO_4	30	(20)	30	(4)	μequiv. l⁻¹
Cl	3	(2)	7	(2)	μequiv. l⁻¹
NO_3	11	(9)	2	(3)	μequiv. l⁻¹
Si			39	(11)	μM
conductivity			7	(1)	μS cm⁻¹
colour			7	(6)	mg Pt l⁻¹
turbidity			0.2	(0.1)	FTU
Al_a			3	(1)	μM
Al_o			2	(1)	μM
Al_i			1	(1)	μM

contribution. The alkalinity was very low (0–4 μequiv. l⁻¹). The pH varied between 4.8 and 5.5 in streamwater (mean 5.2). In snow-columns and precipitation, the pH normally varied between 4.3 and 5.4, with annual mean pH of 4.7. The pH (H_2O) in the upper organic soil layer was 4.1. The greatest drop in streamwater pH was during the first part of the snowmelt, and could partly be due to accumulated dry deposition and an initial wash-out of ions from the snow-pack.

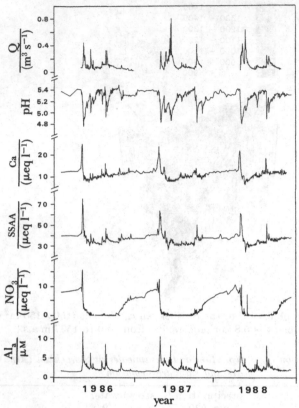

Figure 2. Discharge (Q), pH, calcium, sum of strong acid anions (SSAA), nitrate and aluminium (Al$_a$) in the Storbekken catchment during the period 1986–88.

The concentrations of calcium, magnesium, sodium and potassium varied from 6 to 11 µequiv. l^{-1} in streamwater and from 1 to 3 µequiv. l^{-1} in precipitation. The increase in streamwater concentrations of the base cations compared with precipitation is mainly due to evaporation, weathering, and ion-exchange processes. The highest potassium concentration in streamwater was registered in the first phase of spring flood and could be due to accumulation in the vegetation during summer and decomposition enhanced by freezing–thawing during winter. A similar peak was found for calcium and magnesium but not for silicate, which excludes weathering as a potential causal mechanism. The ammonium concentration was 12 µequiv. l^{-1} in precipitation and 1 µequiv. l^{-1} in streamwater. There was an ammonium peak during snowmelt, but ammonium was consumed by vegetation in the summer and autumn.

The sum of strong acid anions (SSAA $= SO_4 + Cl + NO_3$) was quite low with an average of 38 µequiv. l^{-1} (figure 2). The average sulphate concentration was 30 µequiv. l^{-1} both in streamwater and precipitation (table 2). With an estimated evaporation of approximately 40 %, this indicates a significant loss of sulphate in the catchment because of sulphate reduction or adsorption. A general decrease in sulphate concentrations in precipitation in southern Norway have been observed in recent years (SFT 1989). The average nitrate concentration was 2 µequiv. l^{-1} in streamwater and 11 µequiv. l^{-1} in precipitation. The streamwater nitrate concentration was very low during the summer (< 1 µequiv. l^{-1}), when most is taken up

by the vegetation, but increased towards the end of the winter to 8–9 µequiv. l⁻¹.
There is a growing concern about increasing nitrate concentration as a possible agent
of acidification (Henriksen *et al.* 1988), but in Storbekken, most of the nitrate input
from precipitation seems to be bound in the vegetation in the catchment. Nitrate
does not, as yet, contribute significantly to the water acidification in the area. The
average chloride concentration was 7 µequiv. l⁻¹ in streamwater and 3 µequiv. l⁻¹ in
precipitation. Corrected for evaporation, chloride seems to behave as a conservative
mobile anion, with no net retention in the watershed, in contrast to sulphate. Only
minor, if any, sources of chloride are found in the catchment. The sodium:chloride
ratio in streamwater is 1.26 (seawater 0.86); sodium must therefore be released from
mineral sources. The iron and manganese concentrations were low (< 1 µequiv. l⁻¹).
The average silicate concentration was 39 µM. During snowmelt, the silicate
concentration dropped to 16 µM and increased gradually to 70 µM during the year.
The silicate concentration is unaffected by storm events, compared with pH, which
gives rapid responses. This can possibly be explained by a piston effect whereby
spring meltwater without silicate is flushing out a deeper reservoir, reducing the high
silicate concentration, which gradually builds up during the rest of the year. Acid
reactive aluminium (Al_a) concentrations were low (mean 3 µM). The labile monomeric
aluminium (Al_i), which is considered to be the most toxic Al-fraction, was, in general,
less than 2 µM and probably does not represent any immediate danger to aquatic
biota. In episodes [Al_i] reached 5 µM. As the toxicity of aluminium increases with
decreasing concentration of silicate (Birchall *et al.* 1989) and calcium, these values
may correspond to significant aluminium toxicity during early summer when [Si] and
[Ca^{2+}] are low.

5. Conclusions

Atna is an area thought to be very sensitive to acid deposition. Many streams and
lakes have water quality that can just sustain fish populations. Although the
deposition of acidifying substances is modest, it is therefore important to follow the
development in water quality closely. The importance is enhanced by the large
number of sensitive, alpine catchments in Norway.

References

Birchall, J. D., Exley, C., Chappell, J. S. & Phillips, M. J. 1989 *Nature, Lond.* **338**, 146–148.
Henriksen, A., Lien, L., Traaen, T. S., Sevaldrud, I. S. & Brakke, D. F. 1988 *Ambio* **17**, 259–266.
SFT 1989 The Norwegian monitoring programme for long-range transported air pollutants.
    Annual report. Report no. 375/89.
Stone, A. & Seip, H. M. 1989 *Ambio* **18**, 192–199.

Chemical dynamics of an acid stream rich in dissolved organics

By Harald Grip and Kevin H. Bishop

Department of Forest Site Research, Swedish University of Agricultural Sciences,
S-901 83 Umeå, Sweden

The stream at the Svartberget Forest Research Station (64° 14′ N, 10° 46′ E, altitude 235–310 m) drains a 50 ha† catchment with an 8 ha open mire. The chemistry of atmospheric deposition and streamwater were measured between June 1985 and June 1989.

Wet deposition of sulphate was 40 meq m^{-2} a^{-1}. Total deposition roughly balanced stream output. The catchment was a sink for nitrate, ammonium and hydrogen ions. Marine excess sulphate in precipitation was 95 %. Weekly stream pH varied between 6.8 and 4.2 with a weighted mean of 4.7. Sulphate concentrations in streamwater were negatively correlated with hydrogen ion concentration, with the exception of one, three month winter period.

A balance between major cations and anions was found in the precipitation, but there was an anion deficit of about 50 % in streamwater. The stream carried 5–30 mg C l^{-1}, and unmeasured organic anions may account for the anion deficit. Concentrations of total aluminium were often 0.25 mg l^{-1} but less than 20 % was in the labile monomeric form.

The high concentrations of organic carbon and low fraction of labile monomeric aluminium offer a 'signature' of organic acidity that can be used to distinguish streams acidified by mineral acids. The possibility that anthropogenic acid deposition has altered the complex organic chemistry and pH of the Svartberget stream cannot be dismissed.

1. Introduction

The Svartberget area was chosen for study in the Surface Waters Acidification Program (SWAP) as an area low in pollution load with little marine influence to be compared with the catchments selected in Norway, where the gradient in acid deposition was the main criterion (Seip *et al.*, this symposium). The goal was to gather a data set, for use in testing acidification models under a wide variety of conditions.

The data are now in hand and catchment features specific to the catchment will be discussed below. Although intensive event studies may be the key to understanding processes, longer time series are necessary to quantify input–output balances, to place the studied events in perspective and to gain insight into the cycling of different elements.

Although the data span four years, there are indications that this is a too short

† 1 ha = 10⁴ m².

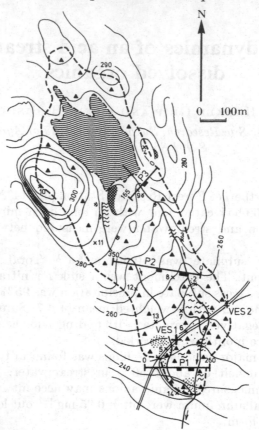

Figure 1. Map of Svartberget research catchment, including soil-sampling pits, seismic profiles and very low frequency (VLF) measuring points; (×11), pit number; (▲), till; (▧), bare bedrock or thin overburden; (▱), wave wash sediment; (▨), till covered with thin wave-wash sediment; (▤), peat; (▨), thin peat cover; (↳⚡), seismic profile with localized fracture zones; (✦), VES point.

time for proper mass balance studies. It is therefore hoped to expand the time series into the future.

The most striking feature of the streamwater at Svartberget is the major imbalance between cations and anions although an ionic charge balance is present in precipitation. The reason was found to be the high content of dissolved organic matter. The pH of the stream was between 6.8 and 4.0 with a volume weighted mean of 4.7. In spite of the acid episodes, fish were still found in other small streams in the vicinity, although none have been reported in this stream (E. Manfredsson, personal communication).

Contrary to many antropogenically acidified sites, there is a negative correlation between sulphate and hydrogen ions in streamwater.

2. Materials and methods

(a) Site description

The 50 ha catchment (figure 1) at the Svartberget Forest Research Park is located 60 km N.W. of Umeå, Sweden (64° 14′ N, 10° 46′ E, altitude 235–310 m). The drainage net consists of two stream channels, deepened during the 1930s to improve forest drainage. The main channel has a several metre deep 8 ha open mire at the head. Along the streams there are another 2.5 ha of peat lenses. The catchment is afforested with mature Scots Pine (*Pinus sylvestris*) on higher ground and Norway Spruce (*Picea abies*) in low lying areas. The field layer is a mix of heather (*Calluna vulgaris*) and berries (*Vaccinium myrtillus* and *Vacc. vitisidaea*) and grasses (mainly *Deschampsia flexuosa*). The bottom layer is a mix of lichen (*Cladonia spp.*) and mesic mosses on the hilltops, mesic mosses (*Hylocomium splendens, Pleurozium schreberi*) on the hillslopes and wet mosses (*Sphagnum spp.* and *Polytrichum spp.*) at the bottom of the hillslopes adjacent to the stream channel.

The gneissic bedrock of Svartberget contains horizons of biotite-plagioclase and graphite-sulphide schists and is overlain by a locally derived glacial till. The sand fraction of the till consists of 50% quartz, 20% plagioclase, 15% K-feldspar, 10% biotite, 3% amphibole and 5% accessories; secondary minerals found were Fe-hydroxide, sericite and chlorite (Miskovsky 1987).

As the catchment is facing southeast it was on the lee side for the dominating ice movement during the last glaciation. The highest shore line was at 255–260 m and thus passes through the central part of the catchment. The hilltops are covered by a compact sandy basal till. On the slopes the basal till is covered by a heterogeneous, less-compacted ablation till, dominated by the sand fraction. Lenses of well sorted material are abundant. The basal till in the lower parts of the catchment is very compacted and has a higher silt content than the upper parts. Mean particle size distribution in 36 samples from 14 pits was 21% gravel, 49% sand and 30% silt and clay (Ivarsson & Johnsson 1988).

The depth of the overburden was investigated by geophysical methods. The average thickness was found to be 10–15 m with a maximum depth of more than 30 m. A top layer with low compaction had a depth of 1–2 m, an intermediate layer with moderate compaction had a depth of 0–5 m, whereas the very compacted bottom layer had a depth of 0–30 m. One or two fracture zones were found in the bedrock in each of the three seismic profiles (Lindqvist *et al.* 1989).

The soil was Orthic podzol (FAO/Unesco 1974) in most of the catchment, but in low-lying parts it was humic podzol.

(b) Data base

Daily precipitation was collected in a Swedish standard gauge and corrected according to Eriksson (1980). Runoff was measured by using a water stage recorder at a 90° V-notch weir in a heated hut to ensure year-round measurements.

During the period June 1986–June 1989 precipitation was sampled daily and the funnel and bottle rinsed with distilled water. Air concentration of acid gases and particles was sampled weekly by means of denuder technique (Brosset & Ferm 1978). Streamwater was sampled weekly and during spring snowmelt daily or twice a day at the V-notch weir. Groundwater was sampled every second week in five groundwater tubes at depths of 2.9–6.0 m; pH was taken and all samples were deep frozen for later chemical analyses.

The analytical procedures were pH on pH-meter with combination electrode; Cl, SO_4 and NO_3: ion-exchange chromatography; Ca, Mg, Na, Fe, Al_{tot}, S_{tot}, total organic carbon (TOC): inductively coupled plasma spectroscopy (ICP); K: Atomic absorption spectrometer; NH_4: spectrometric colorimetry and N_{tot}: the peroxidisulphate method (Emteryd 1989).

The carbon data were treated as total organic carbon (TOC), as the samples were not filtered before analysis. However, no suspended material was observed and there should be little difference between TOC and dissolved organic carbon (DOC).

The positive charges were calculated as the sum of NH_4^+, Mg^{2+}, Ca^{2+}, Na^+, K^+, Fe^{3+} and Al^{3+}, where all Fe and Al were accounted for as having 3 charges, as about 80% of them had organic ligands (Bishop *et al.* 1990; Emteryd 1989). The negative charges were calculated as the sum of Cl^-, SO_4^{2-} and NO_3^-. The anion charge deficit was the difference between these positive and negative charges.

3. Results and discussion

(a) Deposition

The mean air concentrations of SO_2-S and SO_4-S were each 0.1 µg m^{-3} each and of NO_x-N and NO_3-N 0.01 µg m^{-3} each. Deposition rates for the gases were calculated according to Grennfelt *et al.* (1983) and for the particles a rough estimate of 4 mm s^{-1} was used. These values gave a dry deposition of 0.13 kg S ha^{-1} a^{-1} and 0.02 kg N ha^{-1} a^{-1}, which were negligible compared with the wet deposition of 6.0 kg S ha^{-1} a^{-1} and 4.7 kg N ha^{-1} a^{-1}. Of the deposited nitrogen, 23% was organic nitrogen.

The weighted mean pH of precipitation was 4.4, and hence H^+ was the dominating cation together with ammonium (table 1). Sulphate was the dominating anion, followed by nitrate.

(b) Streamflow output

The sulphur output from the catchment as SO_4-S was 6.2 kg ha^{-1} a^{-1}. Total sulphur, as measured by the ICP technique, exceeded SO_4-S by 25%, the residual sulphur probably partly being organic sulphur and sulphides. This sulphur, which is normally not accounted for in ecological studies, was significant and is thus included here. The catchment was leaching 2.4 kg N ha^{-1} a^{-1}, mostly in the form of organic nitrogen. The inorganic nitrogen output was only 0.2 kg N ha^{-1} a^{-1} (table 1).

The weighted mean pH in the stream was 4.7, with extremes of 6.8 and 4.0. The hydrogen ion concentration was positively correlated with streamflow, as is usually the case. Especially, the increase in H^+ was large in the flow interval 0–2 mm per day, where the increase was 20 µeq l^{-1}, or equal to that in the flow interval 2–16 mm per day.

High hydrogen ion concentrations were often, but not always, accompanied by high iron and aluminium concentrations. Maximum iron and aluminium concentrations were about 80 and 35 µeq l^{-1}, disregarding the extreme situation during the winter 1987, when the aluminium concentration rose to about 90 µeq l^{-1}. Iron as well as aluminium were only measured as Fe_{tot} and Al_{tot} in the regular sampling, but as about 80% were complexed by organic acids, it seems justified to give them three positive charges. The stream output of iron was 7.3 kg ha^{-1} a^{-1} and of aluminium 0.85 kg ha^{-1} a^{-1}. The export of organic carbon amounted to 56 kg ha^{-1} a^{-1}, of basic cations 0.81 keq ha^{-1} a^{-1} and of strong acids anions 0.47 keq ha^{-1} a^{-1}.

(c) Charge balance and organic matter

The dissolved organic matter was identified as a key variable to explain cation transport mechanisms and anion charge deficit.

There was a consistently large anion charge deficit ranging between 90 and 330 µeq l^{-1}, with a mean value of 175 µeq l^{-1}.

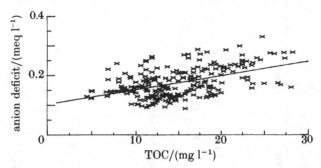

Figure 2. Relation between anion charge deficit (meq l^{-1}) (see text for definition) and TOC (mg l^{-1}) in streamwater at Svartberget in weekly samples between 20 June 1985 and 5 June 1989; $r = 0.50$, $P < 0.001$, $n = 250$.

Table 1. *Deposition and runoff of different elements at Svartberget Research Catchment given as volume weighted-mean concentrations for the period 20 June 1985–5 June 1989*

element	unit	wet deposition	runoff
water	mm a^{-1}	720	370
pH		4.4	4.7
H^+	μeq l^{-1}	36.0	20.0
Ca	μeq l^{-1}	8.6	99.5
Mg	μeq l^{-1}	2.4	53.5
Na	μeq l^{-1}	4.8	52.8
K	μeq l^{-1}	2.5	9.9
NH_4-N	μeq l^{-1}	17.5	1.9
Cl	μeq l^{-1}	10.2	20.8
SO_4-S	μeq l^{-1}	51.6	104.6
NO_3-N	μeq l^{-1}	18.6	1.4
Al_{tot} (as 3+)	μeq l^{-1}	—	25.6
Fe_{tot} (as 3+)	μeq l^{-1}	—	39.1
sum cation	μeq l^{-1}	71.8	302.3
sum anion	μeq l^{-1}	80.4	126.8
anion charge deficit (%)		−12	58
TOC (mg l^{-1})	mg l^{-1}	—	15.2
N-tot	mg l^{-1}	—	0.62
S-tot	mg l^{-1}	—	2.14

	air concentration μg m^{-3}	dry deposition kg ha^{-1} a^{-1}
SO_2-S	0.10	0.05
SO_4-S	0.06	0.08
NO_x-N	0.01	0.004
NO_3-N	0.01	0.01

The yearly minimum values occurred when the spring flood peaked and the yearly maximum values came with the first summer floods in August, except during 1987, when the maximum was in the extremely cold January.

The stream was rich in TOC, with the concentrations being 5–30 mg l^{-1} and the volume weighted mean 14.9 mg l^{-1}. The yearly concentration pattern followed the

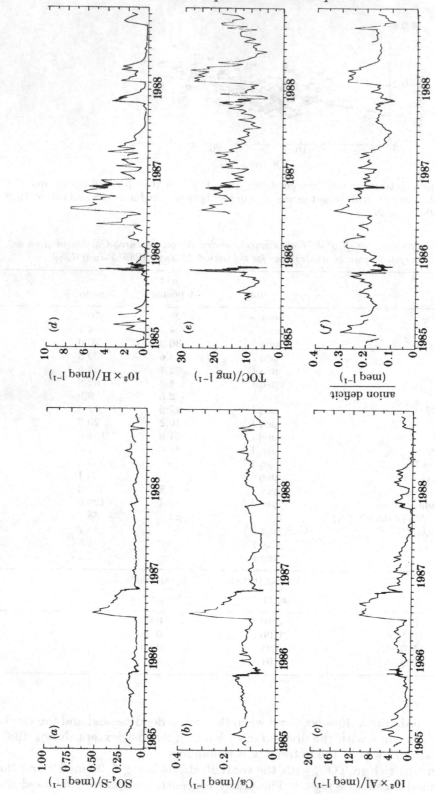

runoff, but the correlation was weak. The change in runoff was more important for the change in TOC concentration than the absolute values. The summer high flow in August–September had the maximum TOC. Decreased flow in October–December was accompanied by a decrease in TOC, but the continued flow decrease throughout January–April was accompanied by an increase in TOC. After spring snowmelt, flow decreased until July, and so did the TOC concentration.

There was a good correlation between TOC and anion charge deficit, the correlation coefficient being 0.5 ($P < 0.001$, figure 2). The conclusion was therefore that the charge imbalance was accounted for by organic anions, e.g. protolysed organic acids or organic anions as ligands in metallic complexes. The annual weighted mean charge density of the organic material at ambient pH was 12 mequiv. g C^{-1}, which was equal to that at microcatchment F3 at Lake Gårdsjön. The carbon content at that microcatchment was only half that observed at Svartberget, and had a weighted DOC of 7.7 mg l^{-1} (Hultberg 1985).

The dissolved organic material contained sulphur and nitrogen. Only during January–September, 1987, or 20 % of the time, were the C/N ratios less than 30. The C/N ratio of microbial biomass and humus is between 5 and 15, and of nitrogen rich plant material, such as sweet clover, about 30 (Alexander 1977), which suggests that the stream TOC is generally not well decomposed or it contains non-nitrogen organic acids. 90 % of the C/S ratios were less than 100. As the C/S ratio of microbial biomass is normally of the order of 100–1000 and of soil organic matter about 100, the low C/S ratio indicates that some sulphides may be present.

To see if the anion charge deficit could be accounted for by organic anions, a streamwater sample was exposed to uv-light while the pH was held constant (Cronan & Aiken 1985; L. O. Öhman & O. Emteryd, personal communication). The sample, which was run in four replicates, contained 34.5 mg C l^{-1}. The cation sum was 412 µeq l^{-1}, the inorganic anion sum 94 µeq l^{-1} and thus the anion deficit was 318 µeq l^{-1}. During oxidation, 287 µeq l^{-1} of H^+ had to be added to keep the pH constant. Obviously a great part of the deficit was accounted for by organic anions, but 31 µeq l^{-1} H^+ less than the anion deficit was added. This could be accounted for by the oxidation of sulphur and nitrogen not in the form of sulphate, ammonium or nitrate.

The sample contained 0.31 mg S l^{-1} that was not sulphate, and it was assumed to be organic sulphur and sulphides and 1 mg l^{-1} organic nitrogen. As the increase in sulphate, ammonium and nitrate was measured and the nitrogen loss to the atmosphere could be calculated, the final balance was found to depend on the proportions of organic sulphur to sulphide, which was not known. This preliminary test gave a standard deviation of 5 % of added H^+ and less than 10 % of the anion deficit remained unexplained and due to the uncertainties involved. This limited experiment showed that the anion deficit could be accounted for by organic acids.

(d) *The winter of 1987*

During the winter of 1987, extremely high concentrations were measured for many chemical constituents in the stream (figure 3). January was the coldest month in more than a 100 years, with a mean temperature of − 19 °C. As the snow was only

Figure 3. Observed time series in the stream at Svartberget between 20 June 1985 and 5 June 1989. From top SO_4-S, Ca, Al_{tot}, H^+, TOC and anion charge deficit (see text for definition). TOC is given in mg l^{-1}, the other variables in meq l^{-1}.

20 cm deep, comparatively deep frost was formed in the soil, even in the forest where there is normally little frost in the ground. Nearly all water in the stream was frozen. The shallow groundwater that flew out of the ground formed bulge ice that overfilled the stream channel as the winter proceeded. The stream flow amounted to only 1.3 mm during the first three months of the year, and only water entering at the bottom of the stream close to the weir had a chance to leave the catchment. As winter drew on and the bulge ice and the snow pack grew thicker, the ice could be melted from below by the outflowing groundwater to form a conducting tunnel; at the beginning of April the base-flow increased.

The concentrations in the deep groundwater (from 6 m depth) were Ca^{2+} 0.34, SO_4^{2-} 0.37, Mg^{2+} 0.18 and Na^+ 0.11 meq l^{-1}, and in streamwater the concentrations were Ca^{2+} 0.36, SO_4^{2-} 0.51, Mg^{2+} 0.18 and Na^+ 0.12 mequiv. l^{-1} in the beginning of January. From this it can be concluded that it was only deep groundwater that contributed to the stream flow. However, the high concentrations might have been the effect of outfreezing as TOC, Al_{tot}, Fe^{3+} and H^+ also had peak concentrations at the same time. The peak concentrations in January were depleted during the following months, indicating that more and more shallow groundwater contributed to streamflow.

The deep groundwater contribution to streamflow was calculated to be 6 mm a^{-1}, if the stream flow during the first three months of 1987 could be a measure. If, on the other hand, the hydraulic conductivity of the compacted deep soil layer (measured to be 8×10^{-7} m s^{-1}), the cross-sectional area of that layer as found by the refraction seismic investigation and the general slope along the catchment were used to calculate the flow, the deep groundwater flow was 15 mm a^{-1} or 5% of the runoff measured in the stream. A part of that groundwater would, however, pass below the measuring dam.

The simple calculation shows that the contribution of deep groundwater to stream flow and thus to streamwater chemistry was minor. Only at very low flow would the deep groundwater chemistry have significant influence on streamwater composition. If the deep compacted soil layer was estimated to have a mean depth of 10 m, a porosity of 40% and deep groundwater flow of 15 mm a^{-1}, the mean residence time for water in that reservoir would be 250 years.

(e) Sulphate interactions

The marked effect of the deep groundwater is an example of how flow pathways effect the chemical dynamics of stream flow. It is clear that flow pathways and other factors interact to generate a very complex pattern of stream chemistry. A striking illustration of this may be seen in the two different relations between sulphate and pH (figure 4). For 90% of the data there is a strong ($P < 0.001$) negative correlation between H^+ and SO_4-S. But during the three month cold spell in the winter of 1987, the relation became positive.

The positive relation between sulphate and calcium or magnesium had a correlation coefficient above 0.88 and for sodium it was above 0.60. However, the relation with potassium was weak and with iron, aluminium or TOC there was no relation at all.

In these relations we can see the strong connection between sulphate and some weathering products. Potassium is dominated by its biological importance and is therefore more connected to superficial water pathways, as are iron and aluminium, which need complexing agents to be mobile.

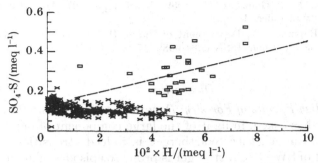

Figure 4. Relations between H⁺ (mequiv. l⁻¹) and SO₄-S (mequiv. l⁻¹) in streamwater at Svartberget in weekly samples between 20 June 1985 and 5 June 1989. Squares are from the extremely cold four first months in 1987; (——): $r = 0.50$ ($P < 0.001$, $n = 275$); (----): $r = 0.42$, ($P < 0.02$, $n = 27$).

4. Conclusion

The Svartberget Research Catchment contains a big water body with a long residence time. On top of that is water with decreasing residence time. This feature endows stream chemistry with two apparently opposite characteristics. On the one hand there is only small yearly variation, and on the other, dramatic shifts may occur in extreme weather situations. The big water body does not allow for proper budget calculations with only a few years of observation. If the time period of the calculations are changed, quite different results may appear, e.g. a net gain of sulphur may be changed to a net loss. Obviously, between-year variations cannot be disregarded.

Much information is yet to be derived from the data; information that may be a key to understanding the dynamics of acid streams rich in organics.

This research was funded by the Surface Waters Acidification Programme. The field observations were made by Elon Manfredsson and the chemical analysis done by Bengt Andersson. Without their skilful work, nothing would have been done. The chief of the laboratory, Ove Emteryd, paid special attention to our problems and helped us to solve them.

References

Alexander, M. 1977 *Introduction to soil microbiology*, 2nd edn., pp. 128–147, 350–367. New York, Chichester, Brisbane and Toronto: John Wiley.

Bishop, K. H., Grip, H. & O'Neill, A. 1990 The origins of acid runoff in a hillslope during storm events. *J. Hydrol.* (In the press.)

Brosset, C. & Ferm, M. 1978 *Atmos. Environ.* **12**, 909–916.

Cronan, C. S. & Aiken, G. R. 1985 *Geochim. cosmochim. Acta* **49**, 1697–1705.

Emteryd, O. 1989 Stencil no. 10, Dept. Forest Site Research, Swedish University of Agricultural Sciences, Umeå, Sweden. ISSN 0280-9168, 181 pp.

Eriksson, B. 1980 SMHI Reports no. RMK 17, Swedish Meteorological and Hydrological Institute, Norrköping, Sweden. ISSN 0347-2116, 150 pp.

FAO/Unesco 1974 Soil map of the world, vol. I, legend, 59 pp.

Grennfelt, P., Fowler, D., Granat, L., Hällgren, J.-E., Person, C. & Richter, A. 1983 PM 1763, Statens natürvårdsverk, Stockholm. ISBN 91-7590-143-9, 101 pp.

Hultberg, H. 1985 *Ecol. Bull. (Stock.)* **37**, 133–157.

Ivarsson, H. & Johnsson, T. 1988 Stencil no. 6, Svartbergets och Kulbäckslidens försöksparker, Sveriges lantbruksuniversitet, Umeå, Sverige. ISSN 0280-4328, 61 pp.

Lindqvist, G., Nilsson, L. & Gonzalez, G. 1989 Draft report 27, Department of Applied Geophysics, University of Luleå, 16 pp.

Miskovsky, K. 1987 Report no. 3, Department of Public Health and Environmental Studies, Umeå University, Umeå, Sweden. ISSN 0284-0588, 17 pp.

Discussion

K. H. BISHOP (*Swedish Faculty of Forestry, Department of Forest Site Research, Umeå, Sweden*). Dr Walker has asked, in light of the large component of organic acidity in the Svartberget stream, whether Svartberget is indeed 'transitional', as it was termed at the outset of SWAP. Svartberget is prone to episodes of acidity, like other transitional SWAP sites. Organic anions, though, often account for as much as half of that acidity. Unfortunately, not enough is yet known about the impact of anthropogenic acid deposition on the cycling of organic material in a catchment to say whether Svartberget's episodic acidity is strictly natural or whether it has been influenced by anthropogenic acid deposition. Dr Grip and myself are looking forward to the results of palaeoecological studies in the area by Dr Renberg's group to give some idea of whether or not surface waters in the area are in transition to greater acidity as a result of increases in acid deposition over the last several decades. We also look forward to conducting further studies at Svartberget to better understand the nature and origin of organic acidity.

Inorganic and organic soil processes and their influence on percolating waters

By T. A. Bruce Walker, Hamish A. Anderson, Alan Hepburn,
Robert C. Ferrier, Ruari G. McMahon, John D. Miller,
Basil F. L. Smith and Ryan G. Main

*Macaulay Land Use Research Institute, Craigiebuckler,
Aberdeen AB9 2QJ, Scotland, U.K.*

The catchment soils at Høylandet, Allt a'Mharcaidh, Loch Chon and Kelty Water form members of a podzolic soil progression, representing the majority of acidic upland soils in the U.K. and Scandinavia. Soils have been analysed by standard methods and when the data are translated into catchment volumetric terms, exchangeable bases, acidity and aluminium can be assessed in terms of soil 'stores'. Both exchangeable and extractable aluminium are predominantly associated with the lower soil horizons, especially Bh and Bs horizons, whereas the exchangeable bases are concentrated in organic topsoils. Sesquioxide accumulations in the Bs horizons are responsible for the adsorption of the mobile sulphate anion, and comparison of sulphate adsorption isotherms for the different sites indicates both the accumulation history of the site and its likely reaction to changing sulphate deposition. Acid-soluble Al decreases with increasing impaction, possibly because of leaching resulting from acidic pollution. Experimental leaching columns demonstrated the feasibility of such a process and the adsorption reactions likely to take place in aluminium-rich Bs horizons.

1. Introduction

The main objective of this part of the Macaulay Land Use Research Institute contribution to the Surface Waters Acidification Programme (SWAP) was the survey and complete characterization of the soils of the Scottish SWAP catchments to provide a basis for comparisons with soil types occurring in Scandinavian experimental catchments, and comparisons of SWAP studies with other similar investigations. In 1987, the pristine site at Høylandet, mid-Norway was instrumented in an identical manner to the Scottish sites and the soils objectives were extended to cover it.

A wide range of soils in Europe are embraced by the terms podzol and podzolic, and in general the so-called Podzol Zone (Muir 1961) is characterized by moderately high to high rainfall and northerly latitudes. This combination provides both leaching and temperature conditions that will podzolize almost any acid or intermediate parent material on freely draining sites. The six instrumented sites established by MLURI in 1986–87 are all underlain by podzolic soils that range across several morphological sub-divisions, from the excessively drained Alpine

Table 1. *Mean soil horizon depths and* pH *values for the experimental plot soils*

site	depth/cm	horizon	pH (H₂O)	site	depth/cm	horizon	pH (H₂O)
Høylandet	0–10	LF	3.94	Mharcaidh	0–20	01	3.86
	10–20	E	4.41		20–40	02	3.87
iron	20–30	Bs1	4.61	peat	40–60	03	3.85
podzol	30–50	Bs2	4.88	(MPt)	60–90	Egh	4.42
	50–	BC	5.08				
(HOYL)	100+						
Mharcaidh	0–5	H	4.10	Chon	0–10	(L)FH	4.23
	5–17	AEh	4.48		10–20	E	4.00
alpine	17–37	Bh	4.90		20–25	Bhs	4.20
podzol	37–67	Bsh	5.03		25–50	Bs	4.22
	67–	C	4.99		50–	BC	4.42
(MAP)	100+				100+		
Mharcaidh	0–10	O	4.20	Kelty	0–30	01	3.70
	10–20	E	4.39		30–50	02	4.06
peaty	20–35	Bh	4.56		50–70	Bgh	4.28
podzol	35–60	Bsh	4.78		70–	BCg	4.61
	60–	C	4.92		100+		
(MPP)	100+						

podzols of the Allt a'Mharcaidh, through the well-drained iron and humus–iron podzols at Høylandet and Chon, respectively, to the poorly drained peaty gleys and peats at Kelty and Mharcaidh. Processes leading to the formation of profiles similar to these, and the genetic relation between them have been described (Romans 1970; Anderson *et al.* 1982).

This paper asks which soil processes may be altering the chemical composition of water percolating through the range of podzolic and peat soils on the MLURI sites and within the surrounding catchments, described elsewhere in this symposium (Ferrier & Harriman, this symposium). On these sites, the central issues dominating such changes are the modifications of incoming acidity, in the form of hydrogen ion associated with excess sulphate, and the subsequent release of Al into surface waters.

2. Soils

In this paper, emphasis has been placed on the soils of the MLURI instrumented plots (Ferrier & Harriman, this symposium), and average soil-horizon thicknesses and pH values are given in table 1. Although the soil pH values for the subsoils at the impacted Loch Ard sites are less than those of the other sites, the Høylandet mineral soils appear equally acid as those at Mharcaidh, whereas the organic topsoil is more acid, approaching the value for the Mharcaidh peat.

The soil samples were recovered from the profiles exposed for the installation of zero-tension lysimeters (Miller *et al.* 1990*a*) and therefore should be related to the throughflow water chemistry (Miller *et al.* 1990*b, c*; Ferrier *et al.* 1990*b*) discussed elsewhere in this symposium (Ferrier *et al.*, this symposium). Disregarding difficulties encountered in identifying plot areas with similar soil profiles, the problem of soil

Table 2. (a) *Exchangeable base cations* (*sum of* Na^+, K^+, Ca^{2+}, Mg^{2+}), *acidity* (*both extracted in* 1 M NH_4OAc *at* pH 7), *and exchangeable* Al (1 M KCl *at* pH 6), *expressed in* mequiv. 100 g^{-1} *air-dried soil* (ND = not detected.)

site			horizons		
HOYL	LF	E	Bs1	Bs2	BC
TBC	25.7	0.4	0.5	0.3	0.2
ExAc	72.5	3.3	17.7	11.6	4.3
ExAl	1.1	1.3	4.7	3.3	0.7
MPP	O	E	Bh	Bsh	C
TBC	3.9	0.3	0.3	0.1	0.2
ExAc	53.0	8.8	19.4	10.8	2.4
ExAl	1.9	6.3	13.4	4.9	ND
MAP	H	AEh	Bh	Bsh	C
TBC	3.3	0.4	0.3	0.3	0.3
ExAc	23.8	12.4	24.9	3.6	1.1
ExAl	3.9	4.4	5.6	1.2	0.6
Chon	(L)FH	E	Bhs	Bs	BC
TBC	14.2	0.5	0.3	0.3	0.3
ExAc	70.4	13.8	16.2	15.6	9.0
ExAl	1.2	3.0	5.4	1.3	0.9
Kelty	01	02	Bqh	BCq	
TBC	2.1	2.7	0.7	0.7	ND
ExAc	81.8	83.7	15.4	8.4	ND
ExAl	10.0	12.7	5.9	2.1	ND
MPt	01	02	03	Eqh	
TBC	18.0	12.5	6.4	0.4	ND
ExAc	114.0	106.2	118.7	12.8	ND
ExAl	1.9	6.3	13.5	4.0	ND

variability is endemic to an exercise such as this (Ball *et al.* 1970), and the average values for soil parameters derived from the four soil pits per site are assumed to give mean values for the soil mapping units concerned.

After air-drying and sieving to less than 2 mm, or in the case of organic soils, milling to less than 1 mm, soils were analysed by standard methods (Soil Conservation Service 1972; Sheldrick 1984).

3. Soil acidity

Table 2a summarizes the contents of major exchangeable cations, acidity and aluminium extracted by neutral salt solutions. As would be expected in leached acidic soils such as these, the base cation contents were low, only the Høylandet, Chon and Mharcaidh peat 0 layers yielding more than 10 mequiv. 100 g^{-1} of total base cations, with the Mharcaidh peat also showing exchangeable bases decreasing down the organic profile. Exchangeable acidity dominated the data on a mequiv. 100 g^{-1} soil basis, and appeared to be related to the organic matter content of the particular soil horizon in question. The exchangeable acidity also increased in the Bh and Bs horizons concurrently with the exchangeable Al, and illustrated the associated of Al and spodic humus noted previously (Anderson 1981; Anderson *et al.* 1982). Three notable features of the data were (i) the high levels of acidity associated with the Mharcaidh peat, (ii) the comparably elevated levels of exchangeable Al in

Table 2. (*b*) *Data from* (*a*) *weighted for bulk density, horizon depth and less than* 2 mm *soil volume per horizon, expressed in* kequiv. ha^{-1} *for* 1 m *deep soil particles*

site			horizons		
HOYL	LF	E	Bs1	Bs2	BC
TBC	30.8	3.0	3.9	4.7	6.3
ExAc	87.0	25.0	136.3	181.0	438.8
ExAl	1.3	9.8	36.2	51.5	22.5
MPP	O	E	Bh	Bsh	C
TBC	5.6	1.7	3.4	1.7	4.3
ExAc	75.5	49.7	216.5	178.2	52.2
ExAl	2.8	35.6	149.5	66.0	ND
MAP	H	AEh	Bh	Bsh	C
TBC	5.5	2.3	2.3	1.4	2.4
ExAc	39.7	71.4	193.5	47.0	13.3
ExAl	6.5	25.3	43.5	15.7	7.5
Chon	(L)FH	E	Bhs	Bs	BC
TBC	19.9	3.3	0.84	4.8	8.6
ExAc	98.6	91.1	44.7	245.8	254.3
ExAl	1.6	19.8	14.9	20.5	24.8
Kelty	01	02	Bqh	BCq	
TBC	8.7	11.8	2.9	14.9	ND
ExAc	343.5	368.2	63.1	168.0	ND
ExAl	42.0	55.8	24.2	42.0	ND
MPt	01	02	03	Eqh	
TBC	36.0	35.0	19.2	9.5	ND
ExAc	228.0	297.4	356.2	153.6	ND
ExAl	3.8	17.6	40.6	96.0	ND

the organic soil horizons at Kelty, and in particular (iii) the similarity between the distributions of the analytes in the four 'podzol' profiles at Høylandet, Mharcaidh and Chon. Apart from more exchangeable Al in the Bh/Bs horizons of the Mharcaidh peaty podzol, there was no obvious pattern, and the Høylandet data indicated that this soil was as acidic as the others.

Assuming that the soil variability shown within the plot soil pits was representative of that in the mapping unit, the data can be weighted for soil bulk densities and horizon thicknesses, to give area data for soil profiles to a depth of 1 m. This procedure allowed estimates to be made of the stores available for the processes that are likely to occur when water percolates through the profiles; at this point, we are obviously disregarding any preferential routing of soil drainage waters. Some weighting procedure like this is necessary if we wish to deduce accurately the relative importance of possible soil reactions within each soil horizon or profile.

As shown in table 2*b*, where the data are expressed in kequiv. ha^{-1}†, the picture of available soil acidity and exchangeable cations has undergone considerable change from table 2*a*. The Bh and Bs horizons now contain the majority of both H$^+$ and Al, whereas the exchangeable bases remain concentrated in the uppermost organic horizons. In these soils of low clay content (average 0.7% in the mineral soils), both exchangeable H$^+$ and Al are obviously bound to non-colloidal fractions, again suggesting the importance of organic matter and, in the case of B horizons, a

† 1 ha = 10^4 m^2

combination of organic matter and sesquioxide fractions. These organic and sesquioxide sites are also available to base cations, but the latter are obviously more firmly retained by the surface organic matter, possibly being supplied to these horizons via input precipitation.

The data in table 2b show clearly that the Høylandet profile had the highest content of available exchangeable acidity, not unexpected from a pristine site where soil acidity is governed by naturally generated organic acidity. Even allowing for the ameliorating effect of amphilitic geology, the exchangeable base cation content was not excessive, being easily exceeded by that of the Mharcaidh peat profile, and being remarkably similar to that shown in the highly impacted Chon and Kelty sites. The Mharcaidh peaty podzol and alpine soils contained the greatest amounts of exchangeable Al to 1 m depth, followed by Kelty and the Høylandet profile, with Chon and the Mharcaidh peat containing the least. Again the unimpacted site appears to occupy a position remarkably similar to that of the supposed transitional and polluted sites. At this level of soil chemical analysis, there was insufficient discrimination between the forms of components to distinguish any effects of anthropogenic acidification on the profiles.

One possible exception lay in the concentration of exchangeable Al in the Kelty 0 horizons, especially when compared with the Mharcaidh peat. In deep organic soils, given the low hydraulic conductivity, precipitation of Al from soil solution is unlikely in upper layers without a distinct upward transport mechanism. Concentrations of aluminium such as those seen in the 01 and 02 horizons at Kelty can only originate in the mineral soil and may arise through erosion followed by deposition (including upward pipe flow from the mineral layers into the organic), or by plant uptake and subsequent deposition in litter. The latter process is probable at Kelty, with the tree roots penetrating through the peat into the upper mineral horizons, allowing uptake of the elevated soil solution concentrations of Al (Joslin *et al.* 1988), and its subsequent return through litterfall (Anderson *et al.* 1990). The roots also alter the hydrology of the peat horizons and allow a more rapid throughflow of canopy interception and stemflow. Immediately outside the forest, under lowland blanket bog vegetation, the Al concentration in the peat is up to ten-times less than under the Sitka, this difference apparently being unrelated to any cultivation at time of planting (Anderson *et al.* 1990). The elevated exchangeable level of Al in the 03 horizon of the Mharcaidh peat is probably also primarily associated with a mixture of plant cycling and mineral inputs.

4. Forms of extractable Al

In podzolized soils, organic matter and aluminium are inter-related, especially in the mineral soil, with the former generally being held responsible for translocation of Al (Duchaufour 1977), although a separate argument favours the deposition of mobile fulvates onto previously precipitated inorganic sesquioxide coatings (discussed in Anderson *et al.* (1982)).

To gauge the relative importance of various solid forms of Al, selective extractions of soil Al were done giving the results shown in table 3. Again, the relative between-site significance of the data increased on conversion to a volumetric output. The most obvious conclusion to be drawn is that the majority of secondary Al in the soil is organically bound. Within the profiles, there is a definite concentration of organic Al and amorphous Al associated with the eluvial horizons (Bh and Bs variants).

Table 3. *Forms of extractable* Al (*Sheldrick 1984; Anderson et al. 1982*) *expressed as tonnes* Al ha^{-1} *to a depth of* 1 m; *data weighted as in table 2b*

(Organic = pyrophosphate soluble; amorphous = acid ammonium oxalate soluble − organic; crystalline = dithionite-citrate soluble − (organic + amorphous).)

	acid-soluble	organic	amorphous	crystalline
HOYL	11.1	24.8	9.0	4.6
MPP	7.7	21.1	12.9	6.0
MAP	4.3	23.9	4.1	7.4
Chon	4.1	23.1	2.3	8.6
Kelty	2.2	9.2	3.5	1.7
MPt	0.9	2.4	0.1	1.1

Table 4. *Mharcaidh catchment soil store* (*to a depth of* 1 m) *of forms of* Al, *expressed as* kt Al *in the catchment*

(Data weighted as in table 3; soil proportions in catchment after Ferrier & Harriman, this symposium.)

	exch.-Al (1 M KCl)	extractable Al acid-sol.	organic	amorphous	crystalline
alpine podzol	0.34	1.66	9.14	1.55	2.81
peaty podzol	0.70	2.36	6.45	3.94	1.84
blanket peat	0.33	0.21	0.56	0.02	0.26
total	1.37	4.23	16.15	5.51	4.81

Expression of the acid-soluble Al on a tonnes per hectare basis reveals a definite pattern. The Høylandet, Mharcaidh peaty, and alpine podzols and Chon showed a decrease in acid-soluble Al with increasing input acidity, although the data for the Mharcaidh peat, and to a lesser extent that for Kelty, were obviously biased by the high proportion of peat in a 1 m profile. It is tempting to speculate that the acid-soluble Al has indeed being leached from the soils by incoming mineral acidity. *In vitro* mineral acid treatment of spodic soil horizons releases not only Al, but also fulvic acids, a major source of the exchangeable acidity associated with these soils. Although the extractable organic-Al shows little variation across the pollution gradient, the exchangeable acidity in the mineral soils decreases HOYL → MPP → MAP, but increases at Chon, especially in the Bs and BC horizons. This pattern is reflected in the acid-soluble humus contents and the exchangeable Al, indicating that these components have been translocated below the natural focus in the Bh horizon, and supporting the argument that leaching of Al and possibly acid-soluble organic matter may be a result of acidic pollution.

At the Allt a'Mharcaidh, the original soil survey indicated that the catchment was dominated by three major soil sub-groups, alpine podzols, peaty podzols and blanket peats, and partition of the catchment into these groups has been used to provide MAGIC input data (Jenkins *et al.* 1988) and to assess catchment deposition inputs (Ferrier *et al.* 1990*a*). A similar exercise with the data for the forms of extractable Al (table 4) suggested that 50% of the potentially available store in the top metre of the catchment soils is in the organic-bound fraction within the alpine and peaty podzol zones. A further 13% of this store occurred as amorphous Al in the peaty podzols, mainly concentrated in the Bs horizons.

These readily extractable forms of Al reflect the severity of the extraction process, and by comparisons, the natural processes would involve very dilute leaching solutions. The concentration of crystalline extractable Al in the Bs horizon of the peaty podzol may reflect the dilute chemistry involved in its deposition from solution, below the major zone for organic complexing reactions (the Bhs). The possibility of easily extractable forms of Al being released into soil solution by acidic input precipitation obviously depends on a wide range of variables, the most important of which will be the accessibility of the solid phase. Concentrations of extractable Al that were noted in deep organic soils, such as the Mharcaidh peat, will be largely inaccessible because of the low hydraulic conductivity of the peat mass. However, if the hydraulic conductivity is increased, e.g. by drainage (either by natural erosion channels or man-made ditches, both of which will lead to shrinkage and cracking), a small proportion of the organic-bound Al will become accessible for desorption, ion exchange and other leaching mechanisms. However, the introduction of extensive and intrusive tree root systems will greatly increase both the hydraulic conductivity and the surface area exposed to leaching processes. Any increase in the rate of humification of these organic soils also allows Al to be released, but predominantly in soluble organic forms. These sources and sinks of Al are discussed elsewhere (Anderson *et al.* 1990).

5. Cation exchange capacities

In terms of exchange processes, podzolic soils are regarded as having a layer depleted in sesquioxides sandwiched between two organic cation exchangers, the surface LFH horizons and the Al/Fe enriched Bh and/or Bs horizons. Both the surface and B horizon organic matter provide numerous sites for both metal complexation and exchange. However, there are differences between the forms of organic matter, involving especially the acidic humic substances (Aiken *et al.* 1985). The surface horizons are dominated by acid-insoluble humic acids and the B horizons contain a high proportion of their organic matter in the form of acid- and water-soluble fulvic acids, in their insoluble form as salts or complexes with sesquioxides (Kononova 1966; Schnitzer & Khan 1972).

The enrichments of sesquioxides and organic matter in podzolic B horizons are responsible for the highly pH-dependent nature of these soil layers (Clark *et al.* 1966). Laverdiere & Weaver (1977) postulated that, at low pH values, positive charges resulted from the protonation of hydrous oxides, whereas desorption of protons from uncomplexed organic functional groups may be the major contributor of negative charges at higher pH values.

As can be seen from figure 1, the cation exchange capacities of the plot soils are critically dependent on pH, and the standard methodology at pH 7 will result in over-estimation, in comparison with the actual field reactivities. Again, as with the extractable-Al estimations, the use of concentrated salt solutions may give unrealistic results, in terms of natural exchange processes. Selectivity of cation retention is noticeable in the Ca/Mg ratios for the plot soils, particularly in the organic upper layers. The availability of these cations is undoubtedly influenced by parent geology, and current weathering rates and cation mobilities dictate the prevalence of Ca and Mg (Bain *et al.*, this symposium). However, in the surface organic horizon at Kelty, the cation retention order is Mg > Ca > K > Na, compared with that (Ca > Mg > K > Na) at all of the other sites, and this reflects the relative importance of Mg over Ca in the Sitka spruce throughfall at this site (Miller *et al.* 1990*b*), in turn related to

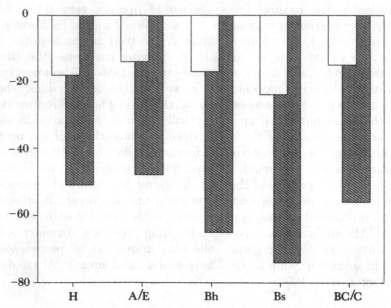

Figure 1. Variation of cation exchange capacity (CEC) with pH, expressed as the percentage decrease from pH 7.0, for mean values of CEC of the five major genetic horizons of the experimental site soils; (□, pH 5; (▨), pH 3.

the interception of greater marine inputs by the canopy, when compared with other sites (Miller & Miller 1980).

6. Sulphate retention

Adsorption of sulphate by soil removes the anion from solution and hence reduces the rate of cation leaching (Johnson & Todd 1983). In a laboratory comparison of the Chon and Høylandet soils, Walker *et al.* (1990) concluded that, at the impacted site, recent reductions in pollution-derived sulphate inputs had led to a desorption of accumulated S, present largely as specifically adsorbed sulphate in the Bs horizons. At the pristine Norwegian site, the soils were in equilibrium with incoming sulphate (predominantly marine in origin), and any acid sulphate inputs would be adsorbed in the Bs, even at very low levels (*ca.* 9 μequiv. l^{-1}). The SO_4^{2-} adsorption isotherms for the mineral soil sequence HOYL-MAP-MPP-CHON (figure 2) demonstrate not only the past history of SO_4^{2-} accumulation, either from 'neutral' marine inputs or from acid deposition, but also the importance of the Bh and Bs horizons in sorption processes, compared with the inactive E horizon.

Site-dependent correlations of the B horizon adsorption capacities with the amorphous and crystalline forms of Al and Fe have been shown by Walker *et al.* (1990) and the sesquioxide surfaces are likely to provide a major SO_4^{2-}-retention compartment in any conceptual model of sulphate retention by soils. Any such model must be able to explain the field-observed retention of sulphate, as noted for Chon and Kelty by Ferrier & Harriman (this symposium), with the experimental findings indicating that these soils are sulphate-saturated. This retention may occur as immobilized-S, either organic or secondary mineral, but the experimental evidence precludes the strongly sorbed S compartment, i.e. that containing the ligand-

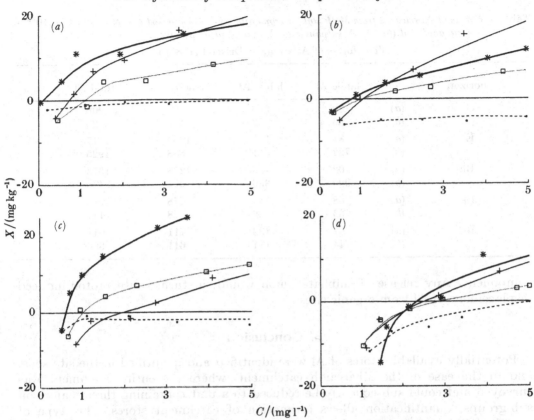

Figure 2. Sulphate adsorption isotherms for the mineral soils at Høylandet, Mharcaidh (alpine and peaty podzols), and Chon. C and X refer to concentration terms in the Freundlich equation: $X = a \times C^b$, where X is the total amount of sulphate retained by the soil, C is the equilibrium concentration of sulphate in solution, and a, b are constants; ($\blacksquare$, E/Ah; (+), BS1/Bh; ($\ast$), BS2/Bs; ($\square$), BC/C. (a) (---), Høylandet; (b) (---), MAP; (c) (---) MPP; (d) (---) Chon.

exchanged S determined by Walker *et al.* (1990). Present studies indicate a possible third mechanism.

7. Organic versus mineral acid leaching

By using 10^{-3} N citric or sulphuric acids, single-horizon columns of soil were leached and the amounts and forms of eluted Al were quantified (McMahon 1990). In the case of the MPP profile, citrate released mainly organic-bound species, contrasting with the predominantly labile form in the sulphur acid leachate (table 5). The increase in the organic-bound Al in the leachate from the Bh horizon corresponds with the acid-soluble nature of spodic horizon humus (Anderson 1981), whereas the decrease in the ratio sulphuric:citric leached Al in the Bs demonstrates the SO_4^{2-}-sorption capacity of this horizon (figure 2).

Similar column experiments are continuing, but using natural input solutions, approximating more closely to the field conditions. In the entire soil, the effect of sulphate retention in the Bs horizon (see §5) would undoubtedly reduce any translocation of inorganic Al in the absence of a replacement anion, but the organic-Al fraction would remain mobile, contributing towards the anionic throughflow

Table 5. *Forms of Al released from MPP soil horizons after six weeks periodic leaching with* (a) 10^{-3} N *citric acid and* (b) 10^{-3} N *sulphuric acid, expressed as* µequiv. Al^{3+} 100 g^{-1} *soil*

(The forms of Al are after Driscoll (1984).)

horizon		labile Al	non-labile Al	acid-reactive Al	total Al
O	(a)	89	278	533	900
	(b)	56	62	111	229
E	(a)	81	147	1089	1317
	(b)	722	371	833	1925
Bh	(a)	102	193	1278	1573
	(b)	900	356	1222	2478
Bs	(a)	68	140	778	986
	(b)	33	36	78	147
BC	(a)	67	133	711	911
	(b)	144	111	611	866

component. Any release of sulphate anion would, in turn, cause proton-induced displacement of Al from organic sites.

8. Conclusion

Potentially available stores of Al were identified and quantified in the site soils, and in the case of the Mharcaidh catchment, where the entire catchment was surveyed and could subsequently be reduced to a unit containing three major soil sub-groups, quantification allows estimation of catchment stores. This type of exercise will be necessary for the characterization of catchments, particularly where nutrient or cation depletion studies are involved, and the cost and logistics may restrict the size of study sites in the future.

In their forms of extractable Al and their capacities to adsorb sulphate, the soils at the MLURI field sites show a definite trend associated with their input acidities. The Høylandet site demonstrates that highly acidic soils can arise by natural processes with their acidity dictated by organic matter. However, the impacted counterpart at Chon is strongly influenced by sulphate, to the extent that organic acidity and its association with Al may have changed. When the comparison is extended to Kelty, the organic acidity is now being released from the soils and intensifies the Al-induced acidity in the streamwaters. Apparently, the transitional Mharcaidh catchment has the potential to adsorb large amounts of sulphate in Bs horizons, but also possesses large stores of readily leached Al. The interactions between sulphate (and possibly nitrate) and the ongoing Al movements in these soils will require further study.

Standard soil analytical methods are necessary to compare current site data with historical examples, but there is a need for milder extraction techniques and the use of fresh soil samples, rather than dried and stored material, to associate analytical data more closely with the natural processes under study.

We are grateful for the essential assistance and support rendered by the Forestry Commission in site selection and preparation, and that given to us by the Land Use and Analytical Divisions of MLURI. Our colleagues at IH, FFL, IC, University of Trondheim and NIVA assisted in various aspects of the work and this is greatly appreciated.

References
Aiken, G. R., McKnight, D. M., Wershaw, R. L. & MacCarthy, P. (eds) 1985 *Humic substances in soil, sediment and water*. New York. John Wiley.

Anderson, H. A. 1981 In *Migrations organominerales dans les sols temperes*. Proc. Colloq. Inter. no. 303, pp. 269–273. Paris: C.N.R.S.

Anderson, H. A., Berrow, M. L., Farmer, V. C., Hepburn, A., Russell, J. D. & Walker, A. D. 1982 *J. Soil Sci.* **33**, 125–136.

Anderson, H. A., Walker, T. A. B., McMahon, R. C., Hepburn, A., Stewart, M. & Main R. G. 1990 *Sci. tot. Envir.* (Submitted.)

Ball, D. F., Williams, W. M. and Hornung, M. 1970 *Welsh soils discussion group report* no. 11, pp. 31–41.

Clark, J. S., McKeague, J. A. & Nichol, W. E. 1966 *Can. J. Soil Sci.* **56**, 161–166.

Driscoll, C. T. 1984 *Int. J. Environ. Anal. Chem.* **16**, 267–283.

Duchaufour, P. 1977 *Pedologie. 1. Pedogenese et classification*, ch. 10. Paris: Masson.

Ferrier, R. C., Jenkins, A., Miller, J. D., Walker, T. A. B. & Anderson, H. A. 1990*a J. Hydrol.* **113**, 285–296.

Ferrier, R. C., Walker, T. A. B., Miller, J. D., Anderson, H. A. & Harriman, R. 1990*b J. Hydrol.* **116**, 251–266.

Jenkins, A., Ferrier, R. C., Walker, T. A. B. & Whitehead, P. G. 1988 *Wat. Air Soil Pollut.* **40**, 275–291.

Johnson, D. W. & Todd, D. E. 1983 *Soil Sci. Soc. Am. J.* **47**, 792–800.

Joslin, J. D., Kelly, J. M., Wolfe, M. H. and Rustad, L. E. 1988 *Wat. Air Soil Pollut.* **40**, 375–390.

Kononova, M. M. 1966 *Soil organic matter*. Oxford: Pergamon Press.

Laverdiere, M. R. & Weaver, R. M. 1977 *Soil Sci. Soc. Am. J.* **41**, 505–510.

McMahon, R. C. 1990 Aluminium speciation in soils and surface waters under impact of acid rain. Ph.D thesis, University of Aberdeen. (Submitted.)

Miller, H. G. & Miller, J. D. 1980 In Proceedings of an international conference on the impact of acid precipitation. Norway: SNSF Project.

Miller, J. D., Anderson, H. A., Ferrier, R. C. & Walker, T. A. B. 1990*a Forestry.* **63**, 251–269.

Miller, J. D., Anderson, H. A., Ferrier, R. C. & Walker, T. A. B. 1990*b Forestry.* (In the press.)

Muir, A. 1961 *Adv. Agron.* **13**, 1–56.

Romans, J. C. C., 1970 *Welsh soils discussion group report* no. 11, pp. 88–101.

Schnitzer, M. & Khan, S. U. 1972 *Humic substances in the environment*. New York: Marcel Dekker, Inc.

Soil Conservation Service 1972 *Soil survey laboratory methods and procedures for collecting soil samples*. Washington, D.C.: U.S. Department of Agriculture.

Sheldrick, B. H. (ed.) 1984 *Analytical methods manual 1984*. Ottawa, Ontario: Land Resource Research Institute.

Walker, T. A. B., McMahon, R. C., Hepburn, A. & Ferrier, R. C. 1990 *Sci. tot. Envir.* **92**, 235–247.

Discussion

J. MULDER (*Center for Industrial Research, Box 124 Blindern, 0314 Oslo 3, Norway*). I think Dr Walker's postulation that acid rain is responsible for the decreased soil contents of acid soluble aluminium when going from impacted to pristine sites is very speculative. The formation of acid soluble (pedogenic) aluminium in soils depends on the intensity of the podzolization process, which may have been different for the studied sites at the onset of acid rain. Could Dr Walker comment on this? Also, (i) what was the spatial variability of the extractable aluminium forms within one catchment and how was this compared with the variation between catchments? (ii) Is Dr Walker's postulation supported by independently measured fluxes of dissolved aluminium from the soil horizons in the respective catchments? Does the summed acid rain induced leaching of aluminium account for the decrease in acid soluble aluminium at the impacted sites?

T. A. B. WALKER. We agree that amounts of pedogenic Al in podzols will depend on the intensity of the podzolization process. However, a more important factor would be the duration of such a process, and we believe that the humus-iron podzol phase has previously passed through an iron podzol stage (Anderson *et al.* 1982). The Chon soil is derived from the same acid schist formation as that at Kelty, and we would expect greater quantities of acid-soluble Al in the podzol B horizon in comparison with a related gley, acid rain inputs being assumed similar for both sites.

The spatial variability of the extractable Al forms for any given soil subgroup within any one catchment differed according to horizon, broadly related to organic matter content. Values for A, E, Bs and B/C horizons lay within $\pm 30\%$ of those shown. Peats, FH and Bh horizons appeared more variable ($\pm 80\%$), but this decreased when profiles were grouped according to vegetation ($\pm 60\%$). Soil solution Al measurements for the different catchments have been made and generally support the postulate (McMahon 1990). Any budgetary approach to long-term Al leaching would have to be attempted by modelling.

E. TIPPING (*Institute of Freshwater Ecology, U.K.*). (i) Can Dr Walker suggest a mechanism for the retention of SO_4^{2-} by organic soils? Organic matter is not usually considered to bear functional groups able to interact with SO_4^{2-}. (ii) Does the observed impoverishment in aluminium of upper soil horizons imply a finite source of aluminium? What implications are there for future soil development?

T. A. B. WALKER. (i) The major functional groups associated with soil organic matter would not, of themselves, be capable of sulphate retention. However, multivalent cations are complexed by these functional groups, and we envisage that these cations provide positively charged sites under the acid conditions encountered in our study areas.

(ii) The Al seen in secondary or tertiary forms in our soils, e.g. organic complexes and other, including exchangeable, forms arises by a number of pedogenic routes. Although these forms of Al may become depleted by anthropogenic acidification, 'natural' pedogenic processes will replace them, albeit slowly, by mineral weathering.

Hydrogeochemical processes in the Birkenes catchment

By Nils Christophersen[1], Michael Hauhs[2], Jan Mulder[3]†, Hans M. Seip[4] and Rolf D. Vogt[4]

[1] Center for Industrial Research, P.O. Box 124 Blindern, N-0314 Oslo 3, Norway
[2] Department of Soil Science and Forest Nutrition, University of Göttingen, Busgenweg 2, D-3400 Göttingen, F.R.G.
[3] Department of Soil Science and Geology, Agricultural University, P.O. Box 37, 6700 Wageningen, The Netherlands
[4] Chemistry Department, University of Oslo, P.B. 1033 Blindern, N-0315 Oslo 3, Norway

Variations in streamwater chemistry at the Birkenes catchment are strongly flow dependent. Typically, the streamwater concentrations of H^+ and inorganic monomeric aluminium (Al_i) increase with flow, whereas the concentrations of Ca^{2+} and Mg^{2+} decrease. On the detailed scale, however, modifications to this picture may be observed depending on the antecedent conditions. The patterns in streamwater chemistry relate to changes in the water flow paths within the catchment. By using both hydrological and chemical techniques, it is shown that during high flow, much of the water generating runoff passes through the soil horizons on the hillslopes, but baseflow originates from deeper layers in the valley bottom where neutralization occurs. The composition of the soil waters in each horizon is strongly controlled by interactions with the solid phase. To a first approximation, the chemically reactive species in streamwater can be explained as a variable mixture of the soil waters from the O/H, E and B horizons on the slopes, and the peat deposits in the valley bottom. The controlling mechanisms for the soil waters in each horizon are discussed.

1. Introduction

Streamwater and precipitation chemistry have been studied at the Birkenes catchment since 1971/72 by the Norwegian Institute for Water Research (NIVA) and the Norwegian Institute for Air Research (NILU). Presently, the site is monitored as part of the programme on 'Long range-transported air pollutants' administered by the Norwegian State Pollution Control Authority (SFT 1989). Long-term monitoring of forest stands and soil chemistry is also done in the catchment.

Based on the precipitation and streamwater data, the hydrochemical Birkenes model was constructed in the 1970s and early 1980s (cf. Christophersen et al. 1982). The model provided a simplified overall picture of the dominant processes controlling streamwater chemistry. A main feature of the model was the difference in flow paths during high flow and low flow; in the former case, most of the water was postulated to pass only upper and acidic soil layers dominated by cation exchange, whereas in

† Present address: Center for Industrial Research, P.O. Box 124, Blindern, N-0314 Oslo 3, Norway.

the latter case baseflow was hypothesized to originate from deeper mineral horizons where neutralization had taken place by base cations released by weathering. Inorganic aluminium was originally assumed for both reservoirs to be controlled by equilibrium with an aluminium hydroxide phase ($Al(OH)_3$, gibbsite). This has later been modified (Stone & Seip 1989) in line with the more recent observations as discussed below.

To test and improve the description of the flow paths and chemical mechanisms in the model, process-oriented studies, which became part of the Surface Waters Acidification Programme (SWAP), were initiated at Birkenes in 1984 (*cf.* Seip *et al.*, this symposium). The emphasis has been on detailed episodic studies in the spring and autumn including sampling of streamwater, soil water, precipitation and melt water, and soils, as well as measurements of the ground-water table and the soil-water potential. The data collected span a wide set of hydrological conditions comprising dry and wet periods (Seip *et al.* 1989; Vogt *et al.*, 1990, this symposium) as well as an extreme sea-salt event (Mulder *et al.* 1990). Much of the effort has been concentrated on a hillslope in the catchment, where detailed hydrochemical plot-scale studies have been done. A distributed two-dimensional hydrological model has been applied to quantify the water-flow patterns on the hillslope. This paper provides a summary of the major results and a discussion of areas where the future hydrogeochemical process studies should be concentrated. The implications for hydrochemical modelling are discussed in a companion paper by Christophersen *et al.* (this symposium).

2. Study site and methods

The Birkenes catchment (41 ha†) is spruce forested and characterized by thin soils on fairly steep hillslopes with deeper organic deposits in the valley bottoms (figure 1; *cf.* Christophersen *et al.* 1982).

Detailed process studies were done on the slope, with plots designated D, E, F and X (figure 1). At these plots, tensiometers and tension lysimeters (Hauhs, 1988; Seip *et al.* 1989; Mulder *et al.* 1990) were installed in the main soil horizons. The tensiometers automatically recorded the soil water potentials at 15 min intervals. Soil sampling was done on several occasions. The most detailed survey took place in the autumn of 1989 in parallel with soil sampling at the other SWAP sites at Høylandet and Birkenes II (Seip *et al.*, this symposium). At Birkenes, soil samples were taken every 10 m in a grid covering 100 m by 100 m overlying the slope (M. Pijpers, personal communication). Soil-water collection has also been done less frequently at plots A, B and C on the other main slope. Piezometers for measurements of the ground-water table were installed at several locations in the main valley.

For streamwater and soil water (200–300 samples per episodic study), inorganic monomeric aluminium (Al_i) and organic monomeric aluminium (Al_o) were determined in the field according to the operationally defined Barnes–Driscoll method (Driscoll 1984). Field analyses also incorporated pH, temperature, conductivity as well as spectrophotometric determination of total organic carbon (TOC) (cf. Sullivan *et al.* 1986; Seip *et al.* 1989). Subsequent laboratory analyses on selected subsets of samples included major cations by atomic absorption (Ca^{2+}, Mg^{2+}, Na^+, K^+), major anions by ion chromatography (Cl^-, SO_4^{2-}, NO_3^-), total fluoride by an ion selective electrode, and silica by the molybdosilicate method. Isotopic analyses of the $^{18}O/^{16}O$

† 1 ha = 10^4 m^2.

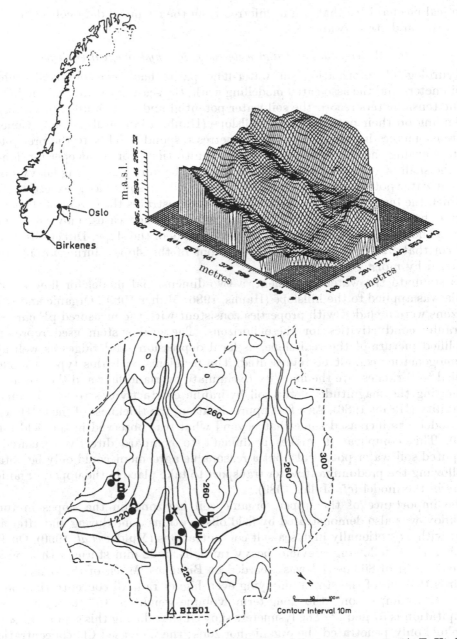

Figure 1. The Birkenes catchment with the weir (BIE 01) and the lysimeter nests (A–F, X). Most of the detailed plot scale work was done on the slope with plots D, E, F and X where tensiometers were also installed.

ratio have also been done for a large number of precipitation, stream and soil water samples.

3. Results and discussion

Here the flow paths will be discussed, starting with the hydrological evidence collected during the episodic studies, and continuing with the hydrological and chemical mechanisms that can be inferred from the chemical data collected on soils, soil water and streamwater.

(a) Water-flow paths and water mixing – hydrological evidence

Hydrological information on water-flow paths has been obtained from the tensiometers and the associated modelling work, the sea-salt event, and the ^{18}O data.

The tensiometers record the soil-water potential and show a differential response depending on their position on the hillslope (Hauhs 1988; Muller 1989). Generally, the tensiometers in the upper few centimetres respond quickly to the precipitation input, showing the effects of direct vertical infiltration. For the deeper tensiometers and the shallow ones downslope, lateral flow can be inferred. During heavy rainfall, the soil water potentials (in centimetres of H_2O) will, in some cases, exceed the depth at which the tensiometers are installed. This demonstrates the extent of the pressure exerted by water upslope. The soil water may, in such cases, move towards the surface and create overland flow on the lower parts of the slope. Direct observations confirm that overland flow occurs at the foot of the slopes during intense events preceded by moist conditions.

To simulate the water movement, a two-dimensional model for flow in porous media was applied to the hillslope (Hauhs, 1986; Muller 1989). Organic and mineral horizons were included with properties consistent with the measured pF curves and hydraulic conductivities for these horizons. The grid system used represents a simplified picture of the real slope, as local depressions and ridges as well as soil inhomogeneities cannot be represented. The problems with this type of model, as applied to Birkenes, are the inability to simulate overland flow and the requirement of limiting the magnitude of the soil hydraulic conductivities to avoid numerical instability (Hauhs 1986). As a first approximation, the thickness of the O/H layer in the model was increased somewhat beyond what was observed in the field (Muller 1989). This complication makes it difficult to compare directly measured and computed soil-water potentials, but a reasonable agreement could only be obtained by allowing the predominant water transport to take place in the upper, organic soil layers in the model (cf. Muller 1989).

The importance of the upper, organic soil horizons on the slopes in routing quickflow was also demonstrated by field observations done during and after a rain event with exceptionally high sea-salt concentrations (Mulder et al. 1990). On 16–17 October 1987, following previous heavy rainfalls, a 22 mm storm with a bulk Cl^- concentration of 862 $\mu eq\,l^{-1}$ was recorded at Birkenes. Much of the sea-salts came towards the end of the storm, implying even higher rainfall concentrations at that time. By comparison, the long-term volume-weighted Cl^- concentration in precipitation is 57 $\mu eq\,l^{-1}$. The lysimeters showed that during this event, the sea-salt 'blanket' only penetrated the organic horizons; the increased Cl^- concentration in the brook being caused by lateral flow in these layers. Furthermore, the sea-salt pulse retained its identity in the O/H horizons until the next major storm (24 mm) occurred almost a week later. The sea-salts were then washed out, partly by lateral flow to the stream and partly by percolation to the mineral horizons. By using chloride as a tracer, calculations indicated that streamwater highflow during the second event was predominantly caused by water from the O/H horizons. After this event, an estimated 50% of the Cl^- that entered the catchment on 16–17 October had left.

Although the upper organic horizons provide important water pathways during

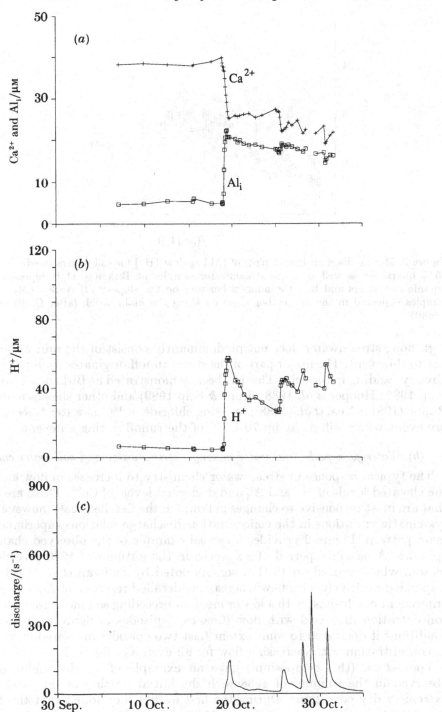

Figure 2. Episodic changes in discharge and streamwater concentrations of Ca^{2+}, Al_i and H^+ for the autumn of 1986.

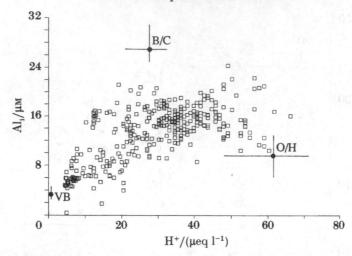

Figure 3. Mixing diagram (EMMA plot) of [Al$_i$] against [H$^+$] for soil waters (median and 75% and 25% quartiles) as well as single streamwater samples at Birkenes. O/H represents the upper organic soil layers and B/C the mineral horizons on the slopes; VB (valley bottom) represents samples collected in the deeper bog deposits along the main brook (after Christophersen *et al.* (1990)).

high-flow, streamwater does not predominantly consist of the rain water that gave rise to the event. The major part of the storm runoff originates from pre-event water already residing in the soils. This has been demonstrated at Birkenes (Christophersen *et al.* 1985; Hooper *et al.* 1988; Stone & Seip 1989) and other sites (see, for example, Rodhe (1981); Neal *et al.* (1988)) by using chloride or ^{18}O as water tracers. Typically, pre-event water will make up 70–80% of the runoff during an event.

(b) *Water-flow paths and water mixing – streamwater and soil-water chemistry*

The typical response in streamwater chemistry to increases in flow at Birkenes is the elevated levels of H$^+$ and Al$_i$ and decreased levels of Ca^{2+}. These are the species that are most responsive to changes in flow. On the detailed scale, however, there are systematic variations in the concentration–discharge relations superimposed on the main pattern. Figure 2 provides a typical example of the observed changes during episodes. A baseflow period of six weeks in the autumn of 1986 preceded the wet season, which started on 19 October. As noted by Sullivan *et al.* (1986, 1987), H$^+$ responded positively with flow whereas the detailed response of Al$_i$ depended on the antecedent conditions; with a low or medium preceding soil moisture content, the Al$_i$ concentration increased with flow (first two episodes in figure 2); under very wet conditions it decreased to some extent (last two episodes measured). Ca^{2+} decreased in concentration with increasing flow for all events in figure 2.

Vogt *et al.* (this symposium) gave an example of another mode of response observed in the summer of 1988 with the initial wetting of the catchment from extremely dry conditions. During the first events they found that the H$^+$ and Al$_i$ concentrations decreased whereas Ca^{2+} (and iron) increased with flow, before the more general pattern was re-established during the subsequent large episode.

The concentration–discharge relations can be explained when combining the soil-water and streamwater chemistry. In figures 3 and 4, two mixing diagrams are given (*cf.* Christophersen *et al.* 1990). Median concentrations and quartiles are indicated for

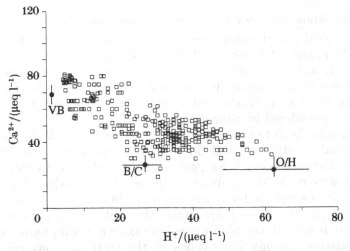

Figure 4. Mixing diagram of [Ca^{2+}] against [H^+] for Birkenes. See legend to figure 3.

three of the major soil water types sampled (called endmembers), and all streamwater samples for the episodic studies from 1984 to 1987 are given. In these figures, high-flow samples are plotted to the right and base-flow samples are to the left. Going from the O/H and B horizons on the slopes to the deeper parts of the main valley bottom (VB), there is a continuous decrease in [H^+] and an increase in [Ca^{2+}]. [Al_l] shows a maximum in the B horizons. An increase in runoff will imply a rise in the water table and contributions from saturated flow higher up in the soil profile (*cf.* Sullivan *et al.* 1987). Disregarding for the moment the initial events following the driest conditions, this picture implies that [H^+] should increase uniformly with increasing flow whereas [Ca^{2+}] should decrease with flow. For [Al_l], with its maximum in the B horizons, it is reasonable to expect (as observed) that a maximum will occur in the stream for events preceded by a low or moderate groundwater table. For the more unusual situations at Birkenes with extremely dry antecedent conditions, it is hypothesized that the initial events will press out deep bog water residing under reducing conditions by a piston effect (Vogt *et al.*, this symposium).

The role of the stream-bed as a source or sink of cations has also been investigated at Birkenes through experimental addition of acids. The results indicated that the stream-bed was not critical in regulating the streamwater chemistry; the most noticeable response to the acid additions was some release of calcium (Vogt 1989).

Under the name of 'end-member mixing analysis' (EMMA), the mixing ideas have been explored in more detail in recent work (Christophersen *et al.* 1990; Hooper *et al.* 1990). A more complete analysis needs to take into account potential effects due to non-conservative mixing (e.g. CO_2 degassing and $Al(OH)_3$ precipitation) (Neal & Christophersen 1989). Also, one must consider the role of temporal changes in the soil water compositions – especially in the O/H horizons – with season (Andersen *et al.*, this symposium), meteorological conditions, and the ionic loading of precipitation and meltwater (Vogt *et al.* 1990, this symposium). Furthermore, the observed soil waters in the O/H and B horizons on the slopes contain too little calcium to explain the concentration in the brook at highflow (figure 4). The stream-bed could potentially supply some calcium, but more sampling will need to be done in the catchment to search for other calcium sources.

(c) Mechanisms governing the soil and soil-water chemistry

Both the hydrological and chemical observations confirm the overall importance of water-flow paths in the catchment. Within the chemical variations observed for the end-members, the soil waters of each horizon seem to have a specific 'fingerprint' composition, different from those of the other horizons. The solid phases, therefore, must exert strong controls over the soil solutions (cf. Rosenqvist 1978). These mechanisms will be considered in some detail.

For the O/H horizons both the sea-salt event of October 1987 (Mulder et al. 1990) and the experimental addition of NaBr in the field (Vogt 1989; Vogt et al. 1990) demonstrated the effects of cation exchange under field conditions. In both cases, H^+, Ca^{2+}, Mg^{2+} and Al_i were released in response to the addition of sodium ions. Also, for the field observations as well as for laboratory experiments on organic soils from Birkenes (Andersen et al., this symposium), the soil solutions were strongly undersaturated with respect to any aluminium hydroxide mineral phase. For the sea-salt event, the relations among the cations in the O/H horizons corresponded approximately to what one would expect from simplified cation exchange equations (Mulder et al. 1990). For example, the slope in a diagram of pAl^{3+} against pH was roughly equal to 3. However, different results were obtained in the other cases (Andersen et al., this symposium; Vogt et al. 1990), and the chemical control mechanisms in the organic horizons need a more thorough investigation (cf. Neal et al., this symposium). Future work will be based on the CHAOS model by Tipping & Hurley (1988), where the aluminium activity is determined by (de)complexation with solid phase organics. This model was applied to the results from the laboratory experiments on the Birkenes soils and the initial results are promising (Andersen et al., this symposium).

There is a uniform increase in Al_i, and to a small extent Ca^{2+}, and a decrease in H^+ from the O/H to the B horizons (cf. figures 3 and 4). Thus during vertical percolation, neutralization of H^+ is mainly accomplished through release of Al_i. The concentration levels in the B horizons show little temporal variations during a given episodic study (Seip et al. 1989). The soil solutions in the B horizons broadly satisfy the necessary (but not sufficient) condition for equilibrium with an aluminium hydroxide mineral (Seip et al. 1989). However, organic Al–Fe complexes, deposited on the mineral surfaces as part of normal soil genesis, could play an important role in controlling the soil solution chemistry. The CHAOS model has also therefore been applied to these horizons (Andersen et al., this symposium).

In the valley bottom both the H^+ and Al_i concentrations are consistently low (figure 3). Proton consumption occurs both through reduction, as evidenced by elevated iron levels and low sulphate values for some lysimeters, and base cation weathering, as evidenced by high base cation concentrations for other lysimeters.

The nature of the strong relations between the solid and solution phases seen in all soil horizons has been under discussion. Based on other studies, Hauhs (1988) suggested that the chemical soil-water interactions predominantly occurred in the inner solution close to the surface. He hypothesized that the soil water would get its chemical 'fingerprint' under unsaturated conditions. As the flow became saturated, mixing processes would dominate and little chemical interaction would take place with the solid phase. An argument in favour of this hypothesis is that soil mineral profiles are characterized by strong vertical chemical gradients (due to unsaturated vertical flow) and not by lateral gradients along slopes.

To look for lateral soil gradients, the detailed grid sampling on the slope was done in the autumn of 1989. The results so far show that there is a systematic variation in the surface soil chemistry with decreased base saturation and increased aluminium saturation in the bogs as compared with the O/H horizons of well drained soils. This was found for the bog along the main brook as well as for small bogs situated in local depressions on the slopes (M. Pijpers, personal communication, *cf.* also Seip *et al.*, this symposium). Retaining the above hypothesis of Hauhs (1988), one could argue that the bogs are fed from the O/H and B horizons and that the upper peat layers will interact with this water during recession periods when conditions become unsaturated. Alternatively, the soil-water chemistry could be controlled to a smaller or larger degree by the solid phase also under saturated conditions. In this case, the streamwater chemistry would be more or less regulated by the soil horizons of last contact before entering the stream.

4. Conclusions

The results from the field work at Birkenes conclusively show the importance of water flow paths in determining streamwater chemistry. Each soil horizon has a typical 'fingerprint' composition, different from that of the others. The most important remaining questions are related to quantification of the chemical mechanisms controlling the soil-water compositions in each horizon, and the role of mixing processes versus reactions along water pathways in determining streamwater chemistry.

This work has been done as part of the SWAP and has also been funded by the Norwegian Department of the Environment and the Norwegian Council for Scientific and Industrial Research (NTNF). Valuable comments were received from C. Neal.

References

Christophersen, N., Seip, H. M. & Wright, R. F. 1982 *Wat. Resour. Res.* **18**, 977–996.

Christophersen, N., Kjærnsrød, S. & Rodhe, A. 1985 *Hydrologic and hydrogeochemical mechanisms and model approaches to the acidification of ecological systems, report no.* 10, pp 29–40. Oslo: Nord. Hydrol. Programme.

Christophersen, N., Neal, C., Hooper, R. P., Vogt, R. D. & Andersen, S. 1990 *J. Hydrol.* (In the press.)

Driscoll, C. 1984 *Int. J. environ. analyt. Chem.* **16**, 267–283.

Hauhs, M. 1996 *Geoderma* **38**, 97–113.

Hauhs, M. 1988 *Report department of soil science and forest nutrition, University of Göttingen*, F.R.G., Germany, 18 pp.

Hooper, R. P., Stone, A., Christophersen, N., de Grosbois, N. & Seip, H. M. 1988 *Wat. Resour. Res.* **24**, 1308–1316.

Hooper, R. P., Christophersen, N. & Peters, N. R. 1990 *J. Hydrol.* (In the press.)

Mulder, J., Christophersen, N., Hauhs, M., Vogt, R. D., Andersen, S. & Andersen, D. O. 1990 *Wat. Resour. Res.* **26**, 611–622.

Muller, D. I. 1989 *M.Sc. thesis, University of Göttingen*, F.R.G., 139 pp.

Neal, C. & Christophersen, N. 1989 *Sci. tot. Env.* **80**, 195–203.

Neal, C., Christophersen, N., Neale, R., Smith, C. J., Whitehead, P. G. & Reynolds, B. 1988 *Hydrol. Process.* **2**, 155–165.

Rodhe, A. 1981 *Nord. Hydrol.* **12**, 21–30.

Rosenqvist, I. Th. 1978 *Sci. tot. Envir.* **10**, 39–49.

Seip, H. M., Andersen, D. O., Christophersen, N., Sullivan, T. J. & Vogt, R. D. 1989 *J. Hydrol.* **108**, 387–405.

SFT, 1989 Annual report 1988, Norwegian state polution control authority, report no. 375/89, 286 pp., Oslo.

Stone, A. & Seip, H. M. 1989 *Ambio* **18**, 192–199.

Sullivan, T. J., Christophersen, N., Muniz, I. P., Seip, H. M. & Sullivan, P. D. 1986 *Nature, Lond.* **323**, 324–327.

Sullivan, T. J., Christophersen, N., Hooper, R. P., Seip, H. M., Muniz, I. P., Sullivan, P. D. & Vogt, R. D. 1987 *International symposium on acidification and water pathways, Bolkesjø, Norway,* 4–8 *May,* pp. 269–286. Oslo: Norwegian Hydrological Committee.

Tipping, E. & Hurley, M. A. 1988 *J. Soil. Sci.* **39**, 505–519.

Vogt, R. D. 1989 M.Sc. thesis, Chemistry department, University of Oslo, 98 pp.

Vogt, R. D., Seip, H. M., Christophersen, N. & Andersen, S. 1990 *Sci. tot. Envir.* (In the press.)

Spate-specific flow pathways in an episodically acid stream

By K. H. Bishop, H. Grip and E. H. Piggott

*Department of Forest Site Research, Swedish University of Agricultural Sciences,
S-901 93 Umeå, Sweden*

Isotope tracer studies indicate that most of the water in acid episodes is pre-event water that was in the catchment before the onset of the episode. This has not been a feature of conceptual models of episodic acidity in which hydrology plays a crucial role. To determine experimentally the pathways that pre-event water takes to the stream during acid episodes and how these pathways affect the chemistry of episodes, a field study was conducted on a hillslope adjacent to an episodically acid stream at Svartberget in northern Sweden. Spatially and temporally intensive measurements of hydraulic potential and soil water chemistry were made during acid episodes to ascertain the major flow pathways and their chemistry.

Groundwater was alkaline before, during and after acid episodes that were caused by increases in organic acidity rather than by increases in mineral acids or decreases in base cations. During episodes, the majority of the runoff moved laterally through 'spate-specific' flow pathways in the upper four decimetres of soil. The large amount of pre-event water held in this zone before the episode led to the large component of pre-event water in the runoff. The organic acidity of the runoff was acquired in the shallow, organic-rich soil horizons intersected by the spate-specific flow pathways.

Hydrologically similar catchments may respond to changes in acid deposition more quickly than would be predicted by models of catchment hydrochemistry that do not adequately represent the localized origins of runoff and acidity.

1. Introduction

A characteristic feature of the headwater streams most susceptible to acidification is a strong correlation between stream acidity and streamflow. This correlation is manifested most clearly in 'acid episodes' during periods of high flow after spring snowmelt or large rainstorms. During acid episodes, the concentration of H^+ can increase by as much as an order of magnitude in a matter of hours. (Davies 1989) The profound biological impact of such sudden declines in pH and deleterious increases in labile inorganic aluminium concentrations are responsible for much of the ecological depredations caused by acidification (Gunn & Keller 1984; Henriksen *et al.* 1984).

The strong correlation between water flow and acidity is thought to be an indication of the fundamental relationship between the movement of water through a catchment and the chemistry of streamwater leaving the catchment. Conceptual models, such as Birkenes and ILWAS, reflect this relation and strive to simulate the essential processes that generate acid episodes. In both models, a key component is the routing of water through a series of discrete reservoirs that are supposed to represent the catchment (Christophersen *et al.* 1982; Chen *et al.* 1982; Gherini *et al.* 1985). In published applications of these models, the water feeding simulated streams at low flow is less acid because it has a residence time in the catchment of weeks or longer. During that time, weathering reactions reduce the acidity in the water. The model runoff that feeds the stream during acid episodes has a lower pH because much of the episode runoff is routed to the stream in a matter of days, moving through shorter residence time reservoirs where there is less chance for weathering and more opportunities to acquire acidity located in the shallower soils of the catchment (Fendick & Goldstein 1987; Seip *et al.* 1985).

The hydrological data used in the calibration of these models have been the daily inputs to and outputs of water from a catchment. However, this information is insufficient to define the routing of the water within the catchment, which is crucial for realistic formulation of the acidification models. Of particular importance to the chemical behaviour of the models is the partitioning of the runoff into the less acid water with a longer residence time in the catchment and the more acid water with a shorter residence time.

Efforts were made to confirm independently this aspect of the Birkenes model by using stable isotope 'labels' to separate the streamflow during acid episodes into an older 'pre-event' component, which was present in the catchment before the onset of the runoff event, and the component of new rain or snowmelt that entered the catchment during an acid episode. This technique, known as isotope hydrograph separation (IHS), has generally indicated that 60–90 % of the storm hydrograph is composed of pre-event water (Rodhe 1987). The episodically acid Birkenes stream, for which the Birkenes model was developed, proved to be no exception as it also has a large component of pre-event water in stormflow, thus contradicting the hydrology used in published applications of the full Birkenes hydrochemical model, which relied on a large component of acid new water to effect the dramatic changes in stream chemistry during acid episodes.

Both Christophersen *et al.* (1985) and Hooper *et al.* (1988) were unable to reproduce satisfactorily even the streamflow and isotopic concentrations at Birkenes with the Birkenes model, though Taugbøl & Bishop (this symposium), report slightly more success by using 'piston-flow' to bypass the model's normal routing of runoff. Efforts to simulate episodic acidity with any model that outputs a large proportion of pre-event water during episodes have yet to be reported.

The large component of pre-event water in the storm hydrograph indicated by isotope hydrology presents a two-fold paradox for models of episodic acidification. The first aspect of the paradox is how 'old' water can comprise such a large portion of the storm hydrograph. This has been a long-standing question for those concerned with runoff generation. The new twist to the paradox in studies of acid episodes is to explain how the dramatic changes in chemistry during episodes can be effected largely with pre-event water when pre-event water was so much less acid in the low flow just hours before the onset of a storm event.

This shortcoming in the critical characterization of catchment hydrology results largely from the dearth of field observations appropriate to the detailed questions about the movement of water raised by efforts to model acidification. At the Svartberget Research Catchment in northern Sweden, the paradox of pre-event water in acid episodes has been addressed by using instrumented hillslopes adjacent to an episodically acid stream to study how flow pathways affect the chemistry of water as it moves through the soil catena on its way to the stream.

2. Site description

The hillslope research area is located on the 50 ha Svartberget research catchment 60 km N.W. of Umeå, Sweden (64° 14′ N, 10° 46′ E). A full description of the catchment can be found in Grip & Bishop (this symposium). The hillslopes in the research area slope at an angle of 5°–10° toward a stream. The hillslope is underlain by some 30 m of till derived from the local gneissic bedrock (Lindqvist 1989; Miskovsky 1987). The soils on this till are well developed iron podzols, blending into humus podzols within 10 m of the stream. The hillslope is afforested with Norway Spruce. The field layer is primarily lichens, grasses and berries on the hillslopes, changing into mosses on the streambanks.

3. Methods

(a) Hillslope observations

To observe the changing flow pathways of water during acid episodes and the chemistry of water moving along those pathways, cross-sections of hydraulic potential and water chemistry were taken along instrumented hillslope transects oriented parallel to the downhill flow of water. Transects on both sides of the stream were monitored, but only the results of the 50 m transect to the west of the stream are reported here.

Zero-tension lysimeters (which doubled as piezometers), groundwater tubes and tensiometers were located in the stream itself and at distances of 2 m, 7 m, 15 m, 25 m and 50 m from the stream (figure 1). Sets of samples were collected before, during and after storm events between June and September in 1986, 1987 and 1988. Cross-sections of hydraulic potential were taken from the piezometers and groundwater tubes in each nest every 3–4 h during the rising limb of a storm event. Sampling frequency tapered off to every 48 h during prolonged periods of low flow.

The key to using these measurements of hydraulic potential to determine the flow pathways of water through the hillslope lay in determining the hydraulic conductivity of the hillslope. With information on both the hydraulic conductivity and the distribution of hydraulic potential, Darcy's Law could be used to calculate flows through the slope. Saturated hydraulic conductivity was measured *in situ* at depths down to 1 m with double-ring infiltrometers. It is, however, notoriously difficult to determine the hydraulic conductivities that actually obtain in the field. Observations of the hydraulic gradient and the flux of water through the hillslope made it possible to use Darcy's law to calculate the effective transmissivity of the hillslope at different times. By analysing the transmissivity of the hillslope when the groundwater was at different levels, it was possible to estimate the saturated hydraulic conductivity of discrete 10 cm layers of the soil (K. H. Bishop, unpublished results).

(b) Stream observations

To assess the effect of changes in flow pathways on streamwater, time-series of the amount and chemistry of runoff from the instrumented hillslope were required. These were calculated from the difference in discharge and chemistry between weirs above and below the instrumented hillslope. These two weirs were 180 m apart and the reach that they delimited drained a subcatchment of roughly 4.8 ha†. Samples

† 1 ha = 10^4 m².

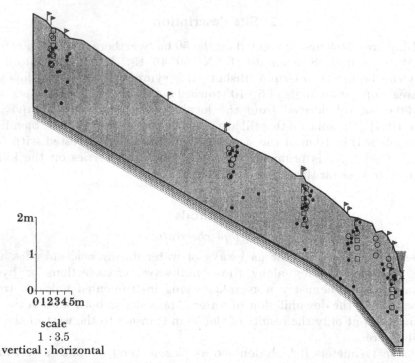

Figure 1. Cross-section of the hillslope transect. The instruments were arranged in six 'nests' to provide measurements of water chemistry and hydraulic potential down to a depth of two metres; (▶), groundwater tube; (□), suction lysimeter; (●), piezometer; (+), stage post; (○), tensiometer.

for analysis were taken from these weirs at the same time as the hillslope hydraulic potential was recorded.

Potential exists for large errors in calculations of hillslope output from the small difference in two large measurements. The errors in calculated discharge and chemistry were on the order of ±30% at low flow, but the relative size of these errors decreases with increasing flow to below ±10%. This should allow for meaningful interpretation of the patterns in hillslope runoff and its chemistry during acid episodes.

(c) Chemical analyses

A thorough description of the techniques used for chemical analysis in this research may be found in Bishop et al. (1990). The large contribution of organic acids to the acidity of the stream was assessed by using the titration technique developed by Lövgren et al. (1987) to determine the concentration and pKa of organic acids. A second technique to determine the quantity of organic anions in streamwater involved the oxidation of organic carbon using ultraviolet light (Grip & Bishop, this symposium).

Cross-sections of hillslope chemistry indicated that the organic acidity entered the runoff as it passed the organic-rich forest mor and streambank. This hypothesis was tested by leaching groundwater through columns of soil taken from the forest mor and streambank.

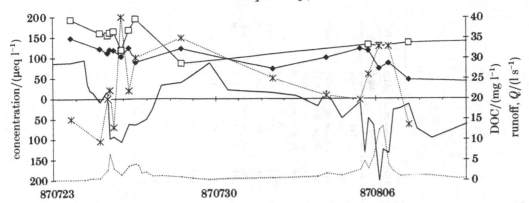

Figure 2. Concentration of principal chemical constituents in hillslope runoff during three acid
episodes; (—), alkalinity; (♦), sulphate; (□), Ca^{2+} and Mg^{2+}; (✳), DOC; (......), runoff.

4. Results

The chemistry of streamwater during acid episodes has been described by Ring
(1989). The relation between hillslope processes and acidity has been considered in
more detail by Bishop *et al.* (1990), whereas the flow of water in the hillslope during
these episodes is treated in Bishop & Rodhe (1989). In this review, the hydrology and
chemistry of three consecutive storm events in 1987 is presented. They convey the
essential features observed in the hydrochemical behaviour of the hillslope during
acid episodes.

(a) *Hillslope runoff chemistry*

On 23 July 1987, after six weeks of dry summer weather, runoff from the hillslope
had fallen to less than 0.5 mm per day. The alkalinity of this runoff had climbed
to 10 µeq l^{-1} as measured by the Swedish standard method of titration to an
endpoint of pH 5.4. Storms with more than 20 mm of rain followed on 25 July, 26
July and 5 August. Each storm resulted in episodes of increased acidity. The
alkalinity of the runoff decreased to between -50 and -100 µeq l^{-1} on 25 July
and 26 July. The alkalinity dropped to almost -200 µeq l^{-1} on 5 August (figure 2).
The concentration of acid sulphate anions did not increase during these pH
depressions. Instead, the concentration fell during successive acid episodes from
150 µeq l^{-1} on 23 July, before the first acid episode, to less than 75 µeq l^{-1}
during the third and most intense acid episode. The concentration of total organic
carbon (TOC) went from 12 mg l^{-1} at low flow to above 30 mg l^{-1} during each acid
episode, generally paralleling the increases in acidity during episodes. The chemistry
of these episodes is summarized in table 1.

Less than 20% of the total aluminium in samples speciated with the Driscoll
method (Driscoll 1984) was in the labile monomeric form (*ca.* 2 µeq l^{-1} of Al^{3+} at
low flow and 8 µeq l^{-1} during episodes.) The concentration of the base cations Ca^{2+}
and Mg^{2+} generally stayed 50–100 µeq l^{-1} higher than the sulphate concentration
during the study period. The sum of the equivalent concentrations of the measured
cations (Ca^{2+}, Mg^{2+}, Na^+, K^+, H^+) was consistently more than 100 µeq l^{-1} greater
than the sum of the equivalent concentrations of the measured anions (SO_4^{2-}, Cl^-,
NO_3^-). By using identical analytical methods, the measured anion deficit in the
precipitation during this period was between 10 and -5 µeq l^{-1}, which suggests that
the streamwater analyses are credible.

Table 1. Chemistry of Svartberget waters

water	pH	alk. µeq l⁻¹	SO₄²⁻ µeq l⁻¹	Cl⁻ µeq l⁻¹	NO₃⁻ µeq l⁻¹	Ca²⁺+Mg²⁺ µeq l⁻¹	Na⁺+K⁺ µeq l⁻¹	NH₄⁺ µeq l⁻¹	TOC mg l⁻¹	anion deficit µeq l⁻¹	amount mm	AL(t) mg l⁻¹	Fe³⁺ mg l⁻¹
1987 precipitation (mean)	4.5	N/A	35	10	16	6	4	14	N/A	-5	744	N/A	N/A
1987 discharge (mean)	4.4	N/A	101	9	1	150	55	1	N/A	135	349	N/A	N/A
stream low flow 870723	5.5	-5	91	8	1	135	80	1	12	119	0.5 per day	0.05	0.87
hydrograph peak 870725	4.3	-120	110	10	1	164	70	1	26	164	20 per day	0.27	1.17
hydrograph peak 870726	4.2	-155	62	9	1	126	56	2	23	175	31 per day	0.23	1.30
hydrograph peak 870806	4.1	-145	49	8	1	96	38	1	28	156	71 per day	0.18	0.82
event precipitation 870725	4.6	N/A	20	5	5	2	6	6	N/A	9	48	N/A	N/A
event precipitation 870726	4.6	N/A	16	14	4	3	8	3	N/A	4	33	N/A	N/A
event precipitation 870806	4.8	N/A	24	6	3	3	6	4	N/A	-4	66	N/A	N/A
hillslope runoff low flow	N/A	10	175	9	1	220	90	2	3	130	0.1 per day	N/A	N.A
hillslope runoff 870725	N/A	-59	117	9	1	156	75	0	12	119	5	0.28	0.96
hillslope runoff 870726 event	N/A	-60	113	8	2	158	76	1	14	138	12	0.61	1.42
hillslope runoff 870806 event	N/A	-85	76	8	1	131	50	1	27	149	29	0.46	1.14
groundwater 30–50 cm	5.8	94	117	20	N/A	147	160	N/A	6	164	65	0.53	0.28
groundwater 50–100 cm	5.9	153	146	20	N/A	164	167	N/A	7	197	230	0.44	0.50
groundwater 100–150 cm	6.1	227	149	20	N/A	216	215	N/A	4	319	205	0.30	0.75

(b) Hillslope chemistry

During acid episodes, the shallow groundwater was very alkaline, with a mean alkalinity of 160 µeq l^{-1}. The runoff was sometimes more than 300 µeq l^{-1} less alkaline. The shallowest groundwater samples in the cross-sections were taken from depths greater than 30 cm. Insufficient volumes for sample analysis were obtained from shallower zero-tension lysimeters. Concentrations of base cations, alkalinity and, to a lesser extent, sulphate, were positively correlated with increasing depth, while total aluminium was negatively correlated (table 1).

(c) *The origins of runoff acidity*

Titrations of the organic acids in the streamwater found them to be adequately described by a monoprotic acid with a pKa of between 4.6 and 5.2. The estimated concentrations of this organic acid changed from 100 µeq l^{-1} at low flow to 200 µeq l^{-1} during storm events (L. O. Öhman, personal communication). Dissociated organic acids, which are not measured in conventional analyses, help to balance the anion deficit. The estimated concentration of organic anions was not enough to establish the necessary balance between anions and cations in solution, but there is considerable uncertainty in current methods for determining the pK$_a$ and concentration of organic acids. Ultraviolet oxidation of organic material in the streamwater succeeded in eliminating the measured anion deficit, strongly suggesting that organic anions were the source of the measured anion deficit (Grip & Bishop, this symposium).

The significance of organic acids in the runoff acidity, the low concentration of TOC in groundwater and the flux of much runoff through organic-rich soils in the shallowest soil horizons prompted a column leaching experiment to assess the capacity of organic-rich soils to acidify passing alkaline groundwater in transit to the stream. That capacity was found to be large and persistent, reducing the alkalinity of groundwater by as much as 150 µeq l^{-1}.

(d) Hillslope hydrology

The volume of runoff from the hillslope during the events of 25 July, 26 July and 5 August was 5 mm, 12 mm and 29 mm, respectively. By using the soil moisture characteristics and the water tensions measured in the soil before the 25 July storm event, the amount of water in the upper 30 cm of soil is calculated to have been 90 mm. Isotope hydrograph separation of the 25 July storm indicated that 80% of the runoff (4 mm) was pre-event water. 60% of the 26 July storm hydrograph (7 mm) was comprised of pre-event water in the catchment prior to 24 July.

During the 1987 storm events, when discharge increased some two orders of magnitude from that at low flow, the maximum increase in the gradient of the water table was *ca.* 45%. The water table rose to within 20 cm of the ground surface between 3 m and *ca.* 50 m from the stream. Analysis of hillslope transmissivity and groundwater levels found the saturated hydraulic conductivity of the hillslope to increase from 10^{-7} m s^{-1} at a depth of 70 cm to 5×10^{-4} m s^{-1} at a depth of 10 cm. This was comparable to the results from the double-ring infiltrometers.

Darcy's law was used to calculate the lateral flux through different 10 cm thick layers of the soil over the course of the four-week period in 1987 when the three acid episodes of late summer occurred. The majority of this flow passed through the upper 15–45 cm of the soil between 3 m and 50 m from the stream (figure 3).

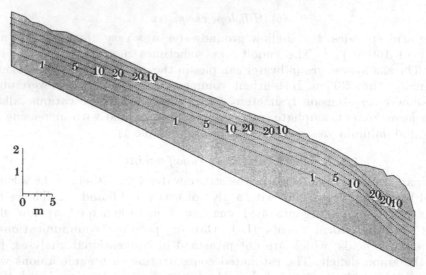

Figure 3. Distribution of lateral water flux through the upper two metres of the hillslope transect between 23 July and 20 August 1987. The flux through each 10 cm layer of soil was reckoned as a percentage of the total flux through the two metre deep soil profile. Less than 2 % of the flux went through any 10 cm layer below a depth of 65 cm.

The flux was not only localized in space to the upper decimetres of the soil, but also in time to the few days when groundwater levels were elevated by the rainstorms that precipitated acid episodes.

5. Discussion

(a) Spate-specific flow pathways

During acid episodes in the Svartberget stream, the principal role of pre-event water in the storm hydrograph is not consistent with the prominent role of new water in existing models of episodic acidification. Thus either the IHS is wrong or models should incorporate processes that can generate acid episodes using pre-event water.

Critical evaluations of the IHS methodology have so far failed to disprove its qualitative validity (Rodhe 1987; Bishop & Richards 1988). To achieve a large increase in the flow of pre-event water to the stream during episodes, the Darcian theory of flow requires a commensurate increase in the hydraulic gradient that forces water through the soil and/or in the transmissivity of the hillslope through which the water moves. As the hydraulic gradient in the hillslope increased by less than 50%, an increase of more than an order of magnitude in the transmissivity of the hillslope is implied. During storm events, the water table rose as much as 80 cm to within a decimetre or two of the soil surface. As there is an exponential increase in saturated hydraulic conductivity in the upper few decimetres of the soil, saturation of these highly conductive soil layers during episodes should provide the indicated increase in the transmissivity of the hillslope. A preponderance of the flow in the hillslope passed through these shallow flow pathways during the brief periods of elevated groundwater level coincident with acid episodes when the stream was in spate.

Storm runoff and snowmelt may follow different flow pathways in other episodically acid catchments. However, there is one feature of the shallow flow pathways at Svartberget that is likely to be common to many different episodically acid catchments. The flow pathways that transport most of the spate flow carry a smaller proportion of the total flow during periods of low flow. The reason why this characteristic is likely to be widespread is that the high saturated conductivity that enables these pathways to conduct spate flow makes those pathways the first to drain when the rainfall or melting of snow stops supplying the catchment with fresh inputs of water. The conductivity of these pathways is dependent upon their saturation, so once drained, their contribution to the hydrological output of the hillslope is greatly diminished. The term 'spate-specific' is used to refer to the shallow flow pathways at Svartberget to emphasize that essential hydrological feature that may be shared by a variety of episodically acid catchments.

Great as the flux is in the spate-specific flow pathways, it did not exceed the amount of pre-event water retained in these pathways before the acid episodes. The large reservoir of pre-event water available for displacement by, or mixing with, the infiltrating precipitation accounts for the large pre-event component of stormflow.

(b) *The chemical dynamics of acid episodes at Svartberget*

The ubiquity of the correlation between streamflow and stream acidity is one of the factors that focused interest on the role of hydrology in acid episodes. The frequency of this correlation cannot, however, be taken to mean that there is a single set of dominant chemical processes common to episodically acid streams. Certain features of the chemistry during the acid episodes at Svartberget distinguish it from many other episodically acid streams reported in the literature. Before considering how flow pathways influence the chemistry of episodes, it is important to determine the precise nature of the chemical changes associated with the acid episodes of July and August 1987.

Grip & Bishop (this symposium) discuss the weekly chemistry of the Svartberget stream that was notable for the prominent anion deficit linked to the high concentrations of organic carbon as well as for the lack of a significant relation between sulphate and stream acidity that is so common in anthropogenically acidified streams. These features of the Svartberget stream are even more pronounced during the acid episodes reported here. There is a negative co-variation between sulphate concentration and stream acidity. There is also a persistent excess of more than 100 μequiv. l^{-1} of cations over measured anions. Both the titration of organic acidity and the ultraviolet oxidation experiment indicated that this anion deficit resulted from unmeasured organic anions.

Changes in mineral acid concentrations or the balance between base cations and mineral acid anions could not explain the dramatic drop in the alkalinity of the runoff water during acid episodes, thereby ruling out the processes of ion exchange or sulphate mobilization often cited in episodically acid streams (Seip 1980; Christophersen & Seip 1983). Increases in TOC did correspond to the peaks in acidity. If TOC is taken as a surrogate for organic anions, then those anions are most likely the key to episodic acidity at Svartberget, either as dissociated acids or as ligands that complex cations. This applies not only to the runoff from the 8 ha peat mire at the head of the catchment, but also to the runoff from predominantly mineral soil hillslopes as well.

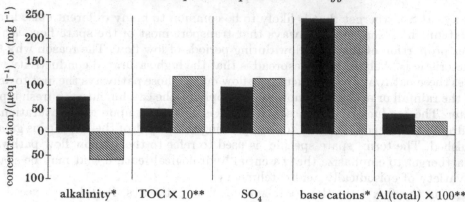

Figure 4. Comparison of hillslope runoff chemistry during the 25 July episode and the chemistry that results if rainwater is mixed with groundwater in the proportions indicated by the isotope hydrograph separation, (*), μequiv. l^{-1}; (**), mg l^{-1}; (■), rain- + groundwater; (▨), hillslope runoff 870725 event.

(c) The role of flow pathways in acid episodes

Often, it has been hypothesized that changes in flow pathways that generate increased runoff also play an essential role in the concomitant pH depressions that distinguish episodically acid streams. If this hypothesis applies to the Svartberget stream, then the essential chemical feature of the episodic acidity in the hillslope runoff that the dynamics of flow pathways should account for is the doubling of TOC during periods of increased flow while the concentration of other major constituents are diluted to a greater or lesser extent and the alkalinity drops precipitously. To resolve the paradox posed by isotope hydrology for models of episodic acidity, these chemical changes must be accomplished by using 60–90 % pre-event water that was roughly 100 μequiv. l^{-1} more alkaline in stream baseflow immediately before an acid episode.

The pre-event component of the storm hydrograph that IHS identified actually encompasses a variety of different sources of water from within the hillslope that can be classified by depth, distance from the stream and mobility within the soil matrix. The groundwater chemistry did not change significantly over the six week period of study, but there were distinct trends with depth between 30 and 200 cm below the ground surface. Flow through the lower 140 cm of this zone amounted to less than 10 % of the total, compared with the large flux of water between 15 and 45 cm depth during acid episodes. Unfortunately, soil-water analyses were not available from the uppermost 30 cm of the soil profile where the flow was concentrated.

To determine what differences in runoff chemistry are caused by the activation of spate-specific flow pathways, a simple mixing model was used. The model combined precipitation and pre-event water (represented by the 30–50 cm deep groundwater) in the proportions specified by the IHS, with the intention of showing the chemistry that would result if the flow pathways remained static and in the deeper soils during the acid episodes. With respect to base cations and sulphate, the simple model satisfactorily reproduces the concentrations seen in runoff during the 25 July episode (figure 4). Neither sources of water with initial concentrations different from that in the 30–50 cm deep groundwater nor rapid reactions along spate-specific flow pathways need to be invoked to explain these aspects of the episode chemistry.

The critical components of runoff chemistry that the mixing model fails to reproduce are the high concentrations of organic carbon and the decline in alkalinity brought about by the increased concentration of organic anions. One way the spate-specific flow pathways could account for the acid episodes is if they mobilized a new source of pre-event water in the shallow soil horizons with high concentrations of TOC and a chemistry which is otherwise similar to pre-event water somewhat deeper in the soil. Spate-specific flow pathways could also contribute to the increase in runoff acidity by moving alkaline pre-event water through the organic-rich soils intersected by the spate-specific flow pathways. In the leaching experiment, the transfer of organic acidity from the soil to alkaline groundwater was rapid enough to be significant in the course of an acid episode (Bishop *et al.* 1990).

6. Conclusions

A field study that integrated intensive observations of hillslope processes with the hydrochemical outputs from that same hillslope has revealed the role of flow pathways in the episodic acidity of a stream. For the area studied, the two-fold paradox posed by isotope hydrograph separation to contemporary models of episodic acidification has been resolved. Both a large component of pre-event water in the storm hydrograph and the increased acidity of that pre-event water during acid episodes resulted from activation of highly conductive flow pathways by a rising water table during storm events. These spate-specific flow pathways hold enough pre-event water and organic acidity to account for the large proportion of acid pre-event water, even though almost all of the groundwater below 30 cm is highly alkaline.

The prominent role of organic anions at Svartberget gives the hillslope runoff a chemical signature that differentiates it from many other episodically acid streams. That signature may be useful in distinguishing naturally acid streams from those acidified by inputs of anthropogenic mineral acidity in which sulphate, or possibly nitrate, accounts for much more of the pH depression during acid episodes. Too little is known about the behaviour of organic anions, however, to categorically state that the organic acidity at Svartberget is unaffected by the 1.5 g m^{-2} of marine-excess sulphate deposited each year on the catchment.

Despite the applicability of the Svartberget chemistry only to waters rich in organic anions, the localization of the spate-specific flow pathways to a narrow band of soil may have implications for the reversibility of acidification in sites acidified by mineral acids. Many acidified or acid sensitive sites throughout northern Europe and northern North America are located on till soils like those of Svartberget. Attempts to identify the flow pathways in such tills have often shown patterns of increased flow in shallow soils during runoff events similar to those identified at Svartberget (Lundin 1982; Mulder *et al.* 1990). The chemistry of these spatially proscribed spate-specific flow pathways determines the chemistry of acid episodes at Svartberget. The episode chemistry of hydrologically similar catchments acidified by the deposition of anthropogenic acidity may also be controlled by spate-specific flow pathways. If the water in those pathways has a short residence time, as seems likely, then changes in the chemistry of precipitation would first affect the chemistry of shallow soil layers that are subject to the most rapid throughput of precipitation. Changes in deposition chemistry would then manifest themselves most rapidly on precisely that portion of the catchment that determines the intensity of the acid episodes so critical to the viability of the aquatic ecosystem. Thus the localization of spate-specific flow pathways may lead to a response to changes in acid deposition that is more rapid than would be predicted by models that use a 'lumped' conceptualization of catchment hydrology.

It may seem ironic that this paper, which began by concentrating on the 'old' pre-event water in acid episodes, should conclude by stressing how 'new' that pre-event water may be in spate-specific flow pathways. Eriksson (1985), however, has emphasized the skewed distribution of water residence times within a catchment. Pre-event water may be old compared with rainwater, but the pre-event water in acid episodes may well be new relative to the mean residence time of water in the catchment or the pre-event water in low flow.

The use of isotope hydrograph separation has done much to stimulate critical consideration of water-flow pathways and residence-time distributions. The challenge now is to identify the residence times of specific flow pathways. Further developments in isotope hydrology promise to figure significantly in this endeavour, and these advances may prove to be of great value in the modelling and prediction of acidification. However, it is now more than twenty years after Svante Odén focussed the attention of the industrialized world on the threat posed by acid deposition with his bold thesis that the long range transport of air pollution was acidifying Scandinavian surface waters (Odén 1968). Much has been learned about the extent and severity of acidification since then. The promise of further discoveries will continue to beckon to scientists, but those responsible for controlling air pollution must balance the uncertain timetable of scientific progress against what is already known and the costs of delaying effective control measures.

This research has been made possible by the financial support of the Anglo–Scandinavian Surface Waters Acidification Programme, the Swedish National Environmental Protection Board, Total Petroleum Inc. and other sponsors of the Cambridge Acid Rain Expedition. The work was conducted by the members of the expedition, Monica Andersson, Enid Irwin, Caroline Jones, Sarah Jones, Andrew O'Neill, Oliver Pollard and Sophie Yangopolous whose extraordinary labours and good humour are deeply appreciated.

References

Bishop, K. H. & Richards, K. 1988 *Consultancy report to the Loch Fleet project*, 50 pp.

Bishop, K. H., Grip, H. & O'Neill, A. 1990 *J. Hydrol.* (In the press.)

Chen, C. W., Dean, J. D., Gherini, S. A. & Goldstein, R. A. 1982 *J. envir. Eng.* **108**, 455–472.

Christophersen, N. & Seip, H. M. 1983 In *Ecological effects of acid deposition*, pp. 129–144. Stockholm: National Swedish Environmental Protection Board.

Christophersen, N., Seip, H. M. & Wright, R. F. 1982 *Wat. Resour. Res.* **18**, 977–996.

Christophersen, N., Kjærnsrød, S. & Rodhe, A. 1985 In *Hydrologic and hydrogeochemical mechanisms and model approaches to the acidification of ecological systems*, report no. 10, pp. 29–40. Oslo: Norwegian Hydrological Programme.

Davies, T. D. 1989 *Eos* **70**, 1122.

Driscoll, C. T. 1984 *Int. J. Envir. Analyt. Chem.* **16**, 267–283.

Eriksson, E. 1985 *Principles and applications of hydrochemistry*. London: Chapman and Hall.

Fendick, E. A. & Goldstein, R. A. 1987 *Wat. Air Soil Pollut.* **33**, 43–56.

Gherini, S. A., Mok, L., Hudson, R. J. M., Davis, G. F., Chen, C. W. & Goldstein, R. A. 1985 *Wat. Air Soil Pollut.* **26**, 425–459.

Gunn, J. & Keller, W. 1984 *Can. J. Fish. aquat. Sci.* **41**, 319–329.

Henriksen, A., Skogheim, O. K. & Rosseland, B. O. 1984 *Vatten* **40**, 255–260.

Hooper, R. P., Stone, A., Christophersen, N., de Grosbois, E. & Seip, H. M. 1988 *Wat. Resour. Res.* **24**, 1308–1316.

Lindqvist, G., Nilsson, L. & Gonzalez, G. 1989 *Draft report, University of Luleå, Department of applied geophysics*, 27 pp.

Lövgren, L., Hedlund, T., Öhman, L.-O. & Sjöberg, S. 1987 *Wat. Res.* 21, 1401–1407.

Lundin, L. 1982 Doctoral dissertation, Uppsala University, UNGI Report no. 56, 216 pp.

Miskovsky, K. 1987 *Umeå University, Department of public health and environmental studies, report no.* 3, ISSN 0284-0588, 17 pp.

Mulder, J., Christophersen, N., Hauhs, M., Vogt, R. D., Andersen, S. & Andersen, D. O. 1990 *Wat. Resour. Res.* 26, 611–622.

Odén, S. 1968 The acidification of air and precipitation and its consequences in the natural environment. *Ecol. Commun. Bull.* no. 1.

Ring, E. 1989 *Report from the division of hydrology, University of Uppsala*, 26 pp.

Rodhe, A. 1987 Doctoral dissertation, Uppsala University, UNGI report series A nr. 41, 260 pp.

Seip, H. M. 1980 In *Ecological impact of acid precipitation*, SNSF-project, pp. 358–365. Oslo: Norwegian Institute for Water Research.

Seip, H. M., Seip, R., Dillon, P. J. & de Grosbois, E. 1985 *Can. J. Fish. aquat. Sci.* 42, 927–937.

Hydrological processes on the plot and hillslope scale

By H. S. Wheater[1], S. J. Langan[1], J. D. Miller[3], R. C. Ferrier[3], A. Jenkins[2], S. Tuck[2] and M. B. Beck[1]

[1] Department of Civil Engineering, Imperial College of Science, Technology and Medicine, London SW7 2BU, U.K.
[2] Institute of Hydrology, Maclean Building, Wallingford, Oxfordshire OX10 8BB, U.K.
[3] Macaulay Land Use Research Institute, Craigiebuckler, Aberdeen AB9 2QJ, U.K.

The importance of hydrological flow paths in determining streamflow chemistry has been widely stated, but in most acidification studies, detailed hydrological processes have not been investigated. As part of the SWAP programme at the Loch Ard and Allt a'Mharcaidh experimental areas in Scotland, soil-water throughflow and three-dimensional soil-water potential have been continuously monitored for representative soils and land-use units in conjunction with precipitation and, in the forested Loch Ard area, with stemflow and throughfall data.

At all sites, soil-water response is spatially very heterogeneous. At the forested Loch Ard sites, spatial variability of surface wetting is influenced by canopy effects, and rapid infiltration to the base of the soil profile (associated with preferred flow paths) generates downslope transmission of spatially discrete wetting fronts in the B/C horizon, which provides the dominant transmission pathway.

At the Allt a'Mharcaidh catchment, observations from representative plots on the alpine podzol, peaty podzol and peat series have been extended to evaluate hillslope response. A consistent but complex pattern of response emerges. Localized induration in the peaty-podzol soils can cause a perched near-surface water table above an underlying water table in the B/C horizon. At the base of the extensive peaty-podzol hillslopes, seeps are generated at points of local topographic convergence, fed by variable saturated areas within the soil profile, but rapid flow paths are also observed. Standing water commonly occurs in the peat areas below the peaty-podzol slopes, but in response to rainfall progressive inter-connection of flow paths can lead to rapid transmission of upslope response.

Characteristic features of the stream hydrograph reflect the observed water flows on the plot scales, the interpretations of which provide the basis for a hydrological and hydrochemical model of the Allt a'Mharcaidh catchment.

1. Introduction

Because hydrological processes provide the water fluxes by which chemical transport in catchments occurs, it is evident that stream hydrochemistry will be dependent on catchment hydrology. This dependence is explicitly defined in many of the first generation of surface water acidification models (see, for example, Christophersen et al. (1982); Chen et al. (1982)) and the exercise of such models has demonstrated the

[121]

importance of the representation of hydrological flow paths in the simulation of stream chemistry. Despite this, few acidification studies have attempted detailed observation of hydrological processes. Commonly, hydrology has been inferred through the application of models to catchment input–output data. This approach does not allow for spatial discrimination and is subject to severe problems of parameter (and process) uncertainty (Wheater *et al.* 1986; Hornberger *et al.* 1985; Beck *et al.* 1990, this symposium). It has been our aim for the U.K. Surface Waters Acidification Programme (SWAP) sites to observe both hydrological and hydrochemical processes, to provide a sound basis in observation for interpretation, conceptualization and modelling of stream response.

To the non-hydrologist, definition of flow paths may appear straightforward. However, as Pearce *et al.* (1986) note, 'despite decades of increasingly more intensive study, storm runoff generation remains a controversial topic'. There are several reasons for this. A fundamental difficulty is the problem of scale. Detailed observations must necessarily be made at small (e.g. experimental plot) scale but, at the scale of interest for prediction (i.e. catchment or subcatchment) non-observed aggregated effects may predominate. A second issue is that of process uncertainty, in particular with respect to preferred flow paths. It is now widely recognized that macropores commonly occur, associated with soil structure, floral and faunal activity and/or internal erosion (Beven & Germann 1982). However, the generation of flows in these pathways, the hydrological interactions between matrix and macropore and the extent to which the chemistry of bypass flows is dependent on that of the soil matrix are largely unknown. A further problem is that of heterogeneity. The hydrologist has available a well-defined spectrum of responses (Chorley 1978), but it is not possible *a priori* to define which will be predominant. Thus Pilgrim *et al.* (1978) note that 'the principal observation from the study was the variability of the runoff processes on the plot, which was selected for its apparent uniformity', and Mosley (1982) concludes that it is not possible to predict hydrological behaviour of the soil without excavation and extensive experimentation.

The primary SWAP sites in Scotland for integrated process studies were at Loch Ard, 40 km north of Glasgow and the Allt a'Mharcaidh catchment, in the Cairngorm mountains, described elsewhere in these proceedings by Ferrier & Harriman. The focus for process studies was a series of experimental plots, nominally 0.2 ha†, which were established on representative soil and land-use units for hydrological and hydrochemical monitoring. On the Allt a'Mharcaidh catchment these were supplemented by additional plots to define hillslope response for characteristic soil sequences and topographic features. Hydrochemical results are reviewed elsewhere (Ferrier *et al.* 1990).

In this paper the hydrological responses are illustrated and summarized. For the Allt a'Mharcaidh catchment, a unified interpretation of plot and hillslope response is presented that forms the basis of hydrological and hydrochemical modelling of the catchment, described in this symposium by Wheater *et al.*

2. Experimental design

As outlined above, experimental plots were established on representative soils at the Loch Ard and Allt a'Mharcaidh sites for combined hydrological and hydro-

† $1 \text{ ha} = 10^4 \text{ m}^2$.

chemical monitoring. Conventional tipping-bucket raingauges were available as part of the basic hydrological instrumentation in both areas. In addition precipitation and occult (mist and fog) deposition were monitored by using assemblies of three replicate gauges for each site mounted at 6 m heights in open areas adjacent to the forested Loch Ard plots, and at 3 m heights above the shorter Mharcaidh vegetation (Ferrier & Harriman 1990).

For the forest sites, canopy effects were quantified through observation of throughfall, stemflow and litter flow (Miller *et al.* 1990*a*, *b*). Throughfall was collected in each plot by using 12 randomly sited collectors (Miller & Miller 1976). Stemflow was collected, by using silicon rubber collars, from four trees in each plot, selected to cover the basal area range. Litterflow was collected by using shallow-tray lysimeters, randomly sited, inserted immediately below the litter layer.

Soil water was monitored by a combination of throughflow pits and distributed tensiometer systems. On each plot, three or four replicate pits, each 0.5 m wide and approximately 1 m deep were established in a similar design to those used by Atkinson (1978). Surface flow and saturated throughflow from selected horizons were intercepted by using polythene sheeting inserted into horizon interfaces and collected in plastic troughs feeding to tipping-bucket recorders. The pit faces were protected by polythene beads, retained by horizontal shuttering (Miller *et al.* 1990*c*). Rainfall, stemflow, throughfall and throughflow were recorded on data-logging systems at 20 min resolution.

At each site a three-dimensional array of tensiometers was used to monitor soil-water potential (Wheater *et al.* 1987; Wheater *et al.* 1990*a*). In an extension of the design of Burt (1978) a 24-port Scanivalve fluid switch was stepped every 2.5 min to connect in turn 22 soil moisture equipment tensiometers and two reference heads to a pressure transducer. The system was logged and controlled by a Campbell Scientific 21X logger, which completes a system scan each hour. At each site the tensimeter arrays were located upslope of one of the throughflow pits. The three-dimensional distribution of tensiometers varied according to site characteristics. Typically, two or three soil horizons were monitored at each of nine locations.

Additional chemical information was obtained from soil-water suction lysimeters and from composite samplers in adjacent streams. Selected throughflow events were also sampled for full chemical analysis by using a combination of automatic and manual sampling.

3. Loch Ard sites

Experimental plots were established on two forested catchments, Loch Chon and Kelty Water, in the Queen Elizabeth Forest Park, Loch Ard. This area has undergone an intensive afforestation programme, initiated in the 1950s, and has been the subject of acidification studies for many years (Harriman & Morrison 1982).

The sites are described by Ferrier & Harriman, this symposium. The underlying geology is mixed, mainly Dalradian metamorphic-igneous rocks including slates, phyllites and mica-schists (Anderson 1947). Soils at both sites are from the same association. At Kelty Water, peaty gley soils are present under Sitka Spruce (Hudson & Hipkin 1985); at Loch Chon the site is on a humus iron podzol under Norway Spruce (Henderson & Campbell 1985). Important differences between the two systems are that Kelty Water is no longer capable of sustaining a viable fish population, whereas the Loch Chon stream retains a reduced population. This is associated with mean streamwater chemistries that are significantly different in some

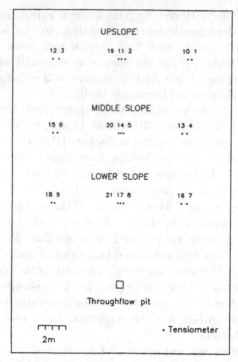

Figure 1. Loch Chon tensiometer layout.

respects. In particular, Kelty Water has lower pH and calcium concentrations and higher concentrations of sulphate and nitrate (Miller *et al.* 1990*b*).

The capability of the experimental systems to detect direct response is briefly illustrated with representative data from June 1987 for Loch Chon, for which the tensiometer layout is presented in figure 1. As noted earlier, measurement of rainfall, stemflow, throughfall and throughflow is available at 20 min resolution. Hourly data from the 22 tensiometers can be analysed with respect to matric potential (i.e. soil-water suction), indicating soil-water state, or total potential energy, which defines flux direction.

Precipitation and throughflow response are shown in figure 2. Throughflow in the 0 soil horizon is initiated simultaneously with the occurrence of intense precipitation, just before 180 h, and ceases with the end of rainfall. B/C throughflow commences some 3 h later, increases rapidly to peak at the end of rainfall and decays quickly before generating a relatively long recession tail. It will be noted that the B/C throughflow peak is two orders of magnitude greater than 0 horizon throughflow. At the onset of this period, soil-water conditions were relatively dry. The low-intensity rainfall in the preceding 130 h is insufficient to generate B/C throughflow. The antecedent flows through the 0 horizons are small and occur in association with bursts of more intense rainfall, following some initial wetting. They are also coincident with periods of significant stemflow response.

The tensiometer data can be interpreted with respect to point profiles, transects or spatial (plan) variability. 0 horizon tensiometers (1–9) show a complex pattern of surface wetting, associated with variations in tree cover. Thus for the June 1987 event, tensiometers 3, 4, 6, 7 and 8 respond with the onset of major rainfall, whereas others respond later (figure 3*a*), in response to lateral migration of the wetting front.

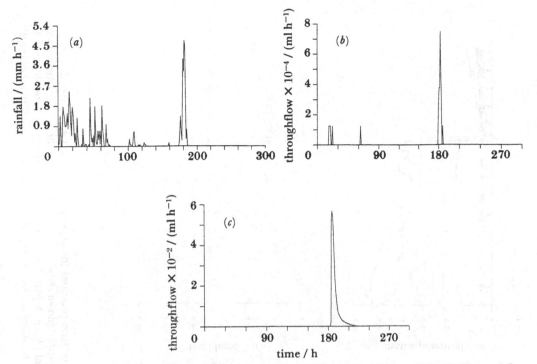

Figure 2. Loch Chon event, June 1987. (*a*) Rainfall, 29 May 1987–10 June 1987. (*b*) Throughflow, 0 horizon, 29 May 1987–10 June 1987. (*c*) Throughflow, B/C horizon, 29 May 1987–10 June 1987.

In general, as would be expected, spatial variability of surface wetting is greatest for small precipitation events.

Downslope migration of wetting fronts at depth can appear to be uniform. Figure 3*b* shows the B/C tensiometers in the downslope centre-line transect. Consistent and relatively rapid downslope movement of a sharp wetting front is apparent, and the onset of saturation (positive matrix potential) at T21 is consistent with the observed initiation of B/C throughflow at the pit immediately downslope. However, data from the right-hand downslope B/C sequence (10, 13, 16) (figure 3*c*) show that uniform downslope migration does not take place. The tensiometer T13 responds considerably later than the lower slope T16, for example. Rapid movement at depth can generate complex wetting patterns. Consideration of total head at the spatial location of tensiometers 8, 17 and 21 (figure 3*d*) shows that in the early part of this period an upwards (evaporative) flux is replaced by downward drainage. However, in response to the major event (180 h), temporary flux reversal occurs at the base of the profile. The mid-profile here is wetted not only from above, but also from below. The wetting front at depth has propagated faster than vertical infiltration, at this location.

A two-dimensional multi-layer finite element model of unsaturated/saturated hillslope drainage has been applied with mixed success to the Loch Chon hillslope (Koide, unpublished results). To match observed tensiometer responses, it was necessary to assume a permeability for the B/C horizon that over-estimated the (large) observed throughflow discharge. Further, to provide appropriate infiltration to depth, regions of high vertical permeability had to be postulated. This strongly suggests that the soils are spatially heterogeneous and that flow is occurring in preferred pathways.

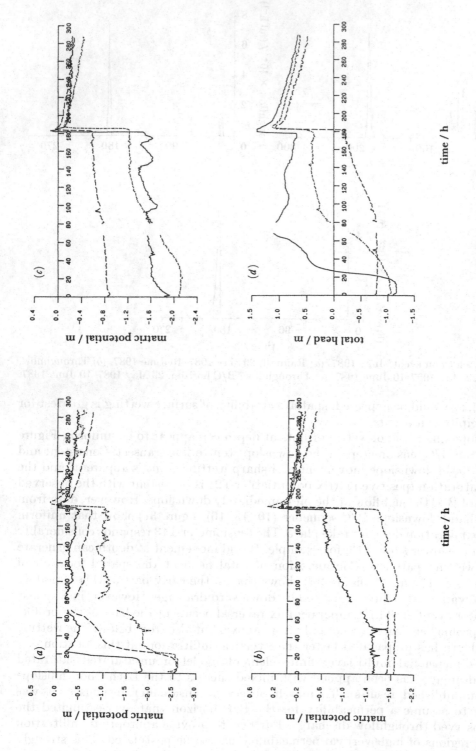

Figure 3. Loch Chon event, June 1987. (*a*) Surface tensiometers, centre-line (matric potential (down slope)); (——), tensiometer 2; (·· - ··), tensiometer 5; (——), tensiometer 8. (*b*) BC tensiometers, centre-line (matric potential (down slope)); (——), tensiometer 19; (·· - ··), tensiometer 20; (——), tensiometer 21. (*c*) BC tensiometers, right-hand line (matric potential (down slope)); (——), tensiometer 10; (·· - ··), tensiometer 13; (——), tensiometer 16. (*d*) Total-head vertical distribution, plot base; (——), tensiometer 8; (·· - ··), tensiometer 17; (——), tensiometer 21.

Data from Kelty Water show a generally similar response, although mean throughflow in the B horizon at Kelty is slightly larger than in the B/C (Miller *et al.* 1990*a*). The conclusion from both of the forested sites (Loch Chon and Kelty Water) is that hillslope response primarily occurs through downslope drainage in the lower soil horizons. However, both the tensiometer data and the variability of throughflow observed in the replicate pits (Miller *et al.* 1990*a*) indicate that there is major heterogeneity in the wetting process. Rapid transmission of vertical infiltration is observed at some locations, but not at others. Sharply defined wetting is observed at depth, but not always in a downslope sequence. It is evident that preferred flow paths are present, as would be expected from research elsewhere (e.g. Tsukamota & Ohta 1988; Mosley 1982; De Vries & Chow 1978) and hence that chemical interactions must be considered in the context of a soil matrix containing preferred flow pathways.

4. The Allt a'Mharcaidh catchment

The Allt a'Mharcaidh catchment in the Cairngorm mountain range has been the subject of an extensive integrated programme of hydrological and hydrochemical observations, as described in this symposium by Harriman & Ferrier. The catchment is relatively large for experimental studies (9.98 km^2) but detailed soil survey (Nolan *et al.* 1985) has shown that from 15 soil-map units, three principal soil groupings can be identified (figure 4 (inset)). These are strongly associated with catchment topography. The highest point in the catchment has an elevation of 1111 m above sea level. At the higher elevations, which are characterized by gently rising plateaux, alpine podzols predominate, colonized by alpine azalea-lichen heath and fescue-woolly fringe-moss heath, and comprise 38% of the catchment area. On the steep valley sides peaty podzolic soils predominate (30% of the catchment) consisting of peaty podzols, humus-iron podzols and peaty rankers with vegetation of lichen-rich boreal heather moor. At the lower altitudes, where the topography is more gently sloping, blanket peat of up to 1 m depth occurs (23% of catchment area) with associated flush communities. The characteristic sequence from catchment divide to stream is thus of alpine plateau giving way to steep peaty-podzol slopes, and, below a break in slope, peat soils. The principal tributary streams are shown in figure 4, however at the base of the steep valley sides several seeps emerge, draining into the peat areas. The peats are predominantly wet, with standing water visible for much of the year. The catchment is underlain by intrusive biotite-granite; thick deposits of boulder clay, derived from local rock, cover much of the valley floor (Nolan *et al.* 1985).

(a) *Plot observations*

The catchment was instrumented from 1985 to 1989 for hydrological and hydrochemical observations. Plot experiments commenced with replicate plots established in representative areas on the dominant soil types (plots AM0, AM1 and AM5, figure 4), supported by three-dimensional tensiometer arrays as described above. This experimental design was subsequently modified in an attempt to characterize hillslope flow on a larger scale. The tensiometer arrays were re-deployed for 1988 to create a sequence of two experimental plots on the eastern peaty-podzol slope (at mid-slope (AM1) as previously and at slope base (AM2)), and a sequence of three plots on the western side, at a point of seep emergence towards the base of the peaty-podzol slope (AM3), in the mid peat area (AM4) and in the peat adjoining the stream (AM5, as previously).

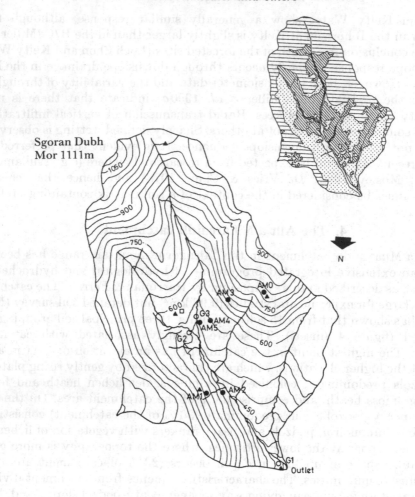

Figure 4. The Allt a'Mharcaidh catchment: (●), experimental plots; (▲), tipping-bucket rain gauges; (□), automatic weather station; (○), stage recorders. Inset, soil types: (□), alluvial; (▨), peat; (▩), peaty podzol; (▤), alpine podzol.

Results from the experimental plots are presented and discussed more fully in Wheater *et al.* (1990 *b*) and Wheater *et al.* (1990 *c*). Salient features of the observed responses are summarized here, first for the original representative plots, and secondly for the hillslope sequences.

(i) *Alpine podzol*

The main feature of the alpine podzol soils was a very limited observed response to rainfall. Tensiometers recorded predominantly unsaturated soil-water conditions, and throughflow, monitored in the organic and Bh horizons, was negligible (figure 5 *a*). The principal organic-horizon responses in summer 1988 occurred in early July (following a long dry period in June), suggesting temporary dessication and/or hydrophillic response. It is concluded that these coarse, poorly developed soils drain freely to depth.

(ii) *Peaty podzol*

The peaty-podzol site (AM1), although located on an apparently uniform section of hillslope, showed marked local heterogeneity. Excavation of the tensiometer locations in later 1988 showed variable occurrence of weak indurated layers. In contrast to the alpine soils throughflow occurred in response to all major events (e.g. figure 5b). The magnitude and initial timing were similar in the 0 and Bh horizons (Bh response occurring slightly later than 0 response), but 0 horizon response was of significantly longer duration. The data appear to suggest that the occurrence of indurated layers is promoting a perched water table and thus generating the longer 0 horizon response. However, it is not immediately apparent from those data how the Bhs response is generated.

The tensiometer data illustrate considerable heterogeneity of point-profile response. At one spatial location with a distinct Bsx horizon, a prolonged perched water table occurs, giving saturation at or close to the soil surface. At other points a damped C-horizon water-table response is observed, with highly transient saturated conditions in the upper profile. A third mode of response is observed in the centre of the plot in which a dynamic response occurs within the Bh horizon above a relatively deep Bsx layer. The vertical hydraulic gradient is temporarily reversed, suggesting a lateral-flow component at depth; the water-table response is similar to that of the 0 horizon throughflow.

It is concluded that the peaty-podzol response is highly heterogeneous. Localized perched water conditions occur, and it is suggested that these can connect with preferred flow paths to generate a rapid response within the profile. The tensiometers provide evidence of a slowly responding underlying water table, not observed in the throughflow response.

(iii) *Peat soils*

The throughflow data for the peat plot AM5 is illustrated in figure 5c. Major rainfall events generate a large and transient response in the Egh horizon, peak flow rates are some three times that of the peaty podzol throughflow, and in general, the largest Egh responses are associated with an isolated and substantially smaller flow in the 0 horizon. The plot throughflow is typically some 2 h later than the peaty-podzol response.

The tensiometer data for this plot are consistent with the throughflow observations. The same response is observed in the Egh tensiometers. Vertical profiles of total soil-water potential show temporary occurrence of an upward flux, which indicates that the 0 horizon response is a result of wetting from below, i.e. the observed throughflow response is transmitted at depth at the base of the organic peat soils.

(b) *Hillslope response*

The plot observations described above define a characteristic response for the principal soils of the catchment. However, the scale problem is apparent. On the basis of these observations alone, response on the hillslope scale is ambiguous; linkages require further investigation, hence the redirection of experimental design in 1988, referred to above.

The response of plot AM3 at the base of the western peaty-podzol slope is particularly interesting. Seep emergence occurs at an incised gulley. The tensiometer

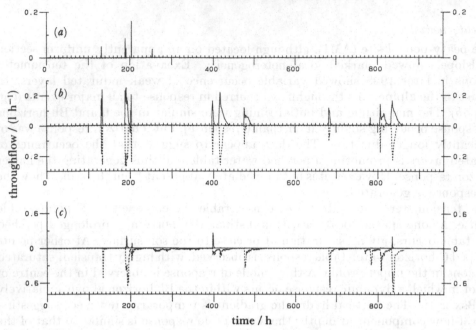

Figure 5. Allt a'Mharcaidh throughflow pits, 27 June–5 August 1988. (*a*) Alpine throughflow: (——), organic; (-----), Bh. (*b*) Peaty-podzol throughflow: (——), organic; (----), Bhs. (*c*) Peat throughflow: (——), organic; (----), Egh.

array showed the gulley to be fed by a water table within the soil profile. In response to a rainfall event, the extent of this saturated area expands and contracts (figure 6), very much in keeping with classical concepts of hillslope response. the temporal response is illustrated in figure 7*a* for tensiometer T16, immediately upslope of the point of seep emergence.

However, 5 m to the north of the seep (T14), a contrasting response is observed (figure 7*b*). A transient pulse is observed at depth, some 2 h after rainfall, associated with a temporary reversal of vertical hydraulic gradient, indicating lateral transmission.

The implications of these data are significant. The extensive western slope is seen to generate two modes of response. Downslope drainage generates water-table conditions, in an area of micro-topographic convergence, which have a relatively damped response. In parallel, a localized preferred pathway gives a rapid response, though it is lagged behind rainfall. Subsequent excavation at this point did not reveal a visually distinct macropore (relatively large natural pipes occur in the peat soils elsewhere on the catchment), but a band of stones located within a sandy-gravel matrix.

Data from the eastern hillslope are somewhat different, but have similar implications. Soil-water conditions at plot AM2 were relatively dry, suggesting that drainage from the extensive upslope area was largely bypassing the plot. A local topographic bench feature caused the generation of a small area of saturation, which propagated upslope in response to rainfall. To one side of the plot, a rapid response indicated the presence of a preferred pathway.

Plot AM4 was established to investigate the hydrological role of the extensive areas of peat and in particular the transmission of response from the upslope peaty-

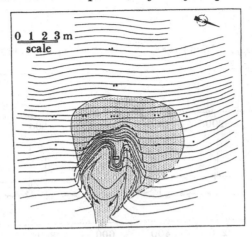

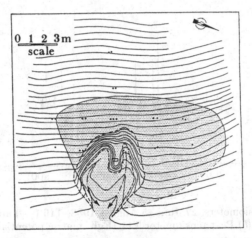

Figure 6. Gulley water-table response: (●), tensiometer locations;
(▨), extent of saturation; (–), 0.2 m contours.

podzol areas. The peat soils are heavily eroded by large gulley features, and, as noted earlier, for much of the year have pools of standing surface water. Flow paths are complex. In response to rainfall, surface pools increasingly interconnect and flows may emerge from contact with the Egh horizon at the base of the eroded gulley to flow overland before re-entering another channelled section. The tensiometer data showed a spatially variable water table, in continuity with the irregular surface topography.

The dynamic response of the peat Egh horizon near the main stream has been noted above, and is illustrated in figure 7c. It is concluded that the peat areas, once wetted, can provide a rapid transmission pathway for flows from upslope. At the timescale of resolution of the tensiometer data (1 h) there is no difference in the timing of the peaty-podzol quick flow-path response (figure 7b) and the peat Egh signal (figure 7c).

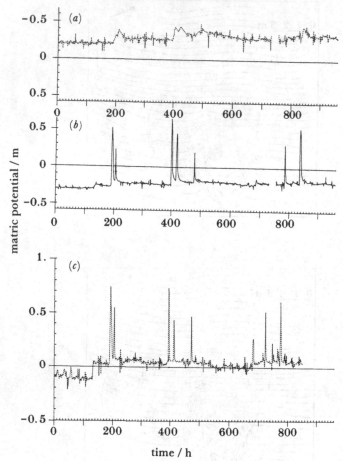

Figure 7. (*a*) Gulley tensiometers, 27 June–5 August 1988, T16 (----) and T14 (——), matric potential. (*b*) Peat tensiometer T11 (Egh, matric potential) (----).

(*c*) *A hypothesis for catchment hydrological response*

The additional data from the hillslope plots provide a consistent set of observations from which hypotheses for hillslope response can be derived. The heterogeneity of the peaty-podzol soil gives rise to locally perched water-table conditions and localized rapid flows in the soil profile, above an underlying water table. These flow responses are consistent with two modes of downslope movement, a relatively damped downslope (matrix drainage) feeding emergent seeps at points of local topographic convergence, and a much more rapid transmission in preferred flow paths. Once wet, the peat areas can provide rapid transmission to the stream, primarily because of flow at the base of the organic soil. Between events, the downslope drainage maintains inflows to the peat so that only in extended dry periods do the surface peat soils dry out.

In figure 8*a* the streamflow response for the main catchment outfall G1 and the two principal tributaries, G2 and G3, is illustrated. It is readily apparent that the streamflow event response has two components. A transient, rapid 'spiked' response generates the hydrograph peak; an underlying 'hump' has a more extended, damped response. These two could be considered to be superimposed on an

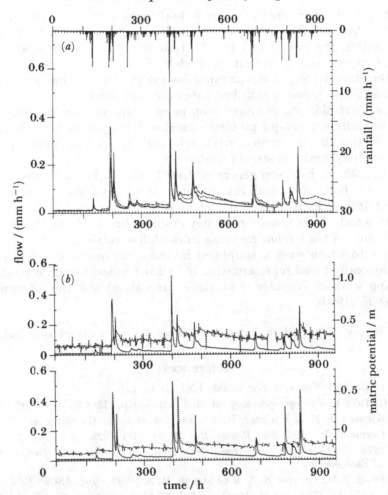

Figure 8. Allt a'Mharcaidh hourly flows and rainfall 27 June–5 August 1988. (*a*) Streamflows for gauging stations: G1 (———), G2 (----) and G3 (——). (*b*) Comparison of streamflow and peaty-podzol gulley tensiometers: (———), flow at G1; (----), tensiometers T16 (upper graph) and T14 (lower graph).

underlying baseflow contribution. In figure 8*b* the two modes of response observed at the peaty-podzol gulley are compared with the G1 hydrograph. The response associated with flow through the preferred flow paths is identical to that of the hydrograph peak; the underlying hydrograph 'hump' probably reflects the response of the water table.

5. Conclusions

The fairly comprehensive hydrological data obtained from the SWAP Scottish experimental sites provide useful insight into the hydrological flow paths within the catchments. An overall feature of the results for both areas is a highly heterogeneous response to precipitation events. For the forested, Loch Ard sites, results are consistent with classical forested hillslope studies done elsewhere (e.g. Whipkey 1965). The major downslope response occurs in the lower soil profile (B/C horizon); large volumes of throughflow were observed. However, the SWAP results have shown that spatial heterogeneity occurs in both surface wetting, and in subsurface transmission. Preferred flow paths are (by implication) responsible for rapid vertical movement within the profile and for the transmission of discrete wetting fronts downslope. Conventional finite element models of porous media flow have not fully simulated the observed response; the rapid downslope transmission is not compatible with observed fluxes unless quick-flow paths are postulated.

The Allt a'Mharcaidh studies have been more comprehensive in scope, and have attempted to address the major problem of scale-related processes in the catchment. A notable feature is the predominant role that quick-flow paths play in the generation of sharp peaks in stream discharges.

A set of hypotheses has been generated that has been shown by inspection to be consistent with both plot- and catchment-scale observations. These have been incorporated into a hydrological model of the catchment (Wheater *et al.*, this symposium), which shows some consistency in simulating both the transient and the slow components of the stream response to rainfall events.

The extent to which such a simplified hydrological model can be extended to include a physically based representation of hydrochemical response is addressed in the modelling work of Wheater *et al.* (this symposium) and the interpretation of Harriman *et al.* (1990).

The financial support of the Surface Waters Acidification Programme is gratefully acknowledged.

References

Anderson, J. G. C. 1947 *Trans. R. Soc. Edinb.* LXI II, no. 18.

Atkinson, T. C. 1978 In *Hillslope hydrology* (ed. M. J. Kirkby), pp. 73–120. New York: John Wiley.

Beck, M. B., Kleissen, F. M. & Wheater, H. S. 1990 *Rev. Geophys.* (In the press.)

Beven, K. & Germann, P. 1982 *Wat. Resour. Res.* **18** (5), 1311–1325.

Burt, T. P. 1978 *Br. Geomorph. Res. Group tech. Bull.*, no. 9. Norwich: Geo-Abstracts Ltd, University of East Anglia.

Chen, C. W., Dean, J. D., Gherini, S. A. & Goldstein, R. A. 1982 *Proc. ASCE, EE3*, 455–472.

Chorley, R. J. 1978 In *Hillslope hydrology* (ed. M. J. Kirkby), pp. 1–42. New York: John Wiley.

Christophersen, N., Seip, H. M. & Wright, R. F. 1982 *Wat. Resour. Res.* **18** (4), 977–996.

De Vries, J. & Chow, T. L. 1978 *Wat. Resour. Res.* **14** (15), 935–942.

Ferrier, R. C., Walker, T. A. B., Harriman, R., Miller, J. D. & Anderson, H. A. 1990 *J. Hydrol.* **115**. (In the press.)

Harriman, R. & Morrison, B. R. S. 1982 *Hydrobiologia* **88**, 251–263.

Harriman, R., Gillespie, E., King, D., Watt, A. W., Christie, A. E. G., Cowan, A. A. & Edwards, T. 1990 *J. Hydrol.* **115**. (In the press.)

Henderson, D. J. & Campbell, C. G. B. 1985 The Surface Waters Acidification Programme; soil survey of the north Loch Chon catchment. Aberdeen: MLURI. (Restricted circulation.)

Hornberger, G. M., Beven, K. J., Cosby, B. J. & Sappington, D. E. 1985 *Wat. Resour. Res.* **21** (12), 1841–1850.

Hudson, G. & Hipkin, J. A. 1985 The Surface Waters Acidification Programme; soil survey of the Kelty Water catchment. Aberdeen: MLURI. (Restricted circulation.)

Koide, S. 1990 Ph.D. thesis, University of London. (In preparation.)

Miller, H. G. & Miller, J. D. 1976 *Lab. Pract. Dec.* 19–20.

Miller, J. D., Anderson, H. A., Ferrier, R. C. & Walker, T. A. B. 1990*a* *Forestry.* **63**, 3.

Miller, J. D., Anderson, H. A., Ferrier, R. C. & Walker, T. A. B. 1990*b* *Forestry.* **63**, 4.

Miller, J. D., Stuart, A. W. & Gsakin, G. J. 1990*c* *Br. Geomorph. Res. Group. Tech. Bull.* No. 37. (In the press.)

Mosley, M. P. 1982 *J. Hydrol.* **55**, 65–92.

Nolan, A. J., Lilly, A. & Robertson, J. S. 1985 Soil and vegetation survey of the Allt a'Mharcaidh catchment. Aberdeen: MLURI. (Restricted circulation.)

Pearce, A. J., Stewart, M. K. & Sklash, M. G. 1986 *Wat. Resour. Res.* **22** (8), 1263–1272.

Pilgrim, D. H., Huff, D. D. & Steele, T. D. 1978 *J. Hydrol.* **38**, 319–341.

Tsukamoto, Y. & Ohta, T. 1988 *J. Hydrol.* **102**, 165–178.

Wheater, H. S., Bishop, K. & Beck, M. B. 1986 *J. Hydrol. Process.* **1**, 89–109.

Wheater, H. S., Langan, S. J., Miller, J. D. & Ferrier, R. C. 1987 The determination of hydrological flow paths and associated hydrochemistry in forested catchments in central Scotland. In *Forest hydrology and watershed management*, pp. 433–449. *Proceedings of the Vancouver Symposium*, IAHS publn no. 167.

Wheater, H. S., Langan, S. J., Brown, A. & Beck, M. B. 1990*a* *J. Hydrol.* (In the press.)

Wheater, H. S., Tuck, S., Ferrier, R. C., Jenkins, A., Kleissen, F. M., Walker, T. A. B. & Beck, M. B. 1990*b* *J. Hydrol.* (In the press.)

Whipkey, R. Z. 1965 *IAHS Bull.* **10** (3), 74–85.

Strong and organic acid in the runoff from peat and woodland

By Olle Westling

Swedish Environmental Research Institute, P.O. IVL Aneboda, S-360 30 Lammhult, Sweden

Distribution of strong acid and organic (weak) acid in runoff from peat- and woodland were studied in two catchments in southern Sweden and in lysimeter studies. Studies of input–output budgets revealed that the strong acid (mineral acid) contribution to pH was 40–50% of the observed free acidity in the runoff from peatland. Dissociation of H^+ from weak-COOH groups in dissolved organic matter (DOM) resulted in 50–60% of the observed acidity, and all of the observed organic acids were dissociated. Dissociated organic acid contributed 60–70% of the free acidity in runoff from woodland (podzol). The contribution from aluminium and carbonic acid was less than 10% in both areas.

Treatment of lysimeters with strong acid (pH 3.6, 3.0 and 2.4) produced increasing leaching of H^+, Ca, Mg, Na, K, S, N, and Al from podzol compared with treatment with normal precipitation of pH 4.2. Corresponding treatment of lysimeters (pH 5.1–3.0) containing oligotrophic sphagnum peat did not change the leaching significantly during the period of investigation (97 weeks), except for a moderate increase of Ca, Mg, S and Al in the treatment with pH 3.0. The treatment with pH 2.4 resulted in chemical mineralization of the peat, and increased leaching of all investigated elements. Input of strong acids to the lysimeters containing peat decreased the leaching of dissociated organic acids. Dissociated organic acid contributed to 100% of the free acidity in treatments pH 5.1, 4.2 and 3.6. Treatment with pH 3.0 resulted in a contribution of 57% from strong acid to free acidity. The low concentrations of dissociated organic acids (7–19 µmol l^{-1}) in the percolated water from the lysimeters containing podzol, were not changed by the treatments, and the free acidity was dominated by strong acid. Dissociated organic acids showed a strong correlation with water-colour index in both output from peat lysimeters and runoff from catchments.

1. Introduction

The majority of Swedish lakes in areas sensitive to acid deposition have a rather high content of dissolved organic matter and are more or less brown coloured (water-colour index > 25 mg Pt l^{-1}). The watersheds often have a high percentage of peat- and wetland, which gives a considerable contribution of organic acids (humus) to runoff. Humus in water influences the acidity, the acid buffering system, and metal species in lakes and running waters (Gjessing 1976).

The acidification process is usually associated with clearwater lakes, often with relatively small catchments, dominated by acidified and shallow mineral soils (Andersson & Olsson 1985). In these kinds of lake, the atmospheric deposition of

Table 1. *Characteristics of the two catchments*

characteristic	Åkhultmire	Aneboda
altitude/m	220	215
area/km²	0.140	0.196
soil horizon	sphagnum/peat	podzol
peatland (%)	100	8
forest cover (%)	10	90
dominating tree species	*pinus sylvestris*	*Picea abies*
ground vegetation dominating species	*Sphagnum* spp.	*Vaccinium myrtillus*

strong acid is the critical factor for the occurrence of high concentrations of free acid in the water. The aim of this study was to investigate the contribution of organic acid to the observed pH in two humic brooks in the central part of southern Sweden, an area with high atmospheric deposition of strong acids.

The effect of increased, or decreased, deposition of strong acids on leaching of organic acids from soil was investigated by lysimeter (soil columns) experiments in the laboratory by determining the input–output of strong and weak acids, and other elements.

2. The study sites

The investigations were done in two small catchments that had different proportions of peatland (see table 1). The mean annual precipitation in the area is 820 mm (corrected), as measured over 30 years. The Åkhultmire is an oligotrophic and ombrogenic bog with sphagnum peat, dominated by open areas that are interrupted by small stands of Scots pine. The Aneboda catchment is 90 % covered by till of moderate depth (3–5 m). The soil type is a weakly developed podzol. The age of the dense spruce forest is 80–100 years.

3. Methods

(a) Catchments

Input of strong acid and other elements was monitored by 10–20 open funnels, randomly placed under the forest canopies, to collect throughfall in the catchments. On the Åkhultmire, most of the funnels were placed so as to collect the wet deposition in the large open areas on the mire. During winter, the funnels were replaced by open buckets to get a better estimate of the snow precipitation. Samples were collected once every month. The throughfall was analysed for pH, Ca, Mg, Na, K, SO_4-S, Cl, NO_3-N, NH_4-N, Kjeldahl-N, tot-P and Mn.

Runoff in the outlets of the catchments was sampled once every two weeks. The same parameters as those for deposition were analysed, with the addition of total Al, monomeric inorganic and organic Al and Fe. All cations were analysed by atomic absorption spectrophotometry (AAS). SO_4-S, Cl and NO_3-N were determined by ion chromatography (IC).

Strong and weak acids in runoff were determined once every month during one year by interpreting titration data. Methods for the analyses of monomeric Al and strong and weak acids are described by Lee (1980, 1985). The specific runoff was monitored continuously at the two measuring weirs.

Table 2. *Mean concentrations in treatments of lysimeters* pH 5.1–2.4

	pH 5.1	pH 4.2	pH 3.6	pH 3.0	pH 2.4
$H^+/(\mu mol\ l^{-1})$	8	77	271	1187	4824
SO_4-$S/(\mu mol\ l^{-1})$	38	53	131	497	1952
NO_3-$N/(\mu mol\ l^{-1})$	47	54	93	276	1004

Element fluxes in deposition and runoff were calculated for three years. Inputs were calculated from the wet deposition and throughfall data. This study covers one year (1985–86), during which strong and weak acids were determined.

(b) *Lysimeters*

The lysimeter studies were done with 10 cells containing oligotrophic peat (pH about 3.8) from the Åkhultmire, and ten with podzol from the Aneboda area. The cells were supplied with pH-adjusted precipitation at the following pH-levels: 2.4, 3.0, 3.6, 4.2 and 5.1. Each pH-level was repeated twice for both the podzol lysimeters and the sphagnum peat lysimeters. All cells were treated with 500 ml per week over 97 weeks and the volume of percolated water was determined. The precipitation was pH-adjusted with H_2SO_4 and HNO_3 (molar ratio 2:1). The mean concentrations of H^+, SO_4-S and NO_3-N are shown in table 2.

Analyses of major ions in percolated water were done after 24, 36, 46, 66, 76 and 97 weeks of treatment. Strong and organic acids were determined after 24, 46 and 76 weeks in lysimeters treated with pH 3.0, 4.2 and 5.1.

4. Results and discussion – catchments

(a) *Atmospheric deposition*

Input–output data from the two catchments are summarized in table 3. Input measured as throughfall represents the sum of wet and dry deposition with addition of uptake or leaching in the canopy. Some elements undergo a net leaching from the canopy (e.g. K, Mn and Ca). Other elements show a net uptake (e.g. nitrogen species, especially NH_4-N). Elements like Na, Cl and SO_4-S are not quantitatively influenced by internal circulation in the trees of areas with high atmospheric deposition (Hultberg 1985; Lövblad *et al.* 1989). The input to the Aneboda catchment was more influenced by dry deposition and uptake or leaching from the dense spruce stands compared with the Åkhultmire with large, open areas, where input was dominated by wet deposition.

The deposition of sulphur from 1985 to 1986 was 38 mmol $m^{-2}\ a^{-1}$ (12.2 kg $ha^{-1}\ a^{-1}$†) and total nitrogen 65 mmol $m^{-2}\ a^{-1}$ (9.1 kg $ha^{-1}\ a^{-1}$) to the Åkhultmire. Corresponding figures for the Aneboda catchment were: sulphur, 70 mmol $m^{-2}\ a^{-1}$ (22.4 kg $ha^{-1}\ a^{-1}$), and total nitrogen, 49 mmol $m^{-2}\ a^{-1}$ (6.8 kg $ha^{-1}\ a^{-1}$), implying a net uptake of nitrogen in the canopy. The deposition of H^+ to the Åkhultmire corresponds to a mean pH of 4.2, whereas the Aneboda catchment showed pH 3.9 as a mean in throughfall.

(b) Elements in runoff

In both catchments, fluxes of nutrients and base cations were low in runoff and the anion flux was dominated by SO_4-S and Cl. For discussion of organic anions see §4 d. Runoff from the Åkhultmire had low pH (4.0 weighted mean), and very low concentrations of Ca, Mg, and inorganic Al compared with the Aneboda catchment (pH 4.5 weighted mean).

Runoff from both catchments was coloured by humic acids. The mean colour index (weighted) from the Aneboda area was 79 mg Pt l^{-1}, and in runoff from the Åkhultmire, 125 mg Pt l^{-1}. The concentrations of elements in the runoff from the Åkhultmire can be compared with a study by Malmer (1962), who investigated the chemistry in groundwater (1–24 cm deep) from the same bog during 1954 to 1960. The results for 1985–1986 are similar to data from the earlier period, except for lower concentrations of Ca, and Mg in 1985–86, indicating small changes over 30 years.

(c) Input–output fluxes

Both areas showed retention of nitrogen and sulphur. The retention of sulphur in the Aneboda catchment was less than in the Åkhultmire, indicating that the organic soil is the cause of retention. In catchments with very low percentage of peatland, input of sulphur normally balances the output (Hultberg 1985). The Aneboda catchment showed retention of H^+ in the soil; the opposite was found in the Åkhultmire. The net release of H^+ from the Åkhultmire was probably caused by ion exchange between deposited base cations and exchangeable H^+ on organic soil particles.

(d) Strong and organic acid in runoff

The study was concentrated on the contribution of H^+ from dissociated organic acids to free acidity observed in runoff. The contribution of H^+ from Al, and H_2CO_3 was calculated to less than 1% in runoff from the Åkhultmire, and less than 10% from the Aneboda catchment.

The dissociation constant of organic acids in runoff from the Aneboda catchment was determined to $pK_a = 4.3$–4.5. The determination of pK_a of organic acids in runoff from the Åkhultmire indicated a very low dissociation constant, $pK_a = 2.5$–2.8. To reach charge-balance in runoff from the Åkhultmire, almost all the organic acids found had to be completely dissociated, resulting in a dissociation constant of $pK_a = 2.7$, which is used in calculations presented later. In runoff from the Aneboda area, dissociated organic acid (organic anions) contributed (on average) 55% of the total organic acid.

The contribution of dissociated H^+ from weak COOH groups in dissolved organic matter resulted in 50–60% of the observed free acidity in runoff from the Åkhultmire (see table 3, H^+ and organic anions). The corresponding figure in runoff from the Aneboda catchment was 60–70%. The explanation for the lower relative contribution in runoff from the Åkhultmire peat land can be found in the seasonal variation of free acid and dissociated organic acid (figure 1). In runoff from the Åkhultmire the strong acid contribution to free acidity was very dominant during periods with high precipitation or snowmelt. On these occasions, most of the precipitation was transported to the brook without contact with the peat under the living sphagnum, across the surface of a very high groundwater table. This was not the case in the Aneboda catchment with a lower groundwater table in the mineral soil.

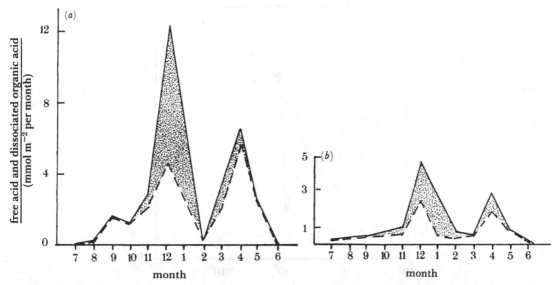

Figure 1. Seasonal variation of free acid and dissociated organic acid in runoff 1985–86 (mmol m^{-2} per month); (——) total free acid; (---) organic anions; (▨) strong acid.

Table 3. *Input and output from catchments during 1985–1986; organic and inorganic* Al *refer to monomeric* Al

	Åkhultmire			Aneboda		
	input	output	output	input	output	output
	mmol m^{-2} a^{-1}	mmol m^{-2} a^{-1}	µmol l^{-1}	mmol m^{-2} a^{-1}	mmol m^{-2} a^{-1}	µmol l^{-1}
H$^+$	38.3	43.9	110	58.8	13.1	33
Ca	4.9	3.6	8.9	21.2	19.2	48
Mg	3.5	5.3	13	12.7	20.7	52
Na	20.6	35	87	52.9	81.0	202
K	5.1	1.8	4.5	37.8	3.9	10
SO$_4$-S	37.7	17.2	43	69.9	50.7	126
Cl	31.9	36.3	91	77.3	72.6	181
NO$_3$-N	31.0	0.95	2.4	22.7	1.6	4.1
NH$_4$-N	19.3	0.3	0.7	4.9	0.3	0.7
org-N	15.0	12.8	32	21.1	10.6	26
Fe	—	1.2	3.0	—	4.7	12
Mn	0.30	0.15	0.4	4.5	0.46	1.1
tot.-Al	—	0.7	1.8	—	4.9	12
org.-Al	—	0.20	0.5	—	2.1	5.3
inorg.-Al	—	0.26	0.7	—	1.3	3.2
tot.-P	0.29	0.06	0.2	0.23	0.08	0.2
org. anions	—	22–26	55–66	—	8–9	20–23

It should be noted that the surface flow at the Åkhultmire did not decrease the pH in runoff, because the throughfall had a higher pH than the normal runoff from the peatland, but it changed the proportions of strong and organic acids.

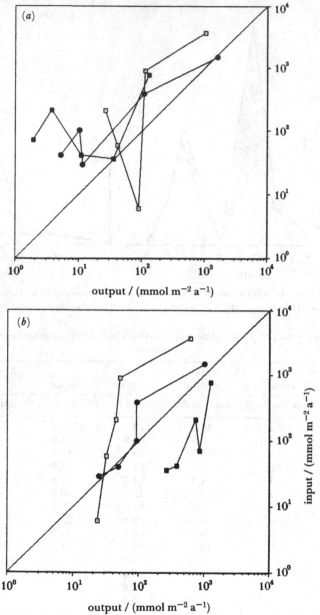

Figure 2. Input–output of H^+, SO_4-S and NO_3-N in (a) peat and (b) podzol lysimeters
(mmol $m^{-2} a^{-1}$); (□) H^+; (●) SO_4-S; (■) NO_3-N; (——) 1:1.

5. Results and discussion – lysimeter experiments

(a) Retention and release of H^+, SO_4-S and NO_3-N

The effect of input of H^+, SO_4-S, and NO_3-N to lysimeters containing peat or podzol on the output of the same elements is shown in figure 2.

H^+ was retained in both peat and podzol except for the lowest input (pH 5.1) to the lysimeters. Sulphur was retained in peat except for the largest input of strong acid (pH 2.4). Podzol showed no retention of sulphur with treatments pH 5.1–3.6, but pH 3.0 and 2.4 resulted in a substantial retention. The probable explanation for the observed sulphur retention is that the normally negatively charged soil particles become oversaturated with H^+, resulting in a capacity to complex anions like SO_4-S. In acid peat this is probably the case even under natural conditions. Podzol does not show this capacity unless the acid load is very high (higher than found in any part of Sweden).

Nitrate was retained in peat except for treatment with pH 5.1. Release of NO_3-N was found in all treated-podzol lysimeters, partly as a result of nitrification, which also released H^+ ions. The vegetation normally assimilating nitrogen (e.g. tree roots) is almost completely absent in the podzol lysimeters.

The mineralization of organic nitrogen starts with ammonification of RNH_2 to NH_4^+, which is a proton consuming process. If NH_4^+ is transformed to NO_3^- by nitrification, a net H^+ contribution is produced (Reuss & Johnson 1987). Ammonification occurred in both peat and podzol lysimeters (see table 4). The nitrification rate in the podzol was high in all treatments, and did not decrease with high acid load, but the nitrification in peat was limited.

(b) *Leaching of base cations*, Fe, Mn *and* Al

The results are shown in table 4. The leaching of base cations was dependent upon the flux of anions. The output of Ca, Mg and K from podzol did not differ between, for instance, treatments with pH 3.6 and 3.0 because of SO_4-S retention. The nitrification and the NO_3-loss in the podzol lysimeters (see §4a caused a high output of base cations in all treatments.

The treatment with pH 2.4 in peat resulted in a chemical mineralization of the peat and very high leaching rates of Ca, Mg and K. The leaching of base cations mentioned above is also demonstrated in figure 3, where the ratios between the different treatments, and treatment pH 4.2 are shown. pH 4.2 (H^+ 59 mmol m^{-2} a^{-1}) is regarded as a normal acid load for the study area.

Iron and manganese did not show any clear reactions to the treatments except for high leaching from peat treated with pH 2.4. The results from podzol indicate different stores of Mn in the soil. Aluminium reacted in the same way as the base cations to acid load. The variation in monomeric organic Al between treatments was small. Monomeric inorganic Al responded well to anion leaching, especially in podzol, where NO_3 loss caused high outflow of aluminium, even with treatment pH 5.1. The leaching of Fe, Mn, and Al from different treatments compared with treatment pH 4.2 is shown in figure 4.

If the leaching of base cations, Fe, Mn, and Al from lysimeters (treatment pH 4.2) is compared with runoff from catchments (see tables 3 and 4), the peat lysimeters and the runoff from Åkhultmire show similarities. The large output of many elements from podzol lysimeters compared with that in the runoff from the Aneboda catchment is mainly caused by the nitrification and NO_3-loss from the lysimeters, because of a lack of vegetation in the lysimeters (see §5a).

(c) *Leaching of strong and organic acids*

The determination of organic acids showed that the difference between the dissociation constant in percolated water from peat and podzol lysimeters was similar to that between runoff from the Åkhultmire and the Aneboda catchments (see §4d). All the organic acids found in output from peat lysimeters were calculated to be dissociated. In output from podzol lysimeters, only 37%, as a mean,

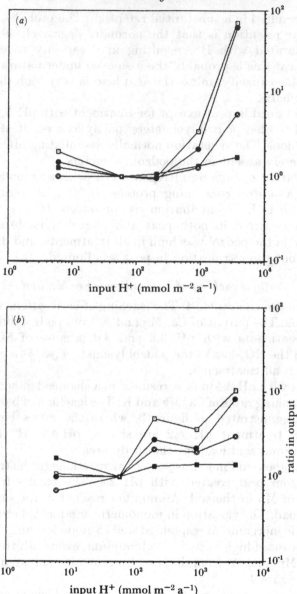

Figure 3. The output of Ca, Mg, Na and K from (a) peat and (b) podzol lysimeters. Output ratio of different treatments and treatment pH 4.2 against H$^+$ input (mmol m^{-2} a^{-1}); ($\square$) Mg; ($\bullet$) Ca; ($\blacksquare$) Na; ($\circ$) K.

was dissociated, and no clear variation could be observed between different treatments.

The results of the evaluation of the contribution from strong and dissociated organic acid to free acidity are shown in figure 5. Some of the results are not measured but calculated from the correlation between organic anions and water-colour index discussed later. Figure 5 includes even results from the two catchments.

The organic acid contributed to 100% of the free acidity in percolating water from peat lysimeters with treatments pH 5.1–3.6.

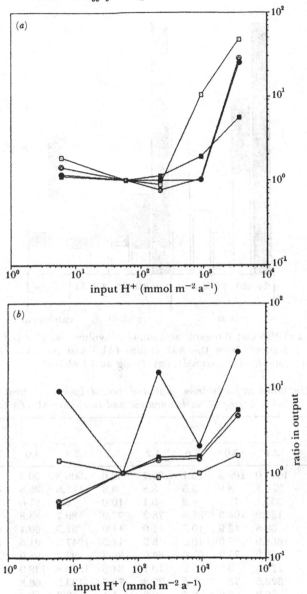

Figure 4. The output of Fe, Mn, tot.-Al and monomeric inorganic Al from (*a*) peat and (*b*) podzol lysimeters. Output ratio of different treatments and treatment pH 4.2 against H$^+$ input (mmol m^{-2} a^{-1}); ($\square$) Fe; ($\bullet$) Mn; ($\blacksquare$) total Al; ($\circ$) inorganic aluminium.

Treatment pH 3.0 resulted in about equally strong and organic acid H$^+$ contributions. In treatment pH 2.4 the organic acid contribution was calculated to be almost zero, both for peat and podzol. No effect of the treatments of the podzol lysimeters on organic anion contribution to free acidity could be observed, except for treatment pH 2.4.

The strong acid contribution from podzol was higher than in the runoff from the Aneboda catchment because of nitrification in the lysimeters and influence of peatland in the catchment. The difference between peat lysimeters (treated with pH

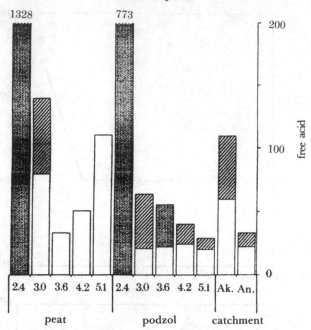

Figure 5. Free acid and dissociated organic acid (µmol l⁻¹, volume weighted). Output from peat and podzol lysimeters and runoff from the Åkhultmire (Åk.) and Aneboda (An.) catchment; (□) organic anions; (▨) strong acid (measured); (▨) strong acid (calculated).

Table 4. *Output (in* mmol m⁻² a⁻¹) *from peat and podzol lysimeters treated with* pH *adjusted precipitation (2.4–5.1) during 97 weeks; organic and inorganic* Al *refer to monomeric* Al

	peat					podzol				
	2.4	3.0	3.6	4.2	5.1	2.4	3.0	3.6	4.2	5.1
H⁺	1042.0	109.5	26.1	39.7	86.6	606.8	50.3	43.9	31.6	22.7
Ca	236.8	8.0	3.5	3.8	5.6	141.6	36.8	46.6	17.1	16.9
Mg	357.2	17.4	4.9	5.1	10.0	76.5	27.0	35.4	7.3	7.9
Na	131.2	104.5	82.6	78.3	97.9	89.7	88.8	84.6	63.9	105.2
K	59.8	13.9	10.8	11.0	11.0	91.3	60.1	63.5	30.5	21.0
SO₄-S	1658.0	107.0	10.2	5.3	11.5	1047.9	91.5	89.6	47.3	24.4
Cl	66.9	77.9	65.6	66.0	139.3	93.2	88.0	116.6	92.0	85.5
NO₃-N	129.0	3.7	1.9	10.7	34.9	1270.8	739.9	854.4	374.6	264.4
NH₄-N	529.6	75.1	25.2	27.2	79.6	133.3	66.8	48.2	17.4	29.3
Fe	26.9	6.0	0.5	0.6	1.1	0.3	0.2	0.2	0.2	0.3
Mn	3.5	0.1	0.1	0.1	0.2	87.4	6.9	50.9	3.3	29.4
tot.-Al	18.4	6.2	3.7	3.3	3.6	815.0	232.8	229.1	148.4	60.4
org.-Al	1.1	1.1	1.9	1.4	1.3	61.7	30.0	41.4	20.4	7.9
inorg.-Al	15.2	0.6	0.4	0.5	0.8	600.3	186.8	181.4	128.0	58.5
org. anions	< 5	62.6	—	101.7	121.6	< 5	8.6	—	11.2	7.8

5.1–3.6) and the Åkhultmire is probably explained by the surface flow of water in the catchment resulting in a strong acid contribution to runoff (see §4d).

The strongest correlation between dissociated organic acids and other analysed elements was found with water-colour index. High water-colour index (greater than 200 mgPt l⁻¹) resulted in different concentrations of organic anions in runoff from the Åkhultmire and Aneboda catchments, and from the output of peat lysimeters because of different dissociation constants of organic acids (see §4d). The water-colour index could also be influenced by different pH in runoff and percolating water (Gjessing 1976). No correlation was found between organic anions and the very low water-colour index (2–10 mg Pt l^{-1}) in output from podzol lysimeters.

This work was funded by the Surface Waters Acidification Programme. Special thanks to Ying-Hua Lee, Hans Hultberg, Johan Knulst and Birgitta Skoglund.

References

Andersson, F. & Olsson, B. (eds) 1985 *Lake Gårdsjön. An acid lake and its catchment. Ecol. Bull.* no. 37, 336 pp.

Brekke, F. H. 1978 SNSF-project, Norway IR 37/38.

Gjessing, E. T. 1976 *Physical and chemical characteristics of aquatic humus.* Michigan: Ann Arbor Science Publications.

Hultberg, H. 1985 *Ecol. Bull.* **37**, 133–157.

Lee, Y. H. 1980 *Wat. Air Soil Pollut.* **14**, 287–298.

Lee, Y. H. 1985 *Ecol. Bull.* **37**, 109–119.

Lövblad, G., Westling, O. & Ivens, W. 1989 Nordic Council of Ministers and the ECE workshop on Mapping.

Malmer, N. 1962 *Opera Botanica* **7**, 1–322.

Reuss, J. O. & Johnson, D. W. 1987 *Ecol. stud.* **59**. New York: Springer-Verlag.

Streamwater, soil-water chemistry, and water flow paths at Birkenes during a dry–wet hydrological cycle

By Rolf D. Vogt[1], Dag Olav Andersen[2], Sjur Andersen[3],
Nils Christophersen[4] and Jan Mulder[4]

[1] Chemistry Department, University of Oslo, P.B.1033 Blindern, N-0315 Oslo 3,
Norway
[2] Chemistry Department, Agder College P.B.607, N-4601 Kristiansand, Norway
[3] Institute for Georesources and Pollution Research, P.B.9, N-1432 Ås-NLH, Norway
[4] Center for Industrial Research, P.B.124 Blindern, 0314 Oslo 3, Norway

Streamwater chemistry at Birkenes can, as a first approximation, be described as a mixture of three soil-water types (end-members), each with its own chemical characteristics. The contributions of the end-members to the streamwater discharge are dependent upon water pathways that vary with precipitation amount and antecedent hydrological conditions. A detailed episode study in the summer of 1988, including a transition from very dry to wet conditions, showed two main features. First, the results highlighted the importance of the bog along the main brook as a source of streamwater both during baseflow and the initial wetting phase. During baseflow, streamwater primarily originated from the water-table level in the bog, whereas the initial wetting caused a piston flow of deeper stagnant and anaerobic bog water. Secondly, a significant dilution occurred in the soil surface layers, and such changes must be taken into account in future applications of the end-member concept.

1. Introduction

Detailed episodic studies at Birkenes have shown the importance of water flow paths in determining streamwater chemistry. High-flow chemistry is characterized by high concentrations of inorganic monomeric aluminium (Al_i) and H^+ and lower concentrations of Ca^{2+} and Mg^{2+}, whereas the reverse is true for baseflow conditions (*cf.* Sullivan *et al.* 1986*a*; Christophersen *et al.* 1990*a*). These observations have been correlated with the soil-water chemistry under the heading of end-member mixing analysis (EMMA; *cf.* Christophersen *et al.* 1990*a*). The present picture is one of mixing where, at high-flow, a major proportion of the runoff originates from the O/H and B horizons on the slopes. At low-flow, water stored in the deeper deposits in the valley bottom contributes most of the runoff.

Superimposed on the flow paths is the effect of the salt levels in the different soil layers in mobilizing aluminium and H^+ with subsequent transport to the stream. This was clearly demonstrated by an episode in October 1987, when an exceptionally high sea-salts event occurred (Mulder *et al.* 1990), and by an *in situ* NaBr manipulation experiment (Vogt 1989; Vogt *et al.* 1990). Also, in earlier work by

Christophersen *et al.* (1982), the role of sulphate mineralization following dry periods was highlighted.

Although it explains the general patterns in streamwater chemistry, the EMMA approach does not take into account temporal changes in the end-member compositions, e.g. caused by changes in the salt level. Another shortcoming is that the calcium concentrations in the O/H and B horizons are too low to explain the streamwater concentrations at high-flow.

In this paper the overall picture is assessed in light of detailed data of soil- and streamwater collected during the summer of 1988. The study started during extremely dry conditions, and was followed through the initial wetting of the catchment during four subsequent rain events.

2. Site description and methods

The Birkenes catchment (see fig. 1 in Christophersen *et al.* (this symposium) is small (0.41 km²) with heterogenous soils in the region of highest sulphur deposition in Norway. During the summer months evapo-transpiration usually exceeds precipitation resulting in summers with little discharge; the autumns are generally wet with high discharge.

Below the organic–humic layer (*ca.* 10 cm), mineral soils, ranging from podzols to acid brown earths, have developed in a layer of shallow glacial till overlaying granitic bedrock. In the valley bottom an extensive area around the brooks is covered by bogs (0.1–2.5 m deep), comprising 7% of the total catchment area. Streamwater is chronically acidic with a volume-weighted mean pH in the main brook of 4.5. Overland flow is common only over the bogs during large events. General soil solution characteristics are presented by Christophersen *et al.* (this symposium) (see their fig. 3).

The episodic studies are characterized by detailed monitoring of soil-water potentials, groundwater tables, precipitation, and discharge as well as the chemistry of streamwater, soil waters, precipitation, and throughfall. The data presented in this paper were collected between 21 June and 9 July 1988. The first event was preceded by more than 3 weeks with no precipitation. At the beginning of the monitoring period, some of the tensiometers dried out whereas others showed soil water potentials of −90 to −250 cmH_2O.

The soil-solution chemistry in the different horizons was monitored by using ceramic-cup suction lysimeters for continuous soil-water sampling using a non-constant tension of maximum 50 kPa. Immediate processing of streamwater grab samples as well as soil waters was done in an on-site laboratory. The Al fractionation was accomplished spectrophotometrically as described in Sullivan *et al.* (1986*b*). Detailed description of sample treatment and analytical procedures may be found in Vogt *et al.* (1990). The determination of Al involves a correction for interference by Fe (as well as Mn). The correction terms are also used as a crude estimate of these elements.

3. Results and discussion

The first three rain events totalling 66 mm precipitation wetted up the catchment but resulted in only minor runoff (figure 1). The fourth, and largest event (total of 89 mm), gave a pronounced discharge response with a water yield generally found during wet conditions. In contrast to the previously described chemical response, the

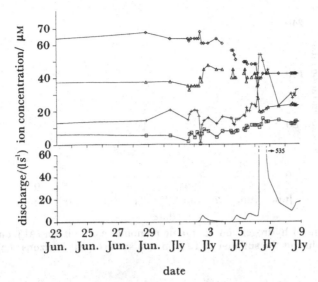

Figure 1. Discharge (litres per second) and concentrations of hydrogen ion, inorganic monomeric aluminum (Al$_i$), and nitrate from dry to wet conditions during the summer of 1988 in the main stream; (+), H$^+$; (□), Al$_i$; (◇), Cl$^-$/2; (△), Ca^{2+}.

first event resulted in a significant decrease in the H$^+$ and Al$_i$ concentrations, whereas the two next events only gave a modest response for these solutes. During the last and major event, both the H$^+$ and Al$_i$ concentrations increased as found previously. For Ca^{2+} and Mg^{2+} as well as NO$_3^-$, an increase in concentration was observed during the first three events followed by a dilution during the fourth episode. Chloride and sodium showed a general dilution over the whole period; no clear response was observed for sulphate. However, because of possible sulphur oxidation during storage, the actual sulphate concentration in the stream may be over-estimated. The estimate of dissolved iron showed an elevated level compared with baseflow, especially during the very first event. After the fourth and largest episode, the iron level decreased below the preceding baseflow concentration.

In contrast to the general picture during events where the high-flow predominantly originates from the O/H and B horizons on the slopes, the observations during the three first storms indicate a different mode of response. This flow-path probably involves interactions with deeper gyttja (organic mud) (DG) layers in the bog. A similar pattern was observed with the initial snow-melt episode in 1987 (unpublished data). The streamwater chemistry during the first three episodes differs from baseflow chemistry that presumably was also fed by bog water. It is hypothesized that this behaviour was due to vertical gradients in the bog-water chemistry. An initial increase in streamwater pH and alkalinity during events has been described for the Allt a'Mharcaidh catchment in Scotland (Harriman *et al.* 1990). A piston effect that pressed out higher alkalinity groundwater was suggested as the probable cause in this catchment.

The dilution observed for Cl$^-$ in streamwater is consistent with earlier observations of decreases in solute concentrations during wet periods preceded by prolonged dry spells (Christophersen *et al.* 1982). However, a typical feature of the previous data from the 1970s was a sharp response of sulphate, which was not observed in 1988. Given the reduced deposition of sulphur and declining SO$_4^{2-}$ concentrations in

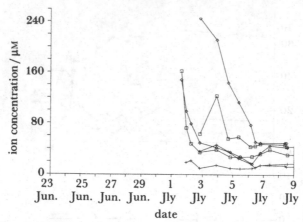

Figure 2. Concentrations of hydrogen ion, inorganic monomeric aluminum (Al$_i$), and nitrate from dry to wet conditions during the summer of 1988 in two surface peat horizons; (□), H$^+$; (+), Al$_i$; (◇), Cl$^-$.

streamwater over the last several years (Christophersen *et al.* 1990*b*), this lack of response could be due to a reduced sulphur pool in the catchment. It is more likely though, that the variations in sulphate concentrations are strongly dependent on the length of the antecedent dry period.

The soil-solution chemistry, though complex, showed definite temporal patterns. For all lysimeters in the bogs, we were able to sample soil water before the first event. Maximum concentrations of H$^+$, Al$_i$, and major anions and cations occurred in the surface peat layers (P) before the onset of the first episode, showing a pronounced drop before levelling out around 6 July (figure 2; table 1).. For the O/H and E horizons on the slope, sampling could not be done before the first event, but similar variations, though less pronounced, were also observed in these horizons. In the B-horizons and the deeper gyttja and mineral (DC) layers of the bog, the chemistry showed significantly smaller temporal variations (table 1).

Note that for Al$_i$ the concentrations were higher for the surface peat layers than for the O/H horizons in the mineral soils. This observation is consistent with the higher exchangeable aluminium levels in the surface peats (M. Pijpers, personal communication). Within the bog, strong vertical gradients existed in the soil-water composition. The concentration of H$^+$ and Al$_i$ decreased, whereas iron and, to a lesser extent, calcium increased with depth. Most likely, neutralization of H$^+$ with depth takes place both through weathering, as evidenced by the higher calcium levels, and by reduction of sulphate as indicated by the higher iron and H$_2$S concentrations (Andersen 1989). Average chloride values did not show a significant trend with depth (table 1). However, the average values mask the underlying temporal changes in the surface layers.

The results show a parallel decrease in Cl$^-$ concentrations both in the surface soil horizons and the stream. However, the temporal trends in Cl$^-$ in deeper gyttja layers (DG) in the bog, postulated as being the major source of streamwater during the first event, are less pronounced. To reconcile the streamwater and the bog-water observations, we hypothesize that the first event pushed out stagnant and H$^+$-depleted water from the deep gyttja layers. This water was not a source of baseflow, which, according to its chemistry, presumably originated from higher aerobic gyttja

Table 1. *Soil-solution chemical characteristics of the Birkenes Catchment.*
The values are based on data from cup lysimeter samples, collected during the summer of 1988 episode study (Vogt 1989). Values in brackets represent total inorganic sulphur because probably high concentrations of sulphide are oxidized before analysis; P, G, DG, DC denotes surface peat, gyttja, deep gyttja (reducing conditions) and deep C (mineral layers below the bogs), respectively.

	soil horizon	H^+/μM	Al_l/μM	Cl^-/μM	NO_3^-/μM	SO_4^{2-}/μM	Ca^{2+}/μM
avg.	H/O	54	6	88	25	55	24
min.	H/O	18	0	56	7	35	12
max.	H/O	83	13	113	93	113	70
std.	H/O	19	4	16	30	25	20
#	H/O	20	15	7	7	7	7
avg.	E	57	19	115	60	61	21
min.	E	18	5	113	7	38	11
max.	E	93	32	118	146	75	26
std.	E	18	6	3	61	17	7
#	E	34	29	3	3	3	3
avg.	B	28	26	109	9	58	15
min.	B	19	20	56	7	50	13
max.	B	34	33	133	21	75	18
std.	B	4	3	22	4	8	3
#	B	42	36	21	21	21	21
avg.	P	52	19	124	69	65	29
min.	P	1	4	28	7	19	10
max.	P	513	46	491	600	165	200
std.	P	65	9	99	101	28	39
#	P	60	58	37	37	37	37
avg.	G	23	14	109	43	42	19
min.	G	11	5	85	9	13	12
max.	G	59	28	141	107	88	28
std.	G	9	7	13	24	22	4
#	G	52	50	20	20	20	20
avg.	DG	0	4	140	9	[15]	34
min.	DG	0	0	127	1	[7]	30
max.	DG	1	17	158	23	[60]	42
std.	DG	0	3	9	6	[18]	3
#	DG	25	25	13	13	[13]	14
avg.	DC	1	5	140	5	[161]	54
min.	D	0	0	127	0	[16]	20
max.	D	2	8	172	22	[216]	84
std.	D	0	2	16	8	[66]	24
#	D	42	38	8	7	[8]	11

(G) levels in the bog close to the water table. The release of the stagnant water in the DG layer also caused an increase in the calcium and iron concentrations and decrease in the H^+ and Al_l levels. Subsequently, the stream was fed from higher bog layers (G) containing diluted and more oxidized solutions and, in the fourth event, the hillslope also contributed significantly.

The authors gratefully acknowledge support from the Surface Waters Acidification Programme (SWAP) and the Norwegian Ministry of Environment. Valuable comments have been given by Colin Neal, Hans Martin Seip and Alex Stone.

References

Andersen, S. 1989 Thesis, University of Oslo. [In Norwegian.]

Christophersen, N., Neal, C., Hooper, R. P., Vogt, R. V. & Andersen, S. 1990a J. Hydrol. (In the press.)

Christophersen, N., Robson, A., Neal, C., Whitehead, P. G., Vigerust, B. & Henriksen, A. 1990b J. Hydrol. (In the press.)

Christophersen, N., Seip, H. M. & Wright, R. F. 1982 Wat. Resour. Res. 18, 977–996.

Harriman, R., Gillespie, E., King, D., Watt, A. W., Christie, A. E. G., Cowan, A. A. & Edwards, T. 1990 J. Hydrol. (In the press.)

Mulder, J., Christophersen, N., Hauhs, M., Vogt, R. D., Andersen, S. & Andersen, D. O. 1990 Wat. Resour. Res. 26, 611–622.

Sullivan, T. J., Christophersen, N., Muniz, I. P., Seip, H. M. & Sullivan, P. D. 1986a Nature, Lond. 323, 324–327.

Sullivan, T. J., Seip, H. M. & Muniz, I. P. 1986b J. environ. analyt. Chem. 26, 61–75.

Vogt, R. 1989 Thesis, University of Oslo.

Vogt, R., Seip, H. M., Christophersen, N. & Andersen, S. 1990 Sci. tot. Envir. 96 (In the press.)

Aluminium solubility in the various soil horizons in an acidified catchment

By Sjur Andersen[1], Nils Christophersen[2], Jan Mulder[2],
Hans Martin Seip[3] and Rolf D. Vogt[3]

[1] Institute for Georesources and Pollution Research, P.O. Box 9,
N-1432 Ås-NLH, Norway
[2] Center for Industrial Research, P.O. Box 124 Blindern, N-0314 Oslo 3, Norway
[3] Department of Chemistry, University of Oslo, P.O. Box 1033 Blindern,
N-0315 Oslo 3, Norway

Most surface water acidification models assume equilibrium with an aluminium mineral phase, usually gibbsite. A uniform aluminium solubility control in the soil profile is also often a common feature. To determine if these assumptions are valid, the aluminium solubility of the four dominating horizons occurring in podzols at the Birkenes catchment, southern Norway, was examined. In air-dried soil samples, the aluminium release from the soils was determined as a function of pH. The aluminium release was highest in the mineral (E and B) horizons and lowest in the organic (O) horizon, implying that there are different reservoirs of soluble aluminium in the various soil horizons. When constructing a plot of aluminium activity (pAl) as a function of pH, the laboratory data could satisfactorily be described by using a linear-regression model. However, the slope varied between 1.7 and 3.1, indicating that neither gibbsite solubility nor cation exchange processes control the aluminium activity in the soil horizons. Other mechanisms, e.g. (de)complexation reactions from soil organics with the equilibrium constant being charge dependent, may better explain the observed behaviour of dissolved aluminium in acidified soils and surface waters. Recently, such (de)complexation reactions were modelled in the chemical equilibrium model CHAOS. Preliminary simulations of our laboratory data with the CHAOS model are presented.

1. Introduction

Elevated levels of aluminium in surface waters and streams, which have harmful effects on the biota, are considered to be one of the most serious impacts of acid deposition. A large effort has therefore been made to evaluate both the aluminium chemistry and the factors that control aluminium solubility in acidified catchments. Computer simulation models have played an important role in the development of the present understanding of acidification processes (Christophersen & Hauhs et al. and Christophersen & Neal et al., this symposium; Stone & Seip 1989). Most hydrogeochemical models generally assume equilibrium with a single gibbsite (Al(OH)$_3$) phase, giving the relation:

$$\{Al^{3+}\}/\{H^+\}^3 = K, \tag{1}$$

where K is a constant and $\{\}$ denote activities. The same equation is used to describe simple cation-exchange reactions where the constant has generally been assigned lower values. Some trends in observed Al concentrations in streamwater may be reproduced assuming equation (1) to hold but, in general, satisfactory agreement between observed and simulated values is not obtained (Stone & Seip 1989). Note that gibbsite minerals are generally not found in temperate forest soils.

Aluminium solubility was examined in all horizons of a podzol profile. The objective was to evaluate both the relation between Al^{3+} and H^+ activities, and the relative size of the aluminium pool in the various horizons. In addition, a first attempt was made to simulate the Al solubility in the various soil horizons by using the CHAOS model (Tipping & Hurley 1988), in which the aluminium activity is determined by (de)complexation reactions with solid-phase organics. Earlier, Bloom *et al.* (1979) already showed the importance of natural organic matter in soils in regulating dissolved Al levels. In addition, Mulder *et al.* (1989) demonstrated that, even in the mineral soil (B horizons) solid-phase alumino-organics may be the major source of dissolved Al. A better insight in the aluminium activity regulation in soils and surface waters is necessary to improve present computer models that describe short- and long-term chemical changes in acidified catchments.

2. Materials and methods

The soils examined were taken from three different plots in the Birkenes catchment, located in southern Norway. The catchment has been extensively studied since 1972 with respect to the impact of acid precipitation (Christophersen & Hauhs *et al.*, Seip *et al.* and Vogt *et al.*, this symposium). Soil samples were collected in the late autumn of 1987 and the summer of 1988 under similar hydrological conditions, with one week without precipitation before sampling. Subsamples were air dried for 48 h; 120 ml of water with a salt matrix of 0.007 M NaCl and varying pH was then added to 24 g of soil. The pH was adjusted with HCl or H_2SO_4 (see §4). NaOH was used as a base throughout the experiment. The mixtures were shaken and sampled after 6, 24, 48 and 196 h. The samples were centrifugated and pH was measured with a Ross-electrode. 25 ml of the supernatant was diluted to 250 ml, and aluminium was fractionated and analysed according to Driscoll (1984), separating non-labile (mainly organically bound) aluminium (Al_o) and labile (mainly inorganic) aluminium (Al_i). Total concentrations of Ca^{2+}, Mg^{2+}, Na^+ were measured by inductive coupled plasma atomic emission spectroscopy (ICP-AES); NO_3^-, SO_4^{2-}, Cl^- by ion chromatography, and total fluoride with an ion-selective electrode. The Al_i fraction was further speciated by using the computer program ALCHEMI (Schecher & Driscoll 1987). Regression analyses were made with pAl^{3+} as a function of pH, where pAl^{3+} denotes the negative logarithm of the activity of Al^{3+}, i.e. $-\log\{Al^{3+}\}$.

The chemical equilibrium model CHAOS is described in detail by Tipping & Hurley (1988). Briefly, the model accounts for Al complexation by solid-phase humic substances as bidentate binding to various functional groups, where the equilibrium constant K depends on the ionic strength in solution as well as on the charge at the organic surface. The version of CHAOS used in this work was programmed in MATHEMATICA (Wolfram 1988).

3. Results

The results of the solubility experiment were plotted in stability diagrams of pAl^{3+} against pH (figures 1 and 2). These plots could satisfactorily be described by a linear regression model for nearly all the mineral soil horizons, with $r^2 = 0.96$ or better

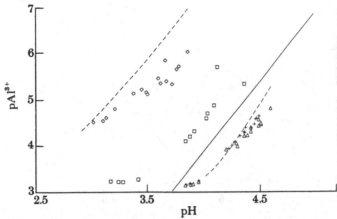

Figure 1. pAl^{3+} against pH for the various soil horizons for the 1987 autumn soil experiment; (--),
CHAOS simulation; (——), gibbsite; (◇) O horizon; (□), E horizon; (△), Bhs horizon; (+), Bs
horizon.

Table 1. *Statistical data for the soil experiment assuming a linear relation,* pAl^{3+} − αpH = pK,
for all series
(α denotes slope.)

soil horizon	autumn			summer		
	α	pK	r^2	α	pK	r^2
organic	1.70±0.15	−0.37±0.15	0.90	3.06±0.28	−4.72±0.24	0.89
eluvial	2.02±0.23	−3.08±0.30	0.89	2.01±0.10	−2.59±0.13	0.96
Bhs	2.10±0.07	−4.64±0.07	0.99	2.06±0.10	−4.58±0.13	0.97
Bs	2.26±0.06	−5.22±0.05	0.99	2.10±0.09	−4.69±0.10	0.97

(table 1). However, a weaker correlation ($r^2 = 0.89$) was found for organic soil and
eluvial soils sampled in the autumn. At the same pH the Al^{3+}-concentrations in the
two mineral horizons were found to be approximately equal, but decreased by a
factor of 10 in the E-horizon and by an additional factor of 10 in the O-horizon. The
levels were not significantly different between soils sampled in the autumn and
summer.

Figures 1 and 2 show that the two mineral horizons were oversaturated, and that
both the E and O-horizons were highly undersaturated with respect to synthetic
gibbsite. All experimental series, even those with sulphuric acid as proton source,
showed an undersaturation with respect to jurbanite.

In figures 1 and 2 the dashed lines indicate the results from CHAOS simulations.
Even though most of the essential model parameters were estimated, reasonable
simulations could be obtained for the laboratory data.

4. Discussion

The use of stability diagrams (e.g. figures 1 and 2) as a test for gibbsite-controlled
Al activity, has been criticized by Neal *et al.* (1987; 1989) because the linearity of the
pAl^{3+} against pH plots may be induced by the dependent axes at pH > 5.5. These
authors argued that, in addition to gibbsite, several aluminium silicates could

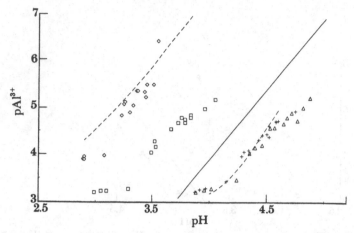

Figure 2. pAl^{3+} against pH for the various soil horizons for the 1988 summer soil experiment; (--),
CHAOS simulation; (——), gibbsite solubility; ($\Diamond$) O horizon; ($\square$), E horizon; ($\triangle$), Bhs horizon;
($+$), Bs horizon.

provide a similar linear pH–pAl response thus making it impossible to decide which
mineral phase controls the Al activity. Although our data show a linear response, the
slope deviates significantly from the generally accepted value of 3. The observed
deviations exceed those expected from the uncertainties in the observations (see Neal
& Christophersen (1989)); therefore, equilibrium with gibbsite or aluminium silicates
must be rejected. At least two hypotheses may explain our observations. First,
different unidentified mineral phases may actually control aluminium solubility in
the various horizons. Secondly, there is no single mineral controlling the Al-
concentrations in each or some of the horizons. The CHAOS calculations demonstrate
that the laboratory data can be modelled reasonably well with an organic Al-
complexation model, suggesting that Al (de)complexation reactions may explain the
observed behaviour better than mineral equilibrium or simple cation exchange, both
in the mineral layers and in the organic layer.

The results discussed were obtained by adjusting pH with HCl for soil collected in
the autumn and with H$_2$SO$_4$ for soil sampled in the summer. It is conceivable that
the strong acid anion may affect the Al solubility, for example if a jurbanite phase
(Al(OH)SO$_4$) is formed. No significant differences were obtained for the two series
(table 1) except for the O-horizon, where the slope was 1.7 for the autumn soil and
3.1 for the summer soil. To test if the difference was due to different acid sources, the
experiment was repeated for the summer soil by using both hydrochloric acid and
sulphuric acid as proton sources. No significant difference was detected.

Variations in slope and pK for the samples from the O-horizon collected in the
summer and autumn are significant. The reason may be a seasonal change in the
humic compounds resulting from, for example, changes in microbiological activity.
Seasonal differences with respect to complexation capacity in aquatic organic
compounds have been observed by Lægreid et al. (1983). A simple ion-exchange
description is thus consistent with our results for the organic summer soil, but not for
the organic autumn soil.

5. Conclusion

The hypothesis that aluminium solubility is controlled by gibbsite, aluminium silicates, or simple ion-exchange reactions does not agree with the results from this experiment. Promising results were obtained when the aluminium activity was assumed to be regulated by Al complexation with organics. However further testing is warranted. The experiment illustrates that acidification models must allow for different aluminium processes in the various soil layers. It may prove necessary to include at least three sub-compartments in the model, each with its own Al-solubility control.

References

Bloom, P. R., McBride, M. B. & Weaver, R. M. 1979 *Soil Sci. Soc. Am. J.* **43**, 488–493.

Driscoll, C. 1984 A procedure for the fractionation of aqueous aluminum in dilute acidic waters, *Int. J. environ. Analyt. Chem.* **16**, 267–283.

Lægreid, M., Alstad, J., Klaveness, D. & Seip, H. M. 1983 *Environ. Sci. Technol.* **17**, 357–361.

Mulder, J., van Breemen, N. & Eijck, H. C. 1989 *Nature, Lond.* **337**, 247–249.

Neal, C., Skeffington, R. A., Williams, R. & Roberts, D. J. 1987 *Earth planet. Sci. Lett.* **86**, 105–112.

Neal, C. & Christophersen, N. 1989 *Sci. tot. Envir.* **80**, 195–203.

Schecher, W. & Driscoll, C. T. 1987 *Wat. Resour. Res.* **23**, 525–535.

Stone, A. & Seip, H. M. 1989 *AMBIO* **18**, 192–199.

Tipping, E. & Hurley, M. A. 1988 *J. Soil Sci.* **39**, 505–519.

Wolfram, S. 1988 *Mathematica, a system for doing mathematics by computer.* New York: Addison-Wesley.

The RAIN project – an overview

By R. F. WRIGHT AND A. HENRIKSEN

Norwegian Institute for Water Research, P.O. Box 69, Korsvoll, 0808 Oslo 8, Norway

International efforts to reduce emissions of sulphur and nitrogen oxides are based in part on the premise that such reductions will restore waters acidified by these emissions. This premise has been tested in Norway by whole catchment manipulations in which acid loading is changed experimentally. The RAIN project (Reversing Acidification In Norway) comprises two experiments: artificial acidification of two pristine catchments in western Norway, and exclusion of ambient acid deposition by means of a roof at an acidified catchment in southernmost Norway. Acid addition has caused major changes in runoff chemistry. Sulphate concentrations have doubled four years after the experiment started. Addition of HNO_3 has so far caused only minor increases in runoff nitrate. The increased sulphate concentrations have been about 50 % compensated by increased concentrations of base cations and about 50 % by decreased alkalinity. Acid exclusion has resulted in lower concentrations of strong acid anions NO_3^- and SO_4^{2-} in runoff relative to control catchments. The decline in strong acid anion concentrations was compensated partly by a decrease in base cation concentrations (55 %) and partly by a decrease in alkalinity (45 %). The input–output budgets indicate that acid exclusion has reversed soil acidification. The effect of organic acids on the pH of runoff has increased in importance as the experiment has proceeded.

1. Introduction

Acid deposition, acidification of surface waters and loss of fish populations occur over large regions of Europe and eastern North America. International efforts to reduce the emissions of SO_2 and NO_x are based in part on the premise that such reductions will restore acidified waters. To test this premise, whole-catchment manipulations are conducted in Norway in which the acid loading is changed experimentally. The RAIN project (Reversing Acidification In Norway) comprises two experiments: artificial acidification of two pristine catchments in western Norway (Sogndal site), and exclusion of ambient acid deposition, by means of a roof, at an acidified catchment in southernmost Norway (Risdalsheia site).

2. The sites

(a) Sogndal: acid addition

The Sogndal site is a pristine but sensitive area in western Norway presently receiving only weakly acidic precipitation (pH 4.8). The site is located 900 m above sea level and on gnesissic bedrock, with patchy, thin (average depth 30 cm) and poorly developed soils having pH (H_2O) 4.5–5.5, and alpine vegetation. Four catchments are being studied: catchment SOG2 (7220 m²) receives H_2SO_4; catchment SOG4 (1940 m²) receives a 1:1 mixture of $H_2SO_4 + HNO_3$; catchments SOG1 (96300 m²) and SOG3 (43200 m²) serve as untreated controls.

Acid addition, which began in April 1984, consists of application to the snowpack of 0.02 mm water at pH 1.9 and four or five events of 11 mm at pH 3.2 during the snow-free months. Acid is mixed with lake water from SOG1 and applied at 2 mm h^{-1} by using commercial irrigation equipment.

Volume and chemical composition of natural precipitation at Sogndal are measured in weekly bulk samples collected at a farm, 500 m above sea level and 3 km from the experimental catchments. Discharge is gauged continuously by weir and stream-level recorders at the outlet of catchments SOG1, SOG2 and SOG4. Runoff samples for chemical analyses are collected at least weekly at the outlet of each catchment; sampling frequency is increased at 2–7 times a week during snow melt. Additional samples from SOG2 and SOG4 are collected every 2 h during and immediately following acid additions, and daily for the five days after addition.

(b) Risdalsheia: acid exclusion

The Risdalsheia site is located in a sensitive area in southernmost Norway currently receiving a high loading of acid deposition (precipitation has a pH of 4.2, sulphate loading, wet and dry, is 100 mequiv. $m^{-2} a^{-1}$). The experimental catchments are situated 300 m above sea level and are characterized by exposed granitic bedrock (30–50 % of the surface) and thin, organic rich, truncated podzolic soils (average depth 10–15 cm, maximum depth 50 cm) with pH (H_2O) 3.9–4.5, and sparse cover of pine and birch.

Acid exclusion is accomplished by using a 1200 m^2 transparent roof that completely covers the 860 m^2 KIM catchment. Incoming precipitation is collected from the roof and pumped through a filter and ion-exchange system. Seawater salts are added back at ambient levels, and the clean precipitation is automatically applied beneath the roof above the canopy at the rate of 2 mm h^{-1}. An adjacent 400 m^2 catchment (EGIL) has also been covered with a roof and receives ambient precipitation from an identical sprinkling system. During the winter, artificial snow is added beneath the KIM roof by using commercial snow-making equipment. Water from a nearby pond is ion-exchanged and sea salts added to make clean snow. At EGIL, ambient acid snow is blown under the roof by using a snowblower. A third uncovered catchment (ROLF) serves as reference. Treatment began in June 1984.

Precipitation volume and chemical composition are measured in bulk samples collected weekly on site. The volume of water applied to the two roofed catchments is measured continuously by meters installed in the sprinkler systems. All runoff from the catchments is collected at fibreglass weirs and conducted in large-diameter hoses to 500 l tanks. Data-logging systems record continuously the total volume of water leaving each catchment. Samples for chemical analyses are collected from each 10–20 mm runoff by automatic samplers. More frequent samples are collected when rain falls off after dry periods.

3. Results

(a) Sogndal

Acid addition at Sogndal caused major changes in runoff chemistry. Each event of acid addition gave a short-term episode of acid aluminium-rich runoff. Recovery between episodes was rapid and full in the first year, but increasingly slower and less complete in subsequent years. Sulphate concentrations in runoff have doubled four years after the experiments started. Addition of HNO_3 caused only minor increases in nitrate runoff until 1989, when signs of nitrate leaching began to show. The increased sulphate concentrations were about 50%, compensated by increased concentrations of base cations and about 50% by decreased alkalinity. About 80% of the sulphur added to the catchment SOG4 over the four-year period 1984–87 was retained and apparently stored in the soil. This addition is sufficient to increase the pool of available sulphate by about 70% of the amount measured in 1984.

Results obtained during the first four years of treatment indicate that these pristine, acid-sensitive catchments respond rapidly to increases of acid deposition. Indeed, a single severe episode of acid precipitation may be sufficient to acidfy runoff to the point at which fish cannot survive. Continued acid addition and loss of base cations at the rate observed at the Sogndal catchments could result in a significant soil acidification within a few decades.

(b) Risdalsheia

Acid exclusion at the KIM catchment resulted in lower concentrations of the strong acid anions NO_3^- and SO_4^{2-} in runoff relative to both the roofed control catchment (EGIL) and the open catchment (ROLF). Nitrate concentrations decreased by 60% within two weeks of starting treatment. Sulphate concentrations showed a general decline beginning about four months after the onset of treatment. After 3.5 years of treatment, these concentrations are about 50% of those at EGIL. At the KIM catchment, sulphate output continues to exceed inputs in wet and dry deposition. Net loss of sulphate during the 3.5 years has been 53 mequiv. m^{-2}. This is equivalent to about 45% of the pool of readily available sulphate in the soil before treatment.

The decline in strong-acid anion concentrations was compensated partly by a decrease in base cation concentrations (55%) and partly by a decrease in alkalinity (45%). The input–output budgets indicate that relative to the control catchments the acid-exclusion catchment is retaining a portion of the base cations deposited in precipitation and released from weathering; the pool of exchangeable base cations is being replenished. The magnitude of this replenishment depends upon weathering rates in these catchments. Weathering rates cannot be obtained directly from input–output budgets at the control catchments because a fraction of the base cations in runoff may come from the pool of exchangeable cations in the soil.

As the acid-exclusion experiment has proceeded, organic acids have become increasingly prominent. The concentrations of organic anions are estimated by difference from the ionic balance. At the acid-exclusion catchment the concentration of organic anions has increased from 22 µequiv. l^{-1} in 1984 to 49 µequiv. l^{-1} in 1987. This increase is due to increased dissociation of organic acids and not to change in total organic carbon (TOC) concentrations. The organic carbon in these acid samples apparently has a maximum charge density of about 4.5 µequiv. mg C^{-1} and pK about 4.

4. Conclusions

At both experimental sites, runoff from the treated catchments now differs substantially in chemical composition from that of the control catchments. New steady-state concentrations have not yet been reached. Runoff at Sogndal has been acidified to levels toxic to fish, and runoff at Risdalsheia has begun to recover to pre-acidification chemical composition. Acid addition at Sogndal has caused soil acidification; acid exclusion at Risdalsheia has initiated reversal of soil acidification.

The changes in runoff chemistry observed at the RAIN project sites can be explained quantitatively by the interaction of key processes such as sulphate adsorption, cation exchange, dissolution of CO_2, mobilization of aluminium and buffering by organic acids. These processes are crucial in understanding soil and water acidification caused by acid deposition.

A process-oriented model, MAGIC (Model of Acidification of Groundwater In Catchments), satisfactorily simulates four-year trends in the runoff chemistry at the two manipulated catchments. The RAIN project provides a unique dataset for the calibration and evaluation of such predictive models. These models can, in turn, serve as the basis for extrapolation of the RAIN results to regional water and soil chemistry.

The RAIN project shows that within a few years, changes in acid deposition cause major changes in surface water chemistry at sensitive sites. These sites are typical for large areas of southern Norway. The reversibility of acidification demonstrated by the RAIN project agrees well with empirical data from Canada, the United States and Scotland. As the RAIN project continues through June 1991, the treated catchments should approach a new 'steady state' and provide information as to whether soil and water acidification is fully reversible.

The RAIN project has received financial support from the Norwegian Institute for Water Research, the Norwegian Institute for Air Research, the Norwegian Ministry of Environment, The Royal Norwegian Council for Scientific and Industrial Research, the Ontario Ministry of the Environment, Canada, the Swedish National Environmental Protection Board, the Central Electricity Generating Board and the Surface Waters Acidification Programme.

RAIN project publications

Wright, R. F. 1985 *Acid Rain Res. Rep.* 7/1985. Oslo: Norwegian Institute for Water Research.

Lotse, E. & Otabbong, E. 1985 *Acid Rain Res. Rep.* 8/1985. Oslo: Norwegian Institute for Water Research.

Wright, R. F. 1985 *Limnos* 1, 15–20. (In Norwegian.)

Wright, R. F., Gjessing, E., Christophersen, N., Lotse, E., Seip, H. M., Semb, A. & Sletaune, B. 1986 *Wat. Air Soil Pollut.* 30, 47–64.

Wright, R. F. & Gjessing, E. 1986 *Acid Rain Res. Rep.* 9/1986. Oslo: Norwegian Institute for Water Research.

Wright, R. F., Gjessing, E., Semb, A. & Sletaune, B. 1986 *Acid Rain Res. Rep.* 10/86. Oslo: Norwegian Institute for Water Research.

Wright, R. F. & Cosby, B. J. 1987 *Atmos. Environ.* 21, 727–730.

Wright, R. F. 1987 In *Reversibility of acidification* (ed. H. Barth), pp. 14–29. London: Elsevier Applied Science.

Hauhs, M. 1986 In *Water in the unsaturated zone* (ed. S. Haldorsen & E. J. Berntsen), pp. 207–217. Oslo: Nordic Hydrologic Programme.

Hauhs, M. 1987 In *Acidification and water pathways*, pp. 173–184. Oslo: Norwegian National Committee for Hydrology.

Wright, R. F. 1987 *Acid Rain Res. Rep.* 13/87. Norwegian Institute for Water Research.

Parmann, G. 1988 *Populærvitenskapelig Magasin* 3/88, 8–11. (In Norwegian.)

Hauhs, M. 1988 *Acid Rain Res. Rep.* 14/88. Oslo: Norwegian Institute for Water Research.

Wright, R. F. 1988 *Acid Rain Res. Rep.* 16/88. Oslo: Norwegian Institute for Water Research.

Wright, R. F., Norton, S. A., Brakke, D. F. & Frogner, T. 1988 *Nature, Lond.* 334, 422–424.

Wright, R. F., Lotse, E. & Semb, A. 1988 *Nature, Lond.* 334, 670–675.

Wright, R. F. 1989 *Wat. Air Soil Pollut.* 46, 251–259.

Wright, R. F., Cosby, B. J., Flaten, M. B. & Reuss, J. O. 1990 *Nature, Lond.* 343, 53–55

Lotse, E. 1989 *Acid Rain Res. Rep.* 18/1989. Oslo: Norwegian Institute for Water Research.
Reuss, J. D. 1989 *Acid Rain Res. Rep.* 19/1989. Oslo: Norwegian Institute for Water Research.
Frogner, T. 1990 *Geochim. cosmochim. Acta* 54, 769–780.

Discussion

I. ROSENQVIST (*Department of Geology, University of Oslo, Blindern 0316, Norway*).
1. We all heard Dr Henriksen say that in the Sogndal area in west Norway, RAIN had added H_2SO_4 and $H_2SO_4 + HNO_3$ in amounts bringing the precipitation acidity up to that of the affected areas in south Norway (i.e. about 40 μequiv. H^+ l^{-1}). Is it not true that Dr Wright and Dr Henriksen added *ca.* 100 μequiv. l^{-1} to the natural level of *ca.* 18 μequiv. l^{-1}, thus bringing the acidity up to about three-times the acidity typical for south Norway? Dr Henriksen said that this addition of strong acids had raised the run-off acidity considerably.

2. Does the addition of 100 μequiv. l^{-1} H^+ only raise the natural H^+ concentration by 2–4 μequiv. l^{-1}, 'mostly, because parts of the strong acids added, fell on the stream and not on the soils'?

3. In Risdalsheia, Dr Henriksen said that the present acid precipitation was replaced with de-ionized water mixed with seawater thus bringing the Cl^- concentration up to the concentration in the present precipitation. Does Dr Henriksen think that the pre-industrial rain had no excess sulphate?

4. Dr Henriksen said that after four years, this treatment had reduced the sulphate and acidity in the run-off. Is it true that four years of treatment (1984–87) had reduced the H^+ concentration from 78 to 77 μequiv. l^{-1} and that the de-acidified water, as well as the acid rain, gave higher acidity in the run-off than in the precipitation?

5. Does Dr Henriksen consider the way he presented the data conformed with scientific ethics?

R. F. WRIGHT. 1. At Sogndal we add 100 mequiv. m^{-2} a^{-1} of acid in about 60 mm of water. Ambient precipitation contains about 17 mequiv. m^{-2} a^{-1} of acid in about 1000 mm of water. Together this gives a total acid loading of 117 mequiv. m^2 a^{-1} in 1060 mm of water, similar to total (wet plus dry acid deposition) at Risdalsheia (118), Birkenes (136) and Storgama (97).

2. Volume-weighted mean concentrations of H^+ in runoff from the two untreated reference catchments at Sogndal averages 2 μequiv. l^{-1} over the five-year period 1984–89. Runoff from the sulphuric-acid treated catchment SOG2 averaged 8 μequiv. l^{-1} and from the sulphuric plus nitric acid treated catchment SOG4 5 μequiv. l^{-1} over this same five-year period. This increase is caused by both direct inputs of added acid on the stream (this also occurs in acidified areas of southern Norway), and by acid-water draining the soils; runoff is acidified for days to weeks following the addition of acid.

3. At Risdalsheia we remove the excess-SO_4, NO_3^-, NH_4^+, and H^+ from incoming wet precipitation. Dry deposition, of course, is only partially reduced by the roofs. Our best estimate is that we have reduced excess-SO_4 deposition (wet plus dry) at Risdalsheia from about 110 mequiv. m^2 a^{-1} to about 15 mequiv. m^2 a^{-1}. The experiment aims to investigate the changes in runoff chemistry as a result of this drastic change in deposition. Whether or not this level reflects pre-industrial deposition is not essential to the interpretation of the results from Risdalsheia.

4. The volume-weighted mean H^+ concentration in runoff from the three catchments at Risdalsheia for the five-year period June 1984–June 1989 are 82 µequiv. l^{-1} at ROLF (no roof, acid rain) (1985–89 only; incomplete data for 1984), 79 µequiv. l^{-1} at EGIL (roof, acid rain), and 73 µequiv. l^{-1} at KIM (roof, clean rain). Concentrations at KIM have decreased from 87 µequiv. l^{-1} in the first year of treatment to 61 µequiv. l^{-1} in 1988–89, the fifth year of treatment.

Wet precipitation at Risdalsheia contains about 60 µequiv. l^{-1}. Dry deposition contributes an additional 30 %. Furthermore, the input–output budgets for inorganic nitrogen components indicate a substantial acid input from NH_4^+ retention in the catchments. Total acid inputs at EGIL and ROLF exceed outputs. At KIM, outputs exceed the greatly reduced inputs because of several factors, incuding release of sulphur stored in the catchment and buffering by natural organic acids. Details are given in Wright *et al.* 1988 (*Nature, Lond.* **334**, 670–675) and Wright (1989) (WASP).

5. Yes.

Chemical effects on surface-, ground- and soil-water of adding acid and neutral sulphate to catchments in southwest Sweden

By Hans Hultberg[1], Ying-Hua Lee[1], Ulf Nyström[1]
and S. Ingvar Nilsson[2]

[1] Swedish Environmental Research Institute, P.O. Box 47086,
S-402 58 Gothenburg, Sweden
[2] Department of Soil Science, Swedish University of Agricultural Sciences,
P.O. Box 7072, S-750 07, Uppsala, Sweden

Experimental treatments that used elemental sulphur (acidic sulphate) and sodium sulphate (neutral sulphate) were carried out with a total of approximately 200 kg S ha^{-1}† in each of two catchments. The catchments were treated in November 1985 and in November 1986. The sodium sulphate treatment showed a rapid response in terms of a sharp increase in sulphate and sodium concentrations in the run-off water. The sulphate concentration then decreased so that 95% of the added sulphate had left the catchment within three years after the last treatment. About 32% of the added sodium was retained. The sodium retention, in combination with the increase in the sulphate flux, caused increased leaching of H$^+$, inorganic Al, Mg and Ca. Whereas leaching of magnesium continued to increase over the whole four-year experimental period, that of Mn and organic Al decreased during the same period. Fe decreased by 40–60% during the three hydrological years 1985–86, 1986–87 and 1988–89.

The treatment with acid sulphate caused pronounced chemical effects in the run-off water, starting in the summer of 1986 and increasing through 1987, 1988 and 1989. The output of Ca, Mg, Al and H$^+$ increased dramatically. The calcium and magnesium concentrations increased by 86% and 128%, respectively during 1987–88 and 1988–89. Sulphate equivalent to 35% of the added sulphur had left the catchment by the end of 1989. Both treatments increased the concentration of several cations in the soil-water and groundwater. In the acid sulphate treated catchment, Mg^{2+} was the cation most clearly affected, followed by Ca^{2+} and total Al. During the experimental period, the concentration of these three elements increased by 30–200%, depending on the location within the catchment. There was a decline in pH in the soil-water, particularly in the B horizon, but no pH change in the groundwater. The sodium sulphate caused a short-term acidification of soil-water and groundwater. The Mg^{2+} and Al concentrations increased by 50–200% immediately after the treatments, but both Mg^{2+} and Al^{3+} had returned to their original levels two years after the last treatment.

The effects on throughfall chemistry were generally small and restricted to the first year after the first application of sodium sulphate to catchment L1. An increased leaching of internally cycled K$^+$, Na$^+$ and Cl$^-$ taken up earlier by the roots was noted.

---

† 1 ha = 10^4 m^2.

1. Introduction

Chemical mass-balance studies began in 1979 in small terrestrial catchments, with the aim of quantifying causes and effects of acidification of Lake Gårdsjön (Hultberg 1985). The fluxes of cations and anions through the catchments were also the basis for assessing the importance of soil acidification for the lake acidification (Nilsson 1985). Budgets of sulphur input and output in coniferous forested catchments in the Lake Gårdsjön area during the period 1979–89 showed that there is little or no retention of sulphur in the soils. Estimated yearly deposition of sulphur, measured throughfall deposition and run-off via streamwater were almost identical at about 26 kg ha^{-1}. The N-deposition (NH_4^+ and NO_3^-) of 10–20 kg N ha^{-1} a^{-1} is almost totally retained in the soil system and only at snow-melt and during autumn/winter are the NO_3^--concentrations elevated in surface waters.

The hypothesis was that there is a fundamental difference between a loading of sulphuric acid and one of a neutral salt such as Na_2SO_4. In the latter case, no further soil acidification takes place. The outcome of a salt input episode is wholly dependent on the previous development of acidification in the soil, and does not cause any long-term change in soil and water chemistry. Its effects are of an episodic and reversible nature. An increased loading of acid sulphate would, on the other hand, cause a long-term change in soil chemistry as well as in water chemistry.

This study was designed to test the anticipated differences between loading of acid sulphate and neutral sulphate by applying elemental sulphur (which is eventually converted to H_2SO_4) and Na_2SO_4, respectively, to two forested catchments in the Gårdsjön area.

2. Description of the study sites and hydrological régime

Three catchments were chosen for the Surface Waters Acidification Programme (SWAP) study. F1 (3.7 ha) is the reference catchment of the Lake Gårdsjön Project; L1 (2.5 ha) was chosen for the Na_2SO_4 treatment and F5 (3.1 ha) for treatment with elemental S (figure 1).

F1 and F5 are situated at the southern end, and L1 east of Lake Gårdsjön, on the Swedish west coast. F1 and F5 drain into Lake Gårdsjön and L1 drains into the nearby Lake Lysevatten. The physiographical, soil and biological characteristics are described by Olsson et al. (1985). The bedrock consists of gneissic granodiorite in catchments F1 and F5; L1 is underlain by greyish red granite. All three catchments are situated between 115 and 140 m a.s.l. and no marine deposits are found in them. The thin till cover is characteristic of the region. The soil type is predominantly iron-humus podzol with a thin cover of till. All three catchments are dominated by Norway spruce with some Scots pine, birch and mountain ash, the forests being 75–90 years old.

Each catchment is drained by a rill surrounded by *Sphagnum* and *Polythricum* on peat. The rill flow is low during the summer, but peaks during high autumn rainfall and spring snowmelt. The run-off during the ten-year period 1979–80 was 609 ± 128 mm a^{-1}. The average precipitation during the same period was 1200 mm.

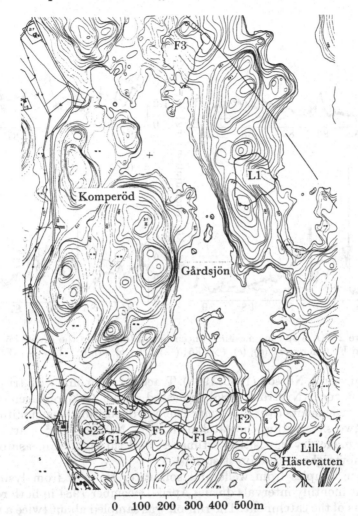

Figure 1. Map of the Lake Gårdsjön area showing the locations of the catchments.

Table 1. *Application rate of sulphur and sodium by elemental S and Na$_2$SO$_4$ treatments in catchments F5 and L1, respectively*

hydrological year	application rate/(kg ha^{-1} a^{-1})		
	sulphur (F5), elemental S	sulphur (L1), Na$_2$SO$_4$	sodium (L1), Na$_2$SO$_4$
1985–86	112.0	90.1	129.5
1986–87	112.0	108.2	155.5
	—	—	—
total	224.0	198.3	285.0

3. Methods

(a) Experimental section

Water discharges were monitored by weirs equipped with water-level recorders. A detailed presentation of the climate and hydrology over the period 1951–1985 is

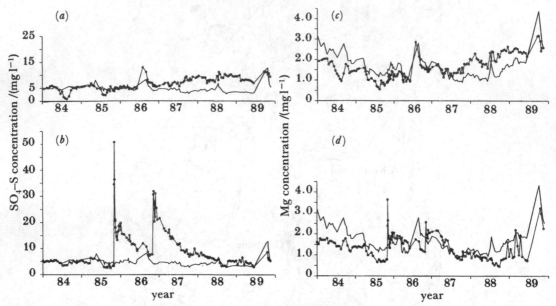

Figure 2. SO_4-S and magnesium concentration (mg l^{-1}) in the runoff F5, L1 and F1 (reference). (*a*), (*c*), (○), F5; (−), F1; (*b*), (*d*), (○), L1; (−), F1.

given by Johansson & Nilsson (1985, 1988). Tracer experiments with tritium injected under the root zone (Nyström 1985) indicated about 40% replacement of water stored in two catchments over one year. The wet deposition was monitored by bulk collectors at two sites within the catchment. SO_2, NO_2 and SO_4^{2-}-particulates were monitored to make dry deposition estimates, and throughfall measurements were made once a fortnight in the forest stands.

Groundwater samples from wells and soil-water samples from lysimeters were taken at about monthly intervals during April–December 1984 in both recharge and discharge areas of the catchment. The run-off was sampled about twice a week for the first six weeks after the treatments and thereafter either weekly or once every two weeks.

Analysis of base cations, iron and manganese were performed by atomic absorption spectrophotometry, whereas sulphate and chloride were analysed by ion chromatography. Total acid reactive aluminium was analysed by using atomic absorption spectroscopy (AAS) with a graphite furnace. The organic Al was determined by the method of Driscoll (1984). Dissolved organic carbon (DOC), phosphorus and nitrogen were analysed at the University of Uppsala (Persson & Broberg 1985).

Catchments L1 and F5 were treated with Na_2SO_4 and elemental sulphur, respectively on 29 and 30 October 1985. A second application was made during early November 1986 in each catchment.

The quantities of chemical, which were evenly distributed over the catchments, are shown in table 1.

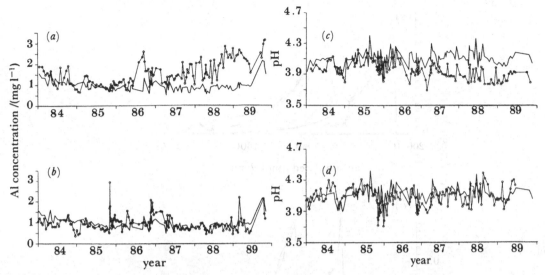

Figure 3. Aluminium (total) concentration (mg l⁻¹) and pH (field) in the runoff F5, L1 and F1 (reference). Symbols as in figure 2.

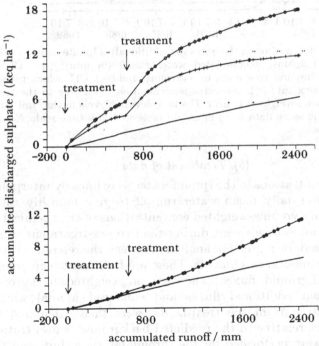

Figure 4. The accumulated discharged sulphate (in keq ha⁻¹) from F5, L1 and F1, and the computed extra amount from L1; (·) and (··) denote the accumulated amount of sulphate added at the first and second treatment, respectively. The monthly totals are marked; 0 mm corresponds to 30 October 1985 (the first treatment occasion) and 613 mm to 30 October 1986 (the second treatment). The F5-curve breaks at 551 mm (31 September 1986) and at 1218 mm (31 August 1987); (○), SO_4-S acc. L1 total (keq ha⁻¹); (+), SO_4-S acc. L1 extra (keq ha⁻¹); (−), SO_4-S acc. F1 (keq ha⁻¹); (●), SO_4-S acc. F5 (keq ha⁻¹); (−), SO_4-S acc. F1 (keq ha⁻¹).

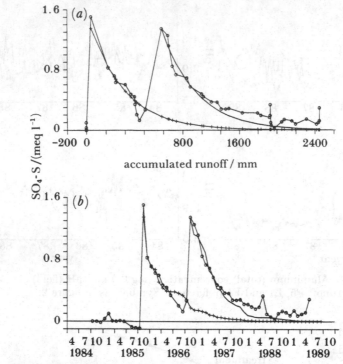

Figure 5. (*a*) The computed decline in the extra (or additional) sulphate concentration due to the treatments (in meq l⁻¹) against accumulated water runoff (in mm); (+), CI, which is the calculation taking only the first treatment into account, and (−), CII, which takes the first and second treatment into account; (○), the extra concentration in L1 due to the treatments. The values are monthly volume-weighted (wmv). The two elevated levels in the 'tail' are dry months with little runoff. (*b*) The same data as in figure (*a*) presented in a time scale. See Figure (*a*) for an explanation of symbols.

(*b*) *Treatment of data*

The chemical concentrations in the run-off water were linearly interpolated in time and multiplied by the daily mean water run-off to give monthly totals of mass transport and monthly volume-weighted concentrations of the chemical substances. The monthly fluxes for each element during the two pre-treatment years, 1983–84 and 1984–85, were used for regression analyses between the reference catchment and each of the two experimental catchments. These pre-treatment regressions were used to predict the 'background fluxes' and 'annual weighted background concentrations'. The annual additional fluxes and enhanced annual weighted mean concentrations resulting from the treatments, were then calculated to give the percentage of changes relative to the predicted background concentrations. Similar regression analyses were performed for the throughfall data, but on a three-month basis.

The *F*-factor:

$$F = (\Delta Ca^* + \Delta Mg^*)/\Delta SO_4^*,$$

where an asterisk superscript denotes the non-marine part (Henriksen 1984), was calculated for soil-, ground- and surface-waters in each of the experimental catchments.

4. Results and discussion

(a) *Effects on run-off water*

(i) *Response of cation and anion concentrations in catchment L1*

Figure 2 shows that the SO_4-S concentration in the two catchments F1 and L1 are almost identical over the years 1984 and 1985. After the Na_2SO_4 treatment in L1, SO_4-S increased immediately from 3.4 to 50.8 mg l^{-1}. The SO_4^{2-} concentration decreased rapidly during the first week and stabilized just below 20 mg l^{-1} until the snow-melt of 1986. After this, a progressive decline occurred until the new application in November 1986, after which a new sharp increase occurred. The Na-concentration before treatment was generally lower in L1 than in F1. The same was true for other sea-salt derived elements, such as K^+, Mg^{2+}, and Cl^-. After the Na_2SO_4 application, changes in Na^+ were almost identical to those in the SO_4^{2-} concentration. Ca, Mg, and K increased in L1 during the first week. Ca and K then declined to pre-treatment concentrations whereas Mg stayed above pre-treatment concentrations over a longer time (figure 2). A decrease was observed in Fe, but the total Al (figure 3) increased in the short-term. No effect was observed in organic Al. The high concentrations in total Al coincided with a decrease in pH from 4.0 to 3.7. The pH was lower than predicted from the reference catchment F1 until January 1986 (figure 3). The Cl-concentration was not affected by the treatment. The second application of Na_2SO_4 in November 1986 resulted in short-term chemical effects very similar to those for the first treatment.

(ii) *Conservative behaviour of sulphate: application of a simple dilution model for L1 run-off*

The background level (keq S ha^{-1}) of sulphate in L1 was computed from:

$$output (L1) = 1.017 \times output (F1) - 0.003,$$

for the pre-treatment period October 1983–September 1985. The conservative behaviour of sulphate is obvious, as the accumulated sulphate output versus accumulated water discharge in F1 is very closely represented by a straight line (see figure 4).

This was also shown earlier by the close 1:1 balance of atmospheric deposited and discharged sulphate at several long-time monitored catchments (Hultberg 1985; Hultberg & Grennfelt 1986). In figure 4, the accumulated discharged sulphate in L1 is shown. The upper curve represents the total amounts in keq SO_4 ha^{-1}. The middle curve shows the accumulated extra sulphate output during the period caused by the treatments. The results of earlier tracer experiments using [3]H (Nyström 1985) and ^{18}O, together with assumptions on the porosity and soil depth, the storage of SO_4 in L1 was estimated to be 300–400 mm. By using a simple one-tank, one-way model with storage 360 mm, the decline in concentration (CI and CII) with respect to accumulated run-off (RA) was calculated as:

$$CI = 5.62/3.60 \times exp(-RA/360) \text{ meq } l^{-1},$$

$$CII = CI + 5.62/3.60 \times exp(-RA/360) + 6.75/3.60 \times exp(-(RA-613)/360) \text{ meq } l^{-1},$$

where CI is valid for the first treatment only and CII for both treatments.

In figure 5b the calculated curves CI and CII are shown in comparison with the monthly volume-weighted extra concentrations from the experiment.

Figure 5b shows the same data as in figure 5a plotted against time. However, the transit times will probably be underestimated as a part of the discharged sulphate ions originate from the continuous input of sulphate. The more lengthy 'tail' of the experimental curve, compared with those calculated from the simple one-tank model, is expected. Thus it is proper to treat the transport of sulphate in the soil as a function of accumulated water run-off rather than time. This illustrates the conservative behaviour of sulphate in soils. The same analysis was done for sodium; this showed a much lower discharge rate than did sulphate, probably because of ion-exchange in the soils.

(iii) *Response in cation and anion concentrations in catchment F5*

During the first months after the treatment, no significant change occurred in run-off. Figure 2, however, shows that a sharp increase in the sulphate concentration occurred nine months after the first treatment. The SO_4-S concentration increased from 4.5 to 13.1 mg S l^{-1}. Oxidation of the applied sulphur had occurred during the summer months with production of sulphuric acid. The acidity increased in run-off and pH decreased from 4.0–4.1 to 3.8–3.9 during the autumn (figure 3). The total Al-concentration increased simultaneously with the decrease in pH (figure 3). An increased leaching of base cations, such as Ca^{2+}, Mg^{2+}, K^+, and metals such as Fe^{3+} and Mn^{2+}, occurred during the sulphate peak in 1986 (figure 2).

The sulphate concentration in the run-off from catchment F5 increased continuously following the initial peak in 1986. The rate of oxidation of the applied sulphur has increased year by year (figure 4).

During 1989, the pH was less than 3.8 in the run-off, and pH values between 3.5 and 3.7 occurred frequently (figure 3). Hence, the H^+ concentration was generally 1.5–2 times higher than in the reference catchment, F1. A similar effect was observed for the inorganic Al, which had increased by about 100% and resulted in a total Al concentration generally around 2.0–2.5 mg l^{-1} during 1988 and 1989 (figure 3). The Ca^{2+} concentration was 2 times higher and Mg^{2+} 2–3 times higher than in the reference catchment. Also, K^+ and Mn^{2+} were 1.5–2.0 times their pre-treatment concentrations during 1989, whereas iron increased by only 10–30%. The concentration of organic Al was reduced during 1987 and 1988 and the sodium and chloride concentrations were unaffected by the elemental sulphur treatments.

(iv) *Effects on cation and anion fluxes*

The regression analyses of the pre-treatment data that were used to predict the background fluxes in F5 and L1 comprised 20–24 pairs of monthly data. The correlation coefficients were generally greater than 0.95 for all elements except for Fe in L1, where it was 0.87.

Figure 6 shows the accumulated monthly fluxes of sulphur and sodium for all the years studied, starting in October 1983. Sulphur and sodium in catchment L1 responded immediately following both treatments. Because of oxidation of the elemental sulphur added in catchment F5, the sulphate leaching increased slowly during 1986. This oxidation was presumably mediated by soil microbes rather than chemical oxidation (Fenchel & Blackburn 1979). An increase of the sulphur leaching occurred during 1987. This probably resulted from an increased oxidation of the added sulphur (see also figure 4). Favourable conditions during these wet and warm years should have enhanced overall microbial activity in the uppermost soil layers where the sulphur was located. The low, initial rate of oxidation and leaching (or both) may be explained either as a lag-phase in the build-up of the microbial

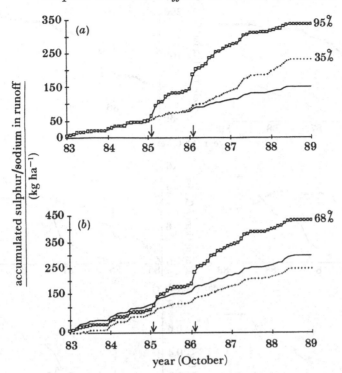

Figure 6. The accumulated monthly fluxes (kg ha^{-1}) for (*a*) sulphur and (*b*) sodium in runoff F5, L1 and F1; (□), sodium sulphate; (–·–), elemental sulphur; (——), reference; arrows indicate applications.

population, or may have been due to sulphur retention. Hence the retention capacity may have been saturated during 1987, resulting in an increased leaching rate of sulphur. However, the sulphur output from neutral sulphate treatment showed no tendency for sulphur retention. On the contrary, as shown earlier by comparing the sulphur decline in L1 with a computation that used a simple dilution model, it was concluded that sulphur behaves very conservatively within the catchments. This means that sulphur retention is of little importance in explaining the input/output fluxes in these sulphur-saturated catchments in southwest Sweden.

Figure 7 *a–e* shows the experimentally induced extra output of sulphate and cations. The high output of neutral sulphur (95% in October 1989) had less effect on Mn, Fe, Al, H$^+$ and K than the lower output (35%) from the elemental sulphur treatment. The loss of Ca and Mg was, however, high after the applications of sodium sulphate, which also produced short-term exchange losses of other cations such as H$^+$ and Al^{3+}. There were three different patterns of effects on the additional output/retention of cations resulting from the neutral sulphate treatment two years after the second treatment: K, Mg, and Al showed little change; Ca, Mn and Fe showed increased retention in soil, but there was a net loss of H$^+$. The effects on Na, K, H, Ca, Mg and Al may be partly explained by the competitive and reversible processes of exchange with sodium ions on soil-exchanged sites and in solution.

The leaching of large amounts of Ca^{2+} and Mg^{2+} following the application of sodium sulphate revealed the presence of a greater pool of Ca^{2+} and Mg^{2+} in the soil of L1 compared with that in F5. The retention of Ca^{2+} that occurred after autumn

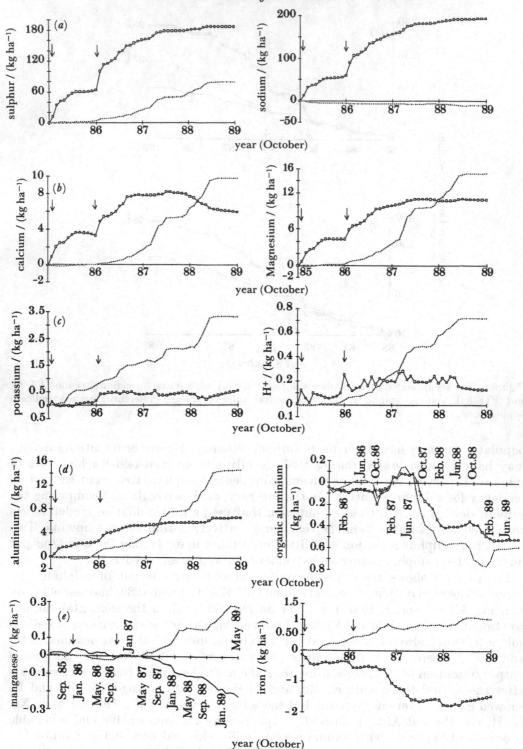

Figure 7. For legend see opposite page.

1988 may have been caused by the increased fraction of Na^+ on the exchange complex and reversal of the exchange reaction described above. A similar but much weaker trend was also observed for H^+.

Figure 8 shows the weighted yearly averages of sulphur, calcium and magnesium concentrations based on observed yearly fluxes in run-off and the extra change in the concentration, as well as the percentage of the observed concentration that is caused by the treatments in F5 and L1. The extra concentration changes, ΔSO_4^*, ΔCa^* and ΔMg^*, were used to calculate the F-factor (table 2). The F-factor ranged from 0.14 to -0.17 in L1 and from 0.11 to 0.4 in F5. Table 2 shows that the increase in acid SO_4 in F5 was neutralized by 40% and 38% during 1987–88 and 1988–89, respectively.

(b) Effects on soil- and groundwater

In the following, average values of soil-water and groundwater data from different lysimeters and wells were used in the evaluation of the effects on Ca, Mg, Na, total Al, Cl and sulphate. Because of the shortage of lysimeter water on some sampling occasions, calculations of the average concentrations in soil-water include different numbers of lysimeter readings on different dates. Evaluation of the sodium sulphate treatment was limited to the period after the second treatment, as very few lysimeter data were available before that period. Therefore, no calibration between the three catchments could be made. Consequently data analysis is restricted to significant time trends which were consistent with the respective treatments.

Figure 9a, b shows the seasonal variation of the average concentration in soil-water and groundwater in catchment F5 (acid sulphate treatment) and in the reference catchment F1. In the treated catchment there was a progressive increase in sulphate, Ca, Mg and total Al. The three cations increased by 30–200% depending on the locations of the lysimeters and groundwater wells within the catchment. Chloride and sodium were not affected. Potassium increased slightly during the last experimental period. The sulphate concentration was practically the same in the water percolating through the B- and C-horizons down to the groundwater, indicating a negligible sulphate retention. This is in accordance with previous budget studies (Hultberg 1985).

The effect of acid sulphate treatment generally began to appear in soil-water after the summer in 1986, and was similar to that on run-off water. For groundwater, the response varied depending on the location of the groundwater well. Figure 9a shows that the increase in Al concentration was particularly great during the autumn of 1986 and the spring of 1988. This can be explained as follows: the organic fraction of Al was probably the by-product of decomposition processes in the organic soil horizon, and increased during summer and autumn (Driscoll 1985). The warm summer of 1986 caused high concentrations of the organic fraction, but also of the inorganic fraction, because of increased oxidation of the applied elemental sulphur that released H_2SO_4 and subsequently inorganic Al by dissolution. Hence a high concentration of total Al was observed in soil-water during the winter of 1986–87. During spring 1988 (snow-melt period), the high Al-concentration was mainly caused by the high inputs of the acid sulphate produced by the oxidation of applied

Figure 7. The accumulated monthly extra fluxes (kg ha^{-1}) of sulphur, sodium, calcium, magnesium, potassium, hydrogen ion, aluminium, organic aluminium, manganese and iron in runoff F5, L1 and F1; ($\square$), sodium sulphate; ($-\cdot-$), elemental sulphur; arrows indicate applications.

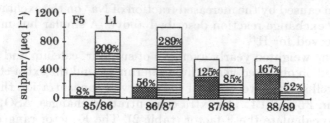

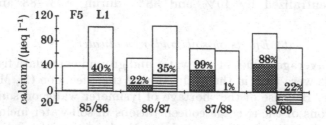

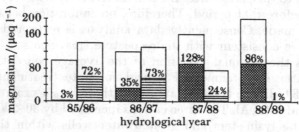

Figure 8. The weighted yearly average concentration (µeq l⁻¹) of sulphur, calcium and magnesium in runoff from F5 and L1, as well as the extra change (µeq per percentage) caused by the treatments.

Table 2. *The Henriksen F-factor from chemistry of runoff*

hydrological year	Na₂SO₄	elemental S
1985–86	0.14	0.11
1986–87	0.11	0.23
1987–88	0.07	0.4
1988–89	−0.17	0.38

elemental sulphur during the mild and wet winter of 1987. The high inputs of H⁺ were only partly neutralized by the available Ca and Mg in the soil, and consequently resulted in a dissolution of a great amount of inorganic Al in the B-horizon.

The increase in sulphate concentration (Table 3) was roughly 0.07–0.14 meq l⁻¹ a⁻¹. The F factor was about 0.2–0.5, in agreement with that for the run-off (table 2). The lower value of the F factor in the B horizon may be due to a lower pool of total calcium compared with the C horizon. Such a trend was also found by Olsson *et al.* (1985) in the nearby control catchment, F1. The increase in aluminium leaching from the B horizon may be caused by several factors. There is a large pool of exchangeable aluminium in the podzolic soils of the Gårdsjö area (Olsson *et al.* 1985; S. I. Nilsson,

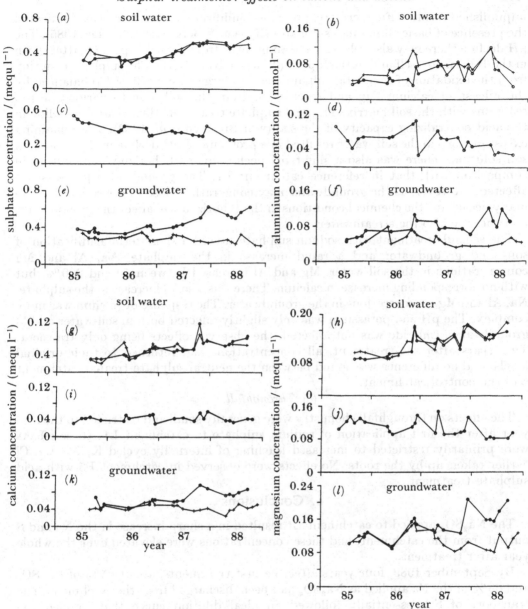

Figure 9. Seasonal variation of the average concentration of sulphate, aluminium, calcium and magnesium in soilwater (B- and C-horizon) and groundwater for catchments F5 and F1; (*a, b, g, h*), (○), F5B; (+), F5C; (*c, d, i, j*), (○), F1C; (*e, f, k, l*), (○), F5; (+), F1.

Table 3 *Mean increase rate ± one standard deviation, given as millimole per litre per year (total Al)*

	SO$_4^{2-}$	Ca^{2+}	Mg^{2+}	total Al
B-horizon	0.118 ± 0.014	0.006 ± 0.004	0.019 ± 0.007	0.018 ± 0.006
C-horizon	0.088 ± 0.013	0.018 ± 0.004	0.026 ± 0.006	0.009 ± 0.005
groundwater	0.108 ± 0.018	0.016 ± 0.004	0.025 ± 0.009	

unpublished results). Furthermore, previous equilibrium calculations have indicated the presence of basic aluminium sulphates (Nilsson & Bergkvist 1983; Lee 1985). The pH decline there may also release some aluminium from the soil organic matter, both in the B horizon itself and in the topsoil (James & Riha 1984). The sulphate resulting from the oxidation of elemental sulphur was to a large extent (60–80%) balanced by the release of calcium, Mg and reactive Al from the soil, due to various proton reactions with the soil matrix. The acid sulphate treatment thus caused a decline in the acid neutralizing capacity of the soil (van Breemen *et al.* 1983). The remaining excess protons in the soil-water resulted in a pronounced pH decline in the B-horizon soil solution. There was also a slight pH decline in the C horizon soil solution in comparison with that in reference catchment F1. The groundwater pH was not affected. However, as the groundwater may come rather close to the soil surface on many occasions, the chemical conditions in the B horizon will affect the groundwater and thereby also the streamwater.

The second treatment with sodium sulphate caused a short-term acidification of soil- and groundwater and a rapid increase in the sulphate, Na, Al and Mg concentrations in the soil-water. Mg and Al increased between 50 and 200%, but with no corresponding increase in calcium. There was also an increase in the sulphate, Na, Al and Mg concentrations in the groundwater. The response of calcium was more complex. The pH and potassium was only slightly affected both in soil-water and in groundwater. Chloride was not affected, the observed effects being only transient. Two years after the treatment, all concentrations had returned to their original levels, and no difference was found between the neutral sulphate-treated catchment and the control catchment.

(c) *Throughfall*

The effects on throughfall chemistry were generally small and restricted to the first year after the first application of sodium sulphate to catchment L1; these effects were primarily restricted to increased leaching of internally cycled K, Na and Cl earlier taken up by the roots. No effects were observed in catchment F5 with acid sulphate treatment.

5. Conclusions

The Na_2SO_4 applied to catchment L1 resulted in a sharp increase in the Na and S output from the catchment, and these concentrations were elevated over the whole year after treatment.

By September 1989, four years after the first treatments, about 95% of the SO_4 and 68% of the Na applied as Na_2SO_4 had been discharged from the catchment. The transport of SO_4 essentially followed an ideal dilution curve that implied the conservative behaviour of sulphate in the catchment. The average concentration of Ca in run-off increased during 1985–86 and 1986–87 by 40% and 35%, respectively, and Mg increased by 72% over this two-year period. However, during 1988–89 there was no further decrease of Mg and the concentrations of Ca and H^+ actually fell to levels some 25% lower than those before the treatment. Total Al increased by about 50% following each of the two applications of Na_2SO_4, whereas Fe, Mn and organic Al decreased. The retention of some 30% of the Na in the catchment resulted in elevated concentrations of Na in soil water and run-off in 1987–89 and may have caused reductions of other cations such as Ca, Mn and H^+ in the soil.

The total transport of S from the F5 catchment treated with elemental sulphur increased progressively over the four years from 1% in the first year, 1985–86, to

35 % in 1988–89. The added acid sulphate produced major effects on ground-, soil- and surface waters. The concentrations of Ca and Mg in run-off doubled, and increased by 20–40 % for K, 30–40 % for H^+, 20 % for Fe, 40 % for Mn and 200 % for total Al. Na and Cl were unaffected, whereas organic Al decreased.

Both acid and neutral sulphate treatments increased the concentration of several cations in the soil- and groundwater. In the acid-sulphate-treated catchment, Mg was the cation most clearly affected, followed by Ca and total Al. During the studied period, the concentration of these three elements increased by 30–200 %, depending on the location within the catchment of the individual lysimeters and groundwater wells. There was a decline in pH in the soil-water, particularly in the B-horizon, but no pH change in the groundwater was measured. The sodium sulphate caused a short-term acidification of soil- and groundwater. The Mg and Al concentrations increased by 50 %, to 200 %, immediately after the treatments, but both Mg and Al had returned to their original levels two years after the last treatment.

The effects on throughfall chemistry were generally small and restricted to the first year after the first application of sodium sulphate to catchment L1.

In ecosystems such as the Lake Gårdsjön catchment with shallow podzolic soils on granitic bedrock, acid deposition increases the loading of proton and strong acid anions (mainly sulphate) on the catchment. This causes an increased leaching of acidic cations (hydrogen ions and aluminium). In the short-term, a neutral sulphate or any other neutral salt input containing a mobile anion would to some extent have the same effect as the acid deposition in terms of acidic cation leaching and an increase of the ionic strength. The salt cations (e.g. Na^+ or Mg^{2+}) will be partly retained in the soil and an increased amount of hydrogen ions and aluminium will be leached out of the soil (exchange acidity).

Acid sulphate input will cause a continuous decline of the acid neutralizing capacity (ANC) of the soil and runoff (Lee & Hultberg, this symposium). The decline could be quite substantial both in terms of the total amount of metal cations lost, and in terms of the depletion of exchangeable basic cations (Ca, Mg, K, Na). These losses will eventually cause a fall in soil pH, an increase of the average aluminium concentration in the soil-water, and consequently increased leaching of acidic cations.

Ingemar Alenäs organized the treatments and installed weirs, groundwater wells and lysimeters together with Ingvar Andersson. Stefan Larsson and Per Leyton have also performed most of the field sampling. The chemical analyses were done by Inger Torbrink and Pia Carlsson. Richard Gould gave linguistic advice and Eva Knudsen assisted in the preparation of the manuscript. Professor Carl Olof Tamm has supported this study through all the years and has made valuable suggestions to improve this paper. Funding for this project was provided from the Scandinavian Waters Acidification Programme (SWAP) through the Royal Swedish Academy of Sciences for the years 1984–85 to 1987–88. The National Swedish Environment Protection Agency provided funding for 1988–89 and 1989–90.

References

Driscoll, C. T. 1985 *Environ. hlth Perspect.* **63**, 93–104.

Driscoll, C. T. 1984 *Int. J. environ. analyt. Chem.* **16**, 267–284.

Fenchel, T. & Blackburn, T. H. 1979 *Bacteria and mineral cycling.* London: AP Press.

Grennfelt, P., Larsson, S., Leyton, P. & Olsson, B. 1985 In *Lake Gårdsjön – an acid forest lake and its catchment* (ed. F. Andersson and B. Olsson), *Ecol. Bull.* no. 37, pp. 101–108. Stockholm: Publishing House of The Swedish Research Councils.

Henriksen, A. 1984 *Verh. Internat. Verein. Limnol.* **22**, 692–698.

Hultberg, H. 1985 In Lake Gårdsjön – an acid forest lake and its catchment (ed. F. Andersson and B. Olsson), Ecol. Bull. no. 37, pp. 133–157. Stockholm: Publishing House of The Swedish Research Councils.

Hultberg, H., Grennfelt, P. & Olsson, B. 1983 Wat. Sci. Tech. 15, 81–103.

Hultberg, H. & Grennfelt, P. 1986 Wat. Air Soil Pollut. 39, 31–46.

James, B. R. & Riha, S. J. 1984 Can. J. Soil Sci. 64, 637–646.

Johansson, S. & Nilsson, T. 1985 In Lake Gårdsjön – an acid forest lake and its catchment (ed. F. Andersson and B. Olsson), Ecol. Bull. no. 37, pp. 86–96. Stockholm: Publishing House of The Swedish Research Councils.

Johansson, S. & Nilsson, T. 1988 In Liming of Lake Gårdsjön – an acidified lake in southwest Sweden (ed. W. Dickson), pp. 9–36. Stockholm: National Swedish Environmental Protection Board, report no. 3426.

Lee, Y. H. In Lake acidification – an acid forest lake and its catchment (ed. F. Andersson and B. Olsson), Ecol. Bull. no. 37, pp. 100–119. Stockholm: Publishing House of The Swedish Research Councils.

Melin, R. 1970 Svenska Utbildningsförlaget Liber, Stockholm, 207 pp. (In Swedish.)

Nilsson, I. S. 1985 In Lake Gårdsjön – an acid forest lake and its catchment (ed. F. Andersson and B. Olsson), Ecol. Bull. no. 37, pp. 311–318. Stockholm: Publishing House of The Swedish Research Councils.

Nilsson, S. I. & Bergkvist, B. 1983 Wat. Air Soil Pollut. 20, 311–329.

Nyström, U. 1985 In Lake Gårdsjön – an acid forest lake and its catchment (ed. F. Andersson and B. Olsson), Ecol. Bull. no. 37, pp. 97–100. Stockholm: Publishing House of The Swedish Research Councils.

Olsson, B., Hallbäcken, L., Johansson, S., Melkerud, P. A., Nilsson, S. I. & Nilsson, T. 1985 In Lake Gårdsjön – an acid forest lake and its catchment (ed. F. Andersson and B. Olsson), Ecol. Bull. no. 37, pp. 10–28. Stockholm: Publishing House of The Swedish Research Councils.

Persson, G. & Broberg, O. 1985 In Lake Gårdsjön – an acid forest lake and its catchment (ed. F. Andersson and B. Olsson), Ecol. Bull. no. 37, pp. 158–175. Stockholm: Publishing House of The Swedish Research Councils.

van Breemen, N., Mulder, J. & Driscoll, C. T. 1983 Pl. Soil 75, 283–308.

Discussion

I. G. LITTLEWOOD (Institute of Hydrology, Wallingford, Oxfordshire, U.K.). Dr Hultberg (and some previous speakers) have analysed mass fluxes to infer processes. However, good estimates of physical and chemical loads transported by streams can be difficult to obtain, especially in small upland catchments where flow changes rapidly and there is a strong relationship between flow and the concentration of the determinand in question. Uncertainty in load estimates increases as sampling interval increases but several other factors are also important. For example, consider two determinands, A and B: concentration A varies inversely with flow (concentration decreases as flow increases) whereas concentration B varies directly with flow as a 'mirror-image' of A. Recent work confirms the findings of previous investigators that for a given period of record, and at any (regular) sampling interval, the uncertainty in the load estimate for B will be greater than for A. Furthermore, some estimation procedures perform better than others.

Some of Dr Hultberg's conclusions depend largely on meaningful comparisons of cumulative fluxes (e.g. detection of significant differences), or on budget calculations. Could Dr Hultberg expand on the sampling frequencies, method of cumulative flux calculations and the uncertainties involved in Gårdsjön work?

[No reply supplied.]

Studies of strong and weak acid speciation in fresh waters in experimental forested catchments

By Ying-Hua Lee and Hans Hultberg

Swedish Environmental Research Institute, P.O. Box 47086,
S-402 58 Gothenburg, Sweden

Strong and weak acids (natural organic acids, aluminium hydroxide complexes and carbonic acid) were specifically quantified in fresh waters of two experimental forested catchments in the Gårdsjön area over three to four years. The aim of this work has been to study the effects of both strong and weak acids on acid neutralizing capacity (ANC), and to evaluate the role of natural organic acids in the acidification of stream water.

Acid sulphate (elemental S) treatment resulted in a long-term acidification in the stream water whereas the neutral sulphate (Na_2SO_4) resulted in an episodic acidification. During the experimental period, the $-ANC$ (strong acidity) level increased by ca. 50% in the stream water of the acid-sulphate-treated catchment and is attributed to the increase of strong acid. The natural organic-acid anions had a very small effect on ANC change due to their interaction with strong acid in the very acidic stream water (pH < 4.2).

No appreciable changes were observed in the concentration of natural organic acids and DOC in streamwaters from the two treatments. The moderately weak organic acidic content of DOC in streamwater was 7.9 ± 2.3 mequiv g^{-1} C ($n = 20$) during autumn and 6.3 ± 1.8 mequiv g^{-1} C ($n = 10$) during summer. The concentration ratio of organic anion to DOC varied from 1 to 4 within the pH range 3.8–4.3.

The acidic contribution from the dissociation of the moderately weak organic acidic sites and organic Al complexes is discussed for these streamwaters.

No appreciable effect on ANC in groundwater was observed in the treated catchments.

---

1. Introduction

Two forested catchments (F5 and L1) within the Lake Gårdsjön area on the west coast of Sweden were chosen for field experiments to study the input effects of acid and sodium sulphate. The aim of this paper has been to study the effects on changes in acid neutralizing capacity (ANC), weak acids (natural organic acids, carbonic acid and aluminium hydroxide complexes), and to evaluate the role of natural organic acids in the acidification of stream water. Quantification of the changes in pH, base cations and anions, and total aluminium in response to acidic and neutral deposition in stream-, soil- and groundwater are reported by Hultberg et al. (this symposium). The effects on phosphorous, nitrogen and DOC are reported by Broberg (this symposium).

In October 1985 and November 1986, catchment F5 was treated with elemental sulphur, and sodium sulphate was added to catchment L1. The elemental sulphur added to the catchment was slowly oxidized and produced H_2SO_4. The untreated catchment F1 was used as a control in the evaluation of the chemical response.

The evaluation of natural organic acids, acid–base properties, their complex formation with aluminium and their contribution to acidification of streamwater in the catchments F1, F5, and L1 has been reported elsewhere (Lee 1990).

2. Concepts

(a) *Acid neutralizing capacity*

The acid neutralizing capacity (ANC), or alkalinity, is an important parameter when evaluating acidification processes in aqueous systems and quantifying the acid–base status of surface waters. Increasing strong acid inputs result in a decrease of ANC. A seriously acidified surface water will have negative values of ANC.

Acid neutralizing capacity is operationally defined as the equivalent sum of the bases that are titratable with the strong acid to an equivalent point. In an aqueous bicarbonate system, ANC can be determined from the titration with strong acid to a proton reference pH ($= 4.5$) as the end point, and conceptually, ANC measures the equivalence of the proton acceptors minus the free H^+ ion concentration (Stumm & Morgan 1981):

$$\text{ANC (ref. 4.5)} = [HCO_3^-] + 2[CO_3^{2-}] + [OH^-] - [H^+], \tag{1}$$

where square brackets denote the concentration of species.

For such systems, the Gran procedure (Gran 1952) has been used for the determination of ANC. In dilute acidified-water systems, weak acid anions other than HCO_3^-, such as natural organic acid anions and aluminium-hydroxide complexes ($AlOH^{2+}$, $Al(OH)_2^+$), are also proton accepters and contribute to the ANC.

For such systems, the end point of the strong-acid titration will be obscure and the conventional titration method will under-estimate the ANC. For estimating the ANC in surface water containing organic acids, the approach used in the literature is based on a charge balance. The ANC was calculated as the equivalence differences between the sum of non-hydrogen cations and the sum of strong acid anions, which can be expressed as follows:

$$\text{ANC} = 2[Ca^{2+}] + 2[Mg^{2+}] + [K^+] + [Na^+]$$
$$+ 2[Mn^{2+}] + [NH_4^+] + n[Al]^{n+} - 2[SO_4^{2-}] - [Cl^-] - [NO_3^-], \tag{2}$$

where n ($= 3$ or 0) is the charge of the aluminium ion.

The accumulated uncertainty associated with the determination of a larger number of ionic concentrations is quite large in the estimation of a low value of ANC.

As ANC measures the net deficiency of protons relative to a reference proton level, it can also be estimated by quantifying the organic weak acidity and the aluminium hydroxide complexes, and calculated using the following equation:

$$\text{ANC (ref. pH} = 3.5) = [HCO_3^-] + [AlOH^{2+}] + 2[Al(OH)_2^-] + [COO^-] + [OH^-] - [H^+], \tag{3}$$

where [COO^-] denotes the dissociated moderately weak acid sites in natural organic acids.

(b) Organic weak acids

The natural organic acids are a complex mixture of polyelectrolytes with several acidic functional groups. Recently, it has been suggested that the organic acids may exhibit a limited number of sites, and the most dominant acidic sites can be described by the acidic functional groups with proton dissociation constants from $pK_a = 1.7$–9.5 (Ephraim *et al.* 1989; Paxéus *et al.* 1985; Lövgren *et al.* 1987; Tipping *et al.* 1988; Lee 1990).

The moderately weak acidic sites were, however, found in all studies with the proton dissociation constant $pK_a = 4.2$–4.5. These moderately weak acidic sites are important proton acceptors for estimating the ANC of environment samples and are therefore included in equation (3).

(c) Strong acidity

In an aqueous system containing both strong and weak acids, where the weak acids are present in undissociated states, the free H^+ concentration is equal to the strong acidity. In dilute acidified water systems containing weak acids (such as natural organic acids and carbonic acid), the weak acids may exist in both dissociated and undissociated states, in which case the free H^+ concentration (free acidity) is the sum of strong acidity and dissociated weak acidity.

As ANC measures the net deficiency of protons (or net base), the negative value of ANC is equivalent to the concentration of strong acidity in the aqueous solution. The natural organic acids may exhibit strongly acidic sites with $pK_a \leqslant 3$, and consequently they will behave like a strong mineral acid and will produce strong acidity and reduce the ANC.

3. Methodology (estimation of ANC)

The concentrations of HCO_3^-, $AlOH^{2+}$, $Al(OH)_2^+$, and COO_w^-, included in equation (2) were estimated by using the following methods. The concentration of Al species (Al^{3+}, $AlOH^{2+}$, $Al(OH)_2^+$) were calculated by using the chemical equilibrium model described elsewhere (Lee 1985). The analytical concentrations of organic weak acids, dissolved inorganic carbon, monomeric inorganic and organic Al, SO_4^{2-}, F^- and pH, as well as the values of the equilibrium constants involved in the Al complexation with ligands and the acidity constants of H_2CO_3, weak organic acids and HF are required for the evaluation.

For determination of the concentrations of HCO_3^- and H_2CO_3, the following equations were used:

$$[HCO_3^-] = T_{CO_2}/(K_2^{-1}[H^+]+1), \tag{4}$$

$$T_{CO_2} = [H_2CO_3^*]+[HCO_3^-], \tag{5}$$

$$[H_2CO_3^*] = CO_2(aq.), \tag{6}$$

where T_{CO_2} denotes the total concentration of inorganic carbon; K_1 and K_2 denote the equilibrium constants of reactions (7) and (8), respectively.

$$H_2CO_3(aq.) = CO_2(aq.)+H_2O, \quad K_1 = 650, \tag{7}$$

$$H_2CO_3^* = H^++HCO_3^-, \quad K_2 = 10^{-6.35}. \tag{8}$$

The total concentration of uncomplexed forms of moderately weak acidic sites (C_{COOH}) was evaluated by using the method described by Lee (1990), whereas a

Figure 1. Site location of minicatchments F5, F1, L1 and the Lake Gårdsjön.

previous method (Lee 1985, 1980), was applied for the earlier data. The concentration of dissociated and undissociated forms of COOH ([COO⁻] and [COOH]) can then be calculated by using the relations:

$$[COO^-] = K_{COOH} \cdot C_{COOH}/([H^+] + K_{COOH}), \qquad (9)$$

$$[COOH] = C_{COOH} - [COO^-]. \qquad (10)$$

4. Experimental section

The sampling sites were the forested catchments F5, L1 and F1 adjoining the Lake Gårdsjön in southwestern Sweden (figure 1). A detailed description of the catchments, field treatment, sampling and analyses of base cations and strong acid anions is given by Hultberg *et al.* (this symposium).

(a) Sampling

Some four collection events per year were performed for sampling of streamwater, groundwater and soil-water. Soil-water was sampled by Cronan lysimeters from two soil horizons (upper B and lower B horizon/C horizon).

(b) Analyses

DOC was analysed according to the procedure of Menzel & Vaceau (1964). The concentration of monomeric Al (Al_m) was determined using the method of Barnes (1975). The non-labile organic Al concentration (Al_o) was determined by the method of Driscoll (1984). The inorganic carbon was analysed by using a modified infra-red gas-analyser system (Schumacher & Smucker 1983). Potentiometric titration was used for determination of C_{COOH}. The detailed procedure is described elsewhere (Lee 1990). To determine C_{COOH}, before titration, the sample was purified by removing the Al^{3+} ion and other cation aluminium complexes, by using a strongly acidic cation-exchange resin (Amberlite 120). The titration was done in a constant ion medium (0.1 M NaClO$_4$) and in a nitrogen atmosphere from pH *ca.* 3–5 at 25 °C by current generation of hydroxide. This system was automatic.

5. Results and discussion

The evaluation of the effect of treatment on ANC and weak acids was focused on streamwater as very few lysimeter and groundwater data were available.

(a) *Effect on streamwater*

(i) *Acid neutralizing capacity*

The ANC has been calculated both with and without including aluminium hydroxide complexes. As the data after the summer of 1988 were not available and the contribution of Al to ANC was very small, part of the ANC data presented in this section does not include Al.

Figure 2*a*–*c* shows the seasonal variation of ANC, in stream-water during the experimental period in catchment F5 (acid sulphate treatment), in catchment L1 (sodium sulphate treatment) and in catchment F1 (reference catchment). All streamwaters have negative values of ANC, indicating net strong acid but no net base.

L1 streamwater showed a fast response to the Na$_2$SO$_4$ treatment as a quick increase in strong acidity followed by an indication of recovery one year after the treatment.

The effect was much less after the second treatment. It may be because the increase of H$^+$ ion was mainly caused by the ion exchange of Na$^+$ with H$^+$ in the upper soil layer and the concentration of Na$^+$ adsorbed on the soil was already high after the first treatment. The effect of the neutral sulphate on the acidification of the catchment was therefore short-term. The S-powder treatment in catchment F5 had a different effect on the acidification of streamwater.

A continued increase in strong acidity occurred after the summer of 1987. The increased acid sulphate deposition caused a progressive increase of acidification. Regression with the strong acidity (−ANC) against time showed a significant increasing trend for strong acidity. The increase in strong acidity was approximately 18 µequiv l^{-1} a^{-1}.

A significant correlation between strong acid and sulphate concentration was also observed for streamwater in catchment F5, thus (−ANC = 0.403·[SO$_4^{2-}$]+0.336; $r = 0.930$, $p = 0$).

Figure 2 illustrates the contribution of weak organic anions and aluminium hydroxide complexes to ANC in streamwater during the experimental period. As shown in figure 2, the aluminium hydroxide complexes made a very small contribution to ANC. The concentration of HCO$_3^-$ was too small to be shown in figure 2. The concentration of weak organic anion varied from 10 to 30 µequiv l^{-1}.

In catchment L1, a clear decrease in weak organic anions after the treatment could be a result of the precipitation of the organic acids in the upper layers of the soil (caused by a great reduction of the pH in soil water). The low, strong acidity during the summer of 1988 may be due to the high concentration of weak organic anions and pH. The results from catchment F5 showed a slight trend for decreasing concentrations of weak organic anions. This was a result of reducing the degree of dissociation of the moderately weak acidic sites due to decreasing pH.

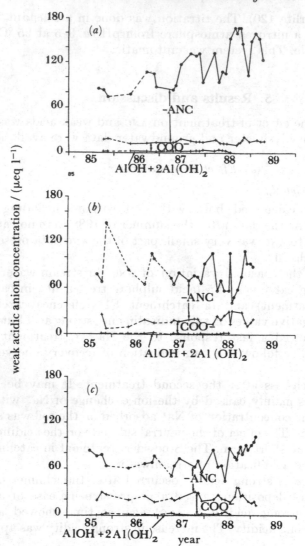

Figure 2. Seasonal variation of H+, −ANC (= strong acidity) and the weak acidic anions (weak organic anion and aluminum hydroxide complexes) in streamwater; (a) F5, (b) L1 and (c) F1 catchments; (○), H+.

The successive increases in acidification in F5 streamwater were therefore attributed to the increase of strong acid. Figure 3 shows that the decrease of strong acidity (−ANC) with increasing pH followed the similar curve of H+ concentration for all three streamwaters. The explanation may be because the streamwaters had the same concentration level of the organic acids that resulted in a strong correlation of nearly 0.99 between −ANC and [H+].

The results of the statistical analysis also implied that when the pH of the streamwater was near 4.56, the strong acidity became almost zero, very similar to the case observed for streamwater L1 in August 1988 (see figure 2b).

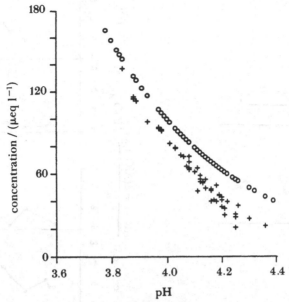

Figure 3. The change of strong acidity $(-\mathrm{ANC}, (+))$ and H^+ (o) concentration varied with pH in streamwater (F5, L1 and F1).

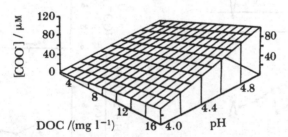

Figure 4. The change of the acid contribution from dissociation of moderately weak acidic sites varied with DOC and pH.

(ii) *Organic acids*

In both treated catchments, the treatments produced no appreciable changes in the concentration of organic acids and DOC. Regression with DOC against C_{COOH} and $T_{COOH} = (C_{COOH} + [Al_o])$ revealed a significant correlation of $r^2 = 0.6$.

The concentration ranges of DOC and T_{COOH} in streamwater are 3.5–13 mg l^{-1} and 30–95 µm, respectively. The concentrations are controlled by the hydrological flow paths and the degree of microbial mineralization in the upper soil profile.

The concentration of dissolved organic substances (as DOC and C_{COOH}) increased between summer and early autumn, dropped in December, and rose again during high-flow periods (early spring). During dry summers, the streamwater mostly consisted of deep soil-water and/or groundwater in which the concentrations of dissolved organic substances were low. At high flow rates, because of the increased contribution of organic horizon leachates to stream discharge, high DOC resulted in the streamwater.

The average number of moderately weak acidic sites per gram carbon in these streamwaters was greater $(7.2 \pm 0.3$ mequiv g^{-1} C) in autumn than in summer

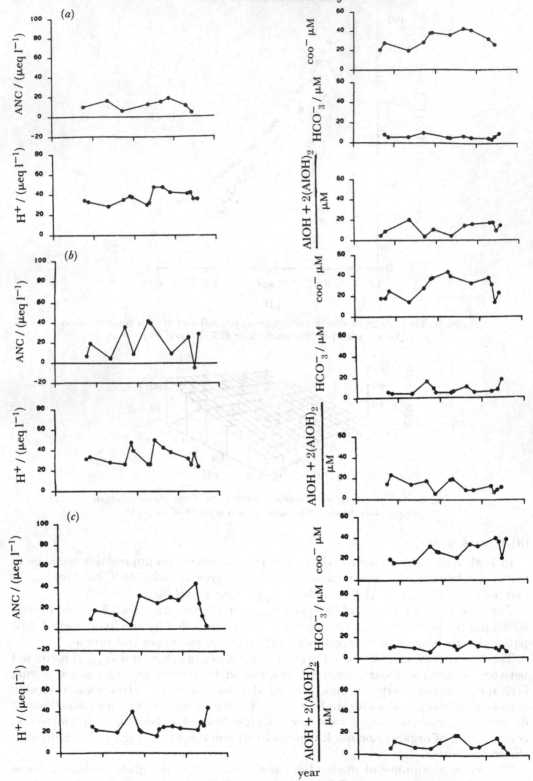

Figure 5. For legend see opposite.

$(6.3 \pm 1.8$ mequiv g^{-1} C). The difference in the moderately weak organic acidic contents of DOC revealed that during dry summers, DOC may contain more fractions of neutral species, bases and hydrophilic acid in soil-water than that during autumn (Cronan & Aiken 1985). Such differences may affect that organic Al complex formation and the acid contribution of DOC.

The estimation of the acid–base characteristics of natural organic acids indicated that the dissolved organic acids contained moderately weak acidic sites with an average pKa of 4.46 ± 0.36 in runoff and groundwaters (Lee 1990).

The average charge of COO$^-$ of DOC (mequiv per gramme carbon) increased with increasing pH in streamwaters and varied from 1 to 4 within the pH range 3.8–4.3.

(b) *The role of organic acids in the acidification of streamwater*

In recent years it has been necessary to include organic acidity for surface water acidification models in order to predict the future trends in surface water quality and the rates of acidification and recovery. The influence of natural organic acids on the acidification of surface water has been documented (Jones *et al.* 1986).

In acidic surface water there are two major processes which contribute to the production of organic acidity – the dissociation of the weak acidic functional group with pK_a less than about 5 and the formation of organic Al (Fe) complexes.

In acidic surface water with pH less than 5.5, the acidic functional groups with pK_a less than 5 have the most influence on the surface water acidification because they are able to partially or completely release protons and acidify waters when they enter the watercourse.

The strongly acidic functional groups with $pK_a < 3$ are believed to form chelate ligands together with adjacent phenolic groups or other moderately weak carboxylic acid groups, COOH, and are involved in the formation of complexes with Al and Fe, (Schnitzer *et al.* 1972; Gamble *et al.* 1980). The results of the statistical analysis indicated that the total concentration of COOH and its content of DOC were the principle factors controlling the concentration of non-labile organic Al (Alorg) in these streamwaters (Lee 1990). The concentrations of monomeric Al and hydrogen ions were less important.

To estimate as a first approximation, the proportion of acidity contributed by the dissociation of organic acids, we only need to consider the COOH as a main acidic group releasing protons by dissociation. Assuming all the strongly acidic carboxylic groups were probably already involved in complex formation, we can therefore quantify the total organic acidity by measuring C_{COOH}, pK_a, pH and $[Al_o]$.

Figure 4 shows that the acid contribution from the dissociation of COOH increases with increasing DOC and pH.

(c) *Effect on groundwater*

Three wells have been chosen to illustrate changes of ANC in groundwater in outflow areas. These wells are assumed to have representative effects within the areas. Figure 5 shows the variation of ANC, H$^+$ and different weak-acid anions in groundwaters of catchment F5, L1 and F1 during the experimental period. It can be seen that there was no appreciable effect on ANC in the two treatments comparing

---

Figure 5. The variation of ANC, H$^+$ and different weak acid anions (moderately weak acid anion, aluminum hydroxide complexes and HCO$_3{}^-$) in groundwaters of (*a*) catchment F5, (*b*) L1 and (*c*) F1 during the experimental period.

them with the reference catchment F1. Among these weak acid anions, the moderate contribution of the weak organic acid anion to ANC was greater than for others.

This study was done with the support of the board of the Scandinavian Surface Waters Acidification Programme (SWAP) and the Research Council of the National Swedish Environmental Protection Board. Thanks to O. Broberg for supply of DOC data, to I. Torbrink, P. Carlsson and Siming Li for performing the remaining analyses, as well as to S. Larsson for sampling; to D. Cooper for checking the English in this paper and to E. Knudsen for her assistance in the preparation of the manuscript.

References

Barnes, R. B. 1975 *Chem. Geol.* **15**, 177–191.

Cronan, C. S. & Aiken, C. R. 1985 *Geochim. cosmochim. Acta* **49**, 1697–1705.

Driscoll, C. T. 1984 *Int. J. environ. analyt. Chem.* **16**, 267–284.

Ephraim, J. H., Boren, H., Petterson, C., Arsenie, I. & Allard, B. 1989 *Environ. Sci. Technol.* **23**, 356–362.

Gamble, D. S., Underdown, A. W. & Langford, C. H. 1980 *Analyt. chem.* **52**, 1901–1908.

Gran, G. 1952 *Analyst* **17**, 661–671.

Jones, M. L., Marmorek, D. R., Reuber, B. S., McNamee, P. J. & Rattie, L. P. (eds) 1986 *Brown waters, LRTAP workshop no. 5.* Toronto: Environmental and Social Systems Analysts Ltd.

Lee, Y.-H. 1980 *Wat. Air Soil Pollut.* **14**, 287–298.

Lee, Y.-H. 1985 In *Lake Gårdsjön – an acid forest lake and its catchment. Ecol. Bull. no. 37* (ed. F. Andersson & B. Olsson), pp. 109–119. Stockholm: Publishing House of the Swedish Research Councils.

Lee, Y.-H. 1990 Natural organic acids in acidic surface water: Acid-base properties, complex formation with aluminum, and their contribution to acidification. In *Proceedings of the international symposium on humic substances in the aquatic and terrestrial Environment.* Linköping, Sweden, August 21–23, 1989.

Lövgren, L., Hedlund, T., Öhman, L. O. & Sjöberg, S. 1987 *Wat. Res.* **21**, 1401–1407.

Menzel, D. W. & Vaccaro, R. F. 1964 *Limnol. Oceanogr.* **9**, 138–142.

Paxéus, N. & Wedborg, M. 1985 *Analytica. chim. Acta* **169**, 87–98.

Schnitzer, M. & Khan, S. U. 1972 *Humic substances in the environment.* New York: Marcel Dekker.

Schumacher, T. E. & Smucker, A. B. 1983 *Pl. Physiol.* **72**, 212–215.

Stumm, W. & Morgan, J. J. 1981 *Aquatic chemistry.* New York: Wiley-Interscience.

Tipping, E., Backes, C. A. & Hurley, M. A. 1988 *Wat. Res.* **22**, 597–611.

Elemental sulphur and sodium sulphate treatment of catchments in the Gårdsjön area, southwest Sweden. Effects on phosphorus, nitrogen and DOC

By Ola Broberg

Institute of Limnology, University of Uppsala, Box 557, 751 22 Uppsala, Sweden

Experimental treatments with elemental sulphur (acid sulphate) and sodium sulphate (neutral sulphate) were done in two catchments in the Lake Gårdsjön area.

The acid-sulphate treatment generated further soil acidification and resulted in increased nitrate concentrations in soil- and groundwater. Further, phosphorus and organic nitrogen decreased in groundwater and a decreased areal loss of DOC was registered. These results support the theory that increased acidification will further decrease phosphorus concentrations and areal losses of DOC leading to a further oligotrophication and increased seccidepth of the lakes in this area.

The effects of the sodium sulphate treatment were less pronounced, resulting mainly in ion-exchange reactions. During the three latest years of this study some changes were registered in nitrogen concentrations, whereas phosphorus and DOC were largely unaffected.

---

1. Introduction

The Lake Gårdsjön catchment area on the west coast of Sweden is one of the sites chosen for a research programme on the effects of acid deposition within SWAP. Two small catchments were selected for treatment with sodium sulphate (L1) and elemental sulphur (F5), respectively. Both catchments were treated in November 1985 and November 1986, each with a total of about 200 kg S ha^{-1}†. The catchments used for 'sulphur' addition are described by Hultberg & Lee (this symposium), Lake Gårdsjön and its surroundings in Andersson & Olsson (1985). Measurements from the 'sulphur' treated catchments were compared with corresponding measurements within the untreated control catchment (F1).

All samples of phosphorus and nitrogen analysed in this sub-project were immediately preserved with $HgCl_2$. Methods are described by Persson & Broberg (1985).

The gradually decreasing phosphorus concentrations in many acidified lakes (Almer *et al.* 1978; Hultberg & Andersson 1982; Jansson *et al.* 1986) have been attributed to decreased phosphorus input with increasing acidification to the surrounding land (Broberg & Persson 1984; Persson & Broberg 1985). Decreasing areal losses due to precipitation of phosphorus by aluminium in the B-horizon were thought to be the main explanation (Jansson *et al.* 1986; Broberg 1988).

Reduced input of organic carbon caused by decreased degradation of organic material (Lohm *et al.* 1984) or precipitation of humic material in the soil profile by the increasing aluminium concentrations occurring upon acidification (Jacks 1982) have been proposed as explanations of the clearance of acidified lakes.

This paper deals with the main effects on phosphorus, nitrogen and dissolved organic carbon (DOC) in throughfall, lysimeters, groundwater and outflow from the catchments.

2. Results

Thoughfall

No significant differences in the concentrations of nitrogen, phosphorus or DOC in the throughfall of the three catchments were detected during the experimental period.

Lysimeters

As the sets of lysimeters closest to the soil surface seldom yielded sufficient volumes of water to give complete time coverage, these will not be evaluated. There is, however, fairly good data on soil-water from the C horizon, which will be presented as a mean of all 'C horizon' lysimeters within each area.

The results from soil-water (C horizon) did not indicate any major differences between the areas regarding concentrations of total phosphorus or organic nitrogen.

DOC concentrations showed a parallel pattern in the areas during the monitored period. Both area F5 and L1 had about twice (8–10 mg DOC l^{-1}) the concentration of the reference area (4–5 mg DOC l^{-1}).

Nitrate concentrations were higher in the soilwater in F5 than in F1 throughout the post-treatment period. In spring 1988 an increase in nitrate concentrations occurred in L1, above the earlier similar levels in F1 and L1.

Groundwater

Three wells have been selected to illustrate changes in groundwater chemistry in outflow areas. These wells are situated close of the outlet weirs, one in each area, and are thought to be representative of cumulative effects within the areas. Furthermore, these wells are assumed to have a chemistry representative of sub-surface input to the lakes.

Finally, mean post-treatment values for all wells will be used as a measure of the 'average' groundwater effect.

Unfortunately, groundwater sampling in the treated areas L1 and F5 before the 'sulphur' additions were too few to give a good comparison with the reference set in area F1. Recorded differences are based on the assumption that no difference existed before treatment. This assumption is likely for catchments F1 and F5 situated next to each other but somewhat more doubtful when results from L1 are compared with the reference area.

The 'average' effect on groundwater concentrations indicates increased concentrations of nitrate in area F5 from spring 1986 and throughout the monitored period (figure 1).

Total nitrogen concentrations were generally lower in F5 than in F1 but similar in L1 compared with F1 until the spring of 1988. At the same time as the increase in nitrate occurred, total nitrogen concentrations decreased in L1 groundwater, indicating that this decrease was due to lower organic-nitrogen concentrations.

The selected wells close to the weirs gave the same general picture as the 'average' groundwater response, with higher nitrate concentrations, and lower total nitrogen concentrations in catchment F5 than in catchment F1.

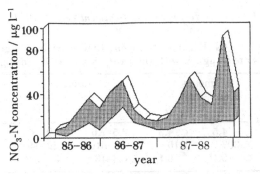

Figure 1. Nitrate concentrations in groundwater in areas F1 (□) and F5 (▦).

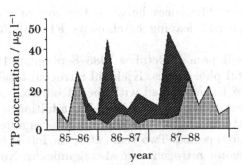

Figure 2. Total phosphorus concentration in groundwater in areas F1 (■) and F5 (▦).

Table 1. '*Average*' *groundwater effect*

(Comparison of groundwater concentrations in areas F1 and F5 during the post-treatment period (Wilcoxon signed rank test); TP, total phosphorus; TN, total nitrogen; NO_3, nitrate nitrogen; DOC, dissolved organic carbon. Concentrations are in µg l^{-1}, with the exception of DOC, in mg l^{-1}.)

	N F1	N F5	means F1	means F5	range F1	range F5	rank test P<
TP	16	17	19.8	8.4	3.5–67.0	2.6–27.5	0.04
TN	16	17	445	285	198–1520	156–740	0.01
NO_3	16	17	10	28	2–28	7–55	0.0006
DOC	16	17	6.0	6.6	2.4–8.7	3.2–9.9	n.s.[a]

[a] Not significant.

DOC concentrations were quite similar in areas F1, F5 and L1, when all wells were included. Groundwater DOC concentrations from outflow areas (single wells), indicated lower concentrations in area F5 (period mean 5.0 mg DOC l^{-1}) than in area F1 (7.9 mg DOC l^{-1}).

Total phosphorus concentrations were lower in groundwater in area F5 ('average' response) from the summer of 1986 to the spring of 1988 (figure 2).

Catchments

During the pre-treatment period (October 1985) there were some differences between the outlet water from area F1 (reference) compared with area L1 (cf. table 2).

Table 2. *Comparison of areas F1 and L1 during the pre-treatment period (1984–85) with the Wilcoxon signed rank test*

(TP, total phosphorus; KJ-N, Kjeldahl nitrogen; DOC, dissolved organic carbon. Concentrations are in $\mu g \, l^{-1}$, with the exception of DOC, in $mg \, l^{-1}$.)

	N F1	N L1	means F1	means L1	range F1	range L1	rank test $p <$
TP	18	21	4.5	4.3	2.5–7.0	1.9–16.0	n.s.[a]
KJ-N	18	21	307	210	188–610	122–540	0.01
DOC	17	20	9.1	6.1	6.3–13.8	4.5–9.3	0.001

[a] Not significant.

There were no significant differences between the concentrations of nitrogen, phosphorus or DOC in the rills leaving catchments F1 and F5 during the pre-treatment period.

During the post treatment period (October 1985–September 1989) there was no significant difference in total phosphorus, Kjeldahl nitrogen, total nitrogen or DOC concentrations in the rill of L1, compared with the F1 outlet.

The most noticeable effect of the elemental sulphur addition was decreased DOC concentration in runoff water from F5. This decrease was significant from 1985–86 and throughout the monitored period. Two years after the initial treatment, lowered runoff concentrations of organic nitrogen were also significant. No significant effects were registered on phosphorus concentrations.

In terms of annual transport, discharge of DOC from area F5 decreased by about $10 \, kg \, ha^{-1} \, a^{-1}$ with a tendency for decreased losses of organic nitrogen to occur during 1987/88–1988/89 compared with the control catchment.

The nitrate–nitrogen and phosphorus discharges were similar in F5 and F1 both before and after the 'sulphur' addition. A small decrease in phosphorus losses from catchment F5 was measured in 1985–86 and 1986–87 ($\Delta - 5 \, g \, P \, ha^{-1} \, a^{-1}$ both years) compared with F1. A closer examination showed that this difference were due to lower losses during high flow (September–October and March–April).

Areal losses from L1 indicated lower output of phosphorus in 1986–87 ($\Delta - 7 \, g \, P \, ha^{-1} \, a^{-1}$), otherwise differences were small compared with L1. There were also a higher discharge of Kjeldahl nitrogen (organic nitrogen + ammonium) from area L1 from 1986/87–1988/89 indicating increased losses of nitrogen of about $0.2 \, kg \, N \, ha^{-1} \, a^{-1}$.

3. Discussion

Both treatments caused clear effects of increased leaching of Ca, Mg and Al (Hultberg & Lee, this symposium). In area F1 those changes were attributed to exchange reactions with sodium ions (neutral sulphate treatment). The acid sulphate treatment caused a pH decrease in B-horizon soilwater, implying a significant soil acidification (Hultberg & Lee, this symposium).

The main purpose of this study was to reveal if an additional acidification of these catchments would further reduce phosphorus and DOC transport through the soils and in the rills leaving them. An additional acidification was evident in catchment F5 (acid sulphate treatment).

The gradually decreasing phosphorus concentrations in many acidified lakes

(Almer *et al.* 1978; Hultberg & Andersson 1982; Jansson *et al.* 1986) has been attributed to decreased phosphorus input with increasing acidification of the surrounding land (Broberg & Persson 1984; Persson & Broberg 1985). Decreasing areal losses due to precipitation of phosphorus by aluminium in the B-horizon was thought to be the main explanation (Jansson *et al.* 1986; Broberg 1988).

Reduced input of organic carbon caused by decreased degradation of organic material (Lohm *et al.* 1984) or precipitation of humic material in the soil profile by the increasing aluminium concentrations occurring upon acidification (Jacks 1982) have been proposed as explanations for the clarification of acidified lakes.

In this study, decreased DOC concentrations were indicated from groundwater in the outflow area of catchment F5 as well as significantly decreased concentrations in the catchment rill. Areal losses of DOC was also reduced by about 10 kg ha^{-1} a^{-1}. No changes were registered as 'average' groundwater effect or in soil water from the C-horizon.

Phosphorus concentrations on the other hand decreased in groundwater and a small decrease was noted in the discharge during 1985–86 and 1986–87 from catchment F5. This decrease was mainly due to lower discharges during (September–October and March–April).

Nitrogen fractions responded to the acid sulphate addition by increased nitrate–nitrogen concentrations in soilwater from the C-horizon and in groundwater accompanied by decreased concentrations of presumably organic nitrogen. Lower concentrations of organic nitrogen were also measured in runoff (rill) water in 1987–88 and 1988–89.

The acid sulphate treatment, followed by further acidification of the soil, seemed to have resulted in decreased degradation of organic material, indicated by decreased concentrations and transport of DOC in outflow and by decreased groundwater concentrations in the outflow area. This change is further emphasized by the significantly lower total nitrogen concentrations in groundwater and runoff.

The registered increase in nitrate concentrations (soil-water C-horizon and groundwater) was probably due to decreased uptake by soil biota, further implicating the damage of the soil biological system by the acid sulphate treatment.

The decreased phosphorus concentrations in groundwater could be due to this probable decrease in degradation and thereby decreased liberation of phosphorus but could also be due to chemical precipitation. The fact that no changes in phosphorus concentrations were registered in water from the C-horizon but were found in groundwater might indicate a time consuming reaction, favouring precipitation reactions.

The neutral sulphate treatment (L1) seems mainly to cause ion-exchange reactions. This treatment did not cause any lasting pH decrease in soil- or groundwater (Hultberg & Lee, this symposium) and was probably more gentle not causing 'any' damage to the soil biological system.

As far as phosphorus, nitrogen and DOC were concerned, effects were only registered on nitrogen. Nitrogen concentrations were altered at the end of the studied period (from the spring of 1988 to the end of the study). A increase in nitrate nitrogen concentrations in soil-water (C-horizon) and groundwater appeared during this period in combination with decreased concentrations of 'organic nitrogen' in groundwater. A slight increase in areal losses of Kjeldahl nitrogen indicated changed leakage of organic nitrogen or ammonium nitrogen. To what extent these changes were due to the sodium sulphate treatment is not known.

If the knowledge of phosphorus, nitrogen and DOC transformations in acidified catchments are going to be better understood in the future, chemical and hydrological measurements must be combined with measurements of the effects on the biological part of the soil ecosystem. Changes in uptake, degradation or micro-biological transformations, such as those caused by the acid sulphate treatment of catchment F5 in this study, are probably as important as 'purely' chemical reations.

References

Almer, B., Dickson, W., Ekström, C. & Hörnström, H. 1978 Pollution and the aquatic ecosystem. In *Sulfur in the environment*, *2* (ed. J. O. Nriago), pp. 271–311. New York: John Wiley.

Andersson, F. & Olsson, B. (eds) 1985 Lake Gårdsjön: an acid forest lake and its catchment, *Ecol. Bull.* no. 37, 336 pp.

Broberg, O. 1988 National Swedish Environmental Protection Board rep. no. 3426, pp. 135–205.

Broberg, O. & Persson, G. 1984 *Arch. Hydrobiol.* **99**, 160–175.

Hultberg, H. & Andersson, I. B. 1982 *Wat. Air Soil Pollut.* **18**, 311–331.

Jacks, G. 1982 *Acidification today and tomorrow*, pp. 176–179. Stockholm: Swedish Ministry of Agriculture and Environment 1982 Committee.

Jansson, M., Persson, G. & Broberg, O. 1986 *Hydrobiol.* **139**, 61–96

Lohm, U., Larsson, K. & Nömmik, H. 1984 *Soil Biol. Biochem.* **16**, 343–346.

Persson, G. & Broberg, O. 1985 *Ecol. Bull.* **37**, 158–175.

The 1000-lake survey in Norway 1986

By A. Henriksen[1], L. Lien[1], T. S. Traaen[1], B. O. Rosseland[1]
and Iver S. Sevaldrud[2]

[1] Norwegian Institute for Water Research, Norway
[2] 3522 Bjoneroa, Norway

The 1000-lake survey was done in Norway in the autumn of 1986 to determine both the chemistry and the status of fish populations in lakes located in areas sensitive to acidic deposition. The survey was also designed to detect possible changes in water quality and fish populations as a follow-up of the extensive regional surveys that were done in 1974–1975. Lakes and watersheds in widespread areas of southern and western Norway are affected by acidic deposition, and thousands of lakes and streams are acidic (acid neutralizing capacity less than 0). The greatest loss in fish populations is to be found in southernmost Norway. The number of barren lakes here and in southwestern Norway has doubled since 1971–1975. The chemical changes in these lakes are characterized by a decrease in calcium and sulphate and an increase in aluminium and nitrate concentrations. There has been little change in pH levels. The nitrate concentrations in most of the lakes in southernmost Norway have in fact doubled over the period, while no significant changes have occurred in the rest of southern Norway. Repeated sampling in the southernmost lakes in 1987–1989 shows that the elevated nitrate concentrations persist. The total land area affected by acidification has increased from 33 000 km^2 in 1974–1979 to 36 000 km^2 in 1986 (ca. 11 % of Norway's total land area). In 1986, over 18 000 km^2 were almost totally damaged compared with 13 000 km^2 in 1974–1979. At present 52 % of the lakes surveyed are endangered. An empirical model predicts that a 30 % reduction in loadings of sulphur compounds would lead to recovery for 28 % of these. A further reduction of 50 % would ensure viable conditions for fish in 40 % of these lakes. These prognoses assume that the content of nitrate and humic compounds in these lakes remain constant. A further increase in nitrate concentration would counteract effects of reduced sulphur deposition. The data from this survey have been used to produce maps for critical loads and where those critical loads are exceeded for southern Norway.

1. Introduction

Norway currently receives considerably more acidic components from air and precipitation than it produces. In 1985 an estimated 185 000 t of sulphur and 88 000 t of nitrogen were deposited. Of these totals, only 10 000 t of sulphur and 22 000 t of nitrogen were traceable to Norwegian sources. Most of the depositions received in southern Norway originate from central and western Europe. Agder and Telemark counties receive the highest loadings. Finnmark in northern Norway also receives considerable loadings of pollutants, largely from smelters in the Soviet Union.

The 1000-lake survey was conducted in Norway in the autumn of 1986 to determine both the chemistry and the status of fish populations in lakes located in areas sensitive to acidic deposition. The survey was designed to establish a baseline of chemical data and fishery status before major planned reductions in emissions of acidic precursors in Europe. The survey was also designed to detect possible changes in water quality and fish status as a follow-up of the extensive surveys that were done in 1974–75.

This survey was done as part of the Norwegian Monitoring Programme for Long-Range Transported Air Pollutants and Precipitation under the Norwegian State Pollution Control Authority (SFT).

2. Methods

The surveys in the 1970s used a sampling grid to produce a statistical unbiased sample of lakes (Wright & Henriksen 1978). Results from these surveys suggested a reasonable procedure for selecting lakes to be included in the present survey: lakes that had been sampled previously in 1974–75 and located in granitic or gneissic bedrock or other bedrocks yielding low ionic strength surface waters, geographical coverage throughout Norway, lake size greater than 0.2 km², and lakes without local sources of pollution or watershed disturbance. A total of 305 lakes included in the 1974–75 surveys that were sampled after the autumn circulation were re-sampled in 1986, when the number of lakes sampled was increased to more than 900. The water samples were collected at the outlets of the lakes shortly after the autumn circulation. Most of the lakes were accessed on foot or by car, although helicopters were used for 375 lakes. All samples were analysed at NIVA (Norwegian Institute for Water Research) by routine procedures. A baseline of fish data was collected in the same manner as in previous surveys (Wright & Snekvik 1978). Questionnaires on fishery status were sent out. Fishery consultants in the various counties were responsible for the collection of data, and the completed questionnaires were returned to NIVA, where they were checked and supplemented before preparation for data registration. Four categories of fish status for the lakes were defined: (1) unaffected; (2) slightly affected: fish still present, but at least one species declining; (3) severely affected: fish still present, but at least one species lost; (4) complete decimation: barren. Altogether, information about the status of fish populations was supplied for 922 lakes. Of these, 31 have never had a fish population. These were excluded in the further analyses.

3. Results

(a) Lake chemistry

Base cations are produced by weathering reactions and ion exchange processes in watersheds. Bedrock and soil are the main sources of base cations, whereas soil thickness and hydrologic flow path are important factors in determining the concentrations. Calcium and magnesium are the principal base cations in most lakes, with some contribution from sodium and potassium. Concentrations of non-sea-salt base cations (non-marine $Ca + Mg$) indicate the normal watershed production through weathering reactions and ion exchange processes. Over 75 % of the lakes had $Ca* + Mg*$ less than 75 μequiv. l^{-1}, half of which had concentrations that were very low (less than 25 μequiv. l^{-1}). Nearly all lakes with concentrations greater than 75 μequiv. l^{-1} were found in areas with thicker soils, or were located below the glaciated marine limit. Fourteen lakes located at elevations above 700 m had

Ca + Mg concentrations less than 10 μequiv. l^{-1}, the lowest measured value was 5.5 μequiv. l^{-1}. At such low concentrations of base cations, very small amounts of strong acids can produce acidic lakes.

Sulphur is considered to be the major contributor to the recent acidification of lakes in Norway (Wright & Henriksen 1978). Sulphate is normally considered to be a mobile anion (Seip 1980) and transports equivalent concentrations of cations (H$^+$, Al^{3+}, Ca^{2+} and Mg^{2+}) to runoff waters. In areas not influenced by acidic deposition, dilute lakes normally have sulphate concentrations near 10–15 μequiv. l^{-1} (Brakke *et al.* 1989). About 50% of the lakes sampled in the survey had SO$_4$ greater than 50 μequiv. l^{-1}. The highest concentrations were observed along the coast in southernmost Norway. High concentrations were also found in eastern Finnmark in northern Norway, largely due to emissions from the Soviet Union.

The pH of a lake is determined by the relative magnitude of the concentrations of base cations and strong acid anions such as sulphate. Forty percent of the lakes in the survey had pH less than 5.0 and 60% had pH less than 5.5. Most of these lakes were found in southernmost and southwestern Norway. The acidic lakes (pH less than 5.0) are dominated by strong acids and aluminium.

Most of the nitrogen compounds received by precipitation and dry deposition are normally taken up by the vegetation in the watersheds, resulting in low concentrations of nitrate in the runoff water. If the deposition of nitrogen exceeds incorporation by vegetation the excess can provide 'mobile' anions and acidify in the same way as sulphate. Nitrate concentrations were highest in lakes in southernmost Norway (100–400 μg N l^{-1}) corresponding with the highest deposition rates of nitrogen. Here nitrate contributes significantly (up to 30%) to the acidity of many lakes.

Elevated concentrations of inorganic (labile) aluminium (toxic to fish) were found mainly in southernmost Norway and were highly correlated with pH.

(b) *Organic carbon in lakes*

In forested and boggy terrain, runoff waters often contain humic compounds imparting a yellowish-brown colour to the water. Total organic carbon (TOC) measurements can be used as an estimate of these humic acids to acidity. Using the simple measure of apparent colour from the lake surveys in Norway in the 1970s (Wright *et al.* 1977), Krug *et al.* (1985) concluded that '88% of the critically acid lakes (pH < 5.0) in Norway were humic colored' on an arbitrary scale ranging from 0 (blue) to 5 (dark brown), rather than instrument measured true colour, available values of TOC, or anion deficit. Both TOC and visual colour were presented for 63 lakes by Wright *et al.* (1977). Of these, 49% had TOC ⩽ 2.0 mg l^{-1}, whereas only 13% had TOC greater than 6 mg l^{-1}. For the 1986 survey, 60% of the lakes had TOC less than 2.0 mg l^{-1} and 10% had TOC greater than 6.0 mg l^{-1}, results very similar to estimates from the 63 lakes sampled in 1976. We thus conclude from the TOC measurements that lakes in Norway generally have a low TOC.

There was no apparent relation between pH and TOC concentrations. This lack of direct relation is also illustrated by a stepwise multiple regression analysis (Henriksen *et al.* 1988). For this purpose we use the parameter strong acid (SA) (Wright & Henriksen 1983), which is operationally defined as:

$$SA = H^+ - HCO_3^- + \Sigma Al^{n+}, \tag{1}$$

where H$^+$ is calculated from the measured pH value, HCO$_3^-$ is the measured alkalinity

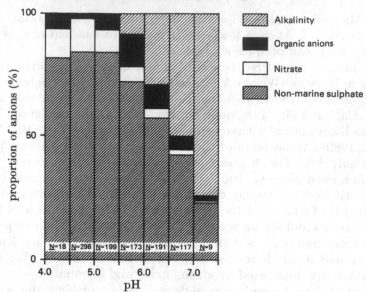

Figure 1. Percentage composition of anions in 1005 lakes sampled in 1986 throughout Norway divided into seven pH classes (4.0–< 4.5, 4.5̈< 5.0, 5.0–<5.5, 5.0–< 6.0, 6.0–< 6.5, 6.5–< 7.0, 7.0–< 7.5).

Table 1. *Stepwise multiple regressions of strong acid (SA) against BC^*, SO_4^*, NO_3, and TOC*

	step 1	step 2	step 3	step 4
variable	BC	SO_4	TOC	NO_3
cumulative percentage	62.6	92.4	92.7	92.8

and ΣAl^{n+} is the sum of all positively charged aluminium ions (measured 'labile' aluminium). In lakes having positive alkalinity the strong acid parameter will be negative, whereas acidic lakes will have positive values. A stepwise multiple regression analysis using the sum of non-marine base cations (BC = Ca + Mg + Na), non-marine sulphate (SO_4^*), NO_3 and TOC as independent variables explained 92.8 % of the variations in SA (table 1). The table gives the variable entered at each step and the cumulative percentage of explained variations. TOC explained only 0.3 % of the variations in strong acid for all lakes sampled. For lakes in eastern Norway, where most of the humic lakes are located, TOC explained 5.9 % of the variations in SA. In southernmost Norway, which had most of the acid lakes, only 1.6 % of the variations in SA was explained by TOC. Of greater importance than the concentrations of TOC is the actual contribution of dissociated strong acids. The anion deficit (difference between the sum of measured cations and measured anions) is a measure of the dissociated organic acids (Henriksen *et al.* 1988 *a*). Organic acids represent less than 10 % of the anions for all pH groups for the full set of lakes sampled in 1986 (figure 1). In southernmost Norway the contribution of organic anions was less than 4 % for any pH range, and NO_3^- had replaced organic anions as the second most important anion. The maximum contribution was found in eastern Norway, where they represented 20 % of the anions in the pH range 5.5–6.0.

It can be concluded that the acidic lakes sampled in southern Norway are mainly inorganic solutions with concentrations of excess sulphate that are high relative to

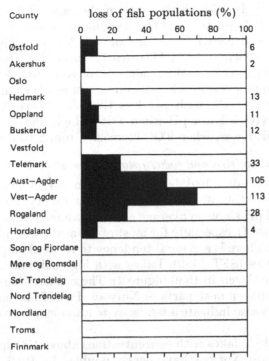

Figure 2. Countywise distribution of lost fish populations expressed as a percentage. The number of lost populations is noted at the side of each bar.

base cations and with little contribution of organic acids to strong acidity. In the acidic lakes in southernmost Norway nitrate has replaced organic anions as the most important anion after sulphate.

(c) *Status of fish populations*

The evaluation of each lake took into account possible causes of any changes in fishery status: regulation measures, disease, competition between species, local sources of pollution, fish management (liming, stocking, rotenone treatment) or other human factors. Fish populations affected by these factors were excluded from further evaluation.

About 40 species of freshwater fish, including eel, have been registered in Norway. In the lakes surveyed, brown trout and perch were the dominant species in eastern and southernmost Norway, whereas Arctic char and brown trout dominated in southwestern Norway and further north. The loss of fish populations varies widely from south to north (figure 2). The most extensive losses, both in number and percentage of populations affected, are found in the counties of Telemark, East and West Agder and Rogaland. In West Agder alone, 70 % of the fish populations in the lakes studied have been lost. Lakes with an unaffected fish population have the highest pH values and the lowest concentrations of inorganic aluminium (figure 3). The pH exceeds 5.0 in 97 % of the lakes, whereas concentrations of aluminium are below 40 µg l⁻¹ in 50 % of the lakes. In lakes where populations are damaged (status 2 and 3) pH values are scattered, but there is a predominance (64 %) of lakes with pH below 5.6, i.e. lakes without bicarbonate buffer capacity. More than 90 % of the lakes with status 4 have pH values below 5.4 and concentrations of inorganic aluminium above 60 μg l^{-1}. In lakes with a pH in the range 4.6–4.8, the calcium content is approximately 40% greater and the aluminium content about 15% lower in damaged lakes than in barren lakes. For lakes in the pH range 4.8–5.0 the corresponding figures are 40% and 30%. This clearly emphasizes the significance of calcium and aluminium for fish survival. In lakes of the same pH there are still fish in lakes with higher calcium levels and lower aluminium levels than barren lakes.

In 117 lakes containing perch, with pH less than 5.5, there is a clear trend indicating that perch survive a lower pH when TOC is high. Perch were present in lakes having a 0.5 pH unit lower, when TOC increased from 2 to 6 mg C l^{-1}.

(d) *Changes in chemistry and fishery status from 1974–75 to 1986*

Three hundred and five lakes had been sampled in earlier surveys after the autumnal circulation. For 254 of the lakes, previous fishery status data were available. Comparisons of pH show on average a slight tendency for higher values in 1986 than in 1974–75 (figure 4), especially for most of the acid lakes (pH < 5.0) found in southernmost Norway. There is a general tendency to reduced sulphur deposition in Norway during the 1980s (SFT 1989). Lakes with high sulphate concentrations showed a tendency to lower levels in 1986 (figure 4). These lakes were mainly located in the eastern and the southernmost parts of Norway. Data from monitored rivers and catchments in Norway also indicate a tendency to lower sulphate concentrations after 1980 (SFT 1989).

Concentrations of nitrate in lakes with concentrations above 100 μg l^{-1} in 1974–75 and located in southernmost Norway have nearly doubled by 1986. The change in nitrogen deposition from the mid-1970s to 1986 is small compared with the increase in runoff concentrations in nitrate. Nitrate contributes about 20–30% of the total acid anion concentrations in the most acidic lakes in southernmost Norway. The concentrations of aluminium in the lakes sampled are also higher in 1986 than in 1974–75. Eighty-eight percent of the 305 lakes sampled both in 1974–75 and 1986 had higher aluminium concentrations in 1986, the corresponding figure for nitrate being 79%. Of the 298 lakes that can be compared with respect to changes in both aluminium and nitrate, 72% of them show increases in both components. Resampling of 100 of the 1000 lakes in 1987, 1988 and 1989 shows that the increased levels of nitrate and aluminium persist. If the lakes continue to show elevated concentrations of these components, this can indicate the beginning of reduced plant uptake of nitrogen in parts of southernmost Norway. If planned reductions of sulphur emissions lead to decreases in sulphate, the increased concentrations of nitrate may cancel out some or all of the effect of the reduced emissions. Of the 254 lakes with fishery status information both in the 1970s and in 1986 30% (77 lakes) were barren between 1971–75, 47% had one or more reduced fish populations, whereas there was no registration as yet of changes in 23% of the lakes. By 1986 the number of barren lakes had risen to 64%, and only 10% of the lakes had not been harmed by acidic water. Thus 48% of the lakes having fish in 1971–75 had become barren by 1986. There has been no major change in water chemistry in the course of the period from 1971–75 to 1986 (Henriksen *et al.* 1989). Also the climatic conditions in 1974, 1975 and 1986 were close to normal (Henriksen *et al.* 1988b). The stress resulting from acidic water is greatest at the fry stage. Loss of catch is first recorded when the fish are from 4–5 to 10–15 years old, in other words 5–15 years after a complete failure in recruitment (Rosseland 1986). We therefore expect effects of the most acidic periods of the 1970s on some fish populations to persist until the late 1990s.

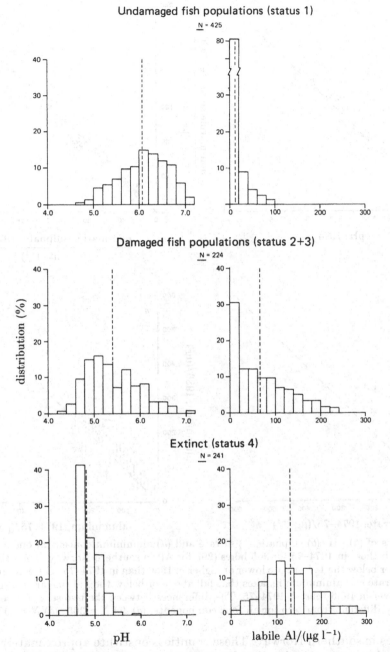

Figure 3. Levels of pH and inorganic (labile) aluminium (μg l^{-1}) in lakes with varying fish status. The dashed line represent the mean value for all lakes in each group.

The surveys done by the SNSF-project (Acid Precipitation – Effects on Forest and Fish) in the period 1974–1979 indicated that the regions with damaged fish populations were about 33000 km². In 1986 the land area had grown by 9% to 36000 km². In 1986, over 18000 km² were almost totally damaged, in contrast to 13000 km² in 1974–1979, an increase of 38%. The regions cover a land area of 22%

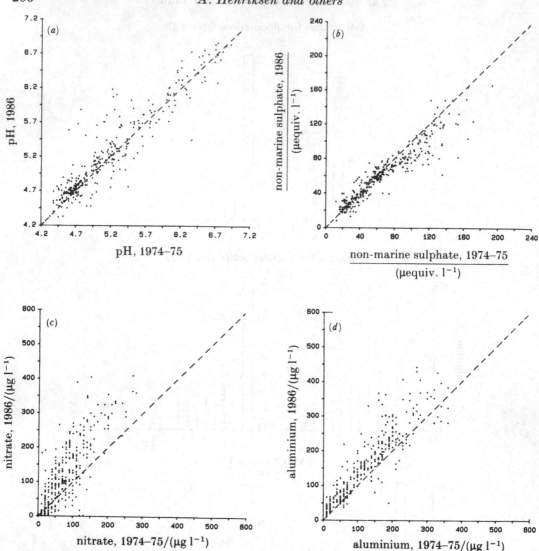

Figure 4. Values of (a) pH, (b) sulphate, (c) nitrate and (d) aluminium concentrations in 1986 in comparison with those in 1974–75 for 305 lakes (299 for Al) in southern Norway. The pH of lakes that fall above or below the 1:1 line are lower or higher in 1986 than in 1974–75. The concentrations of sulphate, nitrate and aluminium in lakes that fall above or below the 1:1 line are considered to be higher or lower in 1986 than in 1974–75. The differences between the two sets of observations are significantly different from zero for all four components. (a)–(c) N = 305; (d) N = 299.

of the counties in southern Norway. These counties constitute approximately half of Norway's total area, so that in all 11 % of the total land area is affected today.

(e) Effect of reduced deposition

Reductions in emissions of acidic precursors have been agreed to by many countries. We have estimated the potential recovery of lakes in southern Norway after a 30 % reduction in deposition, and we have assessed whether further reductions in emissions are necessary to produce suitable water quality for fish populations. For this, a simple empirical model that was used to assess the effects of

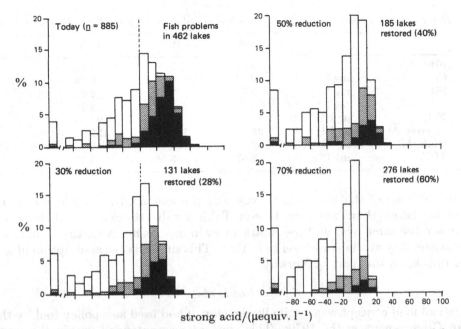

strong acid/$(\mu$equiv. $l^{-1})$

Figure 5. The frequency distribution for content of strong acid and fish status for 885 lakes in 1986, calculated on the basis of a 30, 50 and 70 % reduction in the sulphate concentrations of the lakes. Lakes that are barren (status 4), or with at least one lost species (status 3) are shown in black, lakes with at least one damaged species (status 2) are shaded, and undamaged lakes (status 1) are white.

reduced sulphate inputs on water quality (Henriksen *et al.* 1988) was superimposed on the relation between fish status and water quality derived from the 1986 survey (Henriksen *et al.* 1989). The results are given in figure 5. This figure reveals that a 30 % reduction would mean that 28 % of the lakes that are barren or damaged today would recover. A further reduction in inputs would lead to recovery of more lakes (figure 5). More than 70 % reduction in deposition is necessary to give viable conditions for fish in the most affected areas in Norway. These results are based on conservative estimates, but are strongly dependent on no further increases in nitrate in lakes.

4. Further use of the 1000-lake data

(a) The 100-lake surveys

As a follow up of the 1986 survey, about 100 lakes were selected for yearly sampling after autumnal circulation. Owing to intensified liming activities in Norway during recent years, many lakes that were included in the 1000-lake survey are no longer suitable for monitoring. A base of 94 lakes have been sampled regularly since 1987. Table 2 shows the mean values for some components for the years 1986 to 1988. Particularly the lakes located in eastern and southernmost Norway had lower pH values both in 1988 and in 1987, largely due to high amounts of precipitation in the autumn washing out more organic acids from the watersheds. The sulphate concentrations show a downward trend in the period, again especially in eastern and southernmost Norway. This is most likely due to dilution by high amounts of precipitation. Whether this indicates a general trend will show up in the future monitoring.

Table 2. *Mean values for some components in 94 lakes sampled in the years 1986, 1987 and 1988*

		1986	1987	1988
pH		5.27	5.24	5.17
Ca	mg l^{-1}	0.78	0.80	0.74
SO$_4$	mg l^{-1}	3.5	3.2	2.9
Cl	mg l^{-1}	3.1	3.1	2.7
NO$_3$	µg l^{-1}	88	85	92
reactive Al	µg l^{-1}	121	138	131
labile Al	µg l^{-1}	82	103	95
TOC	mg l^{-1}	2.87	3.56	3.57

A main outcome of the 1000-lake survey was the increased nitrate levels in the southernmost lakes (Henriksen *et al.* 1988*b*). Table 2 indicates clearly that the levels of nitrate are the same over all three years, or even increasing. Also, the aluminium concentrations stay at the same level as in 1986. This supports the assumption of less nitrogen uptake in this part of Norway.

(b) *Critical loads*

The critical load concept was originally developed and used as a policy tool by the Canadian Government in the 1970s. This concept was extended within the work programme under the Convention on Long-Range Transboundary Air Pollution for the development and implementation of control strategies for transboundary air pollutants. First discussed and evaluated within the work of the Nordic Council of Ministers in 1986 (Nilsson 1986), the concept has been elaborated further to become more reliable and to cover a large variety of ecosystems. Some countries have already adopted this concept for their environmental policies. In 1988 the Executive Body for the convention established a Task Force on Mapping under the leadership of the Federal Republic of Germany, as a part of its programme of action for development of the critical loads approach. In March and October 1988 three expert workshops were arranged to evaluate further the critical levels and loads concept and to provide reliable figures. The workshops were held at Bad Harzburg (F.R.G.), Skokloster and Norrtalje (Sweden).

The Skokloster workshop defined critical load as (Nilsson & Grennfelt 1988):

a quantitative estimate of an exposure to one or more pollutants below which significant harmful effects on specified sensitive elements of the environment do not occur according to present knowledge.

The Workshop on Mapping has produced a mapping manual (in the press), which outlines methods for mapping critical loads for sulphur and nitrogen, and areas where they are exceeded. Both static and dynamic modelling methods are given for surface water, groundwater and soils.

We have applied one of the methods given, the steady-state water chemistry method, to surface waters in Norway. In this method sulphate concentrations in runoff are considered to be at a steady state with respect to atmospheric deposition of sulphate, and include nitrate in the runoff water. In its present form, the method can only assess critical loads for sulphur acidity, but not for nitrogen. If data are available to convert concentrations to mass fluxes, quantitative estimates of critical loads can be derived.

Firstly, the base cations and sulphate concentrations are corrected for sea salts, assuming chloride in deposition and runoff originates only from seawater.

From ion balance considerations we have:

$$[BC^*] - [AN^*] = [ANC]. \tag{2}$$

Here, $[BC^*]$ is the sum of non-marine base cations, $[AN^*]$ the sum of nitrate and non-marine sulphate ions and $[ANC]$ the general definition of Acid Neutralizing Capacity:

$$[ANC] = [CO_3^{2-}] + [HCO_3^-] + [OH^-] + [A^-] - [H^+] - [Al^{3+}] - [AlOH^{2+}]$$

$$- [Al(OH)_2^+] - [NH_4^+]. \tag{3}$$

$[A^-]$ is the concentration of organic anions. For softwaters sensitive to acidification, such as in Norway, CO_3^{2-}. OH^- and NH_4^+ are not significant. Adding all inorganic Al ions, we obtain:

$$[ANC] = [HCO_3^-] + [A^-] - [H^+] - [\Sigma Al^{n+}]. \tag{4}$$

A lake's deviation from a 'critical' ANC concentration is:

$$\Delta CC = [BC_0^*] - [AN_t^*] - [ANC]_{\text{limit}}. \tag{5}$$

Here, $[ANC]_{\text{limit}}$ is the critical ANC concentration for a specified biological indicator organism, e.g. trout. $[BC_0^*]$ is the 'pre-acidification' concentration of base cations (Henriksen 1983). Negative values for ΔCC indicate that the critical load of acid inputs is exceeded, whereas positive values show that the deposition of strong acid is below the critical load. Both the ΔCC and the critical load can be quantified by calculating fluxes. Assuming the concentrations given for a lake are equal or close to the weighted yearly average concentration of the actual component, they can be multiplied by the yearly runoff value (Q) to give the annual flux. The critical load (CL) is then given by:

$$CL = ([BC_0^*] - [ANC]_{\text{limit}}) \cdot Q - BC_d^*, \tag{6}$$

and the deviation from the critical load by:

$$\Delta CL = Q \cdot \Delta CC. \tag{7}$$

Here, BC_d^* is the atmospheric deposition of non-marine base cations.

The ECE manual suggests a grid system based on the 0.5° latitude and 1.0° longitude grids and subdivisions. For Norway, we have divided the main grid into 16 subgrids, giving a grid area of about 12 km × 12 km. By means of the 1000-lake survey data and many other data available, we have assessed an average representative water chemistry for each subgrid. The critical load for acidity and the deviation from the critical load have been calculated according to equations (6) and (7) for southern Norway. Figure 6 gives the frequency distributions of CL and ΔCL for southern Norway. For about 30% of the land area the critical load is exceeded. This corresponds well with the figure of 22% obtained for areas with extinct or damaged fish populations in 1986, 22% (Henriksen *et al.* 1989). With a sulphur load of 80 kequiv. km^{-2} per year all over southern Norway, the critical load would have been exceeded in about 75% of the area (figure 6*b*). Using a simple empirical acidification model (Henriksen *et al.* 1988) the effects of different amounts of deposition can be predicted.

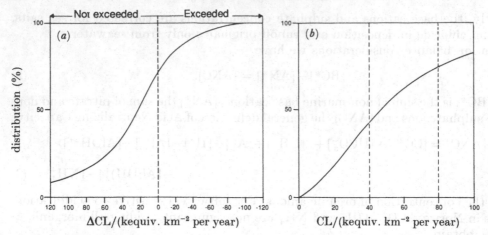

Figure 6. Critical load exceedance (a) and critical loads (b) for southern Norway.

The 1000-lake Survey 1986 received financial support from the Norwegian State Pollution Control Authority (SFT), the Norwegian Institute for Water Research and the Surface Waters Acidification Programme (SWAP).

References

Brakke, D. F., Henriksen, A. & Norton, S. A. 1988 *Water. Res. Bull.* **25**, 247–253.

Henriksen, A. 1984 *Verh. Internat. Verein. Limnol.* **22**, 692–698.

Henriksen, A., Brakke, D. F. & Norton, S. A. 1988a *Environ. Sci. Technol.* **22**, 1103–1105.

Henriksen, A., Lien, L., Traaen, T. S., Sevaldrud, I. S. & Brakke, D. F. 1988b *Ambio* **17**, 259–266.

Henriksen, A., Lien, L., Rosseland, B. O., Traaen, T. S. & Sevaldrud, I. S. 1989 *Ambio* **18**, 314–321.

Nilsson, J. 1986 (ed): Critical loads for nitrogen and sulphur. Report 1986: 11. (232 pages.) The Nordic Council of Ministers.

Nilsson, J. & Grennfelt, P. 1988 (eds) Critical loads for sulphur and nitrogen – Report from a workshop held at Skokloster, Sweden 19–24 March, 1989. (418 pages.) Nord 1988: 15. UN/ECE and Nordic Council of Ministers, 1988.

Rosseland, B. O. 1986 *Water, Air Soil Pollut.* **30**, 451–460.

Seip, H. M. 1980 In *Ecological Impact of Acid Precipitation* (ed. D. Drabløs & A. Tollan), pp. 358–366. SNSF-project, NISK, 1432 ÅS.

SFT (The Norwegian State Pollution Control Authority) 1989 Overvåking av langtransportert forurenset luft og nedbør. Årsrapport 1988. Oslo. (Monitoring of long range transported air pollutants and acid precipitation.) Annual report 1988. (Statlig program for forurensningsovervåking. Rapp. 375/89.) (In Norwegian, English summary.) Oslo.

Wright, R. F. & Henriksen, A. 1983 *Nature, Lond* **305**, 422–424.

Wright, R. F., Dale, T., Henriksen, A., Hendrey, G. R., Gjessing, E. T., Johannessen, M., Lysholm, C. & Støren, E. 1977 Regional surveys of small Norwegian lakes. October 1974, March 1975, March 1976, March 1977. (153 pages.) SNSF-project, NISK, 1432 Ås-NLH, IR 33/77.

Wright, R. F. & Henriksen, A. 1978 *Limnol. Oceanogr.* **23**, 487–498.

Wright, R. F. & Snekvik, E. 1978 *Verh. Internat. Verein. Limnol.* **20**, 765–775.

Discussion

M. CRESSER (*Department of Plant and Soil Science, Aberdeen University, U.K.*). I return to the point raised by Dr Hornung earlier in the meeting. As a chemist turned soil scientist, it worries me that we are concentrating excessively upon soil chemistry, and ignoring the consequences of biological change. Re-analysis of forest soil samples

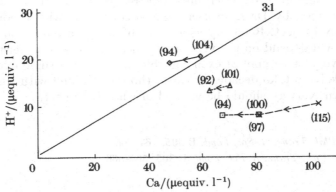

Figure d 1. Mean H^+ versus mean Ca^{2+} in 1974 (tail of arrow) and 1986 (head of arrow) for those lakes between 0 and 249 m which were surveyed both times. The data are divided according to their fish status in 1974: (◇) no fish; (△) sparse fish; (□) good fish; (×) overpopulated. Excess SO_4^{2-} (in microequivalents per litre) is shown in brackets.

collected at a 40 year interval by colleagues at Aberdeen University showed serious acidification to depth, and increases in sulphate saturation and labile aluminium. However, although the soils showed dramatic nitrogen accumulation, the C:N ratio had generally risen rather than fallen because litter accumulation was even more pronounced. We know that the biodegradability of litter from trees in polluted areas may be lower than that from trees in less polluted areas. Presumably too, the acidification of the litter horizons contributes to the increased accumulation rate, and we have shown that respiration rate in turn dilates the available nutrients in the surface horizons. It is imperative that we take such changes into account when attempting to make predictions of long-term water quality change. Unfortunately, we rarely have sufficient information to do so.

R. A. SKEFFINGTON (*National Power Technology and Environmental Centre, Leatherhead, U.K.*). Large data sets like the 1000-lake survey are useful for many purposes, and I present another aspect of the results of this study. This is possible through the generosity of Dr Henriksen, who made the data available. Chester (1984) pointed out that there was a good relation between the ration of Ca^{2+} to H^+ in the lakes of southern Norway and their fish status in 1974: lakes with $Ca^{2+}:H^+$ greater than 4 tended to have good fish, lakes with $Ca^{2+}:H^+$ less than 3 tended to be fishless. How have the lakes in the various fish categories changed in composition between 1974 and 1986?

The answer appears to depend on the altitude. Figure d1 shows the trajectories of the mean chemical composition of lakes in each of the 1974 fish status categories in the altitude band 0–249 m above sea level. In each case the excess SO_4^{2+} concentration has declined by 10–15 %, and the $Ca^{2+}:H^+$ ratio has worsened. Calcium has declined by an average of 11.4 µequiv. l^{-1}, H^+ by 0.9 µequiv. l^{-1}. The situation is very similar in the 250–499 m above sea level band, but at higher altitudes there is no consistent reduction in either excess SO_4^{2-} or Ca^{2+} and H^+. The implication may be that high altitude catchments are better at absorbing SO_4^{2-}, as Sørensen (1984) pointed out. However, at low altitudes, where the majority of lakes are situated, the implications seem clear. A reduction in sulphate deposition has led to worsening conditions for fish in southern Norway, accepting that declining calcium levels are indeed deleterious (see discussion elsewhere in this symposium). This result appears to be borne out by the fish-status described by Dr Henriksen, and is consistent with theory and the results of the RAIN Project. Recovery as a result of emission reduction may come eventually, but it will depend on the very slow process of rock weathering in these areas replacing a very large pool of exchangeable acidity with some exchangeable calcium. This is likely to take decades at least: those concerned with the health of fisheries in southern Norway should be warned of what to expect.

References

Chester, P. F. 1984 *Phil. Trans. R. Soc. Lond.* B **305**, 564–565.
Sørensen, N. A. 1984 *Phil. Trans. R. Soc. Lond.* B **305**, 564–566.

General discussion

C. O. TAMM (*The Swedish University of Agricultural Sciences, Department of Ecology and Environmental Research, S-75007 Uppsala, Sweden*). The discussion of the catchment manipulation experiments has, to a large extent, dealt with the realism of the manipulations and the representativity of the results. Although this type of discussion is necessary for future use of the results, or for models where the results may be incorporated, it fails to emphasize an important point, namely that moderate changes in acid input has led to considerable chemical effects on the outflow, both when the acid input was decreased (at Risdalsheia) and increased (at Sogndal and Gårdsjön). Before SWAP, some scientists refused to accept this possibility. It is also interesting to note that the effects have been clearer on other acid-related characteristics than on the pH, and a further follow-up of these results might give important information on the extent of hysteresis in the acidification processes (a possible difference between the rate of recovery and the rate of acidification).

General discussion

Mineral weathering studies in Scandinavia

By Gunnar Jacks

Department of Land and Water Resources, Royal Institute of Technology,
S-100 44 Stockholm, Sweden

Weathering studies by forest researchers were already underway at the turn of the last century. In connection with acid rain research weathering has attracted renewed attention as it is the major long-term sink for protons in soils.

Weathering rates have been determined with a number of methods among which input–output budgets and the determination of the historical weathering losses in soil profiles are the most commonly adopted. The use of strontium isotopes in conjunction with assuming an analogy between strontium and calcium is another method applied to the problem. Modelling has been done with the MAGIC and PROFILE models.

In spite of the differences in the methods, it is now possible, within fairly narrow limits, to estimate weathering rates for typical Scandinavian landscape types by knowing a few basic facts like mineralogy, soil texture and soil depth. The most sensitive environments are rocky areas with very thin soil covers found commonly in Norway and in coastal tracts of Sweden and Finland. They are likely to have weathering rates of 5–20 mequiv. $m^{-2} a^{-1}$ depending on the mineralogy. Till areas with a soil cover of about 1–5 m have weathering rates of about 20–30 mequiv. $m^{-2} a^{-1}$ if the soil till is derived from acid granites or gneisses. With intermediate rocks as sources for the till and a dark mineral content of around 5% the weathering rate may be found in the range from 30 to 50 mequiv. $m^{-2} a^{-1}$.

Modelling approaches will be very useful in extrapolating results both in time and space. The availability of independent estimates is essential for the development of the models.

Organic acids are important weathering agents. Acidification affects organic matter turn-over. How this in turn affects weathering rates is an important research task.

---

1. Proton sinks and sources in the soil system

The chemistry of surface waters depends on processes in the catchment and processes in the water bodies themselves. The catchment processes are generally the dominating ones. Surface water acidity thus generally depends on the ability of the catchment to produce base cations in exchange for protons added from a variety of sources. Proton sources are acid deposition, base cation uptake by vegetation and oxidation of reduced forms of sulphur and nitrogen. Proton sinks with respect to the soil solution are ion exchange, weathering, and reduction of sulphur and nitrogen species. Reduction–oxidation reactions usually balance each other over timescales of a few years. Long-term effects may be seen in case of drainage or when a drainage system decays. Over a period of a few decades the cation fluxes can be simplified

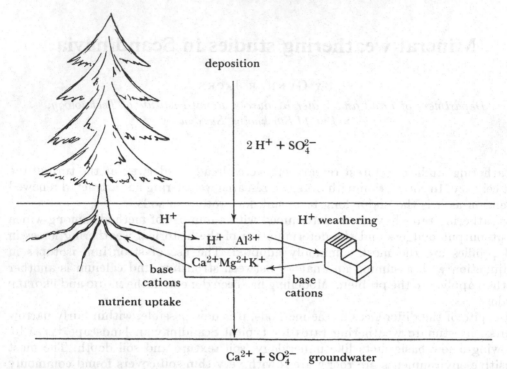

Figure 1. Cation fluxes in a simplified soil system with an exchangeable storage.

Table 1

method	references
historical weathering	Bain *et al.* (this symposium), Teveldal *et al.* (this symposium), Olsson & Melkerud (1989), Wright & Frogner (1989)
input–output budgets; steady state	Rosén (1982), Eriksson (1986), Lundström (this symposium)
input–output budgets; transient processes considered	Pačes (1986), Wright & Frogner (1989)
synthesis of run-off chemistry from weathering reactions	Velbel (1985), Giovanoli *et al.* (1988)
laboratory experiments extrapolated to the field	Frogner (1990), Lundström (this symposium)
strontium-isotope method; analogy with calcium	Åberg *et al.* (1990), Jacks *et al.* (1989)
modelling approaches; MAGIC using individual mineral dissolution rates	Wright & Cosby (1987) Sverdrup & Warfvinge (1988)

according to figure 1. The only long-term sink for protons is weathering. Thus the critical load for acid deposition has been defined on the basis of weathering rate (Nordic Council of Ministers 1988).

The stores of exchangeable cations are large covering the turnover for a few or several decades. In steady-state systems, such as virgin forests in remote areas, the weathering rate could be derived at by a simple input–output budget. In reality, however, we are facing a number of transient processes like modern forestry and acid deposition causing depletion of the exchangeable base cation storage. This justifies the use of many different methods in estimating weathering rates as each may circumvent a particular pitfall.

2. Weathering rate estimation

Table 1 is certainly not complete but covers the most commonly used methods for estimation of weathering rates.

Historical weathering is deduced from the accumulation of conserved elements in soil profiles. Such elements may be zirconium, titanium in some instances, quartz (Teveldal *et al.*, this symposium), or opaque minerals like magnetite when using microscopic techniques (April *et al.* 1986). The soil profile needs to be dated if a mean yearly rate is to be arrived at. In glaciated terrains this is usually rather easy. Especially favourable are the situations when chronosequences of soil profiles can be used, such as shorelines (Bain *et al.*, this symposium) or profiles below a retreating glacier.

Input–output budgets may assume a steady state in which case it is in theory fairly simple to arrive at weathering rate estimates. In practice, however, the steady state is seldom achieved. Such things as oscillations in precipitation climate and after-effects of land use changes may cause problems.

Pačes (1986) has considered exchangeable base cation depletion by assuming that silica and sodium released by weathering are not subject to exchange reactions. He has further assumed that the weathering reactions can be summarized as the difference between the regolith and the bedrock. Velbel (1985) and Giovanoli *et al.* (1988) use a similar starting-point, however, on the basis of a more detailed mineralogical knowledge. They have then made a stochiometric adjustment of the different weathering reactions to the run-off chemistry. It is actually a modern version of the classical calculations by Garrels (1967) on the Sierra Nevada waters.

Laboratory experiments are important in studying processes, but weathering rates are often distorted by artefacts. Batch experiments in general do not give reasonable values, while soil column experiments may do so. However, the high ratio of exchangeable cations to yearly fluxes require lengthy experiments to detect any change in the exchangeable pools (Cronan 1985).

The strontium-isotope method uses an analogy between strontium and calcium based on the chemical resemblance of the two elements. Strontium has stable isotopes which are differently distributed in atmospheric deposition and in old granites and gneisses. ^{87}Sr is often elevated in granites and gneisses and in soils derived from them due to the radioactive decay of ^{87}Rb into strontium with the same mass number. The strontium isotope ratio, $^{87}Sr/^{86}Sr$, is a mixture of the ratio in deposition and in strontium derived from weathering. With a fair knowledge of the deposition rate it is then possible to get an estimate of the weathering rate of calcium using the strontium–calcium analogy.

The MAGIC model gives a weathering rate estimate based on soil chemistry data and time series of run-off chemistry. Catchment manipulations are very useful in providing data for the MAGIC model (Wright & Cosby 1987).

The model developed by Sverdrup & Warfvinge (1988) uses dissolution rates for individual minerals present in the soil profile. The dissolution rates may either be laboratory values or field estimates like those arrived at by Velbel (1985). The model is composed of a number of compartments describing each soil layer. The model requires extensive input data in the form of percentage of individual minerals, specific surface areas, selectivity coefficients for exchange reactions and rate of water percolation.

3. Scandinavian studies

Studies of weathering processes and rates were already started in the beginning of this century by Tamm in his work establishing a link between soil mineralogy and forest site quality. One of his earliest works concerns the historical weathering rate in a lake sediment that had been exposed for about 200 years following a catastrophic tapping of a lake (Tamm 1913). Rosenqvist (1977) pointed to the importance of weathering for the neutralization of protons in the soil system in discussing the effects of acid deposition.

In recent years Eriksson (1986) has published the results of a number of mass-balance studies. He gives the results as alkalinity production after having subtracted the base cations in atmospheric deposition. As most of the studies were performed in the period 1965–75 a variable amount of exchangeable base cation depletion is contained in the results, depending on the geographic position of the respective catchments. For catchment studies in Norway, Sweden and Finland Eriksson found between 33 and 115 mequiv. $m^{-2} a^{-1}$. Values in excess of about 70 mequiv. $m^{-2} a^{-1}$ were found for soils presumably being calcareous.

One of the aims of the SWAP project has been to apply different methods in the same area or catchment. This has been achieved to a considerable extent. In table 2 below is shown a comparison for some catchments.

The scale used for acid deposition is from high to low, covering the gradient from north to south in Sweden and Norway. The mineralogy is described by the terms poor and fair. Poor soils are derived from acid granites and gneisses, while fair means soils derived from intermediate rocks. In such soils the content of dark, rather fast weathering minerals like hornblende and biotite is *ca.* 5–10%.

In general the historical rates are the lowest. Bain *et al.* (this symposium) have found an exponential decrease in weathering rate in a chronosequence in Scotland. That current rates in general are likely to be higher due to induction by the increased acidity found in many soils in southern Sweden and Norway.

The values found with the strontium method for Svartberget are too high. This depends most probably on the special mineralogy of the site where biotite amounts to about 10% and is a major source for cations in run-off water. The biotite has an elevated $^{87}Sr/^{86}Sr$ ratio. Using a citrate extract (Lundström, this symposium) of the soil as one end member of the deposition-weathering mixture yields, about 40 mequiv. $m^{-2} a^{-1}$ which is in level with other estimates.

The soil profile budgets yield higher values than catchment studies as is seen in the studies by U. Lundström (this symposium) and G. Jacks (unpublished work). This may be due to the fact that the drainage depth at least in Buskbäcken-Masbybäcken is very shallow, far less than 2 m (Lundin 1982). According to Lundin most run-off water takes a very shallow pathway.

The effect of the mineralogy is seen in the table as the difference between the Svartberget and Kullarna catchments and Buskbäcken-Masbybäcken. Another example could be mentioned. Gårdsjön in southwestern Sweden has, according to Olsson & Melkerud (1989), a historical weathering rate of 82 mequiv $m^{-2} a^{-1}$. According to Melkerud (1983) the rather thin soils in Gårdsjön have a favourable

Table 2.

site	atm.H$^+$	minera-logy	mequiv. m^2 a	method	references
Svartberget	low	fair	85	budget[a]	Lundström (this symposium)
			31	hist.	Olsson & Melkerud (1989)
			150	Sr	Jacks *et al.* (1989)
Kullarna	med–low	fair	41	budget	Rosén (1982)
			30	hist.	Olsson & Melkerud (1989)
			40	Sr	Jacks *et al.* (1989)
Masbybäcken–Buskbäcken	med.	poor	12	hist.	Olsson & Melkerud (1989)
			15	hist.	G. Jacks (unpublished work)
			20	Sr	Åberg *et al.* (1989)
			55	budget[b]	G. Jacks (unpublished work)
Sogndal	low	fair	56	hist.	Wright & Frogner (1989)
			16	budget	Wright & Frogner (1989)
			7–11	MAGIC	Wright & Frogner (1989)
Risdalsheia	high	poor	10–16	MAGIC	Wright & Frogner (1989)
			4.3	hist.	Wright & Frogner (1989)
Södertörn	med	fair	20–25	budget[c]	Jacks *et al.* (1990)
minicatchments	med	poor	7	budget	Pačes & Jacks (1987)

[a] Budget for lysimeters to 0.8 m depth.
[b] Budget for soil profile 2.0 m deep considering Si conservative.
[c] Budget for data from 1947.

mineralogic composition with a percentage of dark minerals in excess of 5%. Naturally the mild climate has also contributed to the high figure for historical weathering.

The MAGIC modelling for Sogndal and Risdalsheia indicates that this may be a very useful tool in deriving weathering rates. Two studies from similar environments outside Stockholm are included for comparison (Jacks *et al.* 1990; Pačes & Jacks 1987). The Södertörn budget is a lake catchment study on the basis of data from 1947 when sulphur deposition was still rather low. The minicatchments are shallow rock-bowls filled with a few centimetres of mostly organic soil. The budget uses silica as an element unaffected by ion exchange.

The Gårdsjön catchment has been extensively used for modelling by Sverdrup & Warfvinge (1988) yielding weathering rate values in the same order as the historical estimate (Olsson & Melkerud 1989).

4. Weathering rate as affected by acidity

Weathering rate dependence on pH of the solution has been extensively investigated in laboratory tests. Feldspars seem to dissolve at a rather constant rate between pH 4 and 8, the conditions mostly met with in nature (Chou & Wollast 1985). Hornblende seems to dissolve with a rate one order faster for each pH step below pH 6. However, the presence of alumium in solution may even slow down the dissolution rate as pH drops. According to Schnoor & Stumm (1986) the dissolution rate of individual minerals depends on the hydrogen ion activity with an exponent of 0.3–0.7. Overwhelming field evidence indicates that the exponent is less than unity.

Field investigations indicate both increased and decreased base cation release under the influence of acid deposition in nearby sites with similar mineralogy (April *et al.* 1986). In that case the different response of two adjacent catchments may be due to depletion of exchangeable base cations in the catchment with larger soil depth. Teveldal *et al.* are performing laboratory experiments with soil columns indicating that present day rates may be 5–6 times higher than the long-term rates.

A very important aspect dealt with by Lundström (this symposium) is the effect of organic acids in natural soil solutions. The weathering rate is 2.4–2.7 times higher in coloured stream water and peat extract than in distilled water at pH 5.1. If the acidification in some way affects the formation dissolved organic matter, this may have far reaching effects on the weathering. A recent investigation on adsorbed anions in the mineral soil carried out on samples collected from the same site in 1951 and 1989 indicate that the amount of adsorbed organic anions has increased significantly over the almost four decades (Gustafsson & Jacks 1990).

5. Conclusions

There is a fair agreement between different methods used in assessing weathering rate. Deviations can be logically explained. Thus it is now possible to make intelligent guesses about weathering rates for typical Scandinavian landscape types knowing a few basic facts about mineralogy, texture and soil depth. The till terrain characterized by a silty-sandy till to a depth of 1–5 m show a weathering rate of 20–50 mequiv. $m^{-2} a^{-1}$. If the soil is derived from acid granites and gneisses the figures will cover the lower range from 20 to 30 mequiv. $m^{-2} a^{-1}$ while tills formed from intermediate granites and gneisses having dark minerals in percentages of around 5 % will be found in the upper range of the interval, i.e. from 30 to 50 mequiv. $m^{-2} a^{-1}$. Another typical environment common in Norway and in coastal parts of Sweden and Finland is rocky terrains with thin soil covers (less than 0.5 m). In areas with acid rocks such a terrain seems to produce about 5–15 mequiv. $m^{-2} a^{-1}$ while with a better mineralogy the figure may be 10–20 mequiv. $m^{-2} a^{-1}$.

The modelling approaches, MAGIC (Wright & Cosby 1987) and PROFILE (Sverdrup *et al.* 1990), may be very useful in extrapolating results both in time and space. These models, however, do need independent estimates for their verification. The SWAP work on field estimates of weathering will be a good basis for further development of the models.

It is obvious that complex organic acids are important in connection with weathering in the upper portions of the mineral soil. Acidification affects the distribution of different organic carbon fractions in soil solutions (Vance & David 1989). How this in turn will affect weathering rates is an unresolved but important question.

References

Åberg, G., Jacks, G. & Hamilton, P. J. 1989 *J. Hydrology* **109**, 65–78.

Åberg, G., Jacks, G., Wickman, T. & Hamilton, P. J. 1990 *Catena* **17**, 1–11.

April, R., Newton, R. & Truettner Coles, L. 1986 *Bull. Geol. Soc. Am.* **97**, 1232–1238.

Chou, L. & Wollast, R. 1985 *Am. J. Sci.* **285**, 963–993.

Cronan, C. S. 1985 In *The chemistry of weathering* (ed. J. I. Drever), pp. 175–195. Reidel.

Eriksson, E. 1986 In *Critical loads for nitrogen and sulphur* (ed. J. Nilsson). Nordic Council of Ministers, Report 1986:11.

Frogner, T. 1990 *Geochim. cosmochim. Acta* **54**, 769–780.

Garrels, R. M. 1967 In *Researches in geochemistry* (ed. Abelson), pp. 405–420. New York: Wiley.

Giovanoli, R., Schnoor, J. L., Sigg, L., Stumm, W. & Zobrist, J. 1988 *Clays Clay Minerals* **6**, no. 6.

Gustafsson, J. P. & Jacks, G. 1990 Progress report to the Swedish Forest and Agricultural Research Council. (12 pages.)

Jacks, G., Åberg, G. & Hamilton, P. J. 1989 *Nordic Hydrology* **20**, 85–96.

Jacks, G., Stegman, B. & Puke, C. 1990 (Submitted.)

Lundin, L. 1982 Dep. of Physical Geography Univ. of Uppsala, Report UNGI no. 56. (216 pages.)

Melkerud, P. A. 1983 Dept. of Forest Soils, Swedish University of Agricultural Sciences, Report 44.

Nordic Council of Ministers 1988 *Critical loads for sulphur and nitrogen* (ed. J. Nilsson). Miljörapport 1988: 15.

Olsson, M. & Melkerud, P. A. 1989 *Proc. Symp. on Environmental Geochemistry in Northern Europe, Rovaniemi, Finland, October 1989.*

Pačes, T. 1986 *J. Geol. Soc. Lond.* **143**.

Pačes, T. & Jacks, G. 1987 In *Acidification and water pathways. Int. Conf. in Bolkesjø, Norway, May 1987*, vol. 1, pp. 207–222.

Rosén, K. 1982 Dept. of Forest Soils, Swedish University of Agricultural Sciences, Report 41.

Rosenqvist, I. T. 1977 *Sur jord, surt vann.* Ingenjørforlaget, Oslo. (123 pages.)

Schnoor, J. L. & Stumm, W. 1986 *Schweizerische Z. Hydrologie* **48**, 171–195.

Sverdrup, H. U. & Warfvinge, P. G. 1988 In *Critical loads for sulphur and nitrogen.* (ed. J. Nilsson & P. Grennfelt), pp. 81–130. Nordic Council of Ministers, Report 1988: 15.

Sverdrup, H. U., Warfvinge, P. G. & Olausson, S. 1990 *Soil. Sci. Soc. Am. J.* **54**.

Tamm, O. 1913 *Geol. För. Förhandl* **35**.

Vance, G. F. & David, M. B. 1989 *Soil. Sci. Soc. Am. J.* **53**, 1242–1247.

Velbel, M. A. 1985 *Am. J. Sci.* **285**, 904–930.

Wright, R. F. & Cosby, B. J. 1987 *Atmos. Environment* **21**, 727–730.

Wright, R. F. & Frogner, T. 1989 Final report to SWAP for the period 1985–88. (12 pages.)

Discussion

R. A. SKEFFINGTON (*National Power Technology and Environmental Centre, Leatherhead, U.K.*). I do not underestimate the importance of weathering, but it is not the only significant long-term sink for protons in ecosystems. Biologically mediated processes such as denitrification, sulphate reduction where there is a gaseous product, and protonation of organic anions followed by oxidation to CO_2 and water can all be very significant sinks. These are hard to quantify, but should not be forgotten.

G. JACKS. Sulphate reduction is indeed important in catchments with lakes. Denitrification is as yet of minor importance in Scandinavia as most forest ecosystems are able to retain the nitrogen deposition. In wetlands draining agricultural areas denitrification is an important process. Finally the organic matter turnover is indeed difficult to quantify.

R. A. SKEFFINGTON. We have heard a number of good site-specific studies on weathering, and I would like to ask whether it is now possible to deduce weathering rates on a given site from more readily observable properties. I ask in particular because Dr Harald Sverdrup has developed a model that he claims can predict weathering rates based on a knowledge of the content of specific minerals in soils and rocks. Does Dr Jacks feel this approach is valid, and if not what reservations does he have? On the basis of the SWAP studies, could he suggest any further approaches?

G. JACKS. The models developed by Sverdrup and his co-workers are very promising tools. Their data requirement is still quite difficult to meet, but as more experience is gained the models may be simplified in this respect. As is already said other methods are most important for verification of the modelling approaches.

Weathering in Scottish and Norwegian catchments

By D. C. Bain, A. Mellor, M. J. Wilson and D. M. L. Duthie

Macaulay Land Use Research Institute, Craigiebuckler, Aberdeen AB9 2QJ, U.K.

Long-term weathering rates calculated from soil analyses using Zr as an internal immobile element indicate that base cations are lost at rates of 26–72 mequiv. $m^{-2} a^{-1}$ in the three Scottish catchments, the magnitude of the loss being in the order K > Na > Mg > Ca at Allt a'Mharcaidh and Kelty Water, and Na > K > Mg > Ca at Loch Chon. However, Ca and Na are the most mobile cations when abundances in the parent material are taken into account. Current rates calculated from stream chemistry are higher at Allt a'Mharcaidh but lower in the other two. At Høylandet the long-term rate of loss of base cations is 124 mequiv. $m^{-2} a^{-1}$, the values for Ca and Mg being up to 30 and 9 times greater, respectively, than for the Scottish catchments due to the more basic nature of the bedrock. In a chronosequence of seven profiles aged from 80 to 13000 years before present (BP) from Glen Feshie, the weathering rate decreases exponentially with time. Studies of the soils and particle-size fractions have indicated that there are four main weathering processes: (1) decomposition of chlorite with no identifiable weathering product, particularly at Loch Ard and Høylandet; (2) vermiculitization of mica by loss of interlayer K to form interstratified mica-vermiculite with concomitant formation of hydroxy-aluminium in the interlayer space of the vermiculite, particularly in B horizons, in all catchments; (3) weathering of plagioclase feldspar especially in the Allt a'Mharcaidh soils where the effect is most pronounced in the coarse sand fraction; (4) precipitation of imogolitic/allophanic material in B horizons at Allt a'Mharcaidh.

1. Introduction and objectives

Chemical weathering of primary silicate minerals is the principal mechanism by which inputs of acids to soils are counteracted, and the acid neutralizing capacity of soils replenished. It is important, therefore, to understand the processes that occur in soils relating to reactions between soil minerals and percolating waters, particularly those reactions that function as proton sinks and at the same time release base cations (Wilson 1986). It is also important to establish the rate of chemical weathering as increased acidification of soils will occur only if the rate of release of base cations is insufficient to counteract the acidic inputs whether from internal soil processes or from the atmosphere.

The object of this paper is to report the results of a mineral weathering study in the SWAP catchments at Allt a'Mharcaidh, Loch Chon and Kelty Water in Scotland, and at Stor Gronningen, Høylandet, Norway. This study encompasses (1) determination of long-term weathering rates in the catchments from soil chemical data; (2) calculation of current weathering rates from water input/output budgets, and using strontium isotope ratios; (3) studies of the products and processes of weathering of minerals in soil profiles.

[223]

2. Methods

The dried soil was separated into particle-size fractions using standard sedimentation procedures for the clay (less than 2 µm) and silt (2–20 µm) fractions, and sieving for the fine (20–200 µm) and coarse (200–2000 µm) sand fractions.

Chemical analyses of the soils and particle-size fractions were obtained by X-ray fluorescence spectrometry (XRF) using a Philips PW1404 spectrometer and the methods of Norrish & Hutton (1969) for sample fusion and for corrections for interelement effects. In the analysis of Zr, the Kα intensity was corrected for overlap by Sr Kβ, and matrix effects were reduced by an internal ratio technique using the Rh Kα line from the X-ray tube.

X-ray diffraction (XRD) traces were obtained with a Philips 2kW diffractometer with Fe-filtered Co-Kα radiation. For quantitative determinations, the soil and sand fractions were ground to reduce particle size to less than 15 µm. The quantitative XRD method used was essentially that developed by Chung (1974) by which no calibration is necessary as all the absorption and matrix effects are corrected for mathematically.

Soil samples (less than 2 mm) were digested in HF for determination of $^{87}Sr/^{86}Sr$ ratios by thermal ionization mass spectrometry following separation by cation exchange chromatography (Faure 1986).

3. Chemical weathering rates

3.1. *Long-term chemical weathering rates*

The losses of base cations from soil profiles were calculated by comparing the chemical composition of each horizon with that of the C horizon which is taken to represent the unweathered parent material. This requires the use of a reference mineral or element which is considered to be stable and immobile during pedogenic weathering, and Zr was chosen for this purpose. The true quantity X_i of an element X of relative amount x_i in the horizon of interest is obtained from the formula:

$$X_i = x_i(q_c/q_i),$$

where q_c and q_i are the relative amounts of the stable element Q in the C horizon and the horizon of interest. The loss of an element from a horizon was calculated using the formula of Meilhac (1970):

$$\Delta X_i = q_i h_i d_i (X_i - X_c)/q_c,$$

where h_i and d_i are the thickness and bulk density of the horizon of interest, and X_c is the content of element X in the reference horizon C. For the whole profile, the overall loss is the sum of the calculated losses from each horizon.

To convert these losses into weathering rates, all the profiles were assumed to be 10000 years old as they are all post-glacial in age. The calculations are valid only if the soil profiles have developed on uniform parent material, and those profiles which appeared from their chemistry and mineralogy to have developed on non-uniform parent material were excluded.

Table 1 shows the long-term weathering rates for Ca, Mg, K and Na in the three Scottish catchments, and at Høylandet. The figures for the Scottish catchments are

Table 1. *Chemical weathering rates* (mequiv. m^{-2} a^{-1})

catchment		Ca	Mg	K	Na	Total
Mharcaidh:	long-term rate ($n = 6$)					
	range	1–6	2–5	5–68	1–52	14–129
	standard deviation	1.9	1.1	21.8	16.5	39.8
	mean	3	3	21	18	45
	current rate	19	5	1	24	49
Kelty:	long-term rate ($n = 4$)					
	range	1–2	0.1–6	2–18	6–15	12–38
	standard deviation	0.3	2.1	6.1	3.6	11.0
	mean	2	3	11	10	26
	current rate	11	2	< 0.1	6	19
Chon:	long-term rate ($n = 5$)					
	range	3–7	2–23	2–48	11–71	29–133
	standard deviation	1.3	4.2	15.0	20.2	34.7
	mean	5	12	23	32	72
	current rate	23	21	< 0.1	< 0.1	44
Høylandet:	long-term rate ($n = 1$)	57	24	10	33	124

n = number of profiles used in calculation.

Table 2. *Average chemical composition (in wt% air-dry basis) of C horizons used in weathering rate calculations*

	Allt a'Mharcaidh	Kelty Water	Loch Chon	Høylandet	Glen Feshie
SiO_2	73.2	70.5	69.0	68.1	74.8
TiO_2	0.25	0.88	0.99	0.85	0.44
Al_2O_3	14.1	13.1	13.5	12.0	12.3
Fe_2O_3[a]	1.54	6.06	6.08	5.44	2.28
MgO	0.32	1.38	1.37	2.52	0.50
MnO	0.05	0.06	0.10	0.12	0.05
CaO	0.44	0.21	0.28	3.68	0.58
Na_2O	3.10	1.51	2.36	2.48	2.49
K_2O	4.49	2.40	2.18	1.12	3.55
P_2O_5	0.06	0.15	0.12	0.13	0.07
LOI[b]	2.26	3.82	4.19	3.63	3.13
total	99.81	100.07	100.17	100.07	100.19

[a] Total iron as Fe_2O_3.
[b] Loss on ignition at 1000 °C.

based on chemical analyses of 4–6 profiles per catchment, whereas data for only one profile at Stor Grönningen were available for Høylandet. The magnitude of the loss is in the order K > Na > Mg > Ca at Allt a'Mharcaidh and Kelty Water, and Na > K > Mg > Ca at Loch Chon. It is clear that the soils have weathered fastest at Loch Chon (72 mequiv. m^{-2} a^{-1} for total base cations) and slowest at Kelty Water, so the ability of the soils in the catchments to neutralize incoming acidity decreases in the order Loch Chon, Allt a'Mharcaidh, Kelty Water.

The figures for Høylandet are quite different to those for the Scottish catchments. As they are based on one profile only, they may not be typical, but as the Ca, K and Na values are higher by factors of 2–10 than for any individual Scottish profile, it is likely that the rates are genuinely higher at Høylandet. One of the Loch Chon profiles

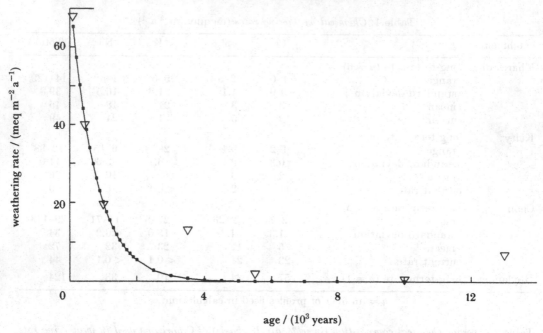

Figure 1. Plot of weathering rate of Ca against age of soil profile for chronosequence of seven profiles from Glen Feshie. The mathematical expression of the calculated curve is $y = 74.35\exp(-0.00131t)$, where y is the weathering rate and t is the age. ■, Calculated points; ▽, measured data points.

has a similar rate of loss of Mg. The much higher rates of loss of Ca and Mg and lower rate of loss of K at Høylandet are due to the more basic nature of the parent material (table 2).

3.2. *Variation in chemical weathering rates with time*

The long-term weathering rates calculated for the catchments are rates of loss averaged over 10000 years. As the rates may vary with time, a chronosequence of soil profiles developed on river terraces in Glen Feshie (near the Allt a'Mharcaidh catchment) was studied. These profiles were dated as 13000, 10000, 5500, 3500, 1000, 450 and 80 years BP by statistical methods on quantitative soil stratigraphic data with several methods of absolute dating control (Robertson-Rintoul 1986), and the chemical weathering rates calculated using Zr as for the catchments. The magnitude of the loss is in the order Na > K > Mg > Ca, the same sequence as at Loch Chon, in all but one profile (450 years BP). The rate of loss decreases exponentially with time for all the base cations, as shown in figure 1 for Ca.

3.3. *Current weathering rates*

Apparent weathering rates can be calculated from analyses of the input water to the catchments and the streams draining the catchment. The figures obtained by subtracting the amount of the base cations in the input from the amount in the output, adjusted to make chloride values constant, are listed in table 1, based on analyses of samples taken over a six-month period. Comparison with the long-term rates (averaged over 10000 years) indicates that the base cations are currently being depleted more quickly at Allt a'Mharcaidh but more slowly at Loch Chon and Kelty

Water. The current figures for Ca and Mg at Loch Chon are distorted by the influence of a dolerite dyke at the head of the catchment; they would be much lower if the only source of cations was quartz-mica-schist.

A modified budget approach can be used whereby the difference in the ratio between the two naturally occurring Sr isotopes in the deposition and the parent material is used to determine the amount of cations derived from the parent material (Jacks & Åberg 1987). The weathering rate can be calculated from the equation:

$$R_w = D(m - a/s - m),$$

where the R_w is the base cation weathering rate, D is the base cation deposition rate, a is the $^{87}Sr/^{86}Sr$ ratio in the deposition, m is the Sr isotope ratio in the run-off, and s is the Sr isotope ratio in the soil matrix. At Loch Chon, the weathering rate for Ca was calculated from the relevant Sr isotope ratios and Ca in the precipitation:

$$R_w = 40.9(0.72051 - 0.70959/0.733797 - 0.72051)$$
$$= 33.6 \text{ mequiv. } m^{-2} a^{-1}.$$

The $^{87}Sr/^{86}Sr$ ratio used for the soil matrix (s) is the average of 21 determinations of horizons in soil profiles in the catchments. Despite the fact that this makes no allowance for the influence of the dolerite dyke, the rate is higher than the value using budgets only (table 3).

4. Mineralogy

4.1. *Quantitative* XRD

To relate long-term weathering rates to mineral weathering in the soils, quantitative determinations were made by XRD for the air-dried soil and the particle-size fractions which comprise the bulk of the soil. The clay fraction is usually less than 1 % at Allt a'Mharcaidh and Høylandet and the sand fraction comprises 80–100 %, with the silt varying from 1–18 % and usually less than 10 % at Allt a'Mharcaidh. The soils at Loch Chon also have a small clay fraction, less than 4 %, and often less than 1 % whereas at Kelty Water the clay varies from 2 to 13 %, the silt fraction comprising 10–27 % at both sites.

The amounts of the minerals in the top and bottom mineral horizons in one profile from each catchment are illustrated in figures 2–5. The long-term weathering rates calculated for these particular profiles are indicated in the legends so that a proper comparison can be made between weathering rates and mineralogy.

At Allt a'Mharcaidh (figure 2), the high loss of Na is clearly related to the decrease in plagioclase feldspar (oligoclase) from 44 % in the C horizon to 31 % in the AEh horizon. Loss of K is associated with weathering of mica and also alkali feldspar which, although increasing in concentration from 23 % in the C to 31 % in the AEh horizon, increases less than the more resistant quartz (17–30 %) and therefore actually decreases in amount. Dissolution features on alkali feldspar grains when examined by scanning electron microscopy confirmed that the feldspar is weathering. Surprisingly, the decrease in plagioclase feldspar from 44 to 23 % in the coarse sand fraction and from 52 to 51 % in the fine sand indicates that weathering is more active in the coarse fractions of this alpine podzol.

The mineralogy of the soils at Kelty Water and Loch Chon is similar. The main changes are the decrease in chlorite and mica, and the increase in interstratified mica-vermiculite in the upper horizons (figures 3 and 4). The vermiculitization of mica by loss of interlayer K and the dissolution of chlorite with no identifiable weathering

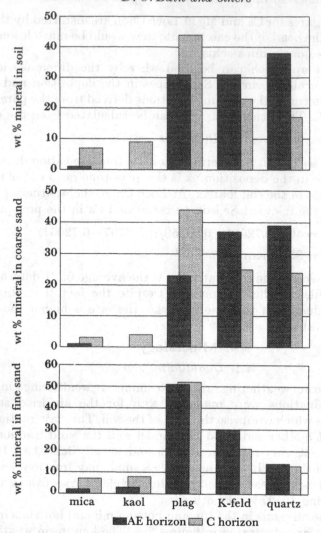

Figure 2. Mineralogy of AEh and C horizons of an alpine podzol from Allt a'Mharcaidh as determined by XRD. Kaol: kaolinite; plag: plagioclase feldspar; K-feld: potash feldspar. Weathering rates calculated from chemical analyses of this profile are: Ca-1, Mg-3, K-10 and Na-13 mequiv. m^{-2} a^{-1}.

product explain the higher losses of K and Mg compared with Allt a'Mharcaidh. Higher losses of Na at Loch Chon compared with Kelty Water probably relate to the weathering of plagioclase feldspar which is more abundant at Loch Chon. Mineral weathering losses occur in all the particle-size fractions in both catchments, but chlorite weathering appears to be particularly active in the silt fractions; amounts of this mineral decrease from 48 to 8% at Kelty Water, and from 64 to 3% at Loch Chon.

The main difference in the mineralogy of the Høylandet soil (figure 5) compared with the soils of the Scottish catchments is the occurrence of 10% amphibole due to the influence of basic material within the granitic gneisses. Quartz increases from 27% in the BC horizon to 39% in the E horizon so that the small decrease in concentration of amphibole (11–10%) will in fact mean a much larger decrease in

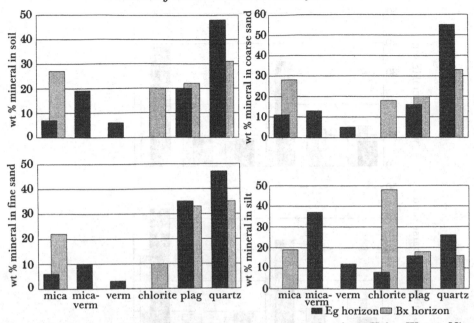

Figure 3. Mineralogy of Eg and Bx horizons of a peaty gley from Kelty Water. Mica-verm: interstratified mica-vermiculite; plag: plagioclase feldspar. Weathering rates calculated from chemical analyses of this profile are: Ca-2, Mg-6, K-13 and Na-13 mequiv. $m^{-2} a^{-1}$.

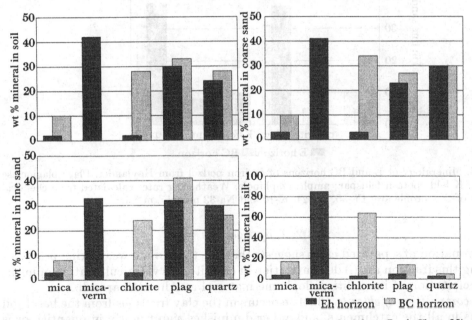

Figure 4. Mineralogy of Eh and BC horizons of an iron humus podzol from Loch Chon. Mica-verm: interstratified mica-vermiculite; plag: plagioclase feldspar. Weathering rates calculated from chemical analyses of this profile are: Ca-4, Mg-16, K-27 and Na-28 mequiv. $m^{-2} a^{-1}$.

actual amount, thereby explaining the high figure for loss of Ca from the profile. Chlorite is absent from the E horizon indicating that it has been weathered out in the soil.

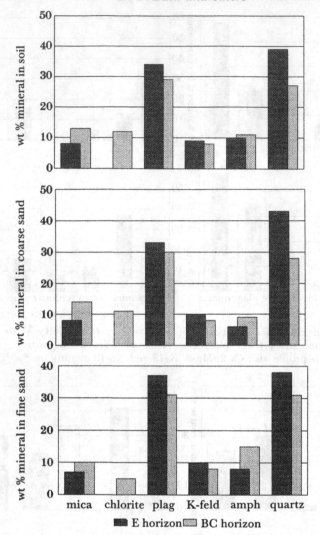

Figure 5. Mineralogy of E and BC horizons of an iron podzol from Høylandet. Plag: plagioclase feldspar; K-feld: potash feldspar; amph: amphibole. Weathering rates calculated from chemical analyses of this profile are: Ca-56, Mg-24, K-10 and Na-33 mequiv. m^{-2} a^{-1}.

4.2. *Clay fractions*

Examination of separated clays suggests that the three main weathering processes affecting this fraction are (i) decomposition of chlorite; (ii) vermiculitization of mica; (iii) precipitation of imogolitic/allophanic material at Allt a'Mharcaidh.

(i) *Decomposition of chlorite*. Chlorite occurs in the clay fractions from the basal soil horizons in all the catchments, and either diminishes significantly in quantity or is completely absent in the E horizons. As there is no detectable product, weathering of the chlorite appears to involve complete breakdown of the mineral structure. Chlorite decomposition is particularly noticeable at the Loch Ard sites where the mineral occurs in significant quantities.

(ii) *Vermiculitization of mica*. Interstratified mica-vermiculite is a common mineral in the clay fractions (and silt fractions) in all the catchments, and forms by the weathering of mica through loss of interlayer K. The interlayer space in the vermiculite component is often filled with varying amounts of polymeric hydroxy-aluminium, the degree of interlayering appearing to be pH-dependent with hydroxyaluminium being removed from the interlayer at pH below about 4.3 (in E horizons usually). A particularly good example of this interstratified mineral is the dominant phase in the clay fraction from the E horizon of a peaty podzol from Kelty Water and has been shown by XRD and XRF to be aluminous in character, and to have formed from a dioctahedral mica (Bain *et al.* 1990).

(iii) *Imogolitic/allophanic material.* The presence of imogolitic/allophanic material has been revealed by infrared spectroscopy in the clay fractions from B horizons of profiles from Allt a'Mharcaidh.

5. Discussion

The above results have highlighted a number of features that require discussion including the different weathering rates obtained by soil analysis compared with budget studies – bearing in mind the temporal variation of weathering rate indicated by chronosequence studies – the marked differences in the rate of release of individual cations, and comparison of the weathering rates with selected rates obtained from the literature.

In the Scottish catchments, the soil analysis weathering rate exceeds the budget weathering rate in Kelty Water and Loch Chon whereas the opposite is found in Allt a'Mharcaidh. The question arises as to whether such differences are real, or whether they are merely a reflection of the rather large uncertainties that are inherent in both methods of estimation. Further consideration of the results inclines the authors to the latter view. For example, the Glen Feshie chronosequence shows an exponential decline with age for the weathering rate of calcium, from an initial rate of 70 mequiv. $m^{-2} a^{-1}$ to less than 5 mequiv. $m^{-2} a^{-1}$ after 2000 years. This result can be explained in terms of initially rapid rates following from the exposure of fresh mineral surfaces and the presence of fine mineral particles in recently laid down sediments. After the highly reactive particles have been depleted, then weathering approaches a steady state. The soil analysis calcium weathering rate at Allt a'Mharcaidh is indeed very low (3 mequiv. $m^{-2} a^{-1}$) but the budget rate is six times higher. Similar discrepancies are noted for the other catchments. This can be accounted for at Loch Chon by the influence on the water chemistry of a base-rich doleritic dyke at the head of the catchment and it may well be that there are additional, so far unidentified, contributions to the outputs of the other catchments. These differences serve to illustrate the difficulty of comparing weathering rates from selected soil profiles (which may not be truly representative of the catchment) as in the soil analysis approach, with a method that integrates the net effect of processes operating over the entire catchment area, and perhaps outwith this area when the complicating effects of groundwater contributions are taken into account.

Other anomalies became apparent when the loss rates of K and Na, as adduced by the two different methods, are compared. According to the budget approach, little or no K is being lost from any of the catchments in the surface drainage whereas soil analysis indicates a mean loss of 11–23 mequiv. $m^{-2} a^{-1}$. These results are, however, not necessarily contradictory. It is well known that in vegetated catchments K is tightly held and cycled within the biomass, at least in the short term. Over geological periods, however, as is represented in the soil analysis approach, K would be removed in drainage waters during the initial phases of soil development and could also be removed by other means such as harvested biomass (trees), offtake by animals, mechanical erosion of K-bearing minerals, removal by aeolian action of K-rich ash following natural fires, etc. For Na, according to budget estimates, there are much smaller amounts being lost from Kelty Water and Loch Chon than from Allt a'Mharcaidh. Probably, this reflects the more Na-rich feldspars that are present in the Allt a'Mharcaidh soils, as well as additional groundwater contributions, although further work is required to confirm these surmises.

The results show that the relative amounts of bases lost from the catchments reflect the abundance of these bases in the parent materials of the soils. For the Scottish catchments, in particular, the highest rates of loss are for K and Na and the lowest for Ca. However, when base weathering rates are adjusted to take abundance in the parent material into account (by dividing the weathering rate by the amount of the element in the basal horizon), then Ca and Na rank as the most mobile cations, certainly for the Kelty Water, Loch Chon and Høylandet catchments. For Allt a'Mharcaidh, there are only slight differences in the mobility of Ca, Na and K, with Mg being the most mobile ion in this instance. Calcium is also the most mobile cation when assessed on the basis of budget weathering rates. It seems likely that plagioclase feldspar will be a key mineral in assessing the rates of Ca and Na weathering and that amphiboles may also be important, as for example in Høylandet.

Despite the difficulties inherent in obtaining weathering rates, it is significant that the results for the Scottish catchments do compare reasonably well with the rates obtained from other catchments developed on similar geological material in northern Europe. For example, the rates of total base cation loss as assessed by soil analysis and budget approaches respectively are 92 and 13 mequiv. $m^{-2} a^{-1}$ for Sogndal in Norway (Wright & Frogner 1987), 10–30 and 40–80 mequiv. $m^{-2} a^{-1}$ for Lake Gårdsjön in Sweden, (Sverdrup & Warfvinge 1988), and 40–110 and 20–230 mequiv. $m^{-2} a^{-1}$ for forest soils in Germany (Fölster 1985). These results compare with weathering rates from the Scottish catchments of 26–72 and 19–49 mequiv. $m^{-2} a^{-1}$.

Finally, it may be noted that when the SWAP catchments are considered individually, the slowest weathering rates are observed for Kelty Water and Allt a'Mharcaidh and the most rapid for Loch Chon and Høylandet, suggesting that the former pair would be more sensitive to acidification. The results for Høylandet also show that the area has a more base-rich geology than do to the Scottish catchments, a point that must be borne in mind when comparisons are made between catchments, particularly when citing Høylandet as an example of a pristine reference site.

The authors are indebted to Dr J. R. Bacon for measurement of strontium isotopes, Dr S. T. Buckland for calculation of the curve in figure 1, Mr J. D. Miller and Dr R. C. Ferrier for the current weathering rates, Dr M. S. E. Robertson-Rintoul for providing the soils and relevant data from Glen Feshie, Dr J. D. Russell for infrared information, and Mr B. F. L. Smith for mechanical analyses. Technical assistance provided by R. L. Perry, L. Thomson, J. J. Harthill and R. Heintze is gratefully acknowledged.

References

Bain, D. C., Mellor, A. & Wilson, M. J. 1990 *Clay Miner.* **25**. (In the press.)

Chung, F. H. 1974 *J. appl. Crystallogr.* **7**, 519–525.

Faure, G. 1986 *Principles of isotope geology*, 2nd edn. New York: Wiley.

Fölster, H. 1985 In *The chemistry of weathering* (ed. J. I. Drever), pp. 197–209. Reidel, Dordrecht: NATO ASI Series.

Jacks, G. & Åberg, G. 1987 In *Surface water acidification programme* (ed. B. J. Mason), pp. 120–131 (Proc. Mid Term Review Conf.). London: Royal Society.

Meilhac, A. 1970 *Thèse Spéc. Univ. Strasbourg.*

Norrish, K. & Hutton, J. T. 1969 *Geochim. cosmochim. Acta* **33**, 431–453.

Robertson-Rintoul, M. S. E. 1986 *Earth Surface Processes Landforms* **11**, 605–617.

Sverdrup, H. & Warfvinge, P. 1988 *Water Air Soil Pollution* **38**, 387–408.

Wilson, M. J. 1986 *J. geol. Soc. Lond.* **143**, 691–697.

Wright, R. F. & Frogner, T. 1987 In *Surface water acidification programme* (ed. B. J. Mason), pp. 106–119 (Proc. Mid Term Review Conf.). London: Royal Society.

Discussion

B. W. BACHE. At the 1983 discussion on 'Ecological Effects of Deposited Sulphur and Nitrogen Compounds', a paper on soil–water interactions pointed out that the composition of water entering streams and lakes is determined mainly by the interplay of the composition of the soil and rock components with the hydrological pathways of the percolating water. Examples were given to illustrate this for contrasting situations, and a plea was made for detailed site investigations to resolve the uncertainties that remain. SWAP has abundantly fulfilled its promise for detailed hydrology, and also in this morning's papers for detailed chemical weathering rates and mechanisms.

With regard to the soil, there are two topics on which we still need some clarification. The first is soil variability, with regard to both the mapped soil units within a catchment, and the chemical variability within a mapped unit; this is likely to be considerable, but is it important? If so, we need to quantify it. The second is to know the most appropriate measurements to make, out of the large number of possibilities, to give predictive ability to our soil chemical work.

May I suggest that in this discussion on weathering (i.e. the decomposition of soil minerals) we address two specific issues: the methodology used in quantitative weathering studies, and how reliable it is; and the extent to which proton consumption by weathering can balance acid inputs. On this second point, we have been shown data that indicates that some weathering reactions do not increase as pH drops from 8 to 4, but this seems contrary to chemical principles; if the concentration of a reactant increases, the position of equilibrium should change.

We can clarify some issues by considering the following simple reaction sequence, where M^+ indicates the alkaline and alkaline earth cations K^+, Na^+, Ca^{2+} and Mg^{2+}:

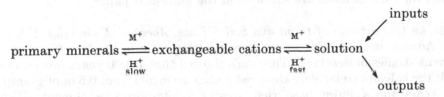

Quantitative weathering studies based in input–output balances are likely to be in error because most of the change will take place in the rapidly equilibrating exchangeable fraction, rather than in the slow mineral decomposition reactions that are the ultimate source of acid neutralization. However, the crucial issues are likely to revolve around the accessibility, in both space and time, of the proton-consuming decomposition reactions. The hydrological studies appear to show that the most damaging water flows occur where there is little possibility for acid neutralization by weathering reactions. If this is indeed so, how relevant is weathering to surface water acidification?

D. C. BAIN. Weathering in the soil is a steady, although slow, process of release of base cations determining the chemistry of streams during base-flow conditions which prevail throughout most of the year. When a short-term hydrological event occurs, changes in stream chemistry are superimposed on the base-flow chemistry, and although the extent to which the stream pH decreases depends on the chemistry of the event water, the actual pH reached will also depend on the initial, base-flow chemistry. Whether the critical pH for fish damage is reached or not will depend on how much the pH falls from the initial base-flow pH. Thus mineral weathering is relevant to surface water acidification even during events in determining the base-flow chemistry on which the event chemistry is superimposed.

M. F. BILLETT (*Department of Plant and Soil Science, University of Aberdeen, U.K.*). I raise two points about Dr Bain's presentation, the first one concerning the homogeneity of C horizons in glaciated areas and secondly a plea for some statistics. Deep mineral soil horizons in glaciated Scottish terrains can be extremely variable and may contain clasts of very different chemical composition from the underlying bedrock. These clasts may be relatively unweathered and easy to recognize or highly weathered and very difficult to recognize in the soil profile. This can lead to a great deal of variability in small areas, particularly in catchments such as Loch Ard and Loch Chon, which also occur on the Dalradian Schists, possibly the most lithologically variable geological sequence in Scotland. How does Dr Bain deal with this problem of soil heterogeneity when comparing long-term weathering rates between SWAP catchments? It would also be interesting to know how much variability there was in the calculation of weathering rates for the individual catchments and if the differences were statistically significant?

D. C. BAIN. One of the assumptions which had to be made in calculating long-term weathering rates was that each soil profile had developed on uniform parent material. There is no doubt that this assumption is not always valid, and where there was evidence from the chemistry and/or mineralogy that a profile had not developed on a uniform parent material, this profile was excluded from the average values calculated from each catchment. As regards statistics, the range and standard deviation for each element are included in the published paper.

M. CRESSER (*Department of Plant and Soil Science, Aberdeen University, U.K.*). Like Dr Bill Adams, we were also assessing weathering rates in the late 1970s but from catchment studies in Scotland. This work showed that 11000 years was required to strip all the calcium (relatively the most mobile element) from 0.5 m of granite, but 22000 years for a more base rich, quartz-biotite-norite catchment. We were impressed by the systematic and random variability within catchments. Weathering and degree of sulphate saturation were more advanced on higher, steeper slopes, and slopes with shallower soils, as soil scientists would expect. Because of chemical variability, and variability in hydrological behaviour such as the irregular distribution of perched water tables, in our research funded by the Department of the Environment over the past seven years we have adopted a rather different approach to that of the SWAP participants. We have concentrated upon prediction of water solute chemistry from soil and catchment characteristics. My colleague, Dr Mike Billett, is having considerable success in using the ratio of exchangeable Ca to C.E.C. to predict river water Ca. A similar approach may be used to predict Na, K, Mg and H^+. A hundred randomly generated points are used to decide upon the contribution to total discharge from each soil horizon close to the river. I feel this is a more useful approach than the highly detailed soil studies that have been presented today, even though these are of excellent quality. The exchange complex effectively provides an overview of the weathering status, all atmospheric and lateral flow inputs and biological effects and hydrological pathway effects. Thus we believe the distribution of cations on the exchange complex provides an integrated overview of the status of all the important processes and contributions.

D. C. BAIN. The distribution of cations on the exchange complex is undoubtedly important in the ability of a soil to neutralize incoming acidity, and the use of the ratio of exchangeable cation to C.E.C. to predict river water cation content seems an interesting approach. What this approach does not seem to address, however, is the ability of the soil matrix as a whole to replenish exchangeable cations which are lost from the exchange complex, i.e. it does not take into account the effect of the different weathering rates of different minerals in catchments of different mineralogy. The influence of mineralogy has been clearly demonstrated in the difference in long-term weathering rates between the Allt a'Mharcaidh and Høylandet catchments.

D. W. JOHNSON (*Desert Research Institute, Reno, Nevada, U.S.A.*). We typically conceptualize soil cations as being either exchangeable or mineral. What about organic pools? Has anyone considered whether soil organic matter is a significant source or sink for base cations? Forest biomass and litter certainly can be.

D. C. BAIN. It is very likely that soil organic matter can be a source or a sink for base cations. It is well known that K is tightly held and cycled within the biomass, at least in the short term.

R. A. SKEFFINGTON (*National Power Technology and Environmental Centre, Leatherhead, U.K.*). We have heard a number of good site-specific studies on weathering, and I would like to ask whether it is now possible to deduce weathering rates on a given site from more readily observable properties. I ask in particular because Dr Harald Sverdrup has developed a model which he claims can predict weathering rates based on a knowledge of the content of specific minerals in soils and rocks. Do speakers feel this approach is valid, and if not what reservations do they have? On the basis of the SWAP studies, could they suggest any further approaches?

I do not underestimate the importance of weathering, but it is not the only significant long-term sink for protons in ecosystems. Biologically mediated processes such as denitrification, sulphate reduction where there is a gaseous product, and protonation of organic anions followed by oxidation to CO_2 and water can all be very significant sinks. These are hard to quantify, but should not be forgotten.

D. C. BAIN. We have some reservations about predicting weathering rates at specific sites from the contents of specific minerals in the soils and rocks because there is very little information available about the weathering rates of these minerals in different environments. The approach needs to be refined by quantifying the weathering rates of the individual minerals in different environments.

J. MULDER (*Center for Industrial Research, Oslo, Norway*). In discussing the aluminium sinks in mineral soils you only consider inorganic forms (interlayer Al, imogolite/allophane) and not organic forms. Why not?

D. C. BAIN. We studied only the mineral horizons in the soil profiles where interlayer Al and imogolite/allophane are likely to be most important sinks for Al. In the overlying more organic horizons the situation will almost certainly be quite different.

Long-term translocation of iron and aluminium in podzols

By MATS OLSSON AND PER-ARNE MELKERUD

Swedish University of Agricultural Sciences, Department of Forest Soils,
S-750 07 Uppsala, Sweden

The long-term translocation and weathering losses of Fe and Al were investigated for podzols on sandy loamy till composed of acid Precambrian basement rocks. No stratification in grain size distribution or mineralogical composition occurred. The different geochemical properties in C horizons and climatic conditions of the studied sites were representative of a large part of Sweden. At each site, 2–3 profiles were sampled at about 10 different levels to a depth of about 1.5 m. Sodium dithionite citrate extractable Fe and Al were determined. These fractions were assumed to correspond to illuvial Fe and Al compounds that were precipitated and accumulated in spodic Bs horizons during soil formation. Total concentrations of elements in soil mineral material of less than 2 mm were analysed by the ICP technique. The results were used for calculations of long-term weathering losses of soil mineral material from the profile. Zr was used as an internal standard. The losses were compared with runoff data from catchment studies and from lysimeter investigations. Long-term weathering rates were determined as the sum of precipitation in Bs horizons and losses from the profile. The weathering so determined was considered to have occurred after deglaciation and exposure to the atmosphere. For this reason it was named historical weathering.

The results showed increased Zr concentrations from 200–400 p.p.m. (parts per million) in C horizons to 250–650 p.p.m. in A and E horizons. This increase was due to a loss of soil material amounting to an average of 2–20 g m^{-2} a^{-1}. The relations between the site properties and the losses of Fe and Al were expressed as regressions. The loss of Fe showed good correlation with the products of temperature and concentration of Fe in the C horizon. For Al, the products of temperature and (Ca + Mg) concentration in C horizons gave the best correlation.

Historical weathering rates of Fe and Al were estimated by adding precipitation in Bs horizons to the historical losses. The amount of annually precipitated Al in Bs constituted only a minor part of the annually weathered Al and was not primarily regulated by weathering rates. This explained why the highest long-term losses were recorded at the best sites, i.e. those with richer mineralogy and warmer climate. A similar situation was true also regarding Fe. The losses of Fe were, however, generally lower than those for Al due to lower weathering rates. Long-term losses of Al were substantially lower than recent leaching from the profile, measured by lysimeter investigations. Fe showed the opposite behaviour.

Table 1. *Location and properties of studied sites*
(Concentrations of Ca, Mg, Al and Fe and base mineral index refer to C horizons.)

	Skv	Gå	Ma	Ri	Ku	Sv
latitude (°N)	56.4	58.2	59.9	60.3	62.1	64.4
altitude (metres a.s.l.)	135	125	335	220	390	290
age (years × 1000)	12	12	9.7	9.5	9	8.8
Ca (%)	1.02	1.61	0.86	0.99	1.14	1.27
Mg (%)	0.25	0.72	0.29	0.52	0.57	0.48
Al (%)	6.25	7.55	5.69	6.31	7.00	6.48
Fe (%)	2.80	4.11	1.24	1.90	3.27	1.78
base mineral index	6.5	14.0	8.7	11.1	8.9	8.9

1. Introduction

Leaching of Al and Fe from podzols on glacial till in Sweden has been recorded through lysimeter investigations (Bergkvist 1986) and catchment studies (Grahn & Rosén 1983). It was shown that leached Al originated from the Bs horizons in the mineral soil (Bergkvist 1986). As the studied soils were podzols one might expect that Fe and Al should be retained in the Bs horizons in organically complexed form (Petersen 1976) or as precipitated proto-imogolite (Farmer *et al.* 1980). Hence, the recent leaching of Al is interpreted as a consequence of increased solubility resulting from atmospheric deposition of acid substances subsequently decreasing pH in the soil. However, the podzolization process does not necessarily result in a complete retention of Fe and Al, as indicated by Petersen (1976). Certain amounts of these elements may pass or might have passed through the Bs horizons. The aim of this study is to quantify translocation and natural long-term losses of Fe and Al.

Because soil processes alter the soil material, present chemical properties reflect loss and translocation taken place since deglaciation and exposure to the atmosphere and biosphere. This enables a quantification of loss and translocation based on chemical properties of soils developed on tills, in a variety of geochemical and climatical conditions. The estimated loss and translocation were characterized as historical and reflect long-term alterations. By comparing historical losses with recent leaching it was possible to elucidate the influence of acid atmospheric deposition.

2. Methods

The study sites were named Skånes Värsjö (Skv), Gårdsjön (Gå), Masbyn (Ma), Risfallet (Ri), Kullarna (Ku) and Svartberget (Sv). They were chosen partly because they represent a variety of different site conditions (table 1) and partly because they were subjected to leaching studies in catchment areas or in lysimeter investigations. All of them are located on sandy loamy till (composed of predominantly archaean acid bedrock material) deposited above the highest coastline. No stratifications in grain-size distribution or mineralogical composition occurred. The study sites have been exposed to the atmosphere and biosphere for different lengths of time after deglaciation (table 1).

Temperature sums (cumulative daily mean air temperatures above 5 °C) were calculated from the latitudinal and altitudinal positions of the sites according to Odin *et al.* (1983). The values were between 940° at the coldest and about 1520° at the warmest site. The sum of Ca and Mg concentrations in soil material less than

2 mm in C horizons (80–150 cm) were in the range 1.5–2.3 % (table 1). The grain-size distributions were fairly similar and characterized by a low clay content amounting to 2–5 %. The soils were characterized as Orthic Podzols except for the Svartberget site, which was a Gleyic Podzol (FAO-Unesco 1974).

At each of the studied sites, two or three profiles were sampled at about ten different levels to a depth of about 1.5 m. From each horizon in each profile, four volumetric samples were taken and put together into a composite sample that was analysed. Base-mineral index was determined according to Tamm (1934). Extractable Al and Fe were analysed by AAS after sodium dithionite–citrate extraction (Mehra & Jackson (1960), as modified by Holmgren (1967)). Total contents of major and trace elements in soil mineral material of less than 2 mm diameter were determined by multielemental ICP analysis. Before analysis, the samples were ignited for 1 h at 900 °C to oxidize organic matter and, after cooling, were ground in an agate mortar.

The mineralogy of the clay fraction (less than 0.002 mm) was determined by X-ray diffraction (XRD) analysis on basal-plane oriented samples after removal of organic matter and sesquioxides. The XRD examination was performed with Philip equipment (PW 1710) by using Ni filtered Cu K_α radiation. The X-ray tube was operated at 40 kV and 24 mA.

The cumulative loss of soil material as well as of major elements were calculated from the present geochemical composition of the different horizons assuming that the original concentrations (immediately after deglaciation) of the studied elements in each one of the sampled horizons were the same as the present concentrations in C horizons.

It is furthermore assumed that the *amount* of Zr (derived from the very stable mineral zircon) in a given soil profile has remained unchanged since deglaciation, but that the concentration of Zr has increased because of weathering of less resistant soil material. Knowledge of present amounts and concentrations of Zr in the A, E, B and C soil horizons allows calculations to be made of the losses of soil material and elements, and thence of weathering rates corrected for estimated changes resulting from formations of new compounds in the soil and deposition from the atmosphere.

3. Results

3.1. *Accumulation in Bs horizons*

In the spodic B horizons extractable Al and Fe amounted to 1300–3000 and 2700–4800 g m^{-2}, respectively (figure 1). These values were considered to correspond to the illuvial and accumulated Al and Fe. During pedogenesis, organic compounds have moved with soil water through the profile and have formed complexes with Fe and Al, released from primary minerals by weathering. The complexes were subsequently partly precipitated and retained in Bs horizons, for instance as coatings. The small variation in accumulated amounts of Al and Fe between the sites was partly caused by the soils having different ages. For this reason the average annual accumulation of Fe and Al in Bs horizons was calculated for the time elapsed since deglaciation and exposure. The average annual accumulation of Fe and Al did not show any correlation with site factors, expressed as geochemical properties of the parent material and climate.

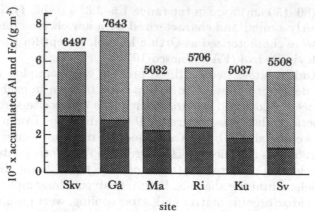

Figure 1. Accumulated Al (⊠) and Fe (⊠) in spodic Bs horizons, extracted by sodium dithionite citrate. Abbreviations are detailed in the text.

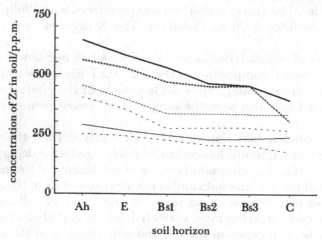

Figure 2. Concentration of Zr (p.p.m.) in soil material (< 2 mm) in different soil horizons; (——), Skv; (-----), Gå; (——), Ma; (-·-·-·), Ri; (-----), Ku; (-·····-), Sv. (Abbreviations are detailed in the text.)

3.2. *Historical losses*

Zr concentrations in C horizons at study sites were between 200 and 400 p.p.m. (parts per million) (figure 2). In the upper part of the soil the concentrations were in the range 250–650 p.p.m. The relative Zr concentrations, i.e. the concentration in individual horizons as a percentage of the concentration in the C horizons, decreased from an average of about 150 % in the upper part of the soil to 100 % in C horizons. The variation between the studied sites was large, indicating different weathering rates. The depth of the weathered part of the soil, that is the horizons in which the relative Zr concentrations were higher than 100 %, showed a variation between 20 cm and 100 cm (figure 3).

Total losses of soil material since deglaciation amounted to an average of 20–240 kg m^{-2}. Standard deviations were generally about 10–20 kg m^{-2}. At two of the studied sites (Skånes Värsjö and Svartberget), standard deviations were substantially higher. This was probably caused by large variations in the levels of the

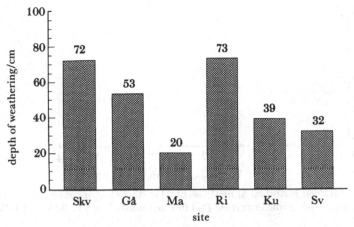

Figure 3. Depth of weathering.

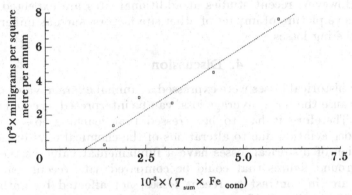

Figure 4. Correlation between historical annual loss of Fe and site factors expressed as the product of temperature sum and Fe concentration (%) in C horizons; $Y = 148.05X - 143.05$; $r = 0.97$.

water table and differences in redox potentials. The mass losses of material roughly corresponded to a 20–200 mm thick layer of soil. The annual average loss of soil material since deglaciation was 2–20 g m^{-2} a^{-1}. Discrepancies between the studied sites were considered being due to a combination of different site properties such as climate, mineralogy and geochemistry. Weathering losses of Ca, Mg, K and Na were calculated by Olsson & Melkerud (1990).

Historical losses of Fe since deglaciation and exposure amounted to 0.02–0.76 g m^{-2} a^{-1} (figure 4). Corresponding values for Al were 0.14–1.67 g m^{-2} a^{-1} (figure 5). The correlation between site properties and loss of Al and Fe, respectively, was expressed as linear regressions (figures 4 and 5). The site factors being tested were temperature sum, and base mineral index and the geochemical composition of the parent material in C horizons. The loss of Fe was best correlated with the product of temperature sum and concentration of Fe in C horizons ($r = 0.97$). Regarding Al, the best correlation was achieved by using the products of temperature sums and the concentration of Ca + Mg in C horizons as an independent variable ($r = 0.90$). This was probably due to easily weatherable Al occurring mainly in minerals rich in Ca and Mg, for instance amphiboles and plagioclases. The calculated weathering losses were slight over-estimations because boulder populations were neglected. The low

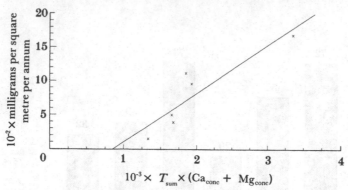

Figure 5. Correlation between historical annual loss of Al and site factors expressed as the product of temperature sum and Ca+Mg concentration (%) in C horizons; $Y = 715.32X - 616.23$; $r = 0.90$.

number of studied sites did not permit the use of more detailed site descriptions in the regression. However, recent studies at additional sites are expected to give a more comprehensive picture of impact of other site factors, such as mineralogy and texture, on weathering losses.

4. Discussion

The calculated historical losses were expressed as annual average values. One may object to this because the term 'average loss' can be interpreted as equal to present leaching losses. Therefore it has to be stressed that leaching losses may have fluctuated after deglaciation due to alterations of the chemical environment in the soil. Determination of historical losses have a fundamental value as they provide long-term background figures that could be compared with recent leaching. The historical losses are, in contrast to recent losses, not affected by anthropogenic impact and short-term climatic fluctuations.

The results clearly show that high losses of Fe and Al have taken place. The losses were higher at sites more prone to weathering. Bergkvist (1986) determined present leaching losses for Al by lysimeter investigations at Skånes Värsjö and Gårdsjön to be 3.14 and 2.15 g m^{-2} a^{-1}, respectively (table 2). These values exceed the historical losses by 2.19 and 0.48 g m^{-2}, respectively. The lysimeter investigation period was restricted to 1979–81 for Skånes Värsjö and 1980–84 at Gårdsjön. The total precipitation sum during these periods exceeded normal values, which may have affected the results. The differences between present and historical losses may also be due to the soils being affected by acid atmospheric deposition and Al^{3+} exchange. The present amounts of extractable Al at Skånes Värsjö and Gårdsjön were 3.0 and 2.8 kg m^{-2}, respectively. At the present leaching rates the accumulated Al should last more than 1000 years.

Fe leaching did not follow the same pattern as Al leaching. Thus historical losses have taken place, though not as high as for Al. Historical losses were roughly equal to leaching losses as determined in the runoff from catchments (Grahn & Rosén 1983; H. Grip and L. Lundin, personal communications). On the other hand, lysimeter investigations at Skånes Värsjö and at Gårdsjön showed very low leaching (table 2). With respect to dry and wet deposition the ecosystems showed slight positive budgets, i.e. +0.14 and +0.01 mg m^{-2} a^{-1}, respectively (Bergkvist 1986). Possible explanations are that historical losses were fossil and that most of the Fe is retained

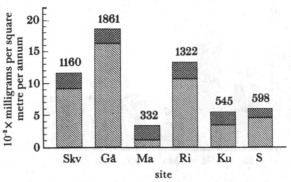

Figure 6. Weathering rate of Al determined as the sum of Al loss (▨) and of illuvial Al (⊠) in Bs horizons minus deposition.

Table 2. *Annual leaching in milligrams per square metre per annum (historical loss within brackets)*

site	Fe	Al	period of investigation
Skv[a]	40 (446)	3140 (950)	1979–81 (3 years)
Gå[a]	140 (760)	2150 (1668)	1980–84 (4 years)
Gå[c]	180 (760)	460 (1668)	1980–81 (1 year)
Ma[b]	210 (20)	250 (139)	1985–87 (2 years)
Ri[b]	42 (300)	44 (1104)	1986–88 (1 year)
Ku[c]	206 (266)	101 (379)	1980–81 (1 year)
Sv[d]	269 (133)	85 (489)	1985–89 (4 years)

[a] Bergkvist (1986), lysimeter investigations.
[b] L. Lundin, personal communication, catchment studies.
[c] Grahn & Rosén (1983), catchment studies.
[d] H. Grip, personal communication, catchment studies.

Table 3. *The mineralogical distribution in the clay fraction*
(Mean values based on XRD-intensity percentage.)

horizon	mixed layers	Smect./ Verm.	Illite	Kaol./ Chlor.	Amph.	Qu.	K-fsp.	Plag.
E	0.8	55.3	7.8	1.3	0.3	13.1	9.3	12.2
Bs	2.2	24.2	15.5	4.9	5.7	18.6	14.1	14.9
C	5.5	4.5	33.4	4.7	6.4	15.4	13.6	16.5

in Bs horizons at well-drained sites by recent processes. The low present leaching through the profile may be due to Fe-rich minerals being consumed by weathering. This is verified by the low content of amphiboles in the clay fraction of E horizons (table 3). Fe may be dissolved and leached at discharge areas by reduction processes due to wet conditions and a rich occurrence of organic matter. This may have resulted in raised concentrations of Fe in runoff water from the catchment areas.

Historical weathering rates for Fe and Al were estimated by adding the accumulation in Bs horizons to the historical losses and by subtracting the deposition. The amount of annually precipitated Al in Bs constituted only a minor part of the annually weathered Al (figure 6). It was in this case apparent that the illuviation in Bs horizons was not primarily regulated by the access of Al^{3+}, i.e. by

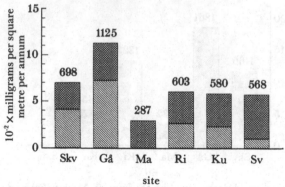

Figure 7. Weathering rate of Fe determined as the sum of Fe loss (◫) and of illuvial Fe (⊠) in Bs horizons minus deposition.

weathering rates. The higher the weathering rate was, the higher was the loss. A similar situation was true also as regards Fe (figure 7). The losses of Fe were generally lower than for Al because the weathering rates were slower.

References

Bergkvist, B. 1986 Ph.D. thesis. Department of Ecology, Plant Ecology. University of Lund, Sweden.

FAO-Unesco 1974 *Soil map of the world, volume 1. Legend*. Paris: Unesco.

Farmer, V. C., Russel, J. D. & Berrow, M. L. 1980 *J. Soil Sci.* **31**, 673–684.

Grahn, O. & Rosén, K. 1983 *Naturvårdsverket rapport, snv pm, 1687*. (In Swedish; summary in English.)

Holmgren, G. G. S. 1967 *Soil Sci. Soc. Am. Proc.* **31**, 210–211.

Lundin, L. 1988 In *Proceedings of the international symposium on the hydrology of wetlands in temperate and cold regions, Joensuu, Finland 6–8 June, 1988*, vol. 1, pp. 197–205. Helsinki. Academy of Finland.

Mehra, O. P. & Jackson, M. L. 1960 *Clays Clay Mineral.* **7**, 317–327.

Odin, H., Eriksson, B. & Perttu, K. 1983 *Reports in Forest Ecology and Forest Soils*, no. 45. Umeå: Swedish University of Agricultural Sciences.

Olsson, M. & Melkerud, P.-A. 1990 In *Proceedings from a symposium on environmental geochemistry in northern Europe, Rovaniemi, Finland, October 1989. The Geological Survey of Finland, Spec. Pap. no. 9*. (In the press.)

Petersen, L. 1976 *Podzols and podzolization*. Copenhagen: DSR forlag.

Tamm, O. 1934 *J. Swed. forest. Soc.* **32**, 231–250. (Summary in German.)

Studies of aluminium species in fresh waters

By E. Lydersen and B. Salbu

Isotope and Electron Microscopy Laboratories, Agricultural University of Norway, P.O. Box 26, N-1432 Ås-NLH, Norway

Size and charge fractionation techniques have been used for determination of Al-species in natural fresh waters and in synthetic solutions. Temperature and pH seriously influence the distribution pattern of Al-species. In identical synthetic solutions, a major fraction of Al is present as amorphous colloids at low temperature; while aggregation of colloids into more crystalline phases takes place at higher temperatures. The formation of $Al(OH)_3(s)$ is rapid, while the dissolution of $Al(OH)_3(s)$ is rather slow.

1. Introduction

In natural water systems Al may be present in different physico-chemical forms. The distribution pattern of Al-species depends on pH, temperature, and the presence of inorganic and organic ligands. Information on Al-species is therefore essential for understanding the transport as well as biological uptake and effect.

In the present work, natural fresh waters and synthetic inorganic Al solutions have been fractionated with respect to size (molecular mass discrimination) and charge (extraction, ion exchange chromatography) properties. Special emphasis has been put on the identification of colloidal or polymeric species in solution. The influence of pH (4.5–6.0) and temperature (2–25 °C) on the distribution pattern of Al-species has been followed during 1 month. Furthermore, the formation and dissolution kinetics of $Al(OH)_3(s)$ have been followed.

2. Experimental

(a) Samples investigated

In situ ultrafiltration has been performed at eight different sites in Norway (Åstadalen (four sites), Lomstjern, Høylandet, Birkenes, Storgamma) in 1987. In addition, synthetic Al solutions (100–800 mg Al l^{-1}) have been prepared from $Al(NO_3)_3 \cdot 6H_2O$ and deionzed water. The influence of pH (in the range 4.5–6.0) and temperature (2, 15, 25 °C) on the distribution pattern of Al-species has been followed during 1 month of storage.

(b) Fractionation techniques

Amicon hollow fibre cartridge H1P10-8 having a molecular mass cut-off level of 10^4 Dalton was used for *in situ* sampling and ultrafiltration (Lydersen *et al.* 1987 *a*, *b*), and for laboratory experiments (Lydersen *et al.* 1990).

The Barnes–Driscoll method was used for separation of Al species according to charge properties (Barnes 1975; Driscoll 1984). By combining the size and charge

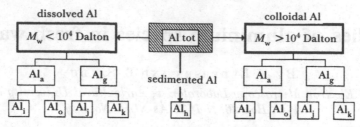

Figure 1. Al-fractions obtained from the combined size and charge fractionation technique
(see text for notation of subscripts).

separation techniques (Lydersen *et al.* 1990) the analytical scheme (figure 1) has been extended as follows.

Al_a: *Total* monomeric Al present in the solution, i.e. HQ/MIBK-extractable Al within 20 s.

Al_g: Non-HQ/MIBK-extractable Al.

Al_i: Inorganic monomeric Al, i.e. HQ/MIBK-extractable Al retained on the cation exchange resin.

Al_o: Non-labile monomeric Al present in the solution, i.e. HQ/MIBK-extractable Al present in the exchange eluate.

Al_j: Non-HQ/MIBK-extractable Al, but retained on the cation-exchange resin.

Al_k: Neither HQ/MIBK-extractable nor exchangeable Al.

Al_h: Precipitated Al.

Hollow fibre ultrafiltration was performed before the Al analysis. Thus, each Al fraction is divided into two molecular mass fractions: $M_w > 10^4$ Dalton, and $M_w < 10^4$ Dalton.

3. Results and discussion

In situ ultrafiltration of natural fresh waters has demonstrated that a significant amount of Al and other trace metals are present as colloids as shown previously (Salbu *et al.* 1985; Lydersen *et al.* 1987*b*; Salbu 1987).

In addition, high molecular mass forms of Al influence the determination of monomeric Al-species. When colloids are removed before analysis, the amount of monomeric Al is significantly reduced, especially in waters high in humic substances and at high pH (> 5) (Salbu *et al.*, this symposium).

To exclude the influence of interfering agents present in natural waters, experiments using synthetic Al solutions were performed. As seen in table 1, the temperature affects the solubility of $Al(OH)_3(s)$, the solubility product formed ($\log *K_s$), the ageing process (decrease in $\log *K_s$), the hydrolysis, and the molecular mass distribution of aqueous Al, as well as the pH of the solution.

At 2 °C (figure 2) a substantial amount of Al is present as colloids ($M_w > 10^4$ Dalton) in solution, having a $\log *K_s$ corresponding to amorphous $Al(OH)_3(s)$ (table 1). At 25 °C, the particle growth (aggregation of colloids) results in precipitation (figure 2) and the solubility product reflects the presence of microcrystalline gibbsite (table 1).

The formation of $Al(OH)_3(s)$ is a rather spontaneous process, while the dissolution of $Al(OH)_3(s)$ is slow (figure 3). After 24 h a substantial amount of $Al(OH)_3(s)$ is still not dissolved, even though the system is far from being saturated with respect to $Al(OH)_3(s)$.

Table 1. pH, log $*K_s$, molar ratio OH:Al, concentration and molecular mass distribution of Al at 2 °C and 25 °C (Final 10 columns are in units of µg Al l⁻¹.)

temp./°C	pH	2 °C log $*K_s$	25 °C log K_s	molar OH:Al	Al_{tot}	Al_a tot.	high[a]	low[b]	Al_g tot.	high	low	Al_j tot.	Al_k tot.	Al_h tot.
2	5.74	11.7	10.3	1.9	400	220	33	187	166	164	2	131	35	14
25	5.02		9.0	1.9	400	69	2	67	110	97	13	75	35	221
2	5.78	11.6	10.2	2.4	400	173	38	135	212	212	0	162	50	15
25	5.07		9.0	2.4	400	60	16	44	102	95	7	56	46	238
2	5.71	11.7	10.3	2.1	600	261	48	213	320	311	9	170	150	19
25	4.95		9.1	2.1	600	102	4	98	185	148	37	100	85	313
2	5.78	11.7	10.3	2.4	600	216	62	154	361	341	20	160	201	23
25	5.07		9.0	2.4	600	53	2	51	174	148	26	93	81	373
2	5.64	11.8	10.4	2.0	800	372	76	296	392	391	1	205	187	36
25	4.83		9.1	2.0	800	201	20	181	272	199	73	119	153	327
2	5.72	11.8	10.4	2.3	800	261	73	188	498	484	14	176	322	41
25	4.95		9.0	2.3	800	106	14	92	302	258	44	150	152	392

[a] $M_w > 10^4$ Dalton.
[b] $M_w < 10^4$ Dalton.

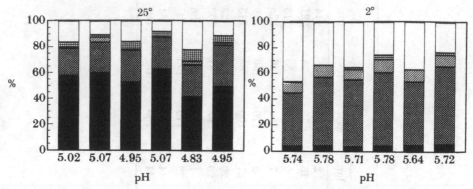

Figure 2. pH and molecular mass distribution of different Al-fractions (%) present in identical solutions stored for 1 month at 2 °C and 25 °C, respectively. ☐, Al_a $M_w < 10^4$ kDa; ▦, Al_g $M_w < 10^4$ kDa; ▨, Al_a $M_w > 10^4$ kDa; ▦, Al_g $M_w > 10^4$ kDa; ■, Al_h: precipitated $Al(OH)_3(s)$.

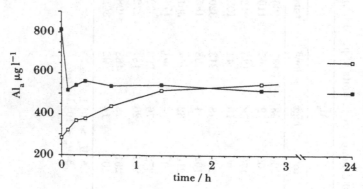

Figure 3. Change in the Al_a concentration during 24 h after a pH change. Experimental temperature: 2 °C. ■, pH adjusted from 4.6 to 5.6 by NaOH; ☐, pH adjusted from 4.6 to 5.6 by NaOH and stored for 3 days before acidifying with HCl to pH 4.3.

4. Conclusions

Especially at low temperature, a substantial amount of Al in solution may be present as colloids thus influencing the determination of monomeric Al. As the present models are based on monomeric Al data, a more accurate Al speciation strategy is needed. The *in situ* hollow fibre fractionation technique is found to be most suitable as colloids are removed from solution during sampling. However, as colloids in solution represent a reservoir of potentially mobilizable Al, colloids should be included in models describing the biogeochemistry of Al.

References

Barnes, R. B. 1975 *Chem. Geol.* **15**, 177–191.

Driscoll, C. T. 1984 *Int. J. Environ. Anal. Chem.* **16**, 267–283.

Lydersen, E., Bjørnstad, H. E., Salbu, B. & Pappas, A. C. 1987*a* In *Speciation of metals in water, sediments and soil systems* (ed. L. Landner). Lecture Notes in Earth Sciences, vol. 11, pp. 85–97. Springer Verlag.

Lydersen, E., Salbu, B., Bjørnstad, H. E., Englund, J. O., Hovind, H. & Rambæk, J. P. 1987*b*
  Proc. Int. Symp. of Acidification and Water Pathways, Bolkesjø, 4 & 5 May 1987, pp. 107–116.

Lydersen, E., Salbu, B., Poleo, A. B. S. & Muniz, I. P. 1990 *Water, air, and soil poll.* (In the press.)

Salbu, B., Bjørnstad, H. E., Lindstrøm, N. S., Lydersen, E., Breivik, E. M., Rambæk, J. P. &
  Paus, P. E. 1985 *Talanta* **32**, 907–913.

Salbu, B. 1987 In *Proc. SWAP Conf., Bergen, June 1987*.

Intercomparison study on the determination of total aluminium and aluminium species in natural fresh waters

BY B. SALBU, G. RIISE, H. E. BJØRNSTAD AND E. LYDERSEN

Isotope and Electron Microscopy Laboratories, Agricultural University of Norway,
P.O. Box 26, N-1432 Ås-NLH, Norway

The results of an intercomparison study for the determination of total Al and Al-species in fresh waters are reported. Unfiltered and *in situ* ultrafiltered water samples from four different sites in Norway were distributed to 19 laboratories (in Norway, Sweden, U.K., U.S.A.) for analysis 3 days after sampling. Wide ranges were obtained for total Al and monomeric Al-species. Outliers were also identified for acid-reactive Al and all the monomeric Al-species. The ranges are reduced when colloids and particles are removed before analysis. Some methodological effects are identified.

1. Introduction

The determination of Al and Al-species as well as other trace elements depends on sampling techniques, pre-analysis handling, methods of analysis and procedures applied. Within the SWAP programme, the determination of Al and Al-species are based on methods described by Dougan & Wilson (1974), Barnes (1975), and Driscoll (1984). However, large differences in the analytical procedures and routines give rise to variable analytical results. Intercalibration of Al and Al-species, major ions, and trace metals has therefore been performed.

2. Experimental

(a) Samples

Water samples from four different sites (Nordmarka, Høylandet, Birkenes, Storgamma) were collected in September and October 1987. The sampling sites were chosen to include water differing in pH, total Al concentration, and total organic carbon (TOC, mg C l^{-1}).

By using a peristaltic pump, water was transferred from a chosen depth directly into a 20 l container. *In situ* fractionation was then performed by hollow fibre ultrafiltration using a cartridge having a nominal molecular weight cut-off value of 10^4 Dalton (Lydersen *et al.* 1987 *a*, *b*).

The ultrafiltrate was transferred directly into 20 l containers. Three days after sampling the subsamples (1 l) were supplied to 19 laboratories. All containers (polyethylene) were carefully cleaned and conditioned with a sample aliquot before collection.

(b) Variables

Analysis of pH, alkalinity, TOC, Al-species (total, Al_{tot}; acid reactive, Al_r; total monomeric, Al_a; organic monomeric, Al_o; inorganic monomeric, Al_i), major cations

Table 1. *Reported results for Al and Al species in unfiltered and ultrafiltered samples*

	unfiltered range	corr. mean	ultrafiltered B range	corr. mean	ratio B/A
Lomstjern					
pH	4.4–4.6	4.52 ± 0.06	4.4–4.7	4.66 ± 0.09	1.03 ± 0.02
TOC/(mg C l^{-1})	12.2–16.9	14 ± 1	5.2–7.0	5.9 ± 0.6	0.4 ± 0.1
Al$_{tot}$/(µg l^{-1})	210–503	424 ± 70	40–460	211 ± 82	0.5 ± 0.2
Al$_r$/(µg l^{-1})	(177)309–505	417 ± 29	125–240	188 ± 35	0.5 ± 0.1
Al$_a$/(µg l^{-1})	(137)235–405	320 ± 51	73–230	162 ± 43	0.5 ± 0.2
Al$_o$/(µg l^{-1})	90–229	158 ± 42	43–88	60 ± 11	0.4 ± 0.1
Al$_i$/(µg l^{-1})	103–202(320)	158 ± 35	49–170(216)	112 ± 34	0.7 ± 0.3
Høylandet					
pH	5.8–6.2	6.0 ± 0.1	5.8–6.2	6.05 ± 0.09	1.00 ± 0.02
TOC/(mg C l^{-1})	4.5–7.1	5.8 ± 0.9	1.1–2.8	1.8 ± 0.5	0.3 ± 0.1
Al$_{tot}$/(µg l^{-1})	20–143	106 ± 24	19–74	35 ± 14	0.3 ± 0.2
Al$_r$/(µg l^{-1})	(19)67–130	94 ± 21	1–57	28 ± 20	0.3 ± 0.2
Al$_a$/(µg l^{-1})	17–100	57 ± 22	5–70	22 ± 15	0.4 ± 0.3
Al$_o$/(µg l^{-1})	33–80	52 ± 12	5–45	15 ± 10	0.3 ± 0.3
Al$_i$/(µg l^{-1})	1–27(115)	6 ± 8	0–25(42)	6 ± 7	1.0 ± 1.8
Birkenes					
pH	4.4–5.2	4.8 ± 0.1	4.7–5.3	5.0 ± 0.1	1.0 ± 0.03
TOC/(mg C l^{-1})	1.3–5	4.4 ± 0.5	2.2–4.2	2.7 ± 0.6	0.6 ± 0.2
Al$_{tot}$/(µg l^{-1})	270–464	410 ± 53	190–360	291 ± 40	0.7 ± 0.1
Al$_r$/(µg l^{-1})	(75)359–462	413 ± 36	(75)282–325	297 ± 15	0.7 ± 0.1
Al$_a$/(µg l^{-1})	156–425	315 ± 66	(113)174–292	242 ± 31	0.8 ± 0.2
Al$_o$/(µg l^{-1})	60–174	101 ± 29	45–99	65 ± 16	0.6 ± 0.2
Al$_i$/(µg l^{-1})	137–276(350)	227 ± 35	118–212(297)	180 ± 24	0.8 ± 0.2
Storgamma					
pH	4.2–4.6	4.55 ± 0.08	4.3–5	4.7 ± 0.1	1.00 ± 0.03
TOC/(mg C l^{-1})	3.9–5.7	4.6 ± 0.6	2.7–3.7	2.9 ± 0.3	0.6 ± 0.1
Al$_{tot}$/(µg l^{-1})	100–492	206 ± 91	70–175(261)	133 ± 24	0.6 ± 0.3
Al$_r$/(µg l^{-1})	(80)155–214	186 ± 17	(75)117–158	137 ± 13	0.7 ± 0.1
Al$_a$/(µg l^{-1})	(74)135–195	153 ± 18	53–137	103 ± 19	0.7 ± 0.4
Al$_o$/(µg l^{-1})	33–84	57 ± 15	18–51	38 ± 10	0.7 ± 0.2
Al$_i$/(µg l^{-1})	75–114(191)	96 ± 12	50–94(142)	69 ± 12	0.7 ± 0.2

(Na, K, Mg, Ca), major anions (SO_4, Cl, NO_3), trace elements (Fe, Mn, Pb, Cu, Zn, Cd) were included in the exercise.

The laboratories used their standard methods of analysis. Time of storage was limited to 5 days.

(c) *Control programme*

To control possible changes in the concentrations of elements and Al-species during storage, separate aliquots were analysed after 1, 3 and 5 days respectively using standardized techniques. No significant changes could be observed during storage.

3. Results and discussion

The ranges and corrected mean values (outliers excluded) for the unfiltered and the ultrafiltered samples are given in table 1. The outlier criterion was considered only if one set of data differed very distinctly from other data. The wide ranges obtained for Al$_{tot}$ seem quite unacceptable. Obviously, different species are included in the analytical results reported. Data obtained by GFAAS seem, however, more reproducible than those obtained by ICP.

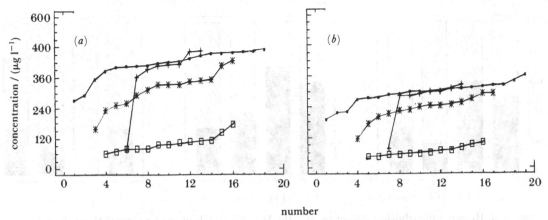

Figure 1. Cumulative distribution for Al_{tot}, Al_r, Al_a, and Al_o. (*a*) Unfiltered sample, Birkenes; (*b*) ultrafiltered sample, Birkenes. ■, Al_{tot}; +, Al_r; ★, Al_a; □, Al_o.

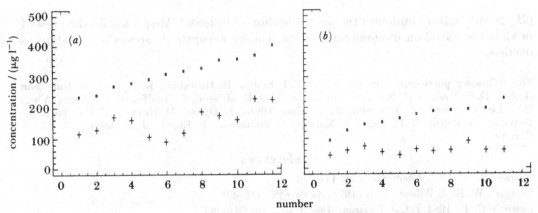

Figure 2. Corresponding results on Al_a (□) and Al_o (+) obtained by 12 different laboratories. (*a*) Unfiltered sample, Lomstjern; (*b*) ultrafiltered sample, Lomstjern.

Outliers are identified for the monomeric Al-species. The wide ranges obtained on Al_{tot}, Al_r, Al_a, and Al_o for Birkenes are given in figure 1*a*, *b*. The concentrations of monomeric Al-species are significantly reduced in the ultrafiltrate, especially for waters high in TOC (figure 2) and pH (table 1), even though the ranges are still wide. For Al_r the wide range can be attributed to the use of different acids. For Al_a the data obtained by using pyrocatechol-violet (Dougan & Wilson 1974) seem more reproducible than those obtained by HQ/MIBK (Barnes 1975). Corresponding data on Al_a and Al_o for Lomstjern are given in figure 2*a*, *b*. Examination of the data illustrates that an increased flow rate through the ion exchange column increases Al_o and thereby reduces the Al_i concentration ($Al_i = Al_a - Al_o$). Furthermore, Al_i is reduced when colloids are removed before analysis.

4. Conclusion

Wide ranges are obtained for Al_{tot} and monomeric Al-species especially in fresh waters high in organic carbon and for pH > 5. The concentrations are reduced when colloids are removed before analysis. Furthermore, high molecular mass Al forms

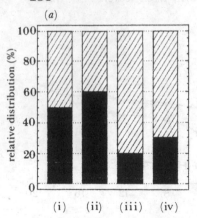
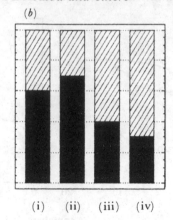
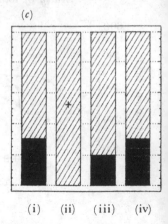

Figure 3. Relative distribution (%) of (a) Al_a, (b) Al_o, and (c) Al_i present as low (▨) and high (■) molecular mass species (10^4 Dalton). (i) Lomstjern, (ii) Høylandet, (iii) Birkenes, (iv) Storgamma. An asterisk denotes a large standard deviation.

($M_w > 10^4$ Dalton) influence the determination of different Al-species. As the present models are based on monomeric Al data, a more accurate Al speciation strategy is needed.

The following participated in our work: F. L. Smith, R. McMahon, E. Tipping, R. Harriman (U.K.); D. C. McAvoy (U.S.A.); U. Lindstrøm, L. G. Danielsson, A. Adolfsson, E. Fröberg, Ying-Hua Lee (Sweden); F. Antonsen, A. R. Selmer-Olsen, G. Ogner, H. Hovind, E. Røgeberg, K. Barland, A. Poleo, R. D. Vogt, V. Martini, O. Johansen, B. Enger, J. E. Hansen, I. Blakar (Norway).

References

Barnes, R. B. 1975 *Chem. Geol.* **15**, 177–191.

Dougan, W. K. & Wilson, A. L. 1974 *Analyst* **99**, 413–436.

Driscoll, C. T. 1984 *Int. J. Environ. Anal. Chem.* **16**, 267–283.

Lydersen, E., Bjørnstad, H. E., Salbu, B. & Pappas, A. C. 1987a In *Speciation of metals in water, sediments and soil systems* (ed. L. Landner). Lecture Notes in Earth Sciences, vol. 11, pp. 85–97. Springer-Verlag.

Lydersen, E., Salbu, B., Bjørnstad, H. E., Englund, J. O., Hovind, H. & Rambæk, J. P. 1987b *Proc. Int. Symp. of Acidification and Water Pathways, Bolkesjø, 4 & 5 May 1987*, pp. 107–116.

Aluminium chemistry in acid environments

By E. Tipping

Institute of Freshwater Ecology/Freshwater Biological Association, Ambleside, Cumbria LA22 0LP, U.K.

Several lines of research have been followed to improve understanding of aluminium chemistry in acid environments and to formulate and test models that can be used to predict the response of aluminium chemistry to changes in conditions. This paper briefly discusses the following topics: (1) evaluation of cation-exchange fractionation for estimating organically complexed Al; (2) modelling the binding of Al by dissolved organic matter (humic substances); (3) modelling ion-binding reactions in organic soils; (4) the conditions required for the precipitation of Al in surface waters; (5) the adsorption of Al by suspended particulates; (6) the uptake and release of Al by immobile streambed components.

Introduction

Mobile aluminium is a characteristic feature of acid environments. The metal has a complicated chemistry, which needs to be understood if the ecological consequences of its mobilization are to be assessed. In surface waters, the distribution of Al among a number of chemical forms – its 'chemical speciation' – is thought to determine the degree of toxicity of Al toward biota. These forms, or species, include Al^{3+} and its soluble complexes with OH^-, F^-, SO_4^{2-} and dissolved organic ligands, together with solid-phase species such as oxyhydroxides, aluminosilicates and adsorbed metal. Only some species – Al^{3+} and its hydroxyl complexes, and freshly precipitated oxyhydroxide – are thought to be toxic (Muniz & Leivestad 1980; Helliwell *et al.* 1983; Dickson 1983). We thus require analytical and theoretical methods of determining which Al species are present in a given water sample. This has been the main subject of our research under SWAP. A related topic is the control of Al concentrations in surface waters. In the first instance, this involves the weathering of primary minerals. Next, solid–solution interactions in soils are important, and our findings have relevance to the description of these interactions. Furthermore, although weathering and soil chemistry, together with hydrology, determine the amounts and forms of Al actually entering surface waters, subsequent events occurring within streams and lakes, notably precipitation–dissolution and adsorption–desorption reactions, may exert additional controls on Al solution chemistry. Some of our work has been directed towards the study of these processes.

Evaluation of cation-exchange fractionation for the determination of organically complexed aluminium (Backes & Tipping 1987)

The cation-exchange technique (Driscoll 1984) relies on the fact that inorganic monomeric species of Al in acid waters are almost exclusively cationic, whereas organic complexes of Al are anionic. Thus the inorganic species are retained by a cation-exchange column while the organic ones are not. Measurement of monomeric Al before and after passage of the sample through the column gives concentrations of total and organic monomeric Al respectively and, by difference, inorganic Al. The method is convenient, reproducible and can be used in the field. However, its application requires that organically complexed Al in the original solution remains complexed throughout passage through the column of exchange resin, and this is questionable, column residence times being in the range 10–50 s. It was therefore considered desirable to evaluate the cation-exchange technique by comparison with an equilibrium analytical method. This was done by estimating organic Al concentrations by cation-exchange fractionation and by equilibrium dialysis, for the same solutions. Experiments were carried out on synthetic solutions containing various amounts of Al, isolated humic substances (HS) and added base, having compositions similar to those in organic-rich acid natural waters,

It was found that concentrations of organically complexed Al estimated by cation-exchange tended to be lower than those obtained by equilibrium dialysis. The extent of the discrepancy increased with the parameter, ν (mol Al bound $(g_{HS})^{-1}$). The results from equilibrium dialysis (ED) and cation exchange (CE) can be related by the empirical equation

$$\nu_{ED} = \nu_{CE} + k\nu_{CE}^2. \tag{1}$$

For a 10 cm³ column, $k = 220\ g_{HS}\ (\text{mol Al})^{-1}$ (13 points, $r = 0.93$); for a 2 cm³ column $k = 400\ g_{HS}\ (\text{mol Al})^{-1}$ (35 points, $r = 0.98$). If values of ν_{ED} are assumed to be the true ones, then equation (1) can be used to correct the results of cation-exchange fractionations, given an estimate of [HS], although application of the equation also embodies the assumption that the values of k do not vary.

A probable explanation for the discrepancies is provided by the results of experiments in which the effect of flow rate was examined as a function of ν. It was found that although satisfactory performances are given by both large and small columns at low ν, at high values it becomes impossible to obtain a plateau region in the plot of $[\text{Al}_m-\text{org}]$ against flow rate. The difficulty could not be overcome simply by increasing the flow rate because of the 'breakthrough' of inorganic Al at flow rates greater than 6 cm³ min⁻¹ cm⁻³ (large column) or 2.2 cm³ min⁻¹ cm⁻³ (small column). The behaviour at high ν most likely reflects Al binding to weaker humic complexing sites, not occupied at lower values. Published data on $[\text{Al}_m-\text{org}]$ and dissolved organic carbon concentrations indicate that in natural waters, ν can be as high as 10^{-3} mol $\text{Al}(g_{HS})^{-1}$: therefore poor column performance at high ν can certainly be of practical significance.

Modelling the binding of aluminium by dissolved organic matter (humic substances)

A series of models (I–IV) have been developed, initially to predict Al–HS interactions in surface waters. Model III was specifically designed for aquatic (dilute) systems, while model IV (also referred to as 'CHAOS') deals with interactions in organic soils (see below).

Model III (Tipping *et al.* 1988*a*) treats humic substances in terms of two types of carboxyl group, together with other more weakly acidic groups. Metal ions (Al^{3+}, AlOH^{2+}, Ca^{2+}) are assumed to bind at bidentate sites. Proton and metal binding can be influenced by the net charge of the humic molecules (macroionic effect). Eight parameters are required to describe the interactions for a given humic sample: total acidity (Ac), which is directly measurable, content of COOH groups (n), four intrinsic equilibrium constants (two for H^+ binding, two for metal binding), and two parameters (P and Q accounting for charge effects at different ionic strengths).

The model has been applied successfully to four samples of aquatic HS. The different equilibrium constants do not vary greatly from sample to sample, which is consistent with the idea that the functional groups in each sample are chemically the same. Equilibrium constants estimated for Al and Ca binding are in reasonable accord with values reported for phthalic and salicylic acids. Parameters accounting for macroionic (charge) effects vary appreciably among the samples; in one case there is no evidence for macroionic behaviour.

The model predicts little effect of Ca^{2+} on Al binding under conditions representative of acid natural waters.

The model predicts that the H^+-buffering capacity in Al-humic solutions is similar to that in humic-only solutions; there is little or no additivity of humic and Al buffering.

Parameter values derived from acid-base titration data give reasonable predictions of Al binding, as judged by comparison with directly determined values in laboratory experiments.

Using parameters derived from the results of experiments with Loch Ard Forest humic substances, reasonable predictions of Al speciation in natural waters can be obtained. Improved prediction is achieved by adjusting one of the model parameters, Ac (see also Tipping & Backes 1988).

Extension of model III to organic soils

In regions sensitive to acid deposition, soils invariably contain large amounts of organic matter, which may be the dominant component of the so-called exchange complex. The need to take into account soil organic matter in assessing the impact of acid deposition has been emphasized by Rosenqvist (1978) and by Krug & Frink (1983). Current models of soil acidification (Birkenes, ILWAS, MAGIC) deal with the ion-binding properties of soil by simple ion-exchange reactions with an exchanger of constant charge. However, soil organic matter has a variable, pH-dependent charge, and it can bind protons and Al species by complexation reactions as well as by 'straightforward' ion-exchange. Thus a quantitative description of organic matter–solution interactions in acid soils would be useful in attempting to understand soil acidification and its reversal. This prompted the extension of model III to soils, and the formulation of model IV also referred to as 'CHAOS' (Complexation by Humic Acids in Organic Soils). The soil is treated as a concentrated suspension of humic substances. Complexation reactions are described by combinations of intrinsic equilibrium constants and electrostatic terms (as in model III), non-specific cation-exchange by a Donnan equilibrium. The two types of reaction are linked via Z, the net humic charge. Details are given in Tipping & Hurley (1988). Model IV was found to account satisfactorily for the results of acid-base titration experiments (pH range 3–5) with soil samples, giving reasonable simultaneous predictions of solution pH and concentration of Al^{3+}. Preliminary predictive calculations with CHAOS suggest that organic soils acidified by acid rain would respond on a timescale of years-to-decades to reductions in rain acidity. An associated effect might be an increase in the concentration of dissolved organic matter in the soil solution.

Conditions required for the precipitation of aluminium in surface waters (Tipping et al. 1988b)

In waters of pH $\leqslant 6$, aluminium in solution is commonly close to saturation with respect to oxide phases or other solid forms (Johnson et al. 1981; Nordstrom 1982; Driscoll 1985). Precipitation of the metal might therefore be brought about by quite small changes in conditions (e.g. pH, concentrations of Al complexing agents, temperature), as might occur when waters of different composition mix, or when groundwater high in CO_2 is exposed to the atmosphere (Norton & Henriksen 1983). It is desirable to know under what circumstances precipitation of Al in natural waters may occur. At present, however, predictions can be made only on the basis of solubility products for pure phases that have not been formed under conditions relevant to natural waters (see, for example, Bache 1986). A study was therefore carried out with the aim of defining more precisely the natural conditions necessary for precipitation, focusing on reactions occurring in dilute solutions of pH 4–6, over short time-periods and at temperature in the range 0–25 °C.

The results showed that if precipitation does occur it involves the formation of Al(oxy)hydroxide, not aluminosilicates or basic aluminium sulphates. The solubility product of the Al(oxy)hydroxide is highly temperature dependent ($\Delta H = -30.5$ kcal mol$^{-1} \approx -130$ kJ mol^{-1}). It is also sensitive to concentrations of SO_4^{2-} and, more markedly, humic substances (HS); both of these decrease solubility, HS by more than an order of magnitude at a concentration of ca. 5 mg l^{-1} (equivalent to ca. 2.5 mg l^{-1} dissolved organic carbon). A semi-empirical equation is proposed that allows the prediction of the effective solubility product at different temperatures and at humic concentrations in the range 0–7 mg l^{-1}:

$$\log_{10} K_S^{\text{eff}} (T, [\text{HS}]) = \log_{10} K_S(T^*) + \frac{\Delta H}{R \ln_{10}} \left\{ \frac{1}{T^*} - \frac{1}{T} \right\} + b[\text{HS}], \tag{2}$$

where T^* is a reference temperature (e.g. 25 °C) and b is an empirical constant, estimated to be -0.25 l(mgHS)$^{-1}$.

Of the 113 natural water samples analysed, only one was calculated to be oversaturated with respect to Al(oxy)hydroxide, suggesting that the removal of Al from surface waters by precipitation is a rare event.

Adsorption of aluminium by suspended particulates (Tipping et al. 1989)

An experimental study was made of the adsorption of aluminium by fine particulates from Whitray Beck, a hill stream in NW England. Adsorption increased with Al^{3+} activity, pH and concentration of particles, and could be quantitatively described by the empirical equation:

$$[Al_{\text{ads}}] = \alpha \{Al^{3+}\}^\beta \{H^+\}^\gamma [\text{particles}], \tag{3}$$

where square brackets indicate concentrations, curly brackets activities, and α, β and γ are constants with values of 5.14×10^{-10} (mol l^{-1})$^{2.015}$ (g particles l^{-1})$^{-1}$, 0.457, and -1.472, respectively. For the experimental data, the equation gave a correlation ratio of 0.99. The equation accounts reasonably well for the adsorption of Al by particulates from seven other streams. In applying the equation, it must be borne in mind that the desorption kinetics of Al depend on pH, and rapid reversibility (less than 15 min) can only be assumed for pH $\leqslant$ 5.5. Calculations using the adsorption equation, and taking competition by dissolved humic substances into account, suggest that adsorbed Al may commonly account for a significant proportion (not less than 10%) of total monomeric Al.

Estimating streamwater concentrations of aluminium released from streambeds during 'acid episodes' (Tipping & Hopwood 1988)

Aluminium that has been mobilized in acid soils enters streams, where some of it accumulates on the stream bed. Manipulation experiments have demonstrated that dissolved Al can be released in substantial amounts when the bed is exposed to acid water. Hall *et al.* (1987) found stream-water concentrations of Al of up to 60 µM after reducing the pH of a New Hampshire stream to 4.0: Henriksen *et al.* (1988) reported a concentration of 90 µM in a Norwegian stream, the pH of which had been reduced to 4.0–4.2. Both groups of authors suggested that streambed release of Al may determine stream-water concentrations of Al during the first stages of the 'acid episodes' that follow rainfall on dry or frozen ground or snowmelt, these periods being thought critical in the toxic action of Al species and H^+ towards fish. The idea is that at the beginnings of such episodes, the water entering stream channels is acid, but very low in Al, having passed into the channel by overland flow or by flow through only the surface layer of soil. Exchange of H^+ for Al^{3+} then occurs at the stream bed.

The above-mentioned experiments were carried out with highly acid input water (pH 3.3–3.9) and at low discharge (4×10^{-4}–4×10^{-3} m^3 s^{-1}). In a natural episode the water might be less acid and the discharge substantially greater. In turn there might be less H^+/Al^{3+} exchange and a greater dilution of released metal. Thus, although streambed release of Al may well take place, its quantitative significance is not clear. We estimated rates of release of Al from streambed components, and used them, to calculate streamwater concentrations under conditions typical of 'acid episodes'.

Rates of release of monomeric aluminium were determined for stream-bed materials placed in a laboratory channel and exposed to acid water, and values of the release rate coefficient, R (µmol Al released per m^2 of bed per second) were calculated. Estimates of R were also made from the reported results of experiments in which streams were artificially acidified. The values of R ranged from 0.1 to 3.2 µmol m^{-2} s^{-1}. They decreased with increase in pH, and were greater for beds containing substantial amounts of the liverwort *Nardia compressa* than for pebble and/or gravel (mineral) beds. Calculations, using the estimates of R, were performed to assess the contribution of bed-derived Al_m to streamwater concentrations, under conditions where acid water enters stream channels having exchanged negligible H^+ for Al^{3+} or base cations in the soil. For typical high-discharge conditions, considerable concentrations of Al_m are possible. For a liverwort-rich bed exposed to water of initial pH 3.5 there may be as much as 60 µM Al_m in the stream water. At the other extreme, a mineral bed subjected to pH 4.5 water would yield less than 1 µM Al_m. Stream bed stores of releasable Al appear sufficient to supply Al to the water for significant lengths of time.

Findings in relation to SWAP objectives

Insofar as mobilized aluminium in solution is a major, perhaps *the* major, toxic agent associated with the deleterious effects of acid deposition, our studies have relevance to all four of the SWAP objectives, as defined by the Management Group (see, for example, Mason & Seip 1985). However, we have kept very much in mind the need to provide quantitative descriptions of Al chemistry that could be used *predictively*; therefore our findings should be of use in testing and extending current models of acid environments (soils as well as waters), leading to more comprehensive accounts of the biogeochemistry of these systems. This is especially relevant to SWAP objective 4, i.e. the assessment of the changes that would be brought about by reductions in sulphur deposition.

The studies reviewed here were carried out in collaboration with C. A. Backes, C. Woof, M. A. Hurley, M. Ohnstad, P. B. Walters and J. Hopwood. Mrs Y. Dickens typed the manuscript. I am grateful to the coordinators of SWAP for funding this work; additional support came from the Commission of European Communities and the Natural Environment Research Council.

References

Bache, B. W. 1986 *J. geol. Soc.* **143**, 699–706.

Backes, C. A. & Tipping, E. 1987 *Int. J. Environ. Anal. Chem.* **30**, 135–143.

Dickson, W. 1983 *Vatten* **4**, 400–404.

Driscoll, C. T. 1984 *Int. J. Environ. Anal. Chem.* **16**, 267–283.

Driscoll, C. T. 1985 *Environ. Health Perspect.* **63**, 93–104.

Hall, R. J., Driscoll, C. T. & Likens, G. E. 1987 *Freshwater Biol.* **18**, 17–43.

Helliwell, S., Batley, G. E., Florence, T. M. & Lumsden, B. G. 1983 *Environ. Technol. Lett* **4**, 141.

Henriksen, A., Wathne, B. M., Rogeberg, E. J. S., Norton, S. A. & Brakke, D. F. 1988 *Water Res.* **22**, 1069–1073.

Johnson, N. M., Driscoll, C. T., Eaton, J. S., Likens, G. E. & McDowell, W. H. 1981 *Geochim. cosmochim. Acta* **45**, 1421–1437.

Krug, E. C. & Frink, C. R. 1983 *Science, Wash.* **221**, 520–525.

Mason, J. & Seip, H. M. 1985 *Ambio* **14**, 45–51.

Muniz, I. P. & Leivestad, H. 1980 *Proc. Int. Conf. on the Ecological Impact of Acid Precipitation, Norway, SNSF Project.*

Nordström, D. K. 1982 *Geochim. cosmochim Acta* **46**, 681–692.

Rosenqvist, I. Th. 1978 *Sci. Tot. Environ.* **10**, 39–49.

Tipping, E. & Backes, C. A. 1988 *Water Res.* **22**, 593–595.

Tipping, E. & Hopwood, J. 1989 *Environ. Technol. Lett.* **9**, 703–712.

Tipping, E. & Hurley, M. A. 1988 *J. Soil Sci.* **39**, 505–519.

Tipping, E., Backes, C. A. & Hurley, M. A. 1988*a* *Water Res.* **22**, 597–611.

Tipping, E., Woof, C., Walters, P. B. & Ohnstad, M. 1988*b* *Water Res.* **22**, 585–592.

Tipping, E., Ohnstad, M. & Woof, C. 1989 *Environ. Pollut.* **57**, 85–96.

Ion-exchange and solubility controls in acidic systems: a comparison of British and Norwegian SWAP sites

By C. Neal[1], J. Mulder[2], N. Christophersen[2], R. C. Ferrier[3],
R. Harriman[4], M. Neal[1], R. McMahon[3] and H. A. Anderson[3]

[1] Institute of Hydrology, Wallingford, Oxfordshire, OX10 8BB, U.K.
[2] Center for Industrial Research, Box 124, Blindern, 0301 Oslo 3, Norway
[3] Macaulay Land Use Research Institute, Craigiebuckler, Aberdeen AB9 2QJ, U.K.
[4] Freshwater Fisheries Laboratory, Pitlochry, Perthshire PH16 5LB, U.K.

A summary of soil and stream water data from catchments at SWAP sites in Scotland and Norway is given. It is concluded that insufficient detail has been established to quantitatively describe the ion-exchange processes in determining stream water chemistry. The importance of simple mixing of soil and ground water, in explaining stream water aluminium–hydrogen ion relationships, is emphasized.

1. Introduction

Predicting the consequences of atmospheric deposition and land-use changes such as conifer afforestation or harvesting, for acidic and acid-sensitive systems, requires a thorough understanding of the major chemical, biological and hydrological processes operating. Among the hydrochemical mechanisms, emphasis is correctly placed on ion-exchange reactions within the soil zone. However, simple and unproven mathematical formulations have been used to describe them. Here, data from the SWAP catchment studies are summarized to provide such an assessment as a prelude to the development of the next generation of acidification models.

2. Reactions within the stream

In catchments with mean pH < 6, stormflow waters are acidic and aluminium bearing while baseflow waters are impoverished in both these components and enriched in base cations (Neal & Christophersen 1989; Seip et al. 1989). Stream water can thus be considered as a mixture of soil and deep percolating 'ground' water components. Baseflow water is derived, essentially, from the ground water areas where aluminium and hydrogen ion concentrations are low while base cation concentrations are high. Stormflow water is essentially characterized as a mixture of groundwater and acidic and aluminium bearing waters derived from within the soil layers (see Harriman et al. (1990) and Vogt et al. (this symposium) for very exceptional cases of low flow waters with mean pH > 6).

Variations in the trends observed for all the catchments relate to the composition of each of the associated end members. For example at a pristine site, Høylandet in mid Norway, preliminary investigations are suggesting that the soil waters are low in aluminium and are of moderate acidity with the consequence that the stormflow waters are not particularly acidic and aluminium bearing ($Al_i < 1$ μM, $H^+ < 20$ μM (Christophersen *et al.* 1990*b*, *c*)). For catchments receiving greater anthropogenic sulphur inputs (6 g S $m^{-2} a^{-1}$), the stream waters are particularly acidic and aluminium bearing like their soil water counterparts. For the Birkenes catchments in southern Norway Al_i rises up to 24 and 4 μM at sites B1 and B2, the corresponding H^+ maxima being 65 and 25 μM (Christophersen *et al.* 1990*b*): B1 represents acidic soils underlying spruce vegetation; B2 represents less acidic soil with deciduous vegetation and more basic underlying geology. Transitional moorland sites such as the Allt A'Mharcaidh in mid Scotland show intermediate, moorland response (4 μM Al_i, 10 μM H^+ (McMahon & Neal 1990; Harriman *et al.* 1990)). In general, the stream waters are undersaturated with respect to aluminium hydroxide for the most acidic waters but saturated to oversaturated under less acidic to alkaline conditions (Seip *et al.* 1989; McMahon & Neal 1990). Similar behaviour is observed for the thousand-lake survey of Norway (Seip *et al.* 1990).

3. Reactions within the soil zone

The acidic soils of concern here (podzols, gleys, brown earths, etc.) have organic rich surface horizons and lower horizons volumetrically dominated by inorganic components (sand, silt and clay sized primary minerals plus secondary components such as complex low crystallinity aluminium/iron oxide/hydroxide/organic materials and the clay minerals). The organic layer provides not only a large cation store, partly bound in organic tissue, which is exchangeable with cations in solution but also release of acidic humic substances to solution. In the inorganic layers the dominant reactions are the breakdown of the oxide/hydroxide and silicate phases by weathering reactions and cation exchange reactions involving inorganic surfaces and organic surface coatings. (N.B. although the term inorganic layer is used as a general description, the importance of the small amounts of organic coatings in controlling cation exchange reactions cannot be overstressed; cf. Mulder *et al.* (1989).)

In the case of three Scottish catchments no relationship is observed between inorganic aluminium and hydrogen ion concentration as the data is highly scattered (McMahon & Neal 1990). At one of these sites, Allt a'Mharcaidh, preliminary analysis of the soil water data revealed sodium, magnesium and the sum of the major anions to be linearly correlated (figure 1): calcium shows no relationships with these other variables. For the soils data at Birkenes, in Norway, spatial variability in trends are observed and evidence for or against specific ion exchange reactions is unclear (Christophersen *et al.* 1990*a*). For this site, a laboratory desorption experiment provides a power term varying from 2 to 3 (Andersen *et al.* 1990). Consequently, examination of soil water data indicates that a simple cation-exchange equilibrium formulation is appropriate for neither the inorganic nor the organic soils: a cubic relationship between trivalent and monovalent ion activities should be observed for the soil waters, for divalent–monovalent cation-exchange reactions the corresponding activity ratio is a squared one (cf. Neal *et al.* 1990*a*). N.B. hydrological and mixing processes come into play within the soil zone producing a complex variety of responses. As a consequence of this, it cannot be conclusively stated that a simple exchange reaction is not occurring. Rather, the results can be viewed as representing simple equilibrium in the micropores and simple conservative chemical mixing in the free draining water supplying the soil solution samplers.

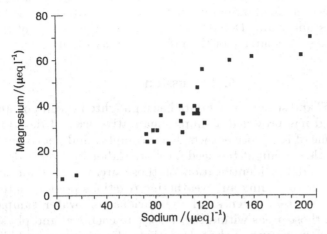

Figure 1. Magnesium–sodium concentration relationships for Allt a'Mharcaidh soil waters (peat site 4, horizons 2 and 3).

None the less, for the ion exchange materials of concern, many of the exchange sites result from ionization of OH groups and as such the assumption of a constant cation exchange capacity is very dubious (Walker *et al.* 1990). Alternative, variable charge, models (see, for example, Tipping & Hurley 1988) can, potentially, explain deviations in the power relationships (Andersen *et al.*, this symposium). However, these models have not been sufficiently developed or tested to show how much improvement can be achieved by their use.

Preliminary comparisons between pristine (Høylandet, mid Norway), transitional (Allt a'Mharcaidh, Scotland) and impacted sites (Birkenes, southern Norway: Plynlimon and Brianne, Wales) suggest that the relative proportion of exchangeable aluminium to exchangeable hydrogen increases with the degree of impaction, although there are large variations in pattern due to the effects of vegetation type and hydrological factors (Christophersen *et al.* 1990 *b*). However, it is difficult to come to any major conclusions as to the variations in the exchangeable cations for different regions: several methods of analysis have been used together with various pretreatments of samples (e.g. drying and sieving soils prior to extraction with unbuffered or buffered salt solutions of various ionic strengths); cf. Walker *et al.* (this symposium.

4. Reactions within the groundwater zone

Despite the stream waters possessing aluminium and hydrogen ion concentrations lower than the associated soil waters, in very few cases have sampling programmes been established to assess where the more alkaline water originates and the nature of the reactions involved. In the exceptional cases at Allt a'Mharcaidh in Scotland (Harriman *et al.* 1990), the Birkenes catchment in southern Norway (Christophersen *et al.* 1990 *a*) and Panola, Georgia, U.S.A., (Hooper *et al.* 1990), less acidic waters have been found under either the hillslope soils or in stream bog areas. These waters, of pH 6 or more, are bicarbonate and alkaline earth metal bearing and inorganic aluminium depleted. Thus in this zone, not only is the acidic water from the overlying soils partially neutralized by weathering and possibly reduction reactions but either aluminium is removed from solution either as aluminium hydroxide/

silicate minerals or the underlying soils produce only organic aluminium: the kinetics of inorganic–organic aluminium interconversions within the soil matrix and soil solution remains uncertain. Detailed evaluation on the nature of the reactions cannot be made given the sparsity of the information available.

5. Discussion

Integrating the soil and stream water data has highlighted the importance of water mixing processes and has provided a simple qualitative way of describing, on the catchment scale, some of the processes which are complex and poorly understood on the plot scale. They also highlight the need for more elaborate models describing ion-exchange reactions and the identification of those areas where waters from the various soil and deeper zones mix and weathering reactions predominate. There is a need to assess the representativeness of the various soil water samples and the delineation between those areas within the soil where chemical and physical mixing reactions take place; the nature of the interactions of aluminium and TOC remains unclear and thus equations linking inorganic and organically bound aluminium in solution also remain uncertain.

Throughout this presentation the importance of the interactiveness of the operative hydrochemical, hydrological, biological processes is recognized: an integrated picture of how catchments respond to anthropogenic influences is required. To obtain this objective an interdisciplinary approach is required and it is insufficient simply to use, *ad hoc*, ideas and methods from the parent disciplines: hydrology, biology, soil science, chemistry, etc. (Christophersen *et al.* 1990 *a*). This has been achieved as new information has accumulated within the surface waters acidification programme and allied research. At the same time these projects have highlighted gaps in our understanding. For example, the case considered here shows a surprising lack of detailed knowledge of aluminium mobilization, given that the importance of this component in acidification impacts was known over 10 years ago. Furthermore, tests for equilibrium processes have been mainly inadequate (Neal 1988; Christophersen & Neal 1989). To capitalize on the field and modelling work over the past decade a new initiative is required to characterize the cation-exchange reactions. On this basis process oriented models along the lines of Tipping & Hurley (1988) deserve encouragement although they must be supported with more rigorous analytical procedures and tests for chemical equilibrium (Neal *et al.* 1990).

References

Christophersen, N. & Neal, C. 1989 *Sci. Total Environ.* **84**, 91–100.

Christophersen, N., Neal, C., Hooper, R. P., Vogt, R. D. & Andersen, S. 1990 *a J. Hydrol.* (In the press.)

Christophersen, N., Neal, C., Vogt, R. D., Esser, J. & Andersen, S. 1990 *b Sci. Total Environ.* (In the press.)

Christophersen, N., Vogt, R. D., Neal, C., Anderson, H. A., Ferrier, R. C., Miller, J. D. & Seip, H. M. 1990 *c Water Resources Res.* (In the press.)

Harriman, R., Gillespie, E., King, D., Watt, A. W., Christie, A. E. G., Cowan, A. A. & Edwards, T. 1990 *J. Hydrol.* (In the press.)

Hooper, R. P., Christophersen, N. & Peters, N. E. 1990 *J. Hydrol.* (In the press.)

McMahon, R. & Neal, C. 1990 *Hydrol. Sci. Bull.* (In the press.)

Mulder, J., Breemen, N. van & Eijck, H. C. 1989 *Nature, Lond.* **337**, 247–249.

Neal, C. 1988 *J. Hydrol.* **104**, 141–159.

Neal, C. & Christophersen, N. 1989 *Sci. Total Environ.* **80**, 195–203.

Neal, C., Reynolds, B., Stevens, P., Hornung, M. & Brown, S. J. 1990*a* *Acidification in Wales* (ed. R. W. Edwards & J. H. Stoner). The Hague: Junk. (In the press.)

Neal, C., Mulder, J., Christophersen, N., Neal, M., Waters, D., Ferrier, R. C., Harriman, R. & McMahon, R. 1990*b* *J. Hydrol.* (In the press.)

Seip, H. M., Andersen, D. O., Christophersen, N., Sullivan, T. J. & Vogt, R. D. 1989 *J. Hydrol.* **108**, 387–405.

Seip, H. M., Andersen, S. & Henriksen, A. 1990 *J. Hydrol.* (In the press.)

Tipping, E. & Hurley, M. A., 1988 *J. Soil Sci.* **39**, 505–519.

Laboratory and lysimeter studies of chemical weathering

By Ulla Lundström

Department of Forest Site Research, Faculty of Forestry, Swedish University of Agricultural Sciences, S-901 83 Umeå, Sweden

The dissolution rates of feldspars in the presence of naturally occurring organics were compared at pH 5.1. Natural silt or ground material of feldspars were suspended for 2500 h in distilled water, streamwater, soil water, water extracts of peat and mor and a citrate solution. pH was maintained at pH 5.1 by using a $CO_2(g)/HCO_3^-$ buffer. The increase of major cations and silicic acid in the aqueous phase was determined and the rates were evaluated from the part of the data showing a linear concentration increase with time.

The dissolution rate for the major cations was 2.7 ± 0.9 ($n = 3$) times as high for streamwater and 2.4 ± 0.4 ($n = 5$) as high for mor and peat extract as for distilled water. For citrate, the rate was enhanced 1.7 ± 0.3 ($n = 3$) times.

By inoculating the suspensions with microorganisms, the weathering rate was decreased to the value for distilled water. This suggests that the action of microorganisms is to consume some 'active' fraction of the organics.

Percolation lysimeters were installed at ten depths in the Svartberget research area in the north of Sweden in a pine forest on a glacifluvial sediment under an undisturbed mor layer.

Basic cations Ca^{2+}, Mg^{2+}, Na^+, K^+ as well as pH, silicic acid, DOC, aqueous Al and Fe were determined in leachates from the lysimeters four times a year (May–October). The leaching of DOC, Al, Fe and Ca from the mor layer was much greater in autumn than in spring and summer. In the upper 10 cm about 80% of the monomeric aluminium occurs as organic complexes.

By considering the net uptake, the precipitation and percolation at different levels, the weathering rates could be evaluated if the pool of exchangeable cations is assumed to be constant. Summarizing the weathering rates of the base cations gives 46 meq m^{-2} a^{-1} for the upper 20 cm of mineral soil profile (2.3 meq m^{-2} a^{-1} cm^{-1}) and 39 meq m^{-2} a^{-1} for the depth 20–80 cm of the profile (0.64 meq m^{-2} a^{-1} cm^{-1}) as an average for three years.

The weathering rate in the upper part of the profile is about three-times higher than in the lower probably because of lower pH, higher waterflow and complex formation of Al and Fe with organics in the upper part.

1. Introduction

Chemical weathering is the only long-term soil process by which hydrogen ions, emanating from growing plants or from acid precipitation, can be neutralized. Chemical weathering is also the dominant natural process delivering all inorganic

[267]

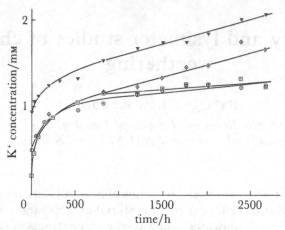

JCFigure 1. Release of K⁺ from K-feldspar in different leachants as a function of time; (▽), mor extract; (◇), streamwater upstream; (□), no organics; (○), streamwater downstream.

nutrients, except nitrogen, to the growing plants. Therefore, the rate at which a soil is weathered determines its sensitivity to acidification as well as its long-term fertility. In this process, the presence of organic substances, able to form complexes, has been claimed to be of major importance (McKeaque *et al.* 1986).

Weathering rates and the effects of organics have been studied in laboratory experiments and in field lysimeter studies.

2. Laboratory experiments

The main purpose of our study was to evaluate if and to what extent naturally sampled organic-containing waters could enhance the weathering rate. This work 'Dissolution of feldspars in the presence of natural, organic solutes' (Lundström & Öhman 1989) is in the press.

(a) Experimental

Three different feldspars, a natural silt, mainly consisting of microcline, oligoclase, quartz and illite, and ground samples of potassium feldspar and plagioclase, were exposed to distilled water, water leaving the humus layer of a pine forest floor, a natural streamwater, water extracts of mor and peat and an ammonium citrate solution; 50 g of silt or mineral were shaken with 950 ml of the solution on a shaker table. To create sterile conditions in these suspensions, 5 ml of formic aldehyde (40 % v/v) was added. During the experiments, carbon dioxide of 1 atmosphere partial pressure was continuously bubbled through the suspensions to maintain a constant pH of 5.1. The experiments were run for 2500 h. At regular time intervals 30 ml aliquots of the homogenized suspension were sampled for determination of pH and cations and silica by ICP.

(b) Results and discussion

(i) Dissolution rates

As an example of the experimental results, figure 1 shows the release of potassium ions from the potassium feldspar.

As seen from this plot, the sample showed much faster dissolution during the

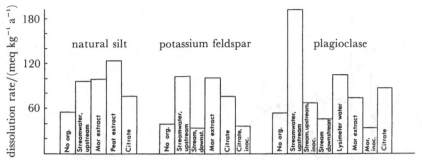

Figure 2. Dissolution rates of feldspars in the absence and presence of different natural organic solutes.

initial part of the experiment, after which the weathering rates attained constant values. A probable explanation for the high initial rates is that freshly ground samples contain ultra-fine particles created through the grinding procedure which are difficult to remove (Holdren & Berner 1979; Stumm *et al.* 1983). The natural silt sample had a constant weathering rate almost from the beginning. This has also been found in earlier investigations with natural soils (Grandstaff 1986; Wright & Frogner 1987). For the long-term weathering process under natural conditions it is, however, evident that this initial nonlinear stage has little relevance. The dissolution rate is therefore evaluated from the stage at which the rate is constant with time. This has been done for each single element from the linear section of the plot. The total dissolution rate for a mineral can be calculated from the formula:

$$\frac{\mathrm{dmin}}{\mathrm{d}t} = \left(\frac{\mathrm{d}[\mathrm{Na}^+]}{\mathrm{d}t} + \frac{\mathrm{d}[\mathrm{K}^+]}{\mathrm{d}t} + 2 \cdot \frac{\mathrm{d}[\mathrm{Ca}^{2+}]}{\mathrm{d}t} + 2 \cdot \frac{\mathrm{d}[\mathrm{Mq}^{2+}]}{\mathrm{d}t} \right),$$

resulting in a figure for the equivalents of cations released per unit mass per unit time. Results from these calculations are presented in figure 2.

From figure 2 it can be concluded that the dissolution of feldspars are significantly enhanced in solutions containing natural fresh organic material. 'Upstream' stream water, coming from a peatland, gives a total dissolution rate of 2.7 ± 0.9 ($n = 3$) times that obtained in a solution containing no organic material, whereas a streamwater taken approximately 1 km downstream had no enhancing effect. Also, when non-sterilized 'upstream' streamwater was inoculated with microorganisms, the weathering-enhancing effect almost disappeared. In experiments done in the presence of mor and peat extract a total dissolution rate 2.4 ± 0.4 ($n = 5$) times that obtained in a solution containing no organic substance resulted. Inoculation of a non-sterilized mor extract with microorganisms resulted in a considerable decrease in dissolution rate. Experiments were also performed in 5.55 mM ammonium citrate at a pH of 5.1 giving an increased dissolution rate 1.7 ± 0.3 ($n = 3$) times compared with a solution containing no organics.

By using the measured specific areas of the solids used, it was also possible to calculate the absolute dissolution rates in units of moles per centimetre per second. In table 1 the linear rate constants at pH = 5.1 in the absence of organic substances are presented.

By comparing the elemental composition of the minerals used with the obtained dissolution rates it appears that the release of aluminium and silicic acid are under-

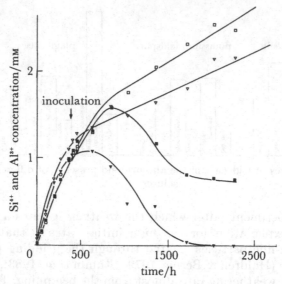

Figure 3. Concentration of aqueous Al ($\triangledown$; $\blacktriangledown$) and Si ($\square$; $\blacksquare$) dissolved from K-feldspar in a sterile citrate solution and in a microbiologically inoculated citrate solution as a function of time; ($\square$, $\triangledown$), citrate sterile; ($\blacksquare$, $\blacktriangledown$), citrate inoculated.

Table 1. *Logarithms of the linear rate constants for the release of the major elements from natural silt,*
K-feldspar and plagioclase in a $CO_2(g)/HCO_3^-$ buffer at pH = 5.1

(The rates are given in units of mole $cm^{-2} s^{-1}$.)

	K	Na	Ca	Si
natural silt	−17.75	−17.66	−17.37	—
K-feldspar	−16.51	−16.65	−16.84	−16.05
plagioclase	−17.38	−16.53	−16.49	−16.49

stoichiometric. This could be explained by the formation of a secondarily formed aluminosilicate phase, containing one or two silicic acid molecules per aluminium ion.

(ii) *Effects of inoculation*

In the weathering experiments using potassium feldspar, an experiment was also conducted where a non-sterilized citrate solution (5.55 mM) was inoculated with microorganisms after 400 hours of leaching. The result of this treatment is illustrated in figure 3.

It may be that when the excess of citrate ions has been decomposed, the soluble aluminium species will start to degrade and, as a consequence, the aluminium will start to precipitate. From a comparison with the behaviour of silicic acid, it is probable that the precipitate formed is an aluminosilicate phase containing approximately one aluminium ion per silicic acid molecule. From this experiment, it is also possible to conclude that the decomposition rate of citrate ions is considerably delayed when in a complex than when occurring as free ions.

2. Lysimeter studies

In our experiments percolation lysimeters were installed under an undisturbed mor layer, so that the uptake of nutrients by the trees was included and artifacts such as mineralization of organic substances because of roots being cut off were minimized. Analysis of soil solution was not undertaken until two years after installation, when the disturbances caused by installation were much reduced.

(a) Experimental

Percolation lysimeters were installed at Heden, in the Svartberget research park, 64° 14′ N, 19° 46′ E, 70 km NW of Umeå in the north of Sweden. The area is a glacifluvial sediment consisting of about 93% fine sand, 6% silt and 1.5% clay of quartz, K-feldspar, plagioclase and some biotite. It is an orthic podzol with a 5 cm thick mor layer (A_0, A_{00}), a *ca.* 3 cm thick elluvial zone (A_2) and an illuvial zone (B) down to about a depth of *ca.* 20 cm. The stand of Scots pine has a mean age of 70 years and a production of 3.9 m³/ha⁻¹ a⁻¹†. The field layer vegetation consists of mosses and dwarf shrubs, *Vaccinium myrtillus* and *Vaccinium vitis-idea*. The mean annual throughfall during the period November 1985–November 1988 was calculated to be 613 mm. The throughfall in the stand was 78% of the precipitation.

Three sets of ten lysimeters were installed under an undisturbed mor layer. Soil cores of only mineral soil profiles were taken and placed in polythene tubes with a filter on its perforated bottom, which was attached to polythene funnels delivering the water to bottles. Tubes were connected to the bottles so that the water could be pumped to the ground surface. The lysimeters were sampled in May, June, August and October. Conductivity, pH, DOC, cations, silicic acid and anions were determined. Data collected during 1986–88 were analysed.

(b) Results and discussion

It was found that the ionic concentration was always higher at a certain depth, when the collected volume of water was small. This could be attributed to the longer residence time of the water during dry periods and thereby longer time for weathering and ion-exchange reactions. This means that the concentrations at a certain depth will not be normally distributed. Therefore weighted means by volume have been calculated for three years for the summer and autumn seasons (figure 4).

As can be seen from the figure, the concentrations of various elements are very similar in the deeper soil throughout the seasons, whereas the upper soil column exhibits a strong seasonal variability. Organic substances, measured as DOC, Ca and Mg are highly leached from the mor layer during the autumn, as well as aqueous Al and Fe. About 80% of monomeric aluminium in the upper 10 cm of the mineral soil has been shown to be in 'organic complexes' by using the method of Driscoll (1984).

Budget studies have been performed for the upper part of the profile down to a depth of 20 cm under the mor layer and for the mineral soil column between a 20 and 80 cm depth. Volume-weighted means were calculated for the ions considered from the three sets of lysimeters between 1986 and 1988. Over the whole profile the volume of water obtained in the lysimeters is about 150 mm a⁻¹ less than either the throughfall precipitation or the run-off measured at the nearby reference station. This is probably because of lower tensions in the lysimeters compared with natural conditions. The volumes used for the budget calculation were corrected for this

† 1 ha = 10⁴ m².

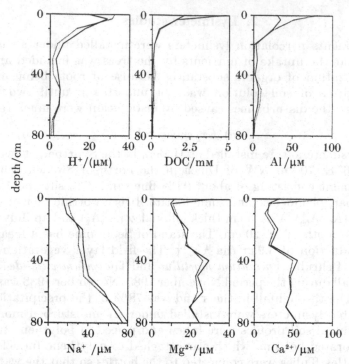

Figure 4. Percolation lysimeters, Svartbergets research area. Means weighted by volume June–August and October–November 1986–88; (—), June–August; (.....); October–November.

Table 2. *Wet deposition* (D), *net uptake by the pine stand* (U), *leaching at 20 cm depth* (L_{20cm}), *weathering rate down to 20 cm under the mor layer* (W_{20cm}), *leaching at 80 cm depth* (L_{80cm}) *and weathering rate 20–80 cm* ($W_{20-80cm}$), *meq m^{-2} a^{-1}, at Heden, the Svartberget research area*

	D	U	L_{20cm}	L_{80cm}	W_{20cm}	$W_{20-80cm}$
			(meq m^{-2} a^{-1})			
Ca	4.0	11.3	16.6	35.2	23.9	18.6
Mg	1.6	3.5	7.6	13.0	9.5	5.4
Na	2.9	0.5	7.1	21.0	4.7	13.9
K	0.9	3.4	5.5	6.1	8.0	0.6
Ca + Mg + Na + K	—	—	—	—	46.1	38.5
					(meq m^{-2} a^{-1} cm^{-1})	
Ca + Mg + Na + K					2.3	0.64

deficiency. The net uptake of elements by the pine stand was calculated by assuming the yearly net increment to be 137 % of the stem increment and that 10 % of this consisted of bark (Marklund 1988). Values of concentration of various elements in stemwood and bark were estimated (A. Albrektsson personal communication). Deposition values were taken from the reference station 1 km from the study area. Dry deposition was not included as data were lacking. Transports considered are presented in figure 5. Input and output data are shown in table 2.

$$W_{20\ cm} = U + L_{20\ cm} - D,$$
$$W_{20-80\ cm} = L_{80\ cm} - L_{20\ cm}.$$

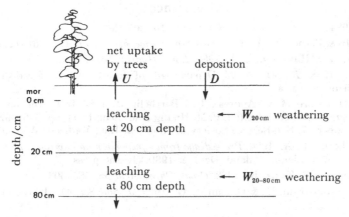

Figure 5. Input–output balance for base cations in the soil.

The symbols in the equations are explained in figure 5.

Considering data in table 2 the weathering rate over the whole profile down to 80 cm below the mor layer was estimated to be 85 mg m^{-2} a^{-1} to keep the pool of exchangeable cations constant. The results also show that the 'weathering rate' in the upper 20 cm of the mineral soil is about three times higher than in the deeper part of the soil profile. This could be because of lower pH, higher flow of water and organic complexation of Al and Fe. In the whole profile, pH is in the range from 4.5 to 5.5, over which the weathering rate is expected to be rather independent of pH (Sverdrup 1989). The water flow is about twice as large in the uppermost part of the mineral soil as in the lower part, which will be of great importance for the weathering kinetics. The dominating factor seems however to be organics, occurring only in the upper part of the profile, forming complexes primarily with aluminium and iron. Comparing the weathering rate for the whole profile with that obtained by Jacks (1989) of 150 meq m^{-2} a^{-1} from an adjacent site, our rate is lower. This however is much higher than that calculated by Olsson (1990), 31 meq m^{-2} a^{-1}, for historical weathering on an adjacent site.

3. Conclusion

The laboratory and field studies have shown that naturally occurring organics influence the weathering rate. In both studies the rate is enhanced 2–3.5 times in waters containing 'active' organic substances. The laboratory measurements show, however, that these 'active' organic substances are probably decomposed by naturally occurring microorganisms and that secondarily formed aluminium silicate phases containing one or two silicic acid molecules per aluminium ion are readily formed.

In the Svartbertget research area in the north of Sweden, the weathering rate was estimated to 85 mg m^{-2} a^{-1} for the upper 80 cm of the profile. The weathering rate obtained by laboratory experiments is about a thousand times higher because of the smaller particle size, the particles being suspended in water, and at higher temperature.

I thank Professor Nils Nykvist and Dr Harald Grip for much valuable advice and for their great interest. The English has been corrected by Mr Kevin Bishop. This work was financially supported by the Surface Waters Acidification Programme.

References

Driscoll, C. T. 1984 *Int. J. environ. analyt. Chem.* **16**, 267–283.

Holdren, G. R. Jr & Berner, R. A. 1979 *Geochim. cosmochim. Acta.* **43**, 1161–1171.

Jacks, G., Åberg, G. & Hamilton, P. J. 1989 *Nord. Hydrol.* **20**, 85–96.

Marklund, L. G. 1988 *Report no. 45* Department of forest survey, Swedish university of agricultural sciences, Umeå.

McKeague, J. A., Cheshire, M. V., Andreax, F. & Berthelin, J. 1986 In *Interactions of soil minerals with natural organics and microbes* (ed. P. M. Huang & M. Schnitzer), pp. 549–592. SSSA Special Publication number 17. Soil Science Society of America, Inc., Madison, Wisconsin, U.S.A.

Olsson, M. & Melkerud, P.-A. 1990. *Proceedings from symposium on environmental geochemistry in northern Europe*, Rovaniemi, Finland, October 1989. (In the press.)

Stumm, W., Furrer, G. & Kunz, B. 1983 *Croat. Chim. Acta* **56**, 585–603.

Sverdrup, H. U., Lundström, U. & Öhman, L. O. 1990 *J. Soil Sci.* **41**. (In the press.)

Aluminium speciation during episodes

By G. S. Townsend, K. H. Bishop and B. W. Bache

Department of Geography, Downing Place, Cambridge CB2 3EN, U.K.

1. Introduction

The hydrochemistry of many acid headwater streams is profoundly modified by short-term storm events. These are characterized by a drop in water pH and changes in aluminium chemistry. This inverse relation between stream discharge and pH has been interpreted as evidence that hydrological pathways play a key role in determining the chemistry of surface waters (Seip 1980; Bache 1984). The effects of acid deposition on organic forest soils depend to a large extent on the processes that occur within surface organic horizons, but these effects are often complicated by the polymeric acids produced naturally by decomposing litter and soil organic matter. To help distinguish these components of stream acidity, a catchment characterized by high natural acidity and relatively low inputs of atmospheric pollutants and marine-derived salts was investigated in detail for a three-week period in August 1988.

2. The site

The detailed hydrochemical survey was done on the streamwater and hillslopes of a 50 ha† catchment at the Svartberget Research Park, Sweden, shown in figure 1. The main stream originates in a 8 ha mire, whose waters were sampled at weir F, and runs 920 m to the catchment outlet at weir A. Precipitation, surface- and soil-water samples were collected over a period of 20 days; this included two storm events on 30 July and 6 August 1988. The first event was comparatively small and occurred after an extensive period of dry weather. The second was considerably larger and provided a discharge of at least an order of magnitude greater that the first event.

3. Experimental methods

The stream hydrograph was recorded by standard equipment. Comprehensive chemical analysis of the water was undertaken, but only the acidity/alkalinity and aluminium data are presented here, together with estimates of dissolved organic carbon (DOC).

Total dissolved aluminium (Al_t) was determined. Within this, the operationally defined fractions total monomeric (Al_m) and organic monomeric (Al_o) were determined by using a modified catechol–violet colorimetric technique based on Driscoll (1984), and inorganic monomeric (Al_i) was calculated as the difference between the Al_m and Al_o concentrations.

Acidity was determined by potentiometric titration to a fixed end-point of pH 5.4.

† $1\ ha = 10^4\ m^2$.

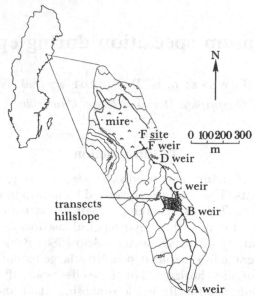

Figure 1. The Svartberget catchment, showing positions of the weirs.

4. Results

In the original poster these were shown in detail in three-dimensional coloured diagrams, in which the contours represented the hydrograph discharge and the colouring represented chemical solute concentrations. Only the key features are discussed here and shown in figures 2 and 3.

During the first event there was no significant increase in DOC concentrations. The pH dropped from the baseflow values of 4.5 and 5.9, down to 4.2 and 5.6, for the F weir and the A weir, respectively. The increase in discharge was accompanied by increases in the concentrations of Al_t and Al_o. The Al_i concentration reached its maximum early on in the rising limb of the hydrograph after which it declined rapidly, even though the discharge continued to rise, as shown in figure 2.

These results for the first event can be interpreted as low intensity rainfall entering a comparatively dry soil apparently inducing a 'piston effect' displacement of some 'old' inorganic Al-rich water into the stream, after it had been in prolonged contact with the mineral horizons. This pool of Al was rapidly depleted indicating either the importance of kinetically slow dissolution reactions in forming the pool, or the short-term mobilization of exchangeable Al as water is temporarily routed through these mineral horizons.

During the second event there was a rapid increase in DOC with discharge (not shown). The water pH dropped to 3.9 at weir F and 4.2 at weir A. Al_i concentrations increased slightly during the rising limb but were a smaller proportion of Al_t than during the first event.

The dramatic increase in DOC, Al_t and Al_o concentrations during the second event may be attributed to water travelling through surface organic horizons when the underlying soils are rapidly approaching saturation as a result of very intense rainfall on soils previously wetted by the first event.

An additional factor in the hydrochemistry of Svartberget is that during base flow conditions, as the water drains the mire and flows toward the catchment outlet, the

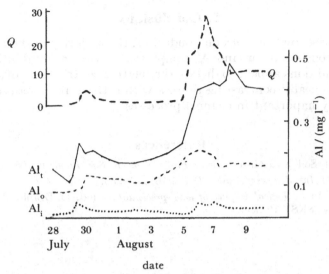

Figure 2. Aluminium fractions in relation to flow at weir *A* over two storm events.

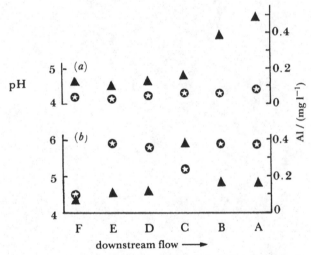

Figure 3. Aluminium and pH from weir *F* to weir *A* during (*a*) storm flow and (*b*) base flow, demonstrating Al mobilization from the streambed during stormflow; (◉), pH; (▲), Al.

pH increases from *ca.* 4.5 to *ca.* 6.0. This increase was shown to be caused partly by degassing of both CO_2 and of H_2S, both of which were formed during a variety of decomposition processes within the mire. This downstream increase in pH appears to encourage the precipitation and adsorption of Al onto stream-bed materials, as shown by the reduction in total and labile Al concentrations in streamwater, which decrease between weirs C and A (figure 3*b*). However, it is probable that during the second event these Al precipitates were rapidly dissolved when acid water was flushed into the stream channel. This source of Al, together with that mobilized from the hillslope during the episode, produced considerably increased Al concentrations with distance downstream, shown for the second storm event in figure 3*a*.

5. Conclusions

During both baseflow and episodic conditions, the majority of the total Al present in the stream consists of organo–Al complexes of varying stabilities. Prevailing hydrological conditions appear to dictate the relative contribution of soil water from a variety of chemically contrasting sources within the soil. Stream composition is then modified by important in-channel processes.

References

Bache, B. W.　1984　Soil–water interactions. *Phil. Trans. R. Soc. Lond.* B **305**, 393–407.

Driscoll, C. T.　1984　*Int. J. envir. Analyt. Chem.* **16**, 267–283.

Seip, H. M.　1980　In *Ecological impact of acid precipitation* (ed. D. Drablos & A. Tollan), pp. 358–366. Norway: SNSF Project.

General discussion

C. O. TAMM (*The Swedish University of Agricultural Sciences, Department of Ecology and Environmental Research, S-75007 Uppsala, Sweden*).
The emphasis within SWAP has been given to two themes within the very broad concept of chemical processes in the soil affecting freshwater quality, namely mineral weathering as a counteracting force to acid deposition, and aluminium chemistry as important for water quality and water biota. Considerable progress has been made within the two themes, within SWAP but also thanks to investigations outside our programme.

The session compared favourably with most earlier discussions of weathering (1) by making the necessary distinctions between important concepts such as current versus historical weathering, and weathering of primary minerals as distinguished from ion exchange processes and formation/dissolution of secondary minerals and amorphous substances, (2) by attempts to use more than one method on the same site. Several methods have been used, some of them recently invented. One such new method used zirconium as internal standard, another one used the differences in strontium isotope ratio in old bedrock and in seawater to distinguish between strontium (and indirectly calcium) derived from bedrock from cyclic strontium in run-off. In addition catchment balances with silicium considered as a conservative element have been used, as well as mineralogical methods, improved with scanning electron microscopy. The potential and the limitation of each method were well recognized.

The results presented show that we now have a much better, although not complete, understanding of rate-determining parameters such as texture, mineral species, and organic ligands. The influence of time was particularly well demonstrated by a study of a Scottish chronosequence of soils. There was general agreement on the rate of weathering, with differences between estimates with different methods often differing by less than a factor of two. There were good indications that the rate of the actual weathering is higher than that of the historical in areas with high acid deposition.

Aluminium chemistry must still be considered as a difficult area, but thanks to the SWAP studies we have now methods intercalibrated between laboratories which make it possible to distinguish in a reproducible way aluminium fractions with widely different reactivity and toxicity. It remains to understand better the rate of transfer between various aluminium fractions in lake and river water, as well as in soil and soil water, and the factors that control aluminium concentrations and speciation.

The next two big challenges in the study of chemical processes affected by acidification of soils and waters, will be, in my mind, the consequences of increased nitrogen deposition and the mobilization of heavy metals by increasing acidity of soils, in addition to direct supply from the polluted atmosphere.

The SWAP Palaeolimnology Programme: a synthesis

By Ingemar Renberg[1] and Richard W. Battarbee[2]

[1] Department of Ecological Botany, University of Umeå, S-901 87 Umeå, Sweden
[2] Palaeoecology Research Unit, Department of Geography, University College London, 26 Bedford Way, London WC1H 0AP, U.K.

Physical, chemical and biological (microfossil) analyses of lake sediment cores enable lake, catchment and atmospheric histories to be reconstructed with high resolution over a range of timescales. In the SWAP Palaeolimnology Programme these techniques have been used to trace the recent (post 1800 A.D.) history of a number of specially selected study sites in Norway, Sweden and the U.K. For this purpose a large calibration data-set of diatoms and water chemistry has been assembled and new statistical techniques of pH reconstruction have been developed. In addition, by comparing temporal trends and spatial patterns within and between sites, various explanations for recent lake acidification have been evaluated.

Studies of complete post-glacial (10000 years) sediment sequences at a number of sites showed that long-term acidification has occurred at some sites but the rate of change before 1800 A.D. at all sites was zero or less than 0.1 pH unit per 1000 years. Tests of the 'land-use change' hypothesis involving studies of past analogues, sites with no decrease in grazing intensity, sites above the treeline and sites with minimal catchments all failed and tended instead to support the acid deposition hypothesis. The only observed land-use effect was an acceleration in the rate of acidification followed afforestation at a site in the Scottish Trossachs. But, since this did not occur at a similar site in an area of low acid deposition in northwest Scotland, this 'forest effect' is probably related more to enhanced scavenging of dry and occult deposition by the forest canopy rather than the effect of base cation uptake.

In contrast to the lack of evidence for alternative causes the results overwhelmingly support the acid deposition hypothesis. Acute acidification always occurs at some date after 1800 A.D., usually after long periods of stable conditions, and sensitive sites in areas of low acid deposition are not acidified. Moreover the uppermost sediments of recently acidified lakes are strongly contaminated by atmospheric pollutants including trace metals and spheroidal carbonaceous particles from fossil fuel burning. When data for all sites are compared it can be demonstrated that there is a space–time and dose–response relationship between acid deposition and lake acidification.

Introduction

The Palaeolimnology Programme of the Surface Waters Acidification Project (SWAP) was designed in 1984 at a time when substantial research on surface waters acidification had already been undertaken in Norway (Overrein et al. 1980), Sweden

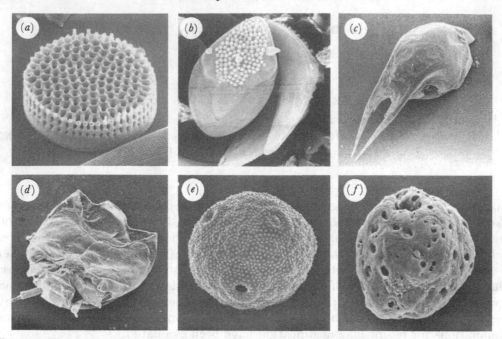

Figure 1. Illustrations of microfossils and particles found in lake sediments used in environmental reconstruction (a) diatom valve; (b) chrysophyte scale; (c) cladoceran head shield; (d) chironomid head capsule; (e) pollen grain; (f) spheroidal carbonaceous particle from oil and coal combustion.

(Ministry of Agriculture 1982; Andersson & Olsson 1985), the United Kingdom (Harriman & Morrison 1982), and North America (Altshuller & Linthurst 1984). However, at that time there was still uncertainty as to the geographical distribution, timing, extent, effects and causes of lake acidification (cf. Mason & Seip 1985). The 'land-use' hypothesis presented by Rosenqvist (1977) was receiving serious consideration as an alternative explanation to acid deposition (cf. Krug & Frink 1983) and the contribution of longer-term natural acidification processes was unclear. In consequence tests of alternative or additional causes for lake acidification became one of the main themes for the Palaeolimnology Programme.

A second major part of the programme concerned detailed integrated palaeo-limnological studies of a number of lakes situated in areas of high and low acid deposition in the United Kingdom, Norway and Sweden. The aim of these projects was to describe trends in lake acidification, catchment change and atmospheric pollution, and to provide background information for other SWAP projects taking place in these regions.

The detailed results of these and other related subprojects were presented at a Discussion Meeting of the Royal Society in August 1989 and have now been published (Battarbee et al. 1990). The Palaeolimnology Programme directly involved more than 40 scientists from the U.K., Norway and Sweden, it organized multinational workshops and meetings which included many scientists external to SWAP, and it benefited from close contacts with several members of the Paleoecological Investigation of Recent Lake Acidification (PIRLA) project in North America (Charles & Whitehead 1986). This paper summarizes the main content of the SWAP Palaeolimnology Programme and discusses the causes and effects of lake acidification using examples drawn from the Programme.

Table 1. *Presentation of projects/techniques, participants, and publications of the SWAP Palaeolimnology Programme*

(a) Integrated site studies

coring, core-handling	UCL-team (UK); Umeå-team (N,S)	Stevenson *et al.* (1987)
pollen	A. C. Stevenson (UK); H. I. Høeg (N); J.-E. Wallin (S)	see Integration and site study presentation
documents	S. Patrick (UK); A. Timberlid (N); J.-E. Wallin (S)	see Integration and site study presentation
dating	P. Appleby, F. Oldfield, N. Richardson, P. J. Nolan (UK); F. El-Daoushy (N,S)	Appleby *et al.* (1990) El-Daoushy (1990)
magnetics	F. Oldfield, N. Richardson	Oldfield & Richardson (1990)
carbonaceous particles	M. Wik (N,S); J. Natkanski, N. Rose (UK)	Wik & Natkanski (1990)
PAH	B. Rippey	Rippey (1990)
trace metals	B. Rippey	Rippey (1990)
sulphur	B. Rippey	Rippey (1990)
chironomids	Y.-W. Brodin	Brodin (1990*a, b*)
cladocera	J. P. Nilssen, S. Sandøy	Nilssen & Sandøy (1990)
chrysophytes	G. Cronberg	Cronberg (1990)
diatoms	V. J. Jones, A. M. Kreiser (UK); F. Berge (N); I. Renberg (S)	see Integration and site study presentation
integration and site study presentation		Kreiser *et al.* (1990) Jones *et al.* (1990) Berge *et al.* (1990) Renberg *et al.* (1990*a*)

(b) pH-reconstruction

diatom data	N. J. Anderson, R. W. Battarbee, F. Berge, S. C. Fritz, E. Y. Haworth, V. J. Jones, A. M. Kreiser, I. Renberg, D. S. Anderson, R. B. Davis, J. C. Kingston	Kreiser & Battarbee (1988) Munro *et al.* (1990)
data handling	M. A. R. Munro, S. Juggins, A. C. Stevenson, H. J. B. Birks, R. W. Battarbee	Munro *et al.* (1990)
statistical analysis	H. J. B. Birks, J. M. Line, S. Juggins, A. C. Stevenson, C. J. F. ter Braak	Birks *et al.* (1990*c*, this symposium) Stevenson *et al.* (1989) Line & Birks (1990)

(c) Hypothesis testing

long-term (U.K.)	E. Y. Haworth, K. M. Atkinson	Atkinson & Haworth (1990)
long-term (S)	I. Renberg	Renberg (1990*a,b*)
hill-top lakes	H. J. B. Birks, F. Berge, J. P. Boyle	Birks *et al.* (1990*a, b*)
Iron-Age analogue	N. J. Anderson, T. Korsman	Anderson & Korsman (1990)
spruce	I. Renberg, N. J. Anderson, T. Korsman	Renberg *et al.* (1990*b*)
afforestation (U.K.)	A. M. Kreiser, P. G. Appleby, J. Natkanski, B. Rippey, R. W. Battarbee	Kreiser *et al.* (1990)
documents, pollen	S. Patrick, A. C. Stevenson (U.K.); A. Timberlid (N)	Patrick *et al.* (1990) Timberlid (1990)

Table 1. (*cont.*)

synthesis	R. W. Battarbee	Battarbee (1990)
	(*d*) Other projects	
diatom ecophysiology	M. A. Smith	Smith (1990)
diatom ecology	F. E. Round	Round (1990)
diatom-MAGIC comparison	A. Jenkins, P. G. Whitehead, B. J. Cosby, H. J. B. Birks	Jenkins *et al.* (1990)
cladocera and predation	J. P. Nilssen & S. Sandøy (1990)	
post-1970 lake changes	R. J. Flower, N. G. Cameron, N. Rose, S. C. Fritz, R. Harriman, A. C. Stevenson	Flower *et al.* (1990)
atmospheric deposition on peat	R. S. Clymo, F. Oldfield, P. G. Appleby, G. W. Pearson, P. Ratnesar, N. Richardson	Clymo *et al.* (1990)

Lake sediments as environmental archives

Palaeolimnologists use the remains of past plant and animal communities in sediments to study how lakes have changed through time, and, because the sediments also contain material derived from the atmosphere and the lake catchment, lake changes can be compared with atmospheric and catchment changes. Figure 1 illustrates a range of microfossils and particles found in lake sediments and used in this project.

Unlike other environmental archives, such as documentary historical sources, lake sediment records are continuous, cover both short and long timescales, and usually accumulate rapidly enough to give a high time resolution (1–2 years). Lake sediments also allow retrospective analyses of environmental changes, many of which are rarely foreseen and accordingly are not monitored during the period of change from pristine to present-day conditions. Palaeolimnology, therefore, has become very popular in the context of lake acidification and a large number of papers on this subject have now been published in the international literature (cf. Battarbee 1984; Charles *et al.* 1989).

The SWAP Palaeolimnology Programme

Table 1 lists the projects and participants in the Palaeolimnology Programme and gives references to papers already published or in press that have been produced with SWAP support.

Integrated studies were undertaken at seven sites: four in Scotland (Round Loch of Glenhead, Loch Tinker, Lochan Dubh, Lochan Uaine); two in Norway (Verevatn, Röyrtjörna) and one in Sweden (Lilla Öresjön) (figure 2). A large number of techniques were applied (table 1). Biological analyses were used to reconstruct lake history included diatom, chrysophyte, cladoceran and chironomid analysis. Documentary sources and pollen analyses were used to study catchment history, and techniques used to trace atmospheric contamination included analyses of trace metals, sulphur, polycyclic aromatic hydrocarbons (PAH), carbonaceous particles and sediment magnetism. Cores were dated by ^{210}Pb and other radiometric methods.

Much developmental work was devoted to techniques, especially to diatom analysis, the most important method for pH reconstruction. A large water chemistry/diatom calibration data-set was created using a standard approach to water chemistry and diatom taxonomy based on the results of workshop exercises and the use of quality control techniques. New methods of pH, DOC and Al

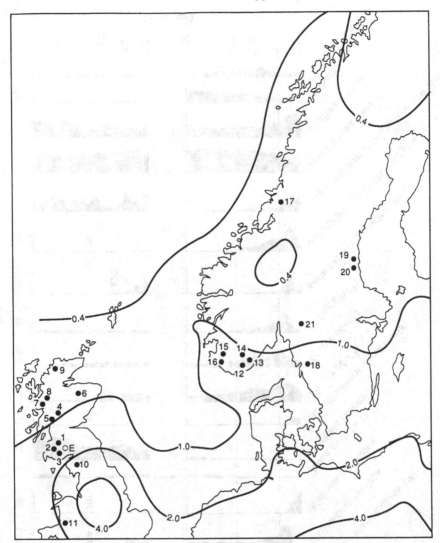

Figure 2. Map of northwest Europe showing the location of study sites in relation to isolines of S deposition (g m^{-2} a^{-1}) (from Eliassen *et al.* 1988). 1, The Round Loch of Glenhead; 2, Loch Grannoch; 3, Loch Fleet; 4, Loch Tinker; 5, Loch Chon; 6, Loch Uaine; 7, Loch Doilet; 8, Lochan Dubh; 9, Loch Sionascaig; 10, Devoke Water; 11, Llyn Hir; 12, Verevatn; 13, Gulspettvatn; 14, Holmevatn; 15, Holetjörn; 16, Ljosvatn; 17, Röyrtjörna; 18, Lilla Öresjön; 19, Sjösjön; 20, Lill Målsjön; E, Ellergower Moss. (From Battarbee & Renberg 1990.)

reconstruction have been developed from this data-set and applied to cores from SWAP sites.

Some or all these techniques were used in a range of projects concerned with the causes of lake acidification (see below). Two projects in the U.K. and Sweden respectively studied long-term acidification and its relationship to recent (post 1800) acidification. Five projects addressed the influence of catchment vegetation and land-use/management on lake acidification. These included studies of afforestation in Scotland, spruce expansion in Sweden, grazing and burning in Norway and Scotland, desettlement in the Swedish Iron Age as an analogue for postwar

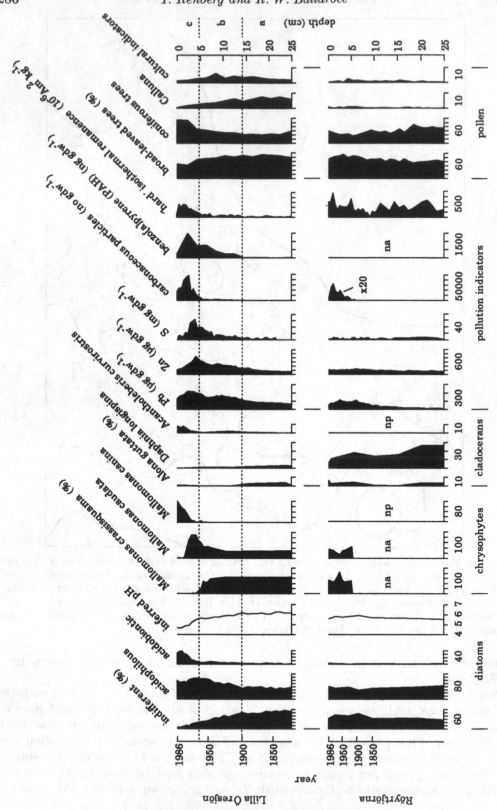

Figure 3. For description see opposite.

agricultural change, and the acidification history of hill-top lakes with very small catchments in south Norway.

The Programme also included a range of projects designed either to aid the interpretation of palaeolimnological data, such as studies of diatom ecophysiology, diatom ecology and the relationship between cladoceran morphology and fish predation, or to complement palaeolimnological reconstructions, such as studies on the use of ombrotrophic peats to record atmospheric contamination and the use of the process-based chemical model MAGIC to reconstruct pH history. A final project considered the lake and lake sediment response to very recent (post-1970) changes, such as liming, forest fertilization and, most importantly, decreasing acid deposition.

Post-1800 A.D. changes at sites with high and low acid deposition

The results of the integrated site studies show that lakes in areas of low acid deposition (northwest Scotland, central Norway) have very low sediment concentrations of air-borne pollutants and have undergone no significant recent change in pH. In contrast the lakes in areas of high acid deposition (southwest Scotland, south Norway, and southwest Sweden) have high concentrations of pollutants and severe recent acidification has taken place. In all cases there is a close correlation between increasing trends in air-borne pollutants and decreasing trends in pH.

Figure 3 illustrates these results using a comparison between Lilla Öresjön and Röyrtjörna (Renberg et al. 1990a: Berge et al. 1990). Lilla Öresjön is situated near the west coast of Sweden (figure 2), an area of high acid deposition and severe surface waters acidification whereas Röyrtjörna is situated close to the coast in central Norway (figure 2) where there are no apparent acidification problems.

In Lilla Öresjön biological changes associated with acidification began gradually about 1900 A.D. (figure 3). A more acute acidification stage followed in the 1960s when pH fell rapidly. Diatom-inferred pH values (based on the weighted averaging method (Birks et al. 1990c)) indicate that pH decreased from 6.2 at the end of the 19th century to the recent values of about 4.6. This reconstruction agrees well with postwar historical data which show that roach and bream occurred in the lake in the 1950s but died out in the 1960s.

The diatom community has changed markedly over the past 100 years; planktonic diatoms with preference for circumneutral waters, such as Asterionella formosa and Cyclotella kuetzingiana disappeared and were replaced by an acidophilous/acidobiontic periphytic flora with high abundance of 'acidification indicators' such as Tabellaria binalis. The scaled chrysophyte flora has changed almost completely. Mallomonas crassisquama has disappeared and the recent assemblage is dominated by M. canina, an acidobiontic taxon.

The cladoceran and chironomid communities have also changed considerably.

Figure 3. Comparison of sediment core data from Lilla Öresjön (SW Sweden, high deposition area) and Röyrtjörna (central Norway, low deposition area). Lilla Öresjön is strongly acidified: a, pre-acidification period; b, pH decrease; c, acute acidification period. Röyrtjörna is not acidified. Hustedt's pH categories for diatoms and diatom-inferred pH using weighted averaging; selected species of chrysophytes and cladocerans; atmospheric pollution indicators (N.B. spheroidal carbonaceous particle scale for Röyrtjörna expanded × 20); pollen (percentages for each category are calculated on the basis of the total pollen sum). na, not analysed; np, not present. Diagram is redrawn from Renberg et al. 1990a) and Berge et al. (1990) with additions from the unpublished results of G. Cronberg, J. P. Nilssen, S. Sandøy, F. Oldfield, N. Richardson and H. I. Høeg.

Several species have disappeared, for example, *Daphnia longispina* (planktonic cladoceran) and *Parakiefferiella bathophila* (chironomid). New acid-tolerant species have appeared and overall diversity has decreased.

In contrast there are few changes in the microfossil composition of the recent sediments in Röyrtjörna. Diatom-inferred pH fluctuates slightly between the relatively high values of 5.6 and 5.9. Although chrysophyte scales are rarer than in Lilla Öresjön the most abundant species in the surface sediments are *Mallomonas crassisquama* and *M. caudata*, the same species that were abundant in the pre-acidification sediments of Lilla Öresjön. The cladoceran community shows no particular change and *Daphnia longispina* is abundant in the recent sediment.

Analysis of atmospheric contaminants in these sediments shows that the concentrations of trace metals and sulphur increased in Lilla Öresjön from the beginning of the 19th century to peak values during the 1960s and 70s. Spheroidal carbonaceous particles and polycyclic aromatic hydrocarbons, indicative of coal and oil combustion, peaked between 1970 and 1980, and measurements of the magnetic mineralogy of the sediment shows that the 'hard' isothermal remanence component, indicative of fly-ash and other ferromagnetic pollutants from fuel combustion and other industrial sources, also increase strongly in the uppermost sediment.

In Röyrtjörna background natural magnetic inputs from the catchment are too high to detect an atmospheric component but, nevertheless, trace metal concentrations are very low and there is no change in the sulphur profile through time. The concentration of spheroidal carbonaceous particles is also very low with a maximum value of $1.6 \times 10^{-3} \, \mathrm{g^{-1}}$ dw (maximum value for Lilla Öresjön $= 48 \times 10^{-3} \, \mathrm{g^{-1}}$ dw). These data clearly confirm that Röyrtjörna has received very little air-borne contamination over the past 100 years or so whereas Lilla Öresjön, as predicted from present-day measurements of S deposition, has been very strongly contaminated over this time period.

The forest vegetation around Röyrtjörna mainly consists of spruce, pine and birch, and large areas of the catchment are covered by blanket mires. There is no agricultural land near the lake. Pollen analysis indicates that conditions have changed little, although there are sporadic indications of small-scale agriculture in the past. For Lilla Öresjön, the pollen analysis and data from documentary sources show that there has been considerable change in vegetation around the lake over the past two hundred years. During the 18th century large areas were covered by *Calluna* heathland. Pine-spruce forest expanded during the 20th century and became increasingly dense until clear-felling started in the 1980s. In 1982 coniferous forest covered about 60% of the catchment (15% clear cut), deciduous woodland 10% and arable land 10%. Agriculture is still practised and no significant area of agricultural land has been abandoned since 1950, but forest grazing decreased during the 1930s and ceased entirely during the 1940s.

The acidification histories and atmospheric contamination record of these two sites are consistent with predictions based on the acid deposition hypothesis. But the acidification of Lilla Öresjön is also correlated in time with land-use/vegetation changes in its catchment. There are many reasons for arguing that the acidification of Lilla Öresjön was caused by acid deposition (Renberg *et al.* 1990*a*), but the simultaneity of acid deposition and land-use change prevents the unambiguous assignment of a cause.

Causes of acidification

Three main explanations have been put forward for the acidification of surface waters. These are, long-term soil acidification over the post-glacial period (10 000 years) (cf. Pennington 1984), soil acidification associated with a decline in pastoral agriculture during the past two centuries (Rosenqvist 1977, 1978), and an increase in S deposition from the combustion of fossil fuels (cf. Odén 1968). In addition, more locally in the U.K., acidification has also been associated with the recent afforestation of upland moorlands (Harriman & Morrison 1982). Because, as noted above, some of these processes may have occurred together over similar timescales, palaeo-limnological tests of the various hypotheses require very careful site selection and research design, including, in some cases, the use of past analogues.

(a) Long-term acidification

The tendency for certain lakes with catchments on base-poor or slow-weathering bedrock to become gradually more acidic during the post-glacial time period was recognized over 60 years ago (Lundqvist 1925). Long-term acidification has been reported from many countries (Charles *et al.* 1989), and has been ascribed to both base-cation leaching and the paludification of catchment soils (Pennington 1984). Following the emergence of interest in acid deposition and its effects there has been an increase in long-term acidification studies (Renberg & Hellberg 1982; Whitehead *et al.* 1986; Jones *et al.* 1989) using pH-reconstruction techniques to quantify the degree and rate of change over this time period.

In SWAP four sites from England, Scotland and Sweden have been studied (Atkinson & Haworth 1990; Renberg 1990*a*; Jones *et al.* 1989; Birks *et al.* 1990*c*). The pH curves for these sites are shown in figure 4.

In the case of Lilla Öresjön Renberg (1990*a*) recognized four pH periods (i) a relatively alkaline period during which pH declined from about pH 7 to 6 following deglaciation around 12 600 before present (BP) to 7800 BP; (ii) a naturally acidic period with gradually decreasing pH from about pH 6.0 to 5.2 between 7800 BP and 2300 BP; (iii) a period of higher pH, mainly above pH 6.0, between 2300 BP and 1900 A.D., the start of which coincides with the expansion of agriculture as indicated by pollen analysis; and (iv) a post-1900 A.D. recent acidification phase divided into two stages, a stage from 1900 to 1950 during which conditions returned to those prevailing before 2300 BP, and an acute acidification stage from 1950 to the present day during which an acidobiontic diatom flora developed and conditions became more acid than at any previous time in the lake's history. Renberg argues that acid deposition is the most reasonable explanation for the recent acidification (Renberg *et al.* 1990*a*).

At Loch Sionascaig and Devoke Water, as at Lilla Öresjön, the basal diatom assemblages indicate pH values of 7 and above (figure 4). At both sites pH declined during the early post-glacial period, stabilizing at about pH 6.5, before 8500 BP in L. Sionascaig and by about 5000 BP in Devoke Water. Unlike Lilla Öresjön there is no tendency for either of these two lakes to become gradually more acidic and pH at both sites remains constant at about 6.2–6.6 until the 20th century when there is a sudden drop in the pH of Devoke Water, but not at L. Sionascaig. The different 20th-century trends between the two sites is related by Atkinson & Haworth (1990) to the much higher levels of sulphur deposition at Devoke Water than at L. Sionascaig.

At the Round Loch of Glenhead the early more alkaline phase is absent. Even the

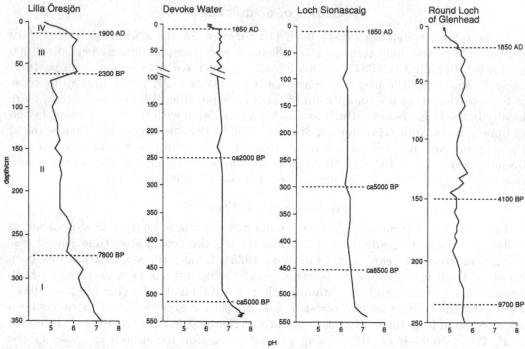

Figure 4. Long-term (10 000 year) pH reconstruction for (a) Lilla Öresjön; (b) Devoke Water; (c) Loch Sionascaig; and (d) The Round Loch of Glenhead (redrawn from Renberg (1990 *a*), Atkinson & Haworth (1990) and Birks *et al.* (1990 *d*).

earliest diatom assemblages at this site indicate pH values as low as 5.5 and although some fluctuations in DOC probably occurred, associated with the development of peatland in the catchment 5000 years ago (Jones *et al.* 1989; Birks *et al.*, this symposium), little pH change occurred throughout its post-glacial history (figure 4) until the acute acidification of modern times from about 1850 onwards (Flower *et al.* 1987), almost certainly caused by acid deposition (Battarbee *et al.* 1989).

The four sites presented here illustrate different individual pH trajectories over the past 10 000 years. Lilla Öresjön shows continuous slow acidification until the expansion of agriculture in the vicinity of the lake, Devoke Water and L. Sionascaig both show early alkaline phases, followed by acidification to pH values between 6 and 7 before stabilizing for the remaining period, while the Round L. of Glenhead shows very low pH values throughout. From these sites and from data from elsewhere it can be seen that long-term acidification does take place, but it is usually very slow suggesting that there is a quasi-equilibrium in most catchment soils between cation generation from weathering and cation loss from leaching. None of the sites has experienced values as low as present pH (below 5) before the influence of acid deposition. In addition, for the Round Loch of Glenhead, Birks *et al.* (1990 *c*) show that the rate of acidification was significantly more rapid between 1850 and the present day than at any previous time.

(b) Land-use/management change

Although long-term acidification processes have undoubtedly increased the sensitivity of many lakes and their catchments to acidification, the acute post-1800 acidification that has occurred cannot be explained in this way, and must instead be a function of environmental changes taking place since 1800 either in the catchment or atmosphere. Rosenqvist (1977, 1978) argued that surface water acidification in Norway has occurred, not as a result of acid deposition, but in response to a decline in the traditional agricultural practices of outfield burning and grazing.

In the SWAP Palaeolimnology Programme tests between these alternatives were carried out by designing studies to eliminate or independently vary the influence of one or the other factor.

(i) *Eliminating catchment influence*

The influence of the catchment can be eliminated or reduced in several ways; by selecting sites where catchment change has not taken place (or sites where catchment changes are not likely to have promoted acidification); by selecting sites where pastoral agriculture has never been practised; or by using sites where the catchment is so small in relation to the lake that its influence is minimal. According to the land-use hypothesis sensitive sites with these characteristics should not be acidified.

In Norway Timberlid (1990) and Patrick *et al.* (1990) tested the land-use hypothesis using data from agricultural censuses, land registrations and reports from farming organizations to assess the degree of land-use change in the acidified southern parts of Norway in comparison with changes in non-acidified parts of western Norway over the past 150 years. In the southern region Timberlid observed that there had been a decrease in grazing intensity only on land below about 600 m. Above this level sheep numbers had changed little since the late 19th century, and since modern sheep consume twice the fodder of earlier breeds grazing pressure in this area had increased. Counter to the prediction of the hypothesis this is the zone in which surface waters acidification is most pronounced (Henriksen *et al.* 1988).

In the western region the grazing of domestic animals and use of summerland has decreased since the 1860s. However, again counter to prediction, there is little evidence of surface water acidification in this area.

In the U.K. there is little support from documentary sources for a decline in moorland grazing. In many regions where lakes are strongly acidified (Battarbee *et al.* 1985, 1988 *a*, 1989) there is evidence for increased grazing and burning and a shift from *Calluna* heathland towards a more grass dominated vegetation, a trend also indicated by pollen analysis (Patrick *et al.* 1990).

Catchment influence can also be eliminated or controlled by considering sites in high elevation montane environments above the treeline and beyond the practice of pastoral agriculture, past or present. Many such sites occur in the Cairngorm Mts of Scotland, close to the SWAP experimental stream catchment on the Allt a'Mharcaidh (Harriman *et al.*, this symposium). The high corrie lakes in this area have steep-sided boulder-strewn catchments with little soil and sparse vegetation. Despite their isolation and lack of catchment change these sites are all acidified, some more than others, and their sediments contain high levels of atmospheric contaminants (Battarbee *et al.* 1988 *a*), a finding which agrees with the expectations of the acid deposition hypothesis but not the land-use hypothesis.

A further test of the land-use hypothesis has been carried out by Birks *et al.* (1990 *a*, *b*) who argued, according to the hypothesis, that lakes perched on hill-tops with very small catchment areas should show no recent acidification since these sites, by their very nature could not have been influenced by land-use change. Two such sites, Holetjörn and Ljosvatn, in Vest-Agder, SW Norway, an area of high acid deposition were studied. Contrary to the prediction both sites showed clear evidence of recent acidification. For example, before about 1900 Holetjørn (figure 5) was acidic with pH values of about 5.0, but then became much more acidic with a pH decline of about 0.4 pH units occurring between about 1915 and the present day and the uppermost sediment is contaminated by carbonaceous particles and has elevated trace metal concentrations. These results represent a falsification of the land-use hypothesis at these sites.

(ii) Eliminating acid deposition

The land-use change hypothesis can be falsified by the above tests since in all cases lake acidification has occurred independently of land-use changes. However, to assess whether land-use changes have played any role at all in lake acidification tests need to be devised which allow the influence of land-use changes to be studied in the absence of acid deposition.

This can be achieved by a study of land-use effects in sensitive systems in an area of very low acid deposition or by looking for past analogues at a time of low deposition. Jones et al. (1986) adopted this latter approach in a study of the Round Loch of Glenhead. They were able to show that the formation of blanket peatland in the catchment of the lake between 3000 and 5000 years ago did not cause a decrease in inferred pH contrary to the prediction of the hypothesis (see figure 4).

There have been two SWAP studies of this kind, both in Sweden; the Iron-Age desettlement project of Anderson & Korsman (1990) and the spruce project by Renberg et al. (1990b).

Anderson & Korsman (1990) argued that a good historical analogue for modern land-use changes is provided by the effects of the depopulation of farms and villages in Hälsingland, N. Sweden during Iron-Age times, an event well-established by archaeologists (Liedgren 1984). Following abandonment of the area from ca. 500 A.D. resettlement did not occur before the Middle Ages (ca. 1100 A.D.). Although evidence for regeneration of forest vegetation during this period can be identified in pollen diagrams from lakes in this region there is no evidence from diatom analysis for lake acidification during this period at the two sites studied, Sjösjön and Lill Målsjön (Anderson & Korsman 1990).

In a similar project Renberg et al. (1990b) assessed the potentially acidifying influence of the colonization of lake catchments by spruce. Since the 1920s the area covered by spruce in Götaland and Svealand, the regions with the most acidified waters in Sweden, has increased by 2.3 million ha. It is known that the pH of forest soils is significantly lower beneath spruce than birch stands and it has been suggested that the increase in the proportion of spruce in forests has been the cause of surface water acidification. To eliminate the acid deposition influences Renberg et al. (1990b) used the natural immigration of spruce into northern and central Sweden about 3000 years ago as an early analogue. Eight lakes that are acidified or sensitive to acidification at the present day and that had minimal amounts of peatland in their catchments were selected. Pollen analysis was used to identify the levels in the sediment cores that corresponded to the arrival of spruce in the region and diatom analysis to identify any consequent acidification. Figure 6 shows results from one site, Sjösjön. It is clear that despite a major shift in vegetational composition towards spruce there was no change in the diatom flora and lake water pH. At other sites slight changes in the diatom assemblages occurred but there changes were not towards an increasing frequency of acid tolerant taxa.

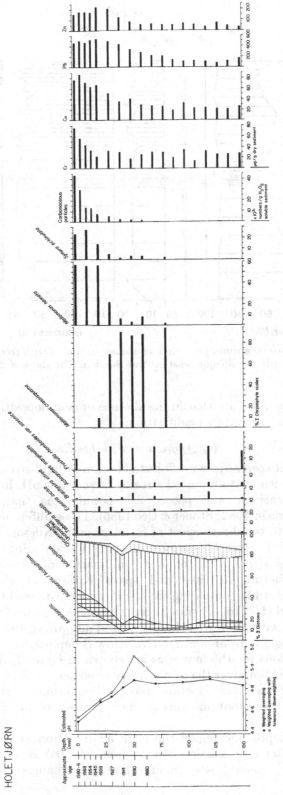

Figure 5. Reconstructed pH and selected diatom, chrysophyte, carbonaceous particle, and trace metal data plotted against sediment depth for a core from Holetjörn (from Birks *et al.* 1990*b*).

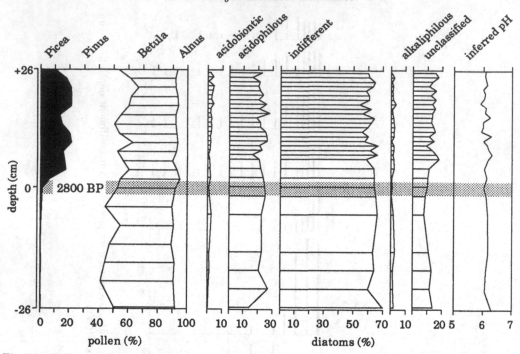

Figure 6. Pollen curves for spruce, pine, birch and alder compared with trends in diatom pH groups and diatom-inferred pH for Sjösjön, east-central Sweden. The date of 2800 BP is based on ^{14}C dating.

The data clearly illustrate that in the absence of acid deposition spruce expansion does not cause surface waters acidification.

(c) Afforestation in the U.K.

Comparisons between adjacent afforested and moorland streams in the uplands of the U.K. have shown that afforested streams have lower pH, higher aluminium and sulphate concentrations, and poorer fish populations than moorland streams (Harriman & Morrison 1982; Stoner & Gee 1985). These differences have been mainly attributed either to the direct effect of forest growth (Nilsson *et al.* 1984) or to the indirect effect of the forest canopy enhancing ('scavenging') dry and occult S deposition (Fowler *et al.* 1989), or to a combination of effects. In an attempt to separate these factors in time and space Kreiser *et al.* (1990) devised a palaeolimnological study comparing the acidification histories of afforested (L. Chon) and non-afforested (L. Tinker) sites in the Trossachs region of Scotland, an area of high S deposition, with afforested (L. Doilet) and non-afforested (Lochan Dubh) sites in the Morvern region, an area of relatively low S deposition.

pH reconstructions for the four sites are shown in figure 7. All four sites showed evidence of acidification over the past century or so and all show contamination over this time period by trace metals and carbonaceous particles although the concentrations of these contaminants in the northwest region are very low (Kreiser *et al.* 1990).

In the high S deposition region the main acidification at the afforested site, L. Chon, occurred after afforestation (from the mid-1950s) at a time when pH at the adjacent moorland 'control' site remained largely unchanged. Since both these sites

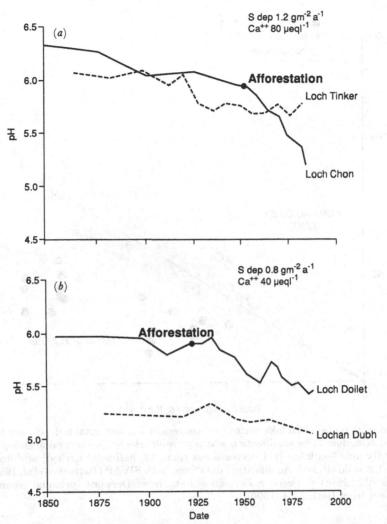

Figure 7. Diatom-inferred pH curves plotted against time for (*a*) two high acid deposition sites, Loch Chon (solid line, afforested 1950s) and Loch Tinker (dotted line, non-afforested) and (*b*) two low acid deposition sites, Loch Doilet (solid line, afforested 1920s) and Lochan Dubh (dotted, non-afforested). (From Kreiser *et al.* 1990.)

have very similar base cation chemistry and both had very similar diatom-inferred pH trends before the afforestation of L. Chon the data strongly suggest that the large difference in present day water chemistry is due to afforestation.

In the region of lower S deposition the main acidification of the afforested site, L. Doilet, also occurs after afforestation but the degree of acidification is slight and not significantly different than for the 'control' site, Lochan Dubh. For these reasons and because L. Doilet is more sensitive to acidification (Ca^{2+} 40 μeq l^{-1}) than L. Chon (Ca^{2+} 80 μeq l^{-1}) Kreiser *et al.* (1990) conclude that any acidification caused by forest growth alone has been minimal and that the data strongly support the 'scavenging' hypothesis. In other words afforestation of sensitive catchments can seriously exacerbate surface water acidification but only in regions receiving high levels of S deposition.

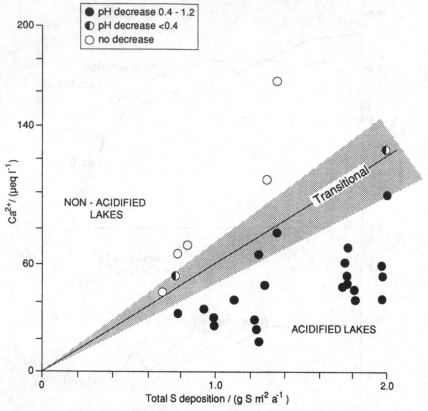

Figure 8. The relation between lake-water Ca^{2+} concentration and total S deposition for core sites in the United Kingdom. The acidification status of each lake is shown: non-acidified lakes (open circles); slightly acidified lakes (pH decrease less than 0.4, half solid circles); acidified lakes (pH decrease 0.4–1.2, solid circles). Acidification data from both SWAP (Battarbee *et al.* 1990) and DoE (Battarbee *et al.* 1988a) projects, S deposition data from Derwent (personal communication). Figure redrawn from Battarbee (1990).

(d) Acid deposition

Excluding the special case of conifer afforestation described above, all palaeo-limnological tests designed to disprove the land-use hypothesis for surface waters acidification have succeeded while those designed to disprove the acid deposition hypothesis have consistently failed, both in Europe (Battarbee *et al.* 1988a; Battarbee 1990) and North America (Charles *et al.* 1989). At individual sites recent acidification always postdates the beginning of major industrialization in the late 18th and early 19th century, diatom responses are always after the first signs of atmospheric contamination in the sediment record and strongly acidified sites are also strongly contaminated by trace metals and carbonaceous particles. Moreover, when the diatom-inferred pH trends from all sites are compared with the regional pattern of S deposition recently acidified sites are found in areas of high S deposition and no recently acidified sites have been reported from areas of very low S deposition (Battarbee *et al.* 1988a; Whiting *et al.* 1989; Charles *et al.* 1989).

For U.K. sites Battarbee (1990) has shown that there is a dose–response relationship between S deposition and lake acidification if lake sensitivity (figure 8) is taken into account. This provisional empirical relationship indicates that recent acidification is not likely at non-afforested sites where the ratio of Ca^{2+} ($\mu eq\,l^{-1}$) to S deposition ($g\,m^{-2}\,a^{-1}$) is greater than about 70:1.

The pH trends inferred from diatom assemblages are also in good agreement with those predicted from chemical models (Jenkins *et al.* 1990) which use S deposition history as the main driving variable. In Galloway, where S deposition has been considerably reduced over the past two decades (cf. Barrett *et al.* 1987) both water chemistry records and diatom trends in the uppermost sediments of acidified lakes indicate either no further acidification or slight recent improvement (Battarbee *et al.* 1988*b*; Flower *et al.* 1990).

Conclusions

Long-term changes have increased the sensitivity of many lakes to acid deposition, and afforestation in areas of high acid deposition has probably altered the timing and intensity of acidification but has not been its direct cause. There is little or no palaeolimnological evidence to support the land-use hypothesis of Rosenqvist. Instead the time–space and dose–response relationships between acid deposition and recent lake acidification noted above, and just as apparent in North America as in Europe (Charles *et al.* 1989) fully substantiate the acid deposition hypothesis.

The data generated by the SWAP Palaeolimnology Programme underline the usefulness of lake sediments not only in providing a high-resolution record of environmental history over a range of short-to-long timescales but also in enabling the causes of environmental change to be evaluated and explained.

References

Altshuller, A. P. & Linthurst, R. A. 1984 Critical assessment review papers EPA/600/8.3/016BF, United States Environmental Protection Agency, Washington, D.C.

Anderson, N. J. & Korsman, T. 1990 In *Palaeolimnology and lake acidification* (ed. R. W. Battarbee, Sir John Mason, F.R.S., I. Renberg & J. F. Talling, F.R.S.) (*Phil. Trans. R. Soc. Lond.* B **327**), pp. 373–376. London: The Royal Society.

Andersson, F. & Olsson, B. (eds) 1985 *Ecological Bull.* **37**, 1–336. Stockholm: Publishing House of the Swedish Research Councils.

Appleby, P. G., Richardson, N., Nolan, P. J. & Oldfield, F. 1990 In *Palaeolimnology and lake acidification* (ed. R. W. Battarbee, Sir John Mason, F.R.S., I. Renberg & J. F. Talling, F.R.S.) (*Phil. Trans. R. Soc. Lond.* B **327**), pp. 233–238. London: The Royal Society.

Atkinson, K. M. & Haworth, E. Y. 1990 In *Palaeolimnology and lake acidification* (ed. R. W. Battarbee, Sir John Mason, F.R.S., I. Renberg & J. F. Talling, F.R.S.) (*Phil. Trans. R. Soc. Lond.* B **327**), pp. 349–355. London: The Royal Society.

Barrett, C. F., Atkins, D. H. F., Cape, J. N., Crabtree, J., Davies, T. D., Derwent, R. G., Fisher, B. E. A., Fowler, D., Kallend, A. S., Martin, A., Scriven, R. A. & Irwin, J. G. 1987 *Acid deposition in the United Kingdom 1981–1985: a second report of the United Kingdom Review Group on acid rain.* Warren Spring Laboratory, Stevenage.

Battarbee, R. W. 1984 *Phil. Trans. R. Soc. Lond.* B **305**, 451–477.

Battarbee, R. W. 1990 In *Palaeolimnology and lake acidification* (ed. R. W. Battarbee, Sir John Mason, F.R.S., I. Renberg & J. F. Talling, F.R.S.) (*Phil. Trans. R. Soc. Lond.* B **327**), pp. 339–347. London: The Royal Society.

Battarbee, R. W. & Renberg, I. 1990 In *Palaeolimnology and lake acidification* (ed. R. W. Battarbee, Sir John Mason, F.R.S., I. Renberg & J. F. Talling, F.R.S.) (*Phil. Trans. R. Soc. Lond.* B **327**), pp. 227–232. London: The Royal Society.

Battarbee, R. W., Flower, R. J., Stevenson, A. C. & Rippey, B. 1985 *Nature, Lond.* **314**, 350–352.

Battarbee, R. W., Anderson, N. J., Appleby, P. G., Flower, R. J., Fritz, S. C., Haworth, E. Y., Higgitt, S., Jones, V. J., Kreiser, A., Munro, M. A. R., Natkanski, J., Oldfield, F., Patrick, S. T.,

Richardson, N. G., Rippey, B. & Stevenson, A. C. 1988a *Lake acidification in the United Kingdom 1880–1986: Evidence from analysis of lake sediments.* London: Ensis Publishing.

Battarbee, R. W., Flower, R. J., Stevenson, A. C., Jones, V. J., Harriman, R. & Appleby, P. G. 1988b *Nature, Lond.* **332**, 530–532.

Battarbee, R. W., Stevenson, A. C., Rippey, B., Fletcher, C., Natkanski, J., Wik, M. & Flower, R. J. 1989. *J. Ecology* **77**, 651–672.

Battarbee, R. W., Mason, J., Renberg, I. & Talling, J. F. (eds) 1990 *Palaeolimnology and lake acidification.* London: The Royal Society.

Berge, F., Brodin, Y.-W., Cronberg, G., El-Daoushy, F., Høeg, H. I., Nilssen, J. P., Renberg, I., Rippey, B., Sandøy, S., Timberlid, A. & Wik, M. 1990 In *Palaeolimnology and lake acidification* (ed. R. W. Battarbee, Sir John Mason, F.R.S., I. Renberg & J. F. Talling, F.R.S.) (*Phil. Trans. R. Soc. Lond.* B **327**), pp. 385–389. London: The Royal Society.

Birks, H. J. B., Berge, F. & Boyle, J. F. 1990a In *Palaeolimnology and lake acidification* (ed. R. W. Battarbee, Sir John Mason, F.R.S., I. Renberg & J. F. Talling, F.R.S.) (*Phil. Trans. R. Soc. Lond.* B **327**), pp. 369–370. London: The Royal Society.

Birks, H. J. B., Berge, F. & Boyle, J. P. 1990b *J. Paleolimnology.* (In the press.)

Birks, H. J. B., Line, J. M., Juggins, S., Stevenson, A. C. & ter Braak, C. J. F. 1990c In *Palaeolimnology and lake acidification* (ed. R. W. Battarbee, Sir John Mason, F.R.S., I. Renberg & J. F. Talling, F.R.S.) (*Phil. Trans. R. Soc. Lond.* B **327**), pp. 263–278. London: The Royal Society.

Brodin, Y.-W. 1990a In *Palaeolimnology and lake acidification* (ed. R. W. Battarbee, Sir John Mason, F.R.S., I. Renberg & J. F. Talling, F.R.S.) (*Phil. Trans. R. Soc. Lond.* B **327**), pp. 295–298. London: The Royal Society.

Brodin, Y.-W. 1990b *National Swedish Environment Protection Board Report.* (In the press.)

Charles, D. F. & Whitehead, D. R. 1986 *Hydrobiologia* **143**, 13–20.

Charles, D. F., Battarbee, R. W., Renberg, I., van Dam, H. & Smol, J. P. 1989 In *Acid precipitation vol. 4. Soils, aquatic processes, and lake acidification.* (ed. S. A. Norton, S. E. Lindberg & A. L. Page), pp. 207–276. New York: Springer-Verlag.

Clymo, R. S., Oldfield, F., Appleby, P. G., Pearson, G. W., Ratnesar, P. & Richardson, N. 1990 In *Palaeolimnology and lake acidification* (ed. R. W. Battarbee, Sir John Mason, F.R.S., I. Renberg & J. F. Talling, F.R.S.) (*Phil. Trans. R. Soc. Lond.* B **327**), pp. 331–338. London: The Royal Society.

Cronberg, G. 1990 In *Palaeolimnology and lake acidification* (ed. R. W. Battarbee, Sir John Mason, F.R.S., I. Renberg & J. F. Talling, F.R.S.) (*Phil. Trans. R. Soc. Lond.* B **327**), pp. 289–293. London: The Royal Society.

El-Daoushy, F. 1990 In *Palaeolimnology and lake acidification* (ed. R. W. Battarbee, Sir John Mason, F.R.S., I. Renberg & J. F. Talling, F.R.S.) (*Phil. Trans. R. Soc. Lond.* B **327**), pp. 239–242. London: The Royal Society.

Eliassen, A., Hor, Ø., Iversen, T., Saltbones, J. & Simpson, D. 1988 *EMEP/MSC–W rep.* no. 1/88.

Flower, R. J., Battarbee, R. W. & Appleby, P. G. 1987 *J. Ecology* **75**, 797–824.

Flower, R. J., Cameron, N. G., Rose, N., Fritz, S. C., Harriman, R. & Stevenson, A. C. 1990 In *Palaeolimnology and lake acidification* (ed. R. W. Battarbee, Sir John Mason, F.R.S., I. Renberg & J. F. Talling, F.R.S.) (*Phil. Trans. R. Soc. Lond.* B **327**), pp. 427–433. London: The Royal Society.

Fowler, D., Cape, N. J. & Unsworth, M. H. 1989 In *Forests, weather and climate* (ed. P. G. Jarvis, J. L. Monteith, F.R.S., W. J. Shuttleworth and M. H. Unsworth) (*Phil. Trans. R. Soc. Lond.* B **324**), pp. 247–265. London: The Royal Society.

Harriman, R. & Morrison, B. R. S. 1982 *Hydrobiologia* **88**, 251–263.

Henriksen, A., Lien, L., Traaen, T. S., Sevaldrud, I. S. & Brakke, D. F. 1988 *Ambio* **17**, 259–266.

Jenkins, A., Whitehead, P. G., Cosby, B. J. & Birks, H. J. B. 1990 In *Palaeolimnology and lake acidification* (ed. R. W. Battarbee, Sir John Mason, F.R.S., I. Renberg & J. F. Talling, F.R.S.) (*Phil. Trans. R. Soc. Lond.* B **327**), pp. 435–440. London: The Royal Society.

Jones, V. J., Stevenson, A. C. & Battarbee, R. W. 1986 *Nature, Lond.* **322**, 157–158.

Jones, V. J., Stevenson, A. C. & Battarbee, R. W. 1989 *J. Ecology* **77**, 1–23.

Jones, V. J., Kreiser, A. M., Appleby, P. G., Brodin, Y.-W., Dayton, J., Natkanksi, J. A., Richardson, N., Rippey, B., Sandøy, S. & Battarbee, R. W. 1990 In *Palaeolimnology and lake acidification* (ed. R. W. Battarbee, Sir John Mason, F.R.S., I. Renberg & J. F. Talling, F.R.S.) (*Phil. Trans. R. Soc. Lond.* B **327**), pp. 397–402. London: The Royal Society.

Kreiser, A. M. & Battarbee, R. W. 1988 In *Proceedings of the Nordic Diatomist Meeting, Stockholm, 10–12 June 1987* (ed. U. Miller & A.-M. Robertsson), pp. 41–44. Stockholm.

Kreiser, A. M., Appleby, P. G., Natkanski, J. A., Rippey, B. & Battarbee, R. W. 1990 In *Palaeolimnology and lake acidification* (ed. R. W. Battarbee, Sir John Mason, F.R.S., I. Renberg & J. F. Talling, F.R.S.) (*Phil. Trans. R. Soc. Lond.* B **327**), pp. 377–383. London: The Royal Society.

Krug, E. C. & Frink, C. R. 1983 *Science, Wash.* **221**, 520–525.

Liedgren, L. 1984 *Archaeol. Environ.* **2**, 93–112.

Line, J. M. & Birks, H. J. B. 1990 *J. Paleolimnology* **3**, 170–173.

Lundqvist, G. 1925 *Sver. Geol. Unders.* C **330**, 1–129.

Mason, J. & Seip, H. M. 1985 *Ambio* **14**, 45–51.

Ministry of Agriculture 1982 Acidification today and tomorrow. A Swedish study prepared for the 1982 Stockholm conference on the acidification of the environment. Stockholm.

Munro, M. A. R., Kreiser, A. M., Battarbee, R. W., Juggins, S., Stevenson, A. C., Anderson, D. S., Anderson, N. J., Berge, F., Birks, H. J. B., Davis, R. B., Flower, R. J., Fritz, S. C., Haworth, E. Y., Jones, V. J., Kingston, J. C. & Renberg, I. 1990 In *Palaeolimnology and lake acidification* (ed. R. W. Battarbee, Sir John Mason, F.R.S., I. Renberg & J. F. Talling, F.R.S.) (*Phil. Trans. R. Soc. Lond.* B **327**), pp. 257–261. London: The Royal Society.

Nilsson, S. I., Miller, H. G. & Miller, J. D. 1984 *Oikos* **39**, 40–49.

Nilssen, J. P. & Sandøy, S. 1990 In *Palaeolimnology and lake acidification* (ed. R. W. Battarbee, Sir John Mason, F.R.S., I. Renberg & J. F. Talling, F.R.S.) (*Phil. Trans. R. Soc. Lond.* B **327**), pp. 299–309. London: The Royal Society.

Odén, S. 1968 *The acidification of air precipitation and its consequences in the natural environment.* Energy Committee Bulletin 1, Swedish Natural Sciences Research Council, Stockholm.

Oldfield, F. & Richardson, N. 1990 In *Palaeolimnology and lake acidification* (ed. R. W. Battarbee, Sir John Mason, F.R.S., I. Renberg & J. F. Talling, F.R.S.) (*Phil. Trans. R. Soc. Lond.* B **327**), pp. 325–330. London: The Royal Society.

Overrein, L. N., Seip, H. M. & Tollan, A. 1980 Final report of the SNSF-project 1972–1980. Oslo-Ås.

Patrick, S. T., Timberlid, J. A. & Stevenson, A. C. 1990 In *Palaeolimnology and lake acidification* (ed. R. W. Battarbee, Sir John Mason, F.R.S., I. Renberg & J. F. Talling, F.R.S.) (*Phil. Trans. R. Soc. Lond.* B **327**), pp. 363–367. London: The Royal Society.

Pennington, W. 1984 *Freshwater Biological Assoc. An. Rep.* **52**, 28–46.

Renberg, I. 1990*a* In *Palaeolimnology and lake acidification* (ed. R. W. Battarbee, Sir John Mason, F.R.S., I. Renberg & J. F. Talling, F.R.S.) (*Phil. Trans. R. Soc. Lond.* B **327**), pp. 357–361. London: The Royal Society.

Renberg, I. 1990*b* *J. Paleolimnology.* (In the press.)

Renberg, I. & Hellberg, T. 1982 *Ambio* **11**, 30–33.

Renberg, I., Brodin, Y.-W., Cronberg, G., El-Daoushy, F., Oldfield, F., Rippey, B., Sandøy, S., Wallin, J.-E. & Wik, M. 1990*a* In *Palaeolimnology and lake acidification* (ed. R. W. Battarbee, Sir John Mason, F.R.S., I. Renberg & J. F. Talling, F.R.S.) (*Phil. Trans. R. Soc. Lond.* B **327**), pp. 391–396. London: The Royal Society.

Renberg, I., Korsman, T. & Anderson, N. J. 1990*b* Spruce and surface water acidification – an extended summary. In *Palaeolimnology and lake acidification* (ed. R. W. Battarbee, Sir John Mason, F.R.S., I. Renberg & J. F. Talling, F.R.S.) (*Phil. Trans. R. Soc. Lond.* B **327**), pp. 371–372. London: The Royal Society.

Rippey, B. 1990 In *Palaeolimnology and lake acidification* (ed. R. W. Battarbee, Sir John Mason, F.R.S., I. Renberg & J. F. Talling, F.R.S.) (*Phil. Trans. R. Soc. Lond.* B **327**), pp. 311–317. London: The Royal Society.

Rosenqvist, I. Th. 1977 *Sur jord – surt vann* (*Acid soil – acid water*) Oslo: Ingeniörsforlaget A/S.

Rosenqvist, I. Th. 1978 *Sci. Total Environ.* **10**, 39–49.

Round, F. E. 1990 In *Palaeolimnology and lake acidification* (ed. R. W. Battarbee, Sir John Mason, F.R.S., I. Renberg & J. F. Talling, F.R.S.) (*Phil. Trans. R. Soc. Lond.* B **327**), pp. 243–249. London: The Royal Society.

Smith, M. A. 1990 In *Palaeolimnology and lake acidification* (ed. R. W. Battarbee, Sir John Mason, F.R.S., I. Renberg & J. F. Talling, F.R.S.) (*Phil. Trans. R. Soc. Lond.* B **327**), pp. 251–256. London: The Royal Society.

Stevenson, A. C., Birks, H. J. B., Flower, R. J. & Battarbee, R. W. 1989 *Ambio* **18**, 228–233.

Stevenson, A. C., Patrick, S. T., Kreiser, A. M. & Battarbee, R. W. 1987 Res. Pap. no. 26. Palaeoecology Research Unit, University College London.

Stoner, J. H. & Gee, A. S. 1985 *J. Inst. Water Engrs Scientists* **39**, 125–157.

Timberlid, A. 1990 *Ökoforsk* **14**. (In the press.)

Whitehead, D. R., Charles, D. F., Jackson, S. T., Reed, S. E. & Sheehan, M. C. 1986 In *Diatoms and lake acidity* (ed. J. P. Smol, R. W. Battarbee, R. B. Davis & J. Meriläinen), pp. 251–274. Dordrecht: Junk.

Whiting, M. C., Whitehead, D. R., Holmes, R. W. & Norton, S. A. 1989 *J. Paleolimnology* **2**, 285–304.

Wik, M. & Natkanski, J. 1990 In *Palaeolimnology and lake acidification* (ed. R. W. Battarbee, Sir John Mason, F.R.S., I. Renberg & J. F. Talling, F.R.S.) (*Phil. Trans. R. Soc. Lond.* B **327**), pp. 319–323. London: The Royal Society.

Discussion

D. W. JOHNSON (*Desert Research Institute, Reno, Nevada, U.S.A.*). It is frequently stated that conifers cause soil acidification. In fact, hardwoods (especially *Quercus* and *Carya* species) can cause considerably more soil acidification because they take up and sequester larger amounts of calcium than conifers as a rule. Thus conifers may produce acid litter and humus, but hardwoods will generally create more acid pressure on mineral soils.

I. RENBERG. Our palaeolimnological studies have not dealt with lakes with hardwood forests in their catchments. Although Dr Johnson's observation is no doubt true, in both Scandinavia and the U.K. such forests are more likely to be found at lower altitudes and on better soils where there is little danger of surface water acidification. We have focused on conifers because these form the forests that surround our acidified lakes. In fact the kind of forest may not matter so much since our evidence supports the view that sulphur scavenging by forest canopies is more important than base cation storage and cycling.

Lake surface-water chemistry reconstructions from palaeolimnological data

By H. J. B. Birks[1], S. Juggins[2] and J. M. Line[3]

[1] Botanical Institute, University of Bergen, Allégaten 41, N-5007 Bergen, Norway
[2] Palaeoecology Research Unit, Department of Geography, University College,
26 Bedford Way, London WC1H 0AP, U.K.
[3] University of Cambridge Computer Laboratory, Pembroke Street, Cambridge
CB2 3QG, U.K.

Reconstruction of past surface-water chemistry from palaeolimnological diatom data involves regression, where responses of modern diatom abundances to selected chemical variables are modelled, and calibration, where these modelled responses are used to infer the variable(s) from diatom assemblages preserved in lake sediments.

When the Surface Waters Acidification Programme (SWAP) modern diatom training set is analysed by canonical correspondence analysis, significant gradients correlated with pH, Ca, alkalinity, and total Al, and with dissolved organic carbon (DOC) emerge. After data screening, weighted averaging regression and calibration are used to reconstruct pH, DOC, and total Al from stratigraphical samples at Round Loch of Glenhead, SW Scotland, for the last 10 000 years. The reconstructions are evaluated in terms of lack-of-fit and analogue measures and root-mean-squared error of prediction (estimated by bootstrapping), and interpreted in terms of changes in vegetation, land-use, soils, and atmospheric inputs during the past 10 000 years.

Introduction

Reconstruction of the history of recent lake acidification by palaeolimnological methods is one aim of SWAP (Mason & Seip 1985). Acidification of surface waters not only results in a decrease in pH and acid-neutralizing capacity but it can also lead to mobilization of Al and other metals and a decrease in DOC. Quantitative reconstructions of lake-water pH and estimation of prediction errors from diatom assemblages preserved in lake sediments have been routinely done within SWAP using the ecologically realistic and theoretically sound approaches of weighted averaging (WA) regression and calibration and bootstrap error estimation (Birks *et al.* 1990). Such reconstructions require a modern training set of diatom assemblages from surficial lake-sediment samples with associated surface-water chemistry data, and counts of fossil diatoms at dated depths in sediment cores.

This paper extends the SWAP reconstruction work to chemical variables other than pH. After describing the data used, the chemical variables that are significantly associated with the major gradients in the modern diatom assemblages are evaluated statistically. These variables are then reconstructed for a complete Holocene (post-glacial) diatom sequence from a lake in the Galloway Hills, SW Scotland that has experienced recent acidification. The reconstructed time-series of pH, DOC, and total Al over the last 10000 years are interpreted in terms of changing vegetation, soils, land-use, and atmospheric inputs.

The data

(a) The modern data

The SWAP training set comprises diatom counts of 178 surface samples from lakes in England (five lakes), Norway (51), Scotland (60), Sweden (30), and Wales (32). It includes all taxa present with 1% or more in any SWAP sample, modern or fossil, and identified to species level or lower. All identifications follow the standardized taxonomy developed by SWAP during its diatom taxonomy workshops (Kreiser & Battarbee 1987; Munro *et al.* 1990). Abundances are expressed as percentages of the total diatom count (*ca.* 500 valves) for each sample. The diatomists who counted these surface samples are D. S. Anderson (Norway), N. J. Anderson (Sweden), F. Berge (Norway), R. J. Flower (Scotland, Wales), E. Y. Haworth (England), and I. Renberg (Sweden).

One or more surface-water pH readings are available for each of the 178 samples (Birks *et al.* 1990); one or more determinations for surface-water pH, conductivity, Ca, Mg, K, SO_4, Cl, alkalinity, and total Al are available for 157 samples; and one or more determinations for these nine chemical variables plus DOC are available for 138 samples (table 1). The 178-, 157-, and 138-sample sets form the pH, Al, and DOC training sets, respectively. The chemical data for each sample are arithmetic means of the available determinations (except for pH, calculated as geometric mean of pH), after preliminary data screening. The data derive from 18 laboratories. Analytical methods generally follow Stevenson *et al.* (1987).

(b) The fossil data

We use the diatom data of V. J. Jones from Round Loch of Glenhead (RLGH), Galloway, SW Scotland for reconstruction purposes. The 101 samples from 0.3 cm to 256.5 cm in core RLGH3 cover the past 10000 years or more (Jones *et al.* 1989). All taxa identified to species level or below using SWAP taxonomy and attaining 1% or more in at least one fossil or modern sample (292 taxa) are included. Some taxa at RLGH are absent from the training sets and vice versa.

RLGH is a small (12.5 ha) lake lying at 300 m. Its catchment is almost entirely *Molina caerulea*-dominated blanket-mire or bare rock. Average lake pH (1977–85) is 4.7 (annual range 1981–82 = 4.6–5.0) (Jones *et al.* 1989; Battarbee *et al.* 1989).

Major directions of variation in the modern and fossil data

(a) The modern data

We used canonical correspondence analysis (cca) (ter Braak 1986, 1987a) to identify statistically significant directions of variation within the 138-sample and 287-taxa training set. cca is a powerful, statistically robust procedure for analysing complex biological data (e.g. diatom counts) in relation to environmental variables (e.g. surface-water chemistry). It provides a simultaneous low-dimensional representation of diatom taxa, samples, and chemical variables. Axes are constrained to

Table 1. *Details of the three SWAP training sets*

training set	number of samples	England	Norway	Scotland	Sweden	Wales	number of diatom taxa	number of chemical variables
pH	178	5	51	60	30	32	294	1
Al	157	3	35	58	29	32	287	9
DOC	138	3	32	44	29	30	287	10

be linear combinations of the environmental variables. Taxon scores are weighted averages of the sample scores that are themselves linear combinations of the environmental variables. The underlying assumption of CCA is that taxon responses to environmental variables are unimodal (ter Braak & Prentice 1988).

We used forward selection of the environmental variables and Monte Carlo permutation tests of the statistical significance of each chemical variable to ensure that only significant or near significant variables were included. All chemical variables were $\log_{10}$ transformed (except pH) and seven 'rogue' samples deleted because their chemical variables had extreme values and exerted a large ($3–5.1 \times$) influence and/or they had high squared residual distances from the first four CCA axes. All chemical variables (except $Kp = 0.06$, 99 permutations) were individually significant ($p = 0.05$, 99 permutations). All ten variables were thus included in the CCA. Computations were done with CANOCO 3.0 a (ter Braak, unpublished work, 1990).

The eigenvalues for CCA axes 1 (0.54) and 2 (0.27) are similar to those for an unconstrained detrended correspondence analysis of the diatom data (0.55, 0.30). The species–environment correlations for CCA axes 1 (0.95) and 2 (0.84) are high. We thus conclude that the measured chemical variables account for the major gradients of diatom composition. The canonical coefficients define the axes and indicate significant contributions (as judged by an approximate t-test) (ter Braak 1987b) to axes 1 and 2 from DOC (axes 1, 2), alkalinity (1), Al (1, 2), pH (2), Mg (2), and Cl (2). Intra-set correlations between environmental variables and the axes show that axis 1 is a gradient of pH ($r = 0.85$), Ca (0.89), alkalinity (0.91), and Al (-0.27) and that axis 2 is primarily a DOC gradient (0.79). Axes 1 and 2 (figure 1) account for 47% of the variance in the weighted averages of the 287 taxa with respect to the ten chemical variables. The statistical validity of CCA axes is, however, best assessed (ter Braak 1987b) by significance tests. Monte Carlo permutation tests (99 permutations of axis 1, 99 permutations of axis 2 with axis 1 as a covariable) show that both axes are significant ($p = 0.01$ axis 1, $p = 0.03$ axis 2).

We conclude that within the 138-sample set there is (1) a significant diatom gradient that is strongly positively correlated with alkalinity and its close correlates Ca and pH, and negatively but less strongly-correlated with total Al, and (2) a significant gradient strongly correlated with DOC. This suggests (1) that there are statistically significant 'signals' associated with alkalinity, Ca, or pH, Al and DOC; (2) that within the measured chemical variables these correlate strongest with the gradients in diatom composition within the training set; and thus (3) that pH (or alkalinity), Al, and DOC are potentially reconstructable from fossil diatom assemblages. The correlation among these variables is 0.17 (DOC–pH), 0.83 (alkalinity–pH), 0.30 (alkalinity–DOC), -0.27 (alkalinity–Al), -0.49 (Al–pH), and

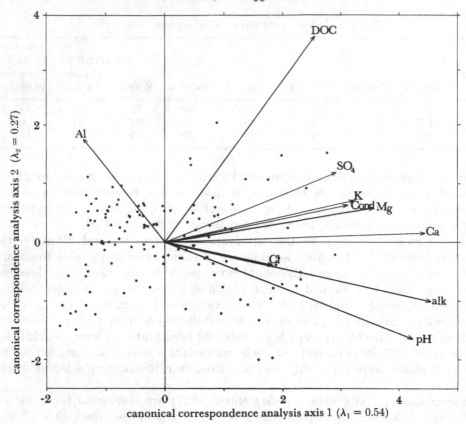

Figure 1. Canonical correspondence analysis of 131 modern samples in the SWAP 138-sample training set in relation to 10 surface-water chemical variables. Variables with significant (t greater than $|1.96|$), canonical coefficients on these two axes have thick biplot arrows. The biplot scores of the chemical variables have been multiplied by 10 for this plot (ter Braak 1987b). Abbreviations: alk, alkalinity; DOC, dissolved organic carbon; Cond, conductivity.

0.37 (Al–DOC). Anderson *et al.* (1986) found similar pH-related and DOC-related gradients in modern diatom samples from southern Norway, as did Huttunen & Meriläinen (1986) in Finland.

CCA with only pH, Al, and DOC as environmental variables produces a nearly identical pattern to figure 1, suggesting that the other seven chemical variables have little independent role in explaining the gradients in diatom composition. CCA of the 157-sample set and its associated nine chemical variables (DOC missing) produces only one significant axis, characterized by strong positive intra-set correlations of pH (0.85), Ca (0.86), alkalinity (0.88), Mg (0.72), and a weak negative correlation with total Al (-0.24).

(b) The fossil data

Samples from RLGH are positioned within the plane of CCA axes 1 and 2 (figure 2) on the basis of their diatom composition, using transition formulae (ter Braak 1987b) to derive scores for the 101 fossil, so-called passive, samples. All fossil samples have a low squared residual distance (2–28) to axes 1 and 2 compared with the modern samples (1–206), indicating that the positioning of the fossil samples in figure 2 is unlikely to be misleading geometrically. The observed patterns of variation in

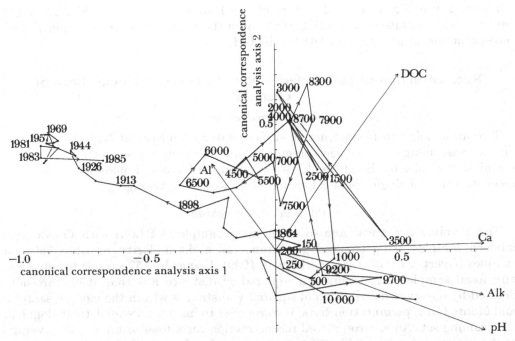

Figure 2. Plot of selected samples from Round Loch of Glenhead positioned as 'passive' samples within part of the CCA plane of figure 1. Not all 101 fossil samples are plotted. All samples for 1985 to 1864 A.D. are plotted, samples at 500 year intervals are plotted for the period 150–6500 BP, and all samples are plotted for the period 6500 BP–10000 BP. Samples are joined up in chronological order. The chemical variables Al, DOC, Ca, alkalinity (alk), and pH are represented as biplot arrows and their scores have been multiplied by 2 for this plot (ter Braak 1987*b*). Abbreviations follow figure 1. Note that the axis scales are different from figure 1.

diatom composition at RLGH (figure 2) can thus be linked to the modern chemical variables depicted as biplot arrows. Arrows for chemical variables point in the general direction of maximum chemical change across the diagram, and their lengths are approximately proportional to the rate of change in that direction (ter Braak 1987*a*).

The positions of the RLGH samples (figure 2) fall into five main groups.

(1) Pre-10000–9200 BP: highish pH, Ca, and alkalinity, and low Al and DOC.

(2) 8700–7000 BP: high DOC, slightly reduced pH, Ca, and alkalinity, and medium Al values. Interestingly this is the time that pine occurred locally near RLGH (Jones *et al.* 1989). *Aulacoseira perglabra* is frequent, a diatom common in lakes today with high DOC values and conifer forests in their catchment in Sweden and Galloway.

(3) 7000–4500 BP: low–medium pH, Ca, and alkalinity, and medium Al and DOC.

(4) 4500–*ca.* 1500 BP: high DOC, medium–low pH, Ca, and alkalinity, and medium Al. This is the time of major deforestation and heath and blanket-mire development of RLGH (Jones *et al.* 1989).

(5) 1864–1985 A.D.: continuously decreasing pH, increasing Al, and low DOC, with some reversal in this trend after 1983 (Battarbee *et al.* 1988). This major acidification during the last 130 years results from the increased atmospheric deposition of strong acids (Flower & Battarbee 1983; Flower *et al.* 1987; Jones *et al.* 1989; Battarbee *et al.* 1989).

Figure 2 provides a general picture of the Holocene and recent history of the surface-water chemistry at RLGH. We turn in the next section to the quantitative reconstruction of pH, Al, and DOC at RLGH.

Reconstruction of past surface-water chemistry at Round Loch of Glenhead

(a) Introduction

Two approaches to reconstructing surface-water chemistry at RLGH from fossil diatom assemblages have been attempted: simultaneous reconstruction of several chemical variables by direct 'analogue' matching of fossil and modern samples, and reconstruction of single chemical variables by weighted averaging.

(b) Analogue matching

This involves numerical comparison of every sample at RLGH with all available modern samples using squared χ^2 distance as a dissimilarity measure between samples (Overpeck *et al.* 1985; Birks *et al.* 1990). Using the 157-sample training set, any fossil sample with a minimum squared χ^2 distance less than 0.59 (threshold derived by observed distribution of squared χ^2 distances within the modern samples and Monte Carlo permutation tests) is considered to have a *close* modern analogue in the training set. An environmental reconstruction for a fossil sample is the average of the environmental variable(s) for modern samples that are close analogues to that fossil sample. Analogue matching and environmental reconstructions were implemented by the program ANALOG 1.4 written by J. M. L.

All RLGH samples below 44.4 cm (*ca.* 900 BP) (except 228.5 cm and 256.5 cm) have squared $\chi^2 > 0.59$ and thus lack *close* analogues. This lack results, in part, from the abundance of *Melosira arentii* at RLGH, a taxon absent from the training set, and, in part, from the scarcity of pristine naturally acid but not acidified lakes in the training set. Such lakes would perhaps provide close analogues to RLGH 900–10000 BP. In the absence of close analogues for much of the RLGH sequence, environmental reconstructions by analogue matching are not presented here.

(c) Weighted averaging regression and calibration

In the absence of reconstructions by analogue matching, quantitative reconstruction of surface-water chemistry from diatoms involves two stages. First, responses of modern diatoms to the chemical variable(s) of interest are modelled. This is a regression problem (ter Braak & Prentice 1988). Secondly, these modelled responses are used to infer past chemistry from the composition of fossil diatom assemblages. This is a calibration problem (ter Braak & Prentice 1988).

Non-linear Gaussian logit regression of diatom responses in relation to two or more environmental variables is, in theory, possible using maximum likelihood (ML) estimation and generalized linear modelling (ter Braak & Looman 1987). Inferring more than one environmental variable simultaneously by ML calibration on the basis of several Gaussian logit response surfaces is much more difficult (ter Braak 1987*c*; ter Braak & Prentice 1988). Moreover, ter Braak & van Dam (1989) and Birks *et al.* (1990) compared the performance of the computationally demanding but formal statistical approach of ML Gaussian logit regression and ML calibration with the computationally simple but heuristic approach of weighted averaging (WA) regression and calibration for inferring a single variable (pH) from diatoms. Birks *et al.* (1990)

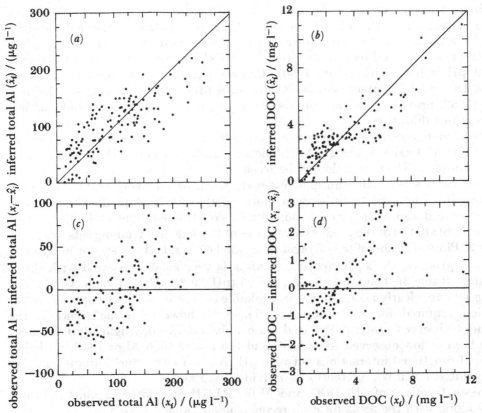

Figure 3. Plots of inferred Al ($\hat{x}_i$) against observed Al (x_i) and of the differences ($x_i - \hat{x}_i$) against observed Al (plots (*a*) and (*c*)) and of inferred DOC ($\hat{x}_i$) against observed DOC (x_i) and of the differences ($x_i - \hat{x}_i$) against observed DOC (plots (*b*) and (*d*)).

Table 2. *Details of the three training sets after data screening*

training set	number of samples after screening	range	mean	median	standard deviation	number of taxa after sample screening
pH	167	4.23–7.25	5.56	5.27	0.77	294
Al (µg l^{-1})	126	13.0–256.3	103.2	104.3	60.04	287
DOC (mg l^{-1})	123	0.12–11.60	3.17	2.50	2.29	283

showed that WA gave superior results in terms of a lower root-mean-square error of prediction (RMSE) in cross-validation than ML. For these reasons we use simple WA regression and calibration to reconstruct separately pH, Al, and DOC at RLGH.

Ter Braak & van Dam (1989) and Birks *et al.* (1990) present the methodology of WA regression and calibration, and ter Braak & Looman (1986), ter Braak & Barendregt (1986), and ter Braak & Prentice (1988) discuss the underlying theory.

The 178-, 158-, and 138-sample training sets (table 1) were used to infer pH (Birks *et al.* 1990), total Al, and DOC, respectively. All sets were subjected to data-screening to detect possible atypical 'rogue', observations using the three procedures of Birks *et al.* (1990). After two to four screenings, no samples appear as obvious rogues within the reduced training sets (see table 2 for details of the screened sets). These screened sets were used for pH, Al, and DOC reconstructions at RLGH and other SWAP sites.

Simple WA without tolerance downweighting was used as it consistently gives lower RMSE as estimated by bootstrapping. Classical regression was used to deshrink inferred pH values (Birks *et al.* 1990). Inverse regression was used to deshrink inferred DOC and Al values because it minimizes RMSE in the training set. Lwin & Maritz (1982) provide a critical analysis of the *pros* and *cons* of inverse and classical regression in calibration.

Root-mean-square of the error $(x_i - \hat{x}_i)$ was calculated for each training set, where x_i is the observed value of chemical variable x in sample i and $\hat{x}_i$ is the inferred value of x for sample i. The correlation (r) between x_i and $\hat{x}_i$ was also calculated. As this apparent RMSE is invariably an underestimate when based solely on training sets, bootstrapping was used to estimate a more reliable RMSE of prediction for each training set and also RMSE for each individual reconstructed value at RLGH (Birks *et al.* 1990). Statistics of the predictive abilities of the screened training sets are given in table 3. Plots of x_i and of $(x_i - \hat{x}_i)$ against x_i for DOC and Al are shown in figure 3.

For all variables, the s_{i1} component of RMSE is very small compared with the s_2 component (table 3). The ratio of s_{i1}/s_2 is 0.23 (pH), 0.27 (Al), and 0.26 (DOC). The training sets are clearly adequate to yield reliable estimates of species optima by WA regression required for WA calibration. There is, however, some trend in the differences (observed − inferred) plotted against observed values (figure 3). Inferences are too high at low observed Al or DOC and too low at high Al or DOC, probably because of functional interactions between pH, Al, and DOC. Such interactions are not accounted for in WA inferences of individual variables.

WA reconstructions of pH, DOC, and Al for RLGH are plotted against depth in figure 4 along with the RMSE for each reconstructed value.

WA regression and calibration, and bootstrapping of WA estimates were implemented by the program WACALIB 3.0 written by J.M.L. in conjunction with C. J. F. ter Braak and H.J.B.B.

(d) *Evaluation of reconstructed* pH, DOC, *and* Al *values*

There is no simple means of evaluating the reliability of individual reconstructed values. Birks *et al.* (1990) introduced two approaches: (1) lack-of-fit measures of fossil diatom assemblages to the variable being reconstructed, and (2) root-mean-square errors for each inferred value.

The squared residual distance of each RLGH sample to the environmentally constrained axis in a CCA of diatom data in relation to the single chemical variable of interest provides a measure of sample lack-of-fit to that variable. Any sample whose residual distance is equal to or larger than the residual distance of the extreme 5% in the training set is marked on figure 4 as having a 'very poor fit', whereas samples with values equal to or larger than the distances of the extreme 10% are indicated as having a 'poor fit' (Birks *et al.* 1990).

Any reconstructed value is likely to be more reliable if the fossil diatom assemblage in that sample has a close modern analogue in the training set. All RLGH samples that lack a close (minimum squared χ^2 distance less than 0.59) or good (less than 0.70) analogue are also shown on figure 4.

The RMSE for individual RLGH samples (figure 4) vary from 0.313 to 0.321 (pH), 1.54–1.59 mg l^{-1} (DOC), and 48.23–49.1 µg l^{-1} (Al). The corresponding RMSE ranges for individual training samples are 0.313–0.492, 1.53–1.89, and 48.19–65.48.

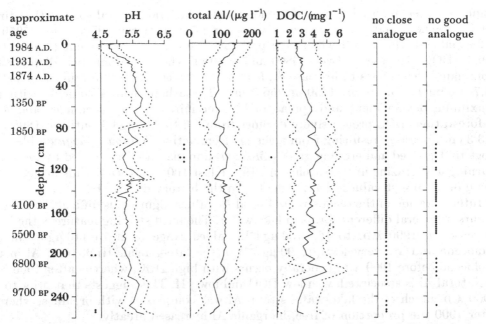

Figure 4. Reconstructed pH, total Al, and DOC values for Round Loch of Glenhead plotted against depth (solid lines) based on weighted averaging. The RMSE of prediction for these estimates are shown as broken lines. Samples with poor (·) or very poor (··) fit to particular chemical variables and those lacking good or close modern analogues in the SWAP training set are indicated.

Table 3. *Apparent root-mean-squared error of prediction (RMSE), RMSE estimated by bootstrapping (RMSE boot) and its error components (s_{t1}, s_2), and correlation (r) between observed and predicted chemistry for the three training set*

	apparent RMSE	RMSE boot	RMSE s_{t1}[a]	RMSE s_2[a]	r
pH	0.297	0.320	0.072	0.312	0.933
Al (µg l⁻¹)	37.820	49.663	12.985	47.935	0.777
DOC (mg l⁻¹)	1.251	1.580	0.403	1.527	0.837

[a] s_t is that part of the prediction error due to estimation error in taxon parameters in the WA calibration formulae, s_2 is that part of the prediction error due to variation in diatom composition at the same water chemistry; see Birks *et al.* (1990).

Discussion

(a) *Reconstructions of* pH, DOC, *and* Al *at Round Loch of Glenhead*

Reconstructed pH varies from 5.4 to 5.8 in the late-glacial and early Holocene, but by 9200 BP (224 cm) it stabilizes to 5.2–5.4 (figure 4). Between 4100 and 1850 BP (142–72 cm) there are short-lived fluctuations, probably due to inwashed material (Jones *et al.* 1989). Acidity changed little (5.3–5.7) until between 1874 and 1931 A.D. (17.3–7.3 cm) when pH dropped by over 0.5 unit. Lake pH never fell below 5.0 until *ca.* 1913 A.D. (10.3 cm).

The magnitude of DOC changes (figure 4) is small relative to the RMSE of prediction (table 3). However, many of the changes are stratigraphically consistent and it is these major trends that we focus on. DOC is low (2.5–3.2 mg l⁻¹) in the late-glacial (figure 4), presumably because of predominantly skeletal mineral soils at that time. DOC reaches maximum values (4.0–4.9 mg l^{-1}) between 9700 and 7050 BP (228–200 cm), the period when pine probably grew in the catchment (Jones *et al.* 1989). DOC decreases between 6800 and 5500 BP when deciduous woodlands were dominant. DOC starts to increase *ca.* 5500 BP (174.5 cm) and maintains high values (3.7–4.0 mg l^{-1}) until *ca.* 1350 BP (56.5 cm). These high values coincide with the maximum development and spread of blanket mires in the catchment and with deforestation. DOC progressively declines from 3.5 to 2.9 mg l^{-1} until *ca.* 1898 A.D. (13.3 cm), possibly resulting from a change from active *Calluna-Sphagnum* blanket-bogs to degraded and eroding grass (?*Molinia*) dominated mires caused by increased burning and grazing in the catchment. In the last 100 years, DOC falls markedly to some of its lowest values (2.4–2.6 mg l^{-1}) in the history of RLGH.

Interpretation of the reconstructed total Al values (figure 4) is difficult because Al occurs in several different forms in lake water. The most striking feature is the large increase *ca.* 1900 A.D. to 125–150 µg l^{-1}. Values range from 75–110 µg l^{-1} in the Holocene and are very low (62–90 µg l^{-1}) in the late-glacial. High total Al in the Holocene before 1900 A.D. generally occurs with high DOC, whereas after 1900 A.D. high total Al is associated with low DOC and low pH. This suggests that prior to *ca.* 1900 A.D. much of the lake-water's total Al was complexed with organics, whereas after 1900 the proportion of free, inorganic Al increased greatly.

RLGH thus underwent major chemical changes in *ca.* 1900 A.D. from being a naturally acid, organic weak-acid dominated lake for much of the Holocene to being an acidified, mineral strong-acid dominated lake in the last 100 years (Battarbee *et al.* 1989). Input of dissolved inorganic Al from the catchment probably increased with the deposition of strong acids. A pH change from 5.5 to 4.5 causes a major change in Al speciation, with a large increase in total Al charge (Charles & Norton 1986). DOC and lake-colour correspondingly decrease because of the increased tendency for organic compounds to aggregate and precipitate because of protonation, and the increased precipitation of organics by acid-mobile metals such as Al (Davis *et al.* 1985). The net results of acidification of RLGH were a loss of surface-water DOC and a decrease in the availability of organics to complex potentially toxic Al, and an increase in toxic inorganic Al (cf. Kingston & Birks 1990).

The null hypothesis that the rate of inferred chemical changes between 1931–1874 A.D. is no different from rates of change in pre acid-deposition times (1874 A.D.–10000 BP) was tested using Kitchell *et al.*'s (1987) time–duration bootstrap test (see Birks *et al.* 1990). Observed rates of chemical change at the time of increased acidic deposition are significantly different ($\alpha = 0.018$, 1000 bootstraps) from expectation. We thus reject the null hypothesis and conclude that the most marked changes in surface-water chemistry during the last 10000 years occurred between 1874 and 1931 A.D. at RLGH (see also figure 2). Acid deposition in the last 100 years has been the major determinant of lake-chemical change at RLGH despite the very important changes in vegetation, land-use, and soils that have occurred in its catchment since 10000 BP (Jones *et al.* 1989).

Deposition of strong acids resulted in a decrease in pH from *ca.* 5.2–5.5 to *ca.* 4.7 and an associated decrease in acid neutralizing capacity. These changes resulted in increased Al concentrations and a decrease in organic acids. Bicarbonate alkalinity was very low or even absent for the previous 9700 years.

(b) Improved reconstruction procedures

This paper is SWAP's initial attempt at multivariable reconstructions from palaeolimnological data. Considerably more statistical and ecological work is needed to refine and improve the reliability of such reconstructions.

Modelling of diatom responses in relation to two or more ecologically critical variables is required before the complex but real interactions between chemical variables influencing diatom abundance can be unravelled. We have constructed response surfaces (Bartlein *et al.* 1986) for several common taxa in relation to pH, DOC, and Al using J. M. L.'s program RSURF. They show interesting and unexpected patterns. Unfortunately, the training sets are still rather small for constructing robust and reliable response surfaces for such highly variable and heterogeneous data. Simultaneous calibration of two or more variables from such surfaces is a major problem (ter Braak & Prentice 1988). One approach, to be implemented, is a direct search of the environmental space to derive estimates of the chemical variables most likely to have produced observed fossil diatom assemblages (see for example, Atkinson *et al.* 1986). Another approach involves diatom–chemical variable dissimilarity surfaces for modern and fossil samples (MacDonald & Reid 1989).

Besides limitations imposed by the size of the SWAP training sets, a further and perhaps more serious limitation is the available chemical data, including inherent (but unknown) inter-laboratory variability and the shortage of many determinations of critical chemical variables such as inorganic monomeric Al, nutrients, dissolved Si, heavy metals, and trace elements.

Future work should include (1) statistical and computing developments that allow for confounding and interacting explanatory environmental variables and improved procedures for inferring variables that can be ecologically important (e.g. DOC, Si, inorganic monomeric Al, heavy metals) but not dominant (e.g. pH); (2) improved, extended, and replicated chemical data subject to similar quality controls as the SWAP diatom data; (3) larger and ecologically more extensive training sets, including naturally acid, 'pristine' lakes that might provide close analogues for pre-acidification assemblages; and (4) improved training set design. Mohler (1983) demonstrates the importance that sampling has on estimating species parameters. With improved palaeolimnological reconstructions, it should then be possible to strengthen links already established between palaeolimnological, biological, and geochemical modelling both for hindcasts (see for example, Jenkins *et al.* 1990) and forecasts (Ormerod *et al.* 1988).

There is also a need for further research on the physiological causes of the marked changes in diatom composition that occur with acidification, along the lines initiated by Smith (1990). Response surfaces for common species provide several hypotheses inviting experimental study. Such future researches would help to reinforce the 'bridges between paleolimnology and aquatic ecology' (Smol 1990).

This research has been supported in part by SWAP, NAVF and IBM (Norway). We are very grateful to our SWAP palaeolimnological colleagues for providing the training set; to Viv Jones for providing the Round Loch of Glenhead data; to Sylvia Peglar and Siv Haugen for invaluable technical assistance, and to Rick Battarbee, Hilary Birks, John Boyle, Viv Jones, and Cajo ter Braak for valuable discussions and commenting on the manuscript.

References

Anderson, D. S., Davis, R. B. & Berge, F. 1986 In *Diatoms and lake acidity* (ed. J. P. Smol, R. W. Battarbee, R. B. Davis & J. Meriläinen), pp. 97–113. Dordrecht: Dr Junk.

Atkinson, T. C., Briffa, K. R., Coope, G. R., Joachim, J. J. & Perry, D. W. 1986 In *Handbook of Holocene palaeoecology and palaeohydrology* (ed. B. E. Berglund), pp. 851–858. Chichester: J. Wiley & Sons.

Bartlein, P. J., Prentice, I. C. & Webb, T., III 1986 *J. Biogeogr.* **13**, 35–57.

Battarbee, R. W., Flower, R. J., Stevenson, A. C. Jones, V. J., Harriman, R. & Appleby, P. G. 1988 *Nature, Lond.* **332**, 530–532.

Battarbee, R. W., Stevenson, A. C., Rippey, B., Fletcher, C., Natanski, J., Wik, M. & Flower, R. J. 1989 *J. Ecol.* **77**, 651–672.

Birks, H. J. B., Line, J. M., Juggins, S., Stevenson, A. C. & ter Braak, C. J. F. 1990 In *Palaeolimnology and lake acidification* (ed. R. W. Battarbee, J. Mason, I. Renberg & J. F. Talling) (*Phil. Trans. R. Soc. B* **327**), pp. 263–278. London: The Royal Society.

Charles, D. F. & Norton, S. A. 1986 In *Acid Deposition: Long Term Trends* (ed. National Research Council Committee on Monitoring and Assessment of Trends in Acid Deposition), pp. 335–506. Washington, D.C.: National Academy Press.

Davis, R. B., Anderson, D. S. & Berge, F. 1985 *Nature, Lond.* **316**, 436–438.

Flower, R. J. & Battarbee, R. W. 1983 *Nature, Lond.* **305**, 130–133.

Flower, R. J., Battarbee, R. W. & Appleby, P. G. 1987 *J. Ecol.* **75**, 797–824.

Huttunen, P. & Meriläinen, J. 1986 *University of Joensuu, Publications of Karelian Institute* **79**, 47–54.

Jenkins, A., Whitehead, P. G., Cosby, B. J. & Birks, H. J. B. 1990 In *Palaeolimnology and lake acidification* (eds. R. W. Battarbee, J. Mason, I. Renberg & J. F. Talling) (*Phil. Trans. R. Soc. Lond. B* **327**), pp. 435–440. London: The Royal Society.

Jones, V. J., Stevenson, A. C. & Battarbee, R. W. 1989 *J. Ecol.* **77**, 1–23.

Kingston, J. C. & Birks, H. J. B. 1990 In *Palaeolimnology and lake acidification* (ed. R. W. Battarbee, J. Mason, I. Renberg, J. F. Talling) (*Phil. Trans. R. Soc. B* **327**), pp. 279–288. London: The Royal Society.

Kitchell, J. A., Estabrook, G. & MacLeod, N. 1987 *Paleobiology* **13**, 272–285.

Kreiser, A. & Battarbee, R. W. 1987 *University of Stockholm, Department of Quaternary Research Report* **12**, 41–44.

Lwin, T. & Maritz, J. S. 1982 *Technometrics* **24**, 235–242. Minor correction in *Technometrics* **27**, 445.

MacDonald, G. M. & Reid, R. T. 1989 *J. Biogeogr.* **16**, 403–412.

Mason, J. & Seip, H. M. 1985 *Ambio* **14**, 45–51.

Mohler, C. L. 1983 *Vegetatio* **54**, 97–102.

Munro, M. A. R. & 15 other authors 1990 In *Palaeolimnology and lake acidification* (ed. R. W. Battarbee, J. Mason, I. Renberg, J. F. Talling) (*Phil. Trans. R. Soc. Lond. B* **327**), pp. 257–261. London: The Royal Society.

Ormerod, S. J., Weatherley, N. S., Varallo, P. V. & Whitehead, P. G. 1988 *Freshwater Biol.* **20**, 127–140.

Overpeck, J. T., Webb, T., III & Prentice, I. C. 1985 *Quat. Res.* **23**, 87–108.

Smol, J. P. 1990 *Proceedings V International Palaeolimnology Symposium.* Ambleside, U.K. (In the press.)

Smith, M. A. 1990 In *Palaeolimnology and lake acidification* (eds. R. W. Battarbee, J. Mason, I. Renberg, J. F. Talling) (*Phil. Trans. R. Soc. Lond. B* **327**), pp. 251–256. London: The Royal Society.

Stevenson, A. C., Patrick, S. T., Kreiser, A. & Battarbee, R. W. 1987 Palaeoecology Research Unit, University College London Research Paper 26, 36 pp.

ter Braak, C. J. F. 1986 *Ecology* **67**, 1167–1179.

ter Braak, C. J. F. 1987*a* *Vegetatio* **69**, 69–77.

ter Braak, C. J. F. 1987*b* Technical Report LWA-88-02, GLW, Wageningen.

ter Braak, C. J. F. 1987c In *Data analysis in community and landscape ecology* (ed. R. H. G. Jongman, C. J. F. ter Braak & O. F. R. van Tongeren), pp. 78–90. Wageningen: Pudoc.

ter Braak, C. J. F. & Barendregt, L. G. 1986 *Math. Biosci.* **78**, 57–72.

ter Braak, C. J. F. & Looman, C. W. N. 1986 *Vegetatio* **65**, 3–11.

ter Braak, C. J. F. & Looman, C. W. N. 1987 In *Data analysis in community and landscape ecology* (ed. R. H. G. Jongman, C. J. F. ter Braak & O. F. R. van Tongeren), pp. 29–77. Wageningen: Pudoc.

ter Braak, C. J. F. & Prentice, I. C. 1988 *Adv. ecol. Res.* **18**, 271–317.

ter Braak, C. J. F. & van Dam, H. 1989 *Hydrobiologia* **178**, 209–223.

Discussion

G. HOWELLS (*Department of Zoology, University of Cambridge, Downing Street, Cambridge CB2 3EJ, U.K.*). While I welcome the attempt to extend the relationship of the diatom record to more than just the pH of lake water, with the objective of inferring historic conditions, I wonder if the three components selected (pH, Al, DOC) are appropriate since in current field data pH and aluminium are dependent, and aluminium toxicity (i.e. speciation) is dependent on DOC. Their dependency, both chemical and biological, should mean that extent and timing of the inferred changes of pH should be paralleled by changes in the inferred aluminium and DOC. Is this shown in the illustration given? Does any discrepancy indicate lack of dependence, or of the errors inherent in the 'back-casting' technique?

H. J. B. BIRKS (*Botanical Institute, University of Bergen, Allégaten 41, N-5007 Bergen, Norway*). In the modern SWAP training set, the correlation between pH and total Al is −0.49, pH and DOC 0.17, and total Al and DOC 0.37. To test the hypothesis that the total Al reconstruction at Round Loch of Glenhead (RLGH) is not a result of the observed correlation between total Al and pH today, we reconstructed total Al indirectly by means of partial canonical correspondence analysis (ter Braak 1988) with pH, Ca, and alkalinity as co-variables and total Al as the sole explanatory variable. The resulting reconstruction of total Al with the effects of pH, Ca, and alkalinity partialled out closely parallels the weighted averaging reconstruction of total Al for RLGH. This suggests that the total Al reconstruction is not a statistical artifact of the significant negative correlation between pH and total Al in our training set. Moreover, response surfaces of the abundance of modern diatom in relation to pH and total Al show that several species differ in their Al optima but have similar pH optima, indicating that total Al diatom-responses are, in part, independent of pH.

Changes in reconstructed pH and total Al at RLGH generally correspond, with total rising whenever pH falls. There is, however, little correspondence between reconstructed DOC and pH or total Al. It is clearly not possible to say whether these patterns result from any dependency in the past. Partial CCA suggests that the pH–total Al correspondence is not an artifact of the correlations in the training set. The errors inherent in our reconstruction procedures are largely captured in our bootstrap error estimation procedure, and are shown on the RLGH reconstruction.

Reference

ter Braak, C. J. F. 1988 In *Classification and related methods of data analysis* (ed. H. H. Bock), pp. 551–558. Amsterdam: North Holland.

Pre-industrial acid-water periods in Norway

By Ivan Th. Rosenqvist

Department of Geology, P.O. Box 1047, Blindern 0316, Olso 3, Norway

Inland fish has always been important in Norwegian rural diets. For this reason, total extinctions of all fish in former good fish brooks and lakes are mentioned in mediaeval folklore. The explanation given was divine punishment for sinful behaviour.

At the beginning of 1900 it was scientifically reported that several good fish lakes in the Agder counties, S. Norway, became barren during the last part of the nineteenth century.

The former theological hypothesis was replaced *ca.* 1920, by geological explanations, as it was found that pH in the barren lakes was lower than in the rain.

The strong acidity was due to biogeochemical reactions between rain and the soil in the catchments. These reactions are complicated and not readily understood by laymen. Between 1960 and 1970, long distance transport of acid industrial air pollutants was established, and a new dogma became popular among politicians and research workers without backgrounds in geochemistry.

Palaeolimnological tests of former acidity have been done in two lakes with the late mediaeval name Fiskeløstjern, i.e. tarn with no fish. The lakes are situated in areas that lost the human population during the plague of 1349. The catchment became woodland until A.D. 1500 when it was used mostly for grazing up until 1940.

Four different methods have been used on sediment cores from lakes that lost their fish after 1950: (i) diatom analysis; (ii) preservation of pine pollen; (iii) quantitative determination of chromium and vanadium; (iv) 6 co-ordinated Al interlayered 2:1 clay minerals.

In both lakes, which now have pH 4.4, the pH was found to be below 4.6 around the years 1400–50. Before 1350, and between 1500 and 1940, pH was above 5.2.

In the westernmost lake, a further acid period with pH 4.4 (perhaps around A.D. 500) was registered. These periods were foreseen and predicted from the vegetation history.

There are several reasons for modern surface water acidity: acid rain is one, but biogeochemical changes in the catchment are the overwhelming factors. Nobody has found a lake in Norway where the acidity of modern precipitation contributes more than 10% of the acid–base budget.

Although the air and precipitation in Norwegian mountains is relatively clean compared with most industrial countries, lakes and brooks may however have very low pH values, down to below pH 4.0. Many watercourses lost their fish populations in the second part of the twentieth century; probably because of low pH and high Al content. Problems with tarns and brooks losing their fish populations are of long standing in Norway. Some lakes are said always to have been acid. This is not true. None of these tarns were acid in preboreal time (9000 years ago) and probably neither in subboreal time 5000–2500 years ago.

[315]

In Norwegian folklore dating back to late mediaeval times, there are stories about good fishing streams and tarns that became barren because of the sinful behaviour of the people. This theological explanation was later replaced by the more rational geological hypotheses.

A. Helland, who was professor in geology at Oslo University, reported in 1900 and 1903 that many good fish lakes in the Agder counties in S. Norway had lost their fish population in the second half of the nineteenth century. At that time, the fossil-fuel consumption in west Europe was less than 20% of the 1975 figure, and very low in Agder rural and alpine areas.

K. Dahl (1926, 1927) was the first Norwegian scientist to explain the mortality of trout fry by the acidity of water. K. M. Ström (1925) investigated 31 mountain lakes over a distance of 60 km with probably the same precipitation and with different gneisses and other metamorphic rocks overlaid by bog-soils and mor; pH in the lakes varied from 3.8 to 5.2 (average, pH 4.5).

Dr K. M. Ström was appointed the first professor in Limnology in Norway. He explained that the chemical composition of surface waters depended mainly upon reactions between the precipitation water and the permeable top- and subsoils in the catchment area.

In Norway, acid lakes are concentrated in areas dominated by shallow mineral soils and topsoils of high cation exchange capacity (CEC) and low base saturation. This may be coniferous forest litter, heather and mosses, forming raw humus (mor) or bog-soil.

Circa 1930, H. Torgersen succeeded in re-establishing the trout population in some lakes by placing blocks of limestone in their inlet brooks. It was suggested that this was not due to the limestone but that the precipitation had become less acid. This was rejected by Torgersen (1934).

The hypothesis of biochemical factors in the catchment area being the dominating factors for the chemistry of lake water, and that changes in these conditions could change pH in either direction, was generally accepted until late 1950 when some biologists (without a background in geochemistry) advanced the hypothesis that increase in rain acidity is the reason for acidification of lakes. It was claimed (without foundation) that the sulphur content in rain water had increased tenfold since pre-industrial times. Later, N. A. Sørensen (1983) by using several methods, including older analyses of river water above the marine limit, maintained that the increase had been around 70% and not 1000%. Instead of the theological dogma we got the geological explanation.

> Geochemical changes in catchments change the runoff water chemistry in brooks, and topsoils of high CEC and low base saturation on sparse mineral soil result in acid runoff independent of the pH of the rain.

However, this explanation was not accepted by many environmentally minded politicians and scientists in Norway and Sweden in the early 1970s, who blamed the acidification of many lakes on pollution from other nations, notably the U.K.

However, in natural sciences nothing can be proved. A hypothesis may only be illustrated or disproved. Hundreds of people were involved and much public money spent to provide illustrations for the new political hypothesis. Thanks to SWAP, money became available for scientific investigations to compare the old and new hypotheses.

In 1981, the institutes of Geology and Limnology at the University of Oslo, compared brook water from three gneiss areas that had undergone severe forest fires in 1965, 1975 and 1976 with brooks from outside the scorched areas. In all three cases the acidities of the 'outside waters' were much higher than those of the 'inside waters'. The maximum H^+ ratio was 1:2000 (Rosenqvist 1985).

The most extreme field was closely examined by T. Hegna (1986), six–nine years after the forest fire. Among other investigations, he compared a lake draining from the scorched area with neighbouring lakes and compared the lake water with precipitation. In the first tarn, the content of Ca^{2+} and Mg^{2+} was markedly higher than in the precipitation.

In the outside tarn, the Ca^{2+} content at all depths (April–October 1983) and particularly in the upper 3 m was even lower than in the rain, whereas the Mg^{2+} content was between three and four times higher than in the rain, probably because this catchment contained some greenstone. This contradicts the concept of 'large-scale titration' but favours the ion-exchange hypothesis in the topsoil (mor).

In A.D. 1349 the great plague hit Norway, reducing the population by two thirds. It took 150 years before the population was back to the pre-plague number. In marginal areas pollen analyses show that agriculture and grazing land had been turned into pine forests for *ca*. 200 years. After resettlement the land was mostly used for grazing and fodder-hay production. After 1930 the land was again afforested.

For palaeolimnological investigations of lake acidity, four methods have been used.

1. *Diatom data* based on the empirical correlation of relative abundance of the different diatom species and the pH values in present lake waters. For palaeoacidity determination Renberg & Hellberg's (1982) methods may be used.

2. *The degree of preservation of pollen*. Empirically, pine pollen are better conserved if they settle in acid lakes rather than in lakes of higher pH. No method for obtaining quantitative values of pH exists.

3. *Vanadium and chromium contents in the sediments*. In mineral soils these elements are present in silicates, and iron oxide–hydroxide minerals as cations. Their solubility increases as the third power of H^+ activity. Above pH 5.2, oxide-hydroxides are practically insoluble. By acid dissolution under oxidizing conditions, vanadium and chromium change into polyvanadate and chromate anions, which may pass through mor and other cation-exchange materials. In reducing organic lake sediments, V and Cr are again precipitated. Under equilibrium conditions pH may therefore be calculated.

4. *Octahedrally co-ordinated inter-layered aluminium in 2:1 clay minerals*. These minerals are non-expanding, unlike other smectites. Below pH 4.8 aluminium in water is only six co-ordinated and concentrations rapidly increase. Once formed, Al interlayer 2:1 minerals do not easily return to expanding smectites by increasing pH. No method for deducing quantitative values of pH exists.

At the SWAP Bergen conference (1987), H. Höeg (1989), B. Stabell (1987) and I. Rosenqvist (1987) each presented their data on pollen, diatoms, vanadium and chromium in sediment cores from Tveitå Fiskeløstjern in Agder. This lake lost its fish in 1950. I. Horstad presented his data on clay minerals two years later (Horstad 1989). All four data-sets show that between 1950 and 1989, and 1350 and 1500, the lake was acid. In the years 1500–1940 and before 1350, pH was higher: above 5.0. The vanadium and chromium data gave pH 4.5 for the years 1400–50, and pH 4.35 for 1975. When these data were presented in Bergen in was objected that results for one lake do not prove anything.

As there was still some SWAP money available, two further cores were collected from two other tarns that had similar vegetation histories during the past 4000 years. These were Ulleberg Fiskeløstjern (20 km), and Ersdal Fiskeløstjern (110 km) to the west of Tveitå. Because of the very high cost of the investigations, only the Ersdal lake has been studied, and that only partly: for pollen, diatoms and chemistry. (Höeg & Stabell have presented their data as a poster at this symposium.) The V and Cr data gave a pH value of 4.4 for the top-most sediment, and pH (V) 4.6, pH (Cr) 4.3 in sediments ca. 500 years old. Between and before the two events, pH was greater than 5.2. An older acid event ca. 1500 years B.P. gave similar values. (The Ulleberg core is kept unopened in the freezer at the University.)

In Sweden, A. M. Robertsson (1988) investigated diatoms in lakes where the Weichsel glaciations had not removed the last interglacial sediments. In one Swedish lake near the border with Norway she found a period some 110000 years ago that coincided with the invasion of pine, spruce and larch. By using Renberg's methods, it can be calculated that the diatom flora had an even lower pH value than they have today.

Two out of two investigated Norwegian lakes and one in Sweden do not prove the geological theory, but they certainly disprove the alternative hypothesis.

In other areas of the world, like New Zealand (South Island), Tasmania and Australia, with no polluted rain, and in the U.S.A., Cape Cod and Adirondacks (partly with more polluted rain than in S. Norway), the same types of changes in pH of the surface water as demonstrated in the two investigated lakes in S. Norway are reported. (It is not possible to quote all of the literature, disproving the dominating effect of industrial pollution upon river and lake water in this short paper. Most it may be found in Edward C. Krug's 350-page 'Assessment report' (1989) containing 258 references.)

In a paper concerning acidity of lakes at Cape Cod, Massachusetts, M. G. Winkler (1985) shows diatom diagrams from Duck Pond, demonstrating the variations in pH between 5.9 and 4.8 during the last 12000 years. The variations before and after the European settlement up to the modern return to forest vegetation, show great similarities to the Tveitå and Ersdal diagrams, from the great plague up to 1988, although the lakes never got quite as acid as the Norwegian lakes.

The effect of thickness of mineral soil in catchment upon the pH of surface waters is demonstrated in a non-acid lake, Maridalen, in Oslo (Rosenqvist 1989). Catchment is above the Weischel III marine limit. The bedrocks are mainly syenites and granites covered by glaciofluvials and tills.

The lake itself was below sea-level for 700 years after glaciation. The pH of precipitation water is pH 4.25. The lake-water is pH 6.1 with 21 µeq l^{-1} Mg^{2+} and 85 µeq l^{-1} Ca^{2+}. The area has mostly been afforested in the twentieth century. When lake-water was filtered through the newly formed mor at an amount corresponding to 60 years of precipitation, pH changed by cation exchange to 4.5 and the contents of Mg^{2+} and Ca^{2+} were reduced to 3.5 and 5 µeq l^{-1}, respectively, which is lower than in the rain.

References

Dahl, K. 1926 Norske Landbruk, Nr. 7.

Dahl, K. 1927 Salmon Trout Magazine 46, 35–43.

Hegna, T. 1986 Ph.D. thesis, University of Oslo.

Horstad, I. 1989 Clay mineralogy in the Tveitå Sediment core during 1000 years. In *Commission of European Communities air pollution research report, 1989* (ed. I. Th. Rosenqvist).

Höeg, H. I. 1989 In *Proceedings of the SWAP mid-term conference, Bergen 1987*, pp. 386–390.

Krug, E. J. G. 1989 *Assessment of the theory of acidification of watersheds*. Illinois: Department of Energy and Natural Resources.

Renberg, I. & Hellberg, T. 1982 *Ambio* **11**, 30–33.

Robertsson, A. M. 1988 *Biostratigraphical studies of interglacial and interstadial deposits in Sweden*, report no. 10. Stockholm: University Department of Quarternary Research.

Rosenqvist, I. Th. 1985 *Acid precipitation and acid soil in freshwater lake chemistry*. Land Use Policy.

Rosenqvist, I. Th. 1987 In *Proceedings of the SWAP mid-term conference, Bergen 1987*, pp. 362–389.

Rosenqvist, I. Th. 1989 *From rain to lake water pathways and chemical changes*. European Geophysical Society XIV General Assembly, Barcelona, 13–17 March 1989.

Sørensen, N. A. 1983 *Miljøvernteknikk* nr. 4, p. 5.

Ström, K. M. 1925 *Nyt Magazin for Naturv.*, vol. LXII.

Stabell, B. 1987 In *Proceedings of the SWAP mid-term conference, Bergen 1987*, pp. 381–385.

Torgersen, H. 1934 Stangfiskeren 1934.

Winkler, M. G. 1985 *Diatom evidence of environmental changes in wetlands, Cape Cod national seashore*. Boston, Massachusetts: U.S.A. Department of Interior National Park Service.

Pre-industrial surface water acid periods in Norway

By Helge Irgens Høeg

Department of Geology, University of Oslo, Blindern, Oslo 3, Norway

It has been scientifically reported that several good fish lakes in the Agder counties in southern Norway became barren during the last part of the 19th and this century, probably because of the lakes being acid. If the lakes can only be acid because of acid rain caused by industrial pollution, the lakes would not have been acid in earlier times. We should look for lakes which in earlier times have had acid phases. If such phases are found we should try to find the cause. In earlier times they could not have been caused by pollution.

A theory put forward by Professor Rosenqvist claims that the acidification of a watercourse is mostly caused by the abandonment of small farms in the outskirts, with the result that forests grew up on earlier pastures and farm land. If this theory is usable, we look for earlier acid phases near farms which have been abandoned after the great plague, around 1350 A.D.

We suppose that the name 'Fiskeløs' (with no fish) indicated that the lake had no fish in the 16th century when the lake got its name. By use of pollen analysis we found a lake with early farming in the vicinity, then a period without farming caused by the plague or other situations. The farming could have resumed in the 16th century and been abandoned once more in modern time, at the same time as the fish died out.

Pollen analysis were carried out from eight bogs near eight lakes with the name Fiskeløs. The analysis showed a development as mentioned above. Two lakes were chosen for further investigations: Tveitå and Ersdal, both in Agder, S. Norway. Complete sediment series from the lakes were gathered. Pollen samples were analysed for each centimetre in the uppermost part, each second centimetre, 5 cm and 10 cm further down. The series have been partly dated.

The pollen analysis should primarily be used to show the vegetational development, especially the gap in farming continuity. Other methods were to be used for investigations of lake acidity (Rosenqvist & Stabell, this symposium).

The preservation of pollen grains, however, depends on pH (Dimbleby 1957; Havinga 1963; Fægri 1971). Pollen grains of *Pinus* have two air bladders, which may fracture. A pollen grain which has lost its one bladder, is called a half *Pinus*. A loose bladder is also called a half *Pinus*. The amount of half *Pinus* in percent of total *Pinus* vary with pH. High pH in a lake cause much half *Pinus* in the top sample of the sediment, low pH small amounts. It is suggested that this can be used when studying pH in the lakes of earlier times.

Tveitå Fiskeløs

The farm Tveitå is referred to in old literature as a waste farm. In this century there were two farms using the area around the lake. One was abandoned 1950 A.D. Grazing and gathering of fodder around the lake continued until 1955 A.D. The other farm was abandoned 1965 A.D., at the same time as the last fish was caught.

[321]

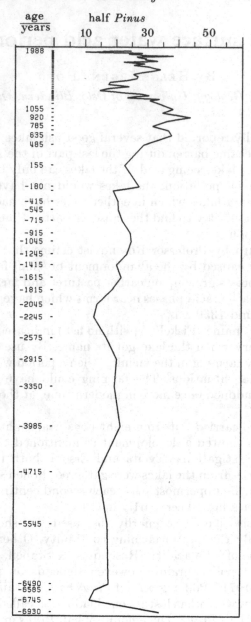

Figure 1. Amount of half *Pinus* in percent of total *Pinus* in the sediment series from Tveitå Fiskeløs. The variations most probably reflect variations in pH. (Small amounts of half *Pinus* means low pH.)

A sediment series was analysed down to 3.04 m, representing the time interval from today and back to shortly after the last ice age, about 9000 years ago. The farming began at 2.80 m, about 2600 B.C. In periods there have been no forest around the lake, because of fodder gathering, vigorous grazing and burning of forest and heather. At 32 cm there were no traces of farming. The age of this level is about 1350 A.D. *Pinus* immigrated and grew in the area for about 200 years, until the farming activity once more became too hard. In 1880 there were neither *Pinus* nor *Picea* in the area according to the owner. After that, i.e. about 10 cm, these trees have been in advance.

The amount of half *Pinus* vary opposite of the amount of *Pinus*, which on the other side vary as a result of the farming activity. At 32 cm there were no farming. *Pinus* increased and half *Pinus* decreased from 27 to 17 %. Farming activity began shortly after, but there were much *Pinus* until 27 cm. Then *Pinus* decreased, and half *Pinus* increased after some time to 32 %. The amount of half *Pinus* was high, up to 10 cm, about 1880 A.D., at the same time as *Picea* began its immigration. From 9 to 7 cm from 1920 to 1940, there were small changes. After that we got a decrease, more or less at the same time as *Pinus* increased and the farms were abandoned. At 3 and 2 cm there were only 14 % half *Pinus*. That was around 1970 A.D. The first increase in the amount of half *Pinus* was around 1 m, at the same time as the farming was intensified. There was also a great change in the amount of half *Pinus* from 3.09 to 3.07 m, when the forest immigrated. The amount of half *Pinus* decreased from 35 to 12 %.

Ersdal

Ersdal is still a farm, but the area around the lake has not been used since 1950 A.D. Before that time the heather was burned regularly, there were gathering of fodder and grazing. The fish died out around 1955 A.D.

A sediment series was sampled from 16.82 to 19.70 m below lake level, representing the time interval from today and back to about 4500 B.C. All the analysed samples contained much charcoal. The lake and the area around must have been an attractive place during the past 6500 years. Some levels contained less charcoal, among others 17.98–17.90 m, 17.54–17.52 m and 17.35–17.33 m. Most probably these levels represent the time intervals 200–100 B.C., 600–700 A.D. and 1350–1450 A.D. At 17.98 m there was a small increase for *Pinus*, and at 17.90 m there was a marked increase. The *Pinus* curve in the diagram shows high values up to 17.72 m. *Pinus* have grown in the area for about 500 years before it again was decimated by fire and cutting.

At 17.54–17.52 m there was no visible changes in the diagram, but from 17.35 to 17.31 m there was a marked increase in *Pinus* and a decrease in *Betula*.

Possibly all three intervals represent periods with reduced farming activity around the lake, even though no changes in farming indicators, such as cereals and *Plantago*, are seen.

The amount of half *Pinus* vary through the series from 12 to 58 %. Below 17.96 m there are five levels with more and 24 levels with less than 25 % half *Pinus*. From 17.96 to 16.82 m there were 63 levels with more and 15 levels with less than 25 % of half *Pinus*. Five of these 15 are between 17.90 and 17.74 m, three from 17.54 to 17.50 m and four from 17.34 to 17.32 m. To that comes 17.44 m, 17.40 m and 16.82 m.

The conclusion is that there are small amounts of half *Pinus* in levels with much *Pinus* and/or small amounts of charcoal, levels which are assumed to represent time intervals with reduced farming activity, and in the top sample, which we know represent a time without farming.

The decrease in half *Pinus* near the top of the series starts at 17.06 m, possibly near 1800 A.D. There is a short increase from 16.93 to 16.88 m, possibly about 1935–50 A.D., from that time the amount of half *Pinus* have been decreasing.

Conclusion

The amount of half *Pinus* looks like it is a pH-indicator. If it is so, the conclusions must be: (1) high pH occurred shortly after deglaciation, before forest establishment; (2) low pH occurred during forest establishment; (3) pH increased during periods with increasing agricultural activity (grazing, burning of forest and heather); (4) pH decreased during periods with reduced farming, dated A.D. 1350 and earlier, as well as A.D. 1950. This agrees well with that have been found by Rosenqvist and Stabell.

References

Dimbleby, G. W. 1957 *New Phyt.* **56**, 12–28.

Fægri, K. 1971 In *Sporopollenin* (ed. J. Brooks, P. R. Grant, M. Muir, P. van Gijzel & G. Shaw), pp. 256–272.

Havinga, A. J. 1963 *Meded. Landbouwhogeschool. Wageningen* **63** (1), 1–93.

Pre-industrial surface water acid periods in Norway: diatom analysis

By Bjørg Stabell

Department of Geology, University of Oslo, Blindern, Oslo 3, Norway

Lake sediments from southern Norway have been investigated to study the development of their diatom flora with regard to changes in acidity. In two lakes, Tveitå Fiskeløstjern and Ersdal Fiskeløstjern, both in Agder, detailed diatom analysis were combined with pollen analysis and geochemical studies.

Results from Tveitå Fiskeløstjern were presented at the SWAP Bergen Conference (Høeg 1987; Rosenqvist 1987; Stabell 1987). It was shown that a major acid period occurred around A.D. 1400–1450 and that pH-values fluctuated highly in the 20th century, with a major acid period after 1945. Both acid periods correlate with vegetational changes; decrease in farming indicators and increased forestation. The diatomological interpretation of an acid period in the 15th century was based on high percentages of *Asterionella ralfsii*, while that of the recent acid period was based on an increase in *Brachysira (Anomoeoneis) serians*, *Navicula subtilissima*, *Semiorbis hemicyclus* and *Tabellaria binalis*.

In addition, Ersdal Fiskeløstjern has been studied. This lake shows the same general development as was found in Tveitå Fiskeløstjern, with acid periods co-occurring with vegetational changes (Høeg and Rosenqvist, this symposium). A third acid period, occurring before A.D. 1200, was observed. Estimated pH-values were based on the Renberg & Hellberg (1982) formula, as there is no database in the area for multivariate statistics. It is evident that Tveitå Fiskeløstjern has been more acid than Ersdal Fiskeløstjern (figure 1), even though the *measured* recent pH-values are equal (Rosenqvist, this volume). Estimated pH-values varies between 4.2 and 4.7 in Tveitå Fiskeløstjern and between 4.7 and 5.4 in Ersdal Fiskeløsvatn.

Another indication of higher pH-values in Ersdal Fiskeløstjern is the occurrence of 12–22 % *Cyclotella kützingiana* in some of the samples. Species belonging to the genus *Cyclotella* was not observed in Tveitå Fiskeløstjern. The acid periods in Ersdal Fiskeløstjern are characterized by increases in the percentages of *Semiorbis hemicyclus* and *Aulacoseira distans* var. *nivalis*.

Selvig (1989) has investigated two additional lakes from southern Norway, Ullstjønn and Stusfu. Ullstjønn is situated quite close to Tveitå Fiskeløstjern. The aim of his study was to determine the diatom development through the whole Holocene, the last millenium was therefore not studied in detail. It is, however, interesting to note that in both lakes a slight decrease in estimated pH-values occur in the 15th century (figure 1). Ullstjønn is less acid (estimated pH-values 4.6–4.9) and has smaller fluctuations in pH-value than Tveitå Fiskeløstjern. Stusfu is situated farther north in an area that is assumed to be less affected by recent acid precipitation. Estimated pH-values range between 5.4 and 5.6. Because of the infrequent sampling it is not possible to infer if a recent acidification occurred in these two lakes.

Even though the exact dating of these cores may be questionable, Tveitå

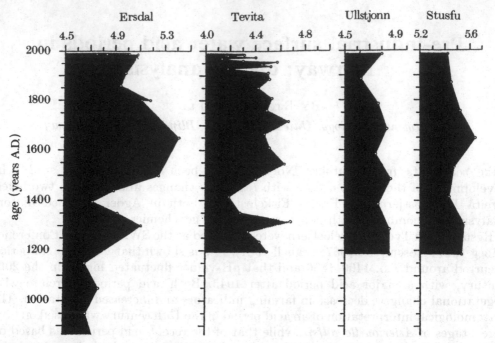

Figure 1. Estimated pH-values against age in four lakes from Southern Norway. Ullstjønn & Stusfu, redrawn from Selvig (1989).

Fiskeløstjern and Ersdal Fiskeløstjern both reacted to a vegetational change (Høeg, this symposium) by becoming more acid. From the estimated pH-values the acidification was equally strong in the 15th century and the late 20th century. The two additional cores from Ullstjønn and Stusfu also show an acid period in the 15th century.

It can be concluded from the present material from southern Norway that the acid lakes in this area have not had large changes in their pH-values, neither of the investigated lakes show fluctuations of more than 0.7 pH-units. However, the estimated pH-values in the two detailed investigated lakes fluctuate repeatedly within this limit through the past eight centuries. In the 20th century the pH of the water in both Tveitå Fiskeløsvatn and Ersdal Fiskeløsvatn seems to have decreased by about 0.4 units, followed by a slight increase in the last decade.

References

Høeg, H. I. 1987 In *Proceedings of the SWAP mid-term Conference, Bergen, Norway*, pp. 386–390.

Renberg, I. & Hellberg, T. 1982 *Ambio* **11**, 30–33.

Rosenqvist, I. Th. 1987 In *Proceedings of the SWAP mid-term Conference, Bergen, Norway*, pp. 362–389.

Selvig, E. 1989 Cand. Scient. thesis, Department of Geology, University of Oslo 1989.

Stabell, B. 1987 In *Proceedings of the SWAP mid-term Conference, Bergen, Norway*, pp. 381–385.

The influence of water quality and catchment characteristics on the survival of fish populations

By I. P. Muniz[1] and L. Walløe[2]

[1] NINA, Division of General Physiology, Institute of Biology, University of Oslo, PO Box 1051, Blindern, Oslo 3, Norway

[2] Departments of Physiology and Informatics, University of Oslo, PO Box 1080, Blindern, Oslo 3, Norway

Data on fish status and water chemistry of lakes in southern Norway have in previous investigations been used to delineate the influence of water quality on the survival of fish populations in acid environments. These analyses generally point to pH and the concentrations of inorganic aluminium species and calcium as the main determinants of fish status. However, these investigations have either been carried out by a series of bivariate statistical methods or not included all potentially important variables.

In the present investigation, data from a number of these earlier studies have been pooled and reanalysed by different bivariate and multivariate statistical methods. The dependent variable in the analyses has been fish status for brown trout, and possible independent variables have been 14 chemical and physiographical variables which include the concentration of all major chemical species in the water, lake size and lake altitude.

The bivariate analyses confirmed previously observed correlations between trout survival and each of the independent variables. In different stepwise multivariate analyses, however, only pH, concentration of inorganic aluminium and lake altitude entered the regression equations, and in that order.

The results also show that low calcium concentration, within the range found in lakes in southern Norway, is neither harmful *per se* for the survival of trout populations, nor does calcium protect the fish from the deleterious effects of inorganic aluminium.

Introduction

Loss of and decline in brown trout populations in some high elevation inland lakes in southernmost Norway, 'Sørlandet', were observed as far back as in the 1920s. Dahl (1920, 1922) then found that these few lakes were acid, had unexpectedly low numbers of fish, and were dominated by large and old individuals suggesting recruitment failure as the cause of the change in fish status. Simple experiments with alevins in low pH solutions were consistent with these field observations (Dahl 1927) as was experienced in local hatcheries. Here previous mass mortalities among egg and alevins of brown trout reared at acid pH could be significantly reduced by letting the water pass through shell sand filters and thereby raise the pH above 6.0 (Sunde 1926).

Continued observations in the 1930s of lake pH and fish status, experiences from stocking of fish to counteract loss of recruitment and to try to restore barren lakes and an occasional fish kill in 1948, all strongly indicated pH as a cardinal variable. As in the case of limed hatchery water, the geographical pattern of extinction also indicated the importance of the geological environment in that areas with higher content of weatherable Ca and Mg minerals proved favourable for fish survival (cf. Muniz 1984).

A series of hatching experiments by Snekvik in the years 1966–71 also showed that addition of Ca and Mg salt and NaCl as well as higher pH, could significantly reduce egg and post-embryonic mortalities of native salmonids. This was further supported by results from his field studies initiated in the late 1960s.

In the regional inventories he collected information on past and present fish status, compiled local observations of pH, and arranged for collections of water samples for chemical analyses, all through local informants and authorities. Such data from a high number of municipalities in our four southernmost counties, Rogaland, Aust-Agder, Vest-Agder and Telemark, showed that both pH and the salt content of the water were important factors determining the fish status. In this region, some 700 lakes, about one-third of the lakes in his material, had lost their fish, mostly during the past 20–30 years, and an additional high number of lakes probably were in the process of extinction (Jensen & Snekvik 1972).

In 1974, Snekvik's regional inventories were continued by the Norwegian SNSF project to include additional lakes. In addition, available information from Snekvik in an area in southern Norway south of the 62nd latitude was revised and the geographical extent of the fish decline was mapped (Muniz *et al.* 1976).

In 1977, water samples were collected from 684 'Sørlandet' lakes and analysed for several chemical components (pH, conductivity, total hardness (CaO), NO_3, SO_4, Cl, Na, K, Ca, Mg and (total) Al. These data were coupled with the Snekvik/SNSF-material on fish status for these lakes and analysed by multiple linear stepwise regression methods. The results pointed to pH and Ca as the two most important factors to explain the fish status (Wright & Snekvik 1978). This was largely in accordance with Snekvik's earlier results, but there was no indication that the sea-salt contribution from NaCl was of any importance for population maintenance. Nor did it appear that the concentration of total Al in the range 50–300 µg l^{-1} observed in approximately 90 % of the lakes was particularly harmful.

When the regional SNSF material on past and present fish status for brown trout was analysed with reference to both chemical and physiographical variables, it was found that the frequencies of extinct brown trout populations increased with decreasing pH and decreasing salt content (conductivity). It was also found that trout from small- and medium-sized lakes were more affected than those from larger lakes and that high-elevation lakes were more affected than lakes at lower altitude. The high-altitude inland lakes were the first to become devoid of trout, and the loss of this species had over the years gradually spread downstream towards the coast (Muniz & Leivestad 1980). This confirmed the general observation of regional fish decline from the affected areas of southernmost Norway (Jensen & Snekvik 1972).

In 1983–84, Sevaldrud & Skogheim (1986) resampled lakes in Aust and Vest-Agder for water chemistry and revised fish status information from many of those localities which according to the SNSF data still had fish in the period 1974–78 (Sevaldrud & Muniz 1980). In 1986, Henriksen *et al.* (1989) conducted a regional survey of 1005 lakes from sensitive areas in Norway, and from 1985 and onwards Hesthagen and coworkers have collected data on fish status and water chemistry from previously unsurveyed areas both at 'Sørlandet', Rogaland and Telemark, as well as from the remaining counties in southern Norway (Hesthagen *et al.* 1989).

The number of water-chemical variables have been increased over the years relative to the Wright–Snekvik protocol, and now also include labile (toxic) and non-labile aluminium fractions (Driscoll *et al.* 1980). The analysis of Sevaldrud & Skogheim pointed to pH, Ca and labile aluminium as being important factors for the survival of fish populations (Sevaldrud & Skogheim 1986; Henriksen *et al.* 1989). The material from Hesthagen and co-workers has not been analysed in this way, and also for the other data materials one might question the choice of variables included in the analysis and the effectiveness of the multivariable method used. Since a large data base on fish population status and a number of chemical and physiographical variables in the lakes now is available, admittedly from different sources and quite heterogenous, we wanted to explore this material using a number of multivariate statistical methods.

In this paper we mainly present results for brown trout, *Salmo trutta* L., to avoid some of the difficulties of analysing multispecies fish communities. The brown trout is more or less uniformly distributed over the entire country, is an important species for recreational and commercial purposes, is a severely affected species, and is probably one of the best studied species in relation to acid aluminium-rich water (Grande *et al.* 1978; Muniz & Leivestad 1980; Brown 1982).

The analysis is carried out for unlimed lakes in southern Norway (south of approximately 62° N). All catchments in this area are to some extent affected by 'acid rain', but the material includes lakes both within and outside the area clearly affected by surface water acidification.

Material

The pooled material is based on three data-sets. The first one (1) is the material of Sevaldrud & Skogheim (1986) from Aust and Vest-Agder where fish status was updated and water samples collected from lakes which according to local informants still had fish in the period 1974–78. The interview methods of Sevaldrud & Muniz (1980) were used and water samples were collected during a week in October 1983 and analysed for the same chemical components as chosen by Wright & Snekvik. In addition, measurements of the concentrations of labile and non-labile aluminium fractions were included. The former fraction is operationally defined as the inorganic monomeric form, the latter as the organically complexed monomeric Al-species.

The second source of information (2) is Henriksen *et al.*'s (1989) material from lakes distributed over the entire country and selected on the *a priori* expectation that the bedrock would give poorly buffered run-off water. The majority of the selected lakes were from the four southernmost most affected counties and the water samples were collected during the autumn in 1986. Information about fish populations was collected from most of the lakes by the county environmental authorities through questionnaires that were handed out and subsequently subjected to checking and supplementation before final compilation. The analytical chemistry protocol comprised macro and microconstituents including the aluminium species.

The third data source (3) is the material which results from the continued national efforts to map and monitor the regional extent and the effect of surface water acidification on freshwater fish stocks in Norway (Hesthagen *et al.* 1989). In recent years, and particularly since 1985, this work has provided data from previously

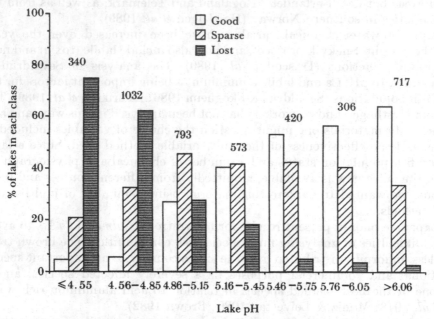

Figure 1. Population status for brown trout for seven different pH categories. Data from 4181 Norwegian lakes south of the 62nd latitude including lakes from areas not appreciably affected by surface water acidification.

unsurveyed lakes in southern Norway. In this material, the data on fish status has been collected as mentioned above, i.e. using local informants and questionnaires and checking the information through test fishing some lakes. The test fishing results are generally in accordance with the information in the questionnaires (Rosseland *et al.* 1980). As before, the water samples were taken from or near to the outlet of lakes at some time during the ice-free season and analysed for chemical components including two aluminium fractions. Information on lake altitude (in metres) and surface area of the lakes (in hectares) were taken from 1:50000 topographical maps (Statens Kartverk M711-series). The three categories of fish status, good, sparse and lost, have been defined as (1) good if there was no indication of a perturbed population, (2) sparse when the population were low in number often reflected as a decline towards low catches and (3) empty when the former population of trout was virtually extinct.

The data from the different investigations appeared on six computer files, but with differing format and map references and often with different units for the same chemical variable. All files were therefore transformed to the same format and units and combined in one large data file. The resulting file contained 9708 lake instances, but the number of individual lakes was only 7533, since there were 2175 instances of repeated data collections from the same lakes. The fish status information was available in the primary files for up to three fish species for each lake. Information of 15 different species could be found in the material, but only brown trout, perch (*Perca fluviatilis* L.) and perhaps char (*Salvelinus alpinus* L.) were present in a sufficiently high number of lakes to be subject to statistical analysis. The information about brown trout is also considered to be more precise because of the high economic and recreational value of this species. Every lake which according to the local informants has had a good and sustained trout population within the memory span

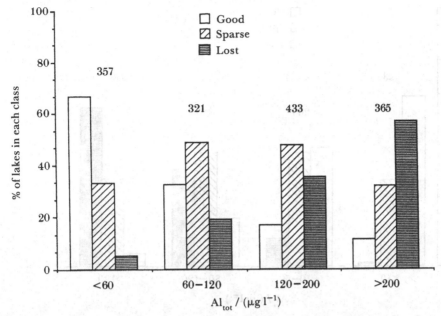

Figure 2. Population status for brown trout for four different categories of total monomeric aluminium, Al_{tot}. Data are for 1476 lakes in southern Norway.

of the informants, is in this investigation called a trout lake. For most lakes the available information about fish status dates back to the 1930s. For some lakes the information given could also be checked against contemporary written sources: agricultural statistics, fishermen's and farmers' diaries, etc.

The total data material contained information about fish status for 4181 different trout lakes, but for many of these the only available chemical information was pH and conductivity. Only about 1200 trout lakes south of 62° N had values for all the variables which in the present investigation have been used in the multivariate analyses: fish status, lake altitude, lake size, pH, conductivity, HCO_3, SO_4, NO_3, Cl, Na, K, Mg, Ca, Al_{tot} (total monomeric aluminium) and Al_i (inorganic monomeric aluminium).

Methods

The relationship between trout population status and the 14 other variables have been analysed both by bivariate and multivariate methods.

The bivariate method used was to display the data in two-way frequency tables and to use the Kendal's correlation coefficient (corrected for ties) as a test statistic. The fish status variable has three categories (good, sparse, lost); the other variables have at least four categories, corresponding to the four quartiles in the marginal distribution of the variable. The results of such analyses is shown in figures 1 to 4.

Since many of the fourteen variables which could possibly influence fish status were strongly intercorrelated, it was considered important also to analyse the material by multivariate statistical methods. What one could hope to obtain was to get rid of some of the uninteresting, spurious correlations in the material and focus the attention on a few variables which would be closer to the causal chain. However, there are difficulties and pitfalls in all multivariate methods, and in the present data

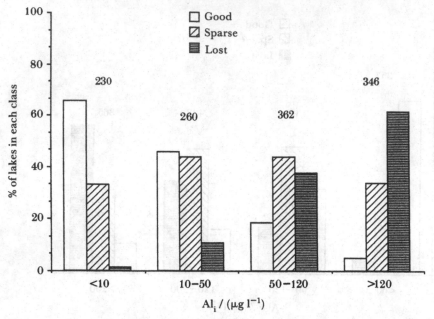

Figure 3. Population status for brown trout for four different categories of inorganic
monomeric aluminium, Al_i. Data are for 1198 lakes in southern Norway.

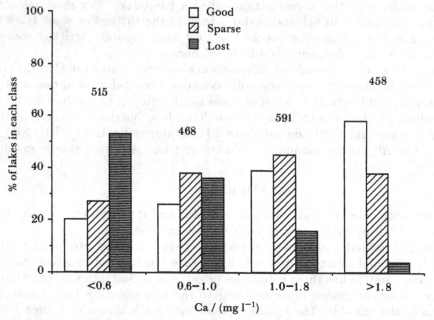

Figure 4. Population status for brown trout for four different levels of calcium.
Data from 2032 lakes in southern Norway.

material there is necessarily a number of false data points. Thus it was necessary (1)
to use robust statistical methods, (2) in each step in the analysis to check the
distribution of residuals and (3) not to rely too much on small differences in the

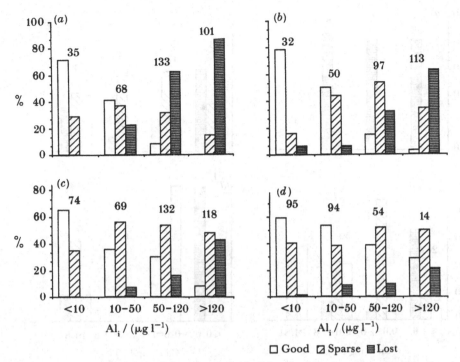

Figure 5. Population status for brown trout for four levels of inorganic monomeric aluminium (Al_i) classified at four levels of calcium concentration: (a) < 0.6 mg l^{-1}, (b) 0.6–1.0 mg l^{-1}, (c) 1.0–1.8 mg l^{-1}, (d) > 1.8 mg l^{-1}. Data from 1279 lakes in southern Norway.

results. In addition, we have used three different multivariate methods with a somewhat different set of assumptions: multiple linear regression, logistic regression (fish status: present, lost), and Cox regression (dependent variable: year of extinction, censoring at year of investigation for lakes with trout populations present). The regressions have been performed by the BMDP system, especially the programs 2D, 5D, 6D, 1R, 2R, 2L and LR (Dixon 1988). Since the results in all main findings are similar by all three regression methods, we present only the results of the linear regression here.

Results

The bivariate analyses showed that trout population status was correlated with many of the 14 variables investigated. Trout populations survived better in lakes at lower altitudes, in larger lakes, at higher pH, at higher conductivities, at lower concentrations of SO_4, NO_3, Al_{tot} and Al_i, and at higher concentrations of HCO_3, Cl, Na, K, Mg and Ca. The results for lake pH are shown in figure 1, for total monomeric aluminium in figure 2, for inorganic monomeric aluminium in figure 3 and for Ca in figure 4. However, all the four independent variables shown in these four figures are strongly correlated with each other. Some of this interdependence can be sorted out by means of multiway classification of the lakes. One example is shown in figure 5. Here fish status has been classified according to level of inorganic aluminium for four different calcium categories. Note that the grouping of both inorganic aluminium and calcium in figure 5 is the same as in figures 3 and 4. Figure 5 illustrates that inorganic aluminium and calcium are negatively correlated. The figure also

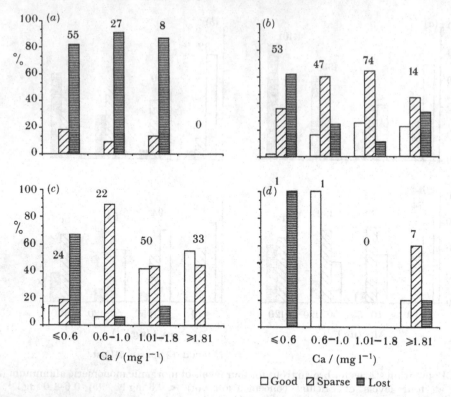

Figure 6. Population status for brown trout for four levels of calcium (Ca) classified at four levels of acidity: (*a*) pH ≤ 4.75, (*b*) pH 4.76–5.15, (*c*) pH 5.16–5.75, (*d*) pH ≥ 5.76. Only lakes with concentrations of inorganic monomeric aluminium between 50 μg l^{-1} and 120 μg l^{-1} have been used.

demonstrates that low calcium is not a critical factor for trout population survival in lakes in southern Norway, as long as the concentration of inorganic aluminium is low. For higher concentrations of inorganic aluminium, however, the figure indicates that high calcium may protect the fish against the deleterious effects of inorganic aluminium. Again, the problem is that high calcium concentration is correlated with high pH in the lake water. We know that low pH is harmful to fish. The possibility therefore exists that the apparently protective effect of high calcium against inorganic aluminium is caused by the corresponding high pH and not by calcium itself. To investigate this point we have classified the lakes in figure 5 also in four pH-categories. The results are shown in figure 6, but only for the inorganic aluminium category 50–120 μg l^{-1} which according to figure 5 is within a range of Al_i where dramatic shifts in the frequencies of fish status occur as we move from low to higher calcium concentrations. The number of lakes in some of the categories now becomes too small to be meaningful, but figure 6 clearly suggests that the protective effect of calcium to a large extent disappears when pH is controlled in the analysis. It is difficult to carry this kind of analysis further. A more appropriate approach is to use multivariate methods.

A stepwise linear regression analysis was carried out with the population status of brown trout as the dependent variable ('good': 1, 'sparse': 2, 'lost': 3) and with lake altitude, lake area, pH, conductivity, concentrations of HCO_3, SO_4, NO_3, Cl, Ca, Mg, K, Na, Al_{tot} and Al_i as possible independent variables. Since many of these variables

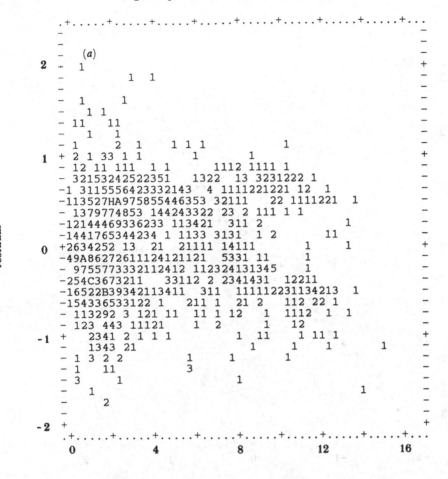

Lake altitude / (10^2 m)

Figure 7(*a*). For description see p. 336.

had skewed marginal distributions with tails towards higher values, logarithmic transformations of these variables were also tried. In qualitative terms, the results were similar whether the original or the transformed variables were used. However, the distribution of the residuals around the regression line was closer to a normal distribution when the original (untransformed) values of the variables were used. To avoid undue influence in the regression of a few lakes with very high values, for instance of calcium, the regression was also performed on truncated distributions. The calcium concentration was truncated both at 7 mg l^{-1} and at 2 mg l^{-1} without changing the results. The regression was carried out for a number of subgroups of the total material, for instance for lakes above 200 m (which is the upper limit for marine sediments) and for lakes in the eastern and western part of southern Norway separately. Again, the results were very similar. In each step and also for the final result, the distribution of residuals was carefully checked. Figure 7 shows one example.

The main result from this multivariate linear regression exercise is that only three of the fourteen possible independent variables has regression coefficients which are

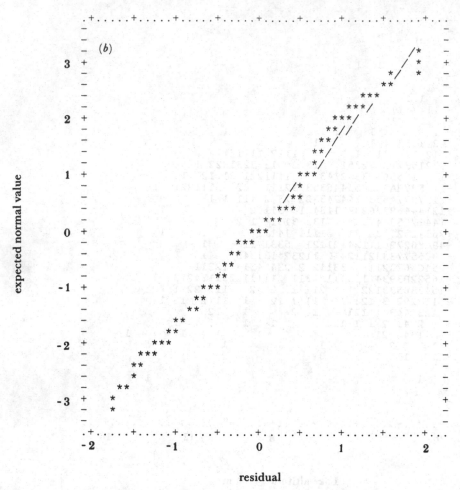

Figure 7. (a) Scatter plot of residuals around the regression line (table 1) against one of the variables in the equation (lake altitude). (b) Normal probability plot of the residuals. Values from normal distribution would lie on the line indicated by the symbol /. (a) and (b) display ordinary printout from the program BMDP1R. The residuals are in units of fish population status.

Table 1. *Results from a stepwise linear regression run with brown trout population status as the dependent variable and with 14 possible independent variables*

(*Multivariate method*: stepwise linear regression. *Independent variables*: lake area (in hectares), lake altitude (in metres), pH, conductivity, $[HCO_3]$, $[SO_4]$, $[NO_3]$, $[Cl]$, $[Ca]$, $[Mg]$, $[K]$, $[Na]$, total monomeric Al $[Al_{tot}]$, inorganic monomeric Al $[Al_i]$. *Dependent variable*: status brown trout: 1 (good), 2 (sparse), 3 (lost). Multiple R: 0.66; multiple R^2: 0.44; $N = 1192$.)

	coefficient	standardized regression coefficient	order of importance
pH	-0.51	-0.42	1
Al_i	0.37×10^{-2}	0.33	2
altitude	0.52×10^{-3}	0.22	3

significantly different from zero at the 5% level; that is pH, inorganic aluminium and lake altitude. On the other hand, these three are all significant at the 0.001 level. In the stepwise regression pH enters the regression equation first, then the concentration of inorganic aluminium, then lake altitudes (table 1). Neither in the analysis of the total material, nor in any of the analyses of subdivisions of material, nor when only lakes with calcium concentrations below 2 mg l^{-1} are included in the analysis, does calcium enter the regression equation at the 5% level.

Discussion

The results of the multivariate statistical analysis are unambiguous. Of the 14 possible independent variables investigated, 11 had no significant relation to the status of brown trout in lakes in southern Norway. Only three variables remained in the regression equation: pH, concentration of inorganic monomeric aluminium and lake altitude. As far as pH and inorganic aluminium are concerned, the results are not surprising, and corroborate earlier results both from the field and the laboratory. There is now good evidence that both low pH and/or high concentrations of inorganic aluminium is detrimental for some or all life-stages of brown trout (for updated review cf. Baker *et al.* 1990).

The strong influence of lake altitude was unexpected. It is not possible that high altitude *per se* is harmful. High altitude must be correlated with some other detrimental variable, a variable that is not already included in the analysis. We can offer some possibilities. The first is that the effect is related to snow accumulation and the nature of snow melt episodes. At inland localities, lakes freeze over annually and snow accumulates throughout late autumn and winter. During the annual spring thaw, deposited acids accumulated in the intact snow pack, leach out in high concentrations to produce the annual snow melt acid episode critical for fish reproduction and survival (Henriksen & Wright 1977; Dickson 1980). Catchments at lower altitudes, and particularly in coastal situations with mild winters, typically exhibit less snow accumulation owing to many alternating periods of melting and deposition. Here only a fraction of the acid deposition is accumulated in the snow pack and, consequently, the trout will not experience the major spring thaw episode (Andersen *et al.* 1984). Another possibility is that the ecosystem in high-altitude lakes contains fewer species and therefore is more vulnerable than lakes at lower altitude. For instance, lack of ecological analogues of key food items for a fish species living in an acidifying lake at high altitude may lead to an insufficiently strong food base or otherwise weaken the overall fitness and performance of the species through food chain perturbations. Although there is currently no data to show that acid-stressed brown trout populations in high mountain lakes become extinct because of lack of suitable food items (Rosseland *et al.* 1980), situations are known where the populations finally converges towards starvation and death of fishes (Mills *et al.* 1987). One possibility is also that the trout populations of high-altitude lakes may be more sensitive because of population dynamics responses to acid stress. If the major mechanism of population extinction under such circumstances is ageing (scenescense), e.g. dominance of older fish as a result of total (or partial) recruitment failure, then recent modelling and quantitative analysis of fish populations suggest that such populations will have shorter extinction times than juvenilized trout populations commonly found at lower altitudes (Bravington *et al.*, this symposium).

The lack of influence of the concentration of calcium on brown trout status was surprising. This result has therefore been scrutinized by different methods and in different subgroups of the total material. It is, however, clear, both from the contingency tables and from the multivariate analysis, that brown trout populations have no problems even with calcium concentrations far below 1 mg l^{-1} as long as the pH is sufficiently high and the concentration of inorganic aluminium sufficiently low. Contrary to current apprehension, calcium does not in our material protect brown trout populations against the damaging effects of inorganic aluminium. In our analysis this is a clear enough result, but it will probably need corroboration from other studies before it will be generally accepted, since it is at variance with results obtained in the laboratory (Brown 1982). However, it is not really at variance with the results obtained by Wright & Snekvik (1978) in their multivariate regression analysis. First, they did not include the physiographical variables lake area and lake altitude in their set of possible independent variables. Since lake altitude and lake calcium concentration in our material is negatively correlated, the calcium concentration in their analysis probably takes care of some of the altitude effect. Secondly, the dependent variable in their analysis was a mixed species fish status variable, containing also perch and some other species in addition to brown trout. Our preliminary analysis shows perch to be much more dependent on high concentration of ions in the lake water than brown trout. Thirdly, most of their lakes were situated in the four most affected counties in southern Norway. Our material contains a higher proportion of lakes from less affected areas of southern Norway, among them a good many with both low calcium concentration and good brown trout populations.

Thanks are due to Trygve Hesthagen, Hans Mack Berger, Iver Sevaldrud, Odd Skogheim and Arne Henriksen and his co-workers for giving us access to their data files, to Kjetil Otter Olsen for help with the data management, to Lars Erikstad for help with the illustrations and to Bjarne A. Waaler for comments on the manuscript.

References

Andersen, R., Muniz, I. P. & Skurdal, J. 1984 *Rep. Inst. Freshwat. Res. Drottningholm* **61**, 5–15.

Baker, J. P., Bernhard, D. P., Christensen, S. W., Sale, M. J., Freda, J., Hetcher, K., Rowe, L., Scanion, P., Stokes, P., Suter, G. & Warren-Hicks, W. 1990 State-of-Science/Technology Report 13. National Acid Precipitation Assessment Program, Washington.

Brown, D. J. A. 1982 *Wat. Air Soil Pollut.* **18**, 343–351.

Dahl, K. 1920 *Norsk Jaeg.-og Fiskeforen. Tidsskr.* **49**, 249–267.

Dahl, K. 1922 *Norsk Jaeg.-og Fiskeforen. Tidsskr.* **51**, 64–66.

Dahl, K. 1927 *Salm. Trout Mag.* **46**, 35–43.

Dickson, W. T. 1980 In *Proc. Int. Conf. Ecol. Impact Acid Precip. Norway* (ed. D. Drabløs & A. Tollan), pp. 75–83. Oslo-Ås, Norway: SNSF project.

Dixon, W. J. (ed.) 1988 *BMDP statistical software manual*, vols 1, 2. Berkeley. University of California Press.

Driscoll, C. T., Baker, J. P., Bisogni, J. J. Jr & Schofield, C. L. 1980 *Nature, Lond.* **284**, 161–164.

Driscoll, C. T. 1980 Ph.D. thesis, Cornell University, New York.

Grande, M., Muniz, I. P. & Andersen, S. 1978 *Verh. Int. Verein. Limnol.* **29**, 2076–2084.

Henriksen, A., Lien, L., Rosseland, B. O., Traaen, T. S. & Sevaldrud, T. S. 1989 *Ambio* **18**, 314–321.

Henriksen, A. & Wright, R. F. 1977 *Nord. Hydrol.* **8**, 1–10.

Hesthagen, T., Berger, H. M., Blakar, I., Enge, E., Fjeld, E., Hansen, L. P., Hegge, O., Larsen, B. M., Sevaldrud, I. H., Strand, R. & Tysse, O. 1989 In *Acid deposition: sources, effects and controls* (ed. J. W. S. Longhurst), pp. 117–142. British Library, Science Reference and Information Service and Technical Communications.

Jensen, K. W. & Snekvik, E. 1972 *Ambio* 1, 223–225.

Mills, K. H., Chalanchuk, S. M., Mohr, L. C. & Davies, I. J. 1987 *Can. J. Fish. Aquat. Sci.* 44 (Suppl. 1), 114–125.

Muniz, I. P. 1984 *Phil. Trans. R. Soc. Lond.* B 305, 517–528.

Muniz, I. P., Sevaldrud, I. H. & Lindheim, A. 1976 SNSF project, TN 21/76. 41 pp.

Muniz, I. P. & Leivestad, H. 1980 In *Proc. Int. Conf. Ecol. Impact Acid Precip. Norway* (ed. D. Drabløs & A. Tollan), pp. 84–92. Oslo-Ås, Norway: SNSF project.

Rosseland, B. O., Sevaldrud, I. H., Svalastog, D. & Muniz, I. P. 1980 In *Proc. Int. Conf. Ecol. Impact Acid Precip. Norway* (ed. D. Drabløs & A. Tollan), pp. 336–337. Oslo-Ås, Norway: SNSF project.

Sevaldrud, I. H. & Muniz, I. P. 1980 SNSF project, IR 77/80 (95 pp., 61 tables and 6 pp. appendix).

Sevaldrud, I. H. & Skogheim, O. K. 1986 *Wat. Air Soil Poll.* 30, 381–386.

Sunde, S. E. 1926 In *Fiskeri-inspektørens innberetning til Landbruksdepartmentet for året 1926*. (*Annual report from the fishery inspector to the Department of Agriculture for the year 1926*.)

Wright, R. F. & Snekvik, E. 1978 *Verh. internat. Verein. theor. angew. Limnol.* 20, 765–775.

Discussion

D. J. A. BROWN (*National Power Technology and Environmental Centre, Kelvin Avenue, Leatherhead KT22 7SE, U.K.*). The recently completed regression analysis reported by Muniz & Walløe suggests that pH, aluminium concentration, and altitude are the major determinants of fishery status in Norwegian lakes, and, contrary to the earlier analysis of data by Wright & Snekvik (1978), calcium concentration is not shown to be an important factor in this regard.

I feel, however, it is premature to eliminate calcium from the deliberations at this time for several reasons, including the following.

(*a*) Many chemical factors in the field are cross-correlated; e.g. low pH is associated with high aluminium and also with low calcium; low calcium is also more likely to be found at higher altitudes, etc. Therefore, the absence of calcium from the results of such a mathematical treatment does not necessarily mean that it is of no importance.

(*b*) To evaluate the real importance of these variables that are cross-correlated in the field, it is sensible to take the problem to the laboratory. There are numerous reported experiments in the literature of where calcium has been shown to be very important in determining survival of fish in acid waters, particularly their early life history stages.

(*c*) These experiments have further demonstrated that concentrations of calcium around 1 mg l^{-1} are crucial, and that at concentrations greater than 2 mg l^{-1}, calcium effects are relatively unimportant. This is another possible reason why regression analysis of a sample of lakes, with many lakes having calcium concentrations in excess of this value, as I believe was the case in the analysis presented by Muniz, may not indicate that calcium is important. However, in the oligotrophic lakes, whose response to a sulphate reduction we are trying to predict, the calcium concentration is going to be in the critical range.

On a related matter, the proposal by Dr Exley that silica is a component of acid waters of greater importance than calcium, in that, contrary to calcium which ameliorates aluminium toxicity, silica can totally remove its toxicity by complexation, is interesting and worthy of further investigation. However, I cannot agree with him that such a hypothesis, if proven to be correct, would negate the importance of calcium. On the contrary, because the survival of freshly fertilized trout eggs is dependent upon pH and calcium concentration, and largely independent of aluminium, it does not matter if this aluminium is complexed with silica or not. If the aluminium, which is toxic to subsequent life history stages, is indeed complexed and rendered non-toxic, the toxicity of hydrogen ion to the especially sensitive freshly fertilized egg, which is most definitely ameliorated by calcium, assumes greater rather than less significance.

L. WALLØE. (*a*) It is certainly true that many of the chemical and physiographical variables in lakes in southern Norway are strongly inter-correlated. Multivariate regression methods have been developed just to enable scientists in natural as well as social sciences to cope with such complex multidimensional information. When the methods are properly used, as we think we have, they may help the scientists to get rid of some of the uninteresting, spurious correlations in the data matrix, and to focus the attention on a few variables which are closer to the causal chain. Therefore, as our results from both the contingency tables (figures 5 and 6 in our paper) and the three multivariate regression methods are consistent and clear-cut, we do indeed draw the conclusion that differences in calcium concentration within the range found in the lakes in southern Norway is of no importance for the survival of populations of brown trout.

(*b*) We do not question the results obtained in the physiological laboratory, but only draw the conclusion that in the field in southern Norway the different life stages of brown trout manage to get sufficient calcium to grow and reproduce.

(*c*) As can be seen from our figures 4 and 5, about half of the lakes in our material have calcium concentrations below 1 mg l^{-1} and about one quarter below 0.6 mg l^{-1}. As long as the concentration of inorganic aluminium is sufficiently low and pH sufficiently high, the populations of brown trout have no problems. As we knew the results from the experiments referred to by Dr Brown, we did indeed perform the regression analyses on the subset of the material with calcium concentrations below 2 mg l^{-1}. As stated in our paper, the results were quite similar to results given in table 1 in our paper. (More precisely, the regression coefficients for pH, inorganic aluminium and altitude in the linear regression analysis were: -0.68, 0.33×10^{-2}, 0.61×10^{-3}, respectively, statistically significant at the 0.001 level. Calcium concentration did not enter the regression equation at the 0.05 level, but the point estimate for the regression coefficient was 0.07 (note wrong sign), $N = 1006$.)

Our results are not contrary to the results obtained by Wright & Snekvik for three reasons. (1) They did not include the physiographical variables lake area and lake altitude in their multivariate analysis. (2) Only total aluminium concentration and not inorganic aluminium concentration was available to them. (3) Most important, they did not exclusively analyse brown trout populations, but multispecies fish populations. The main species in addition to brown trout in their material was perch, which in our preliminary analysis turn out to be dependent on high concentrations of ions (not only Ca, but Mg, K, Na, Cl, etc.) in the lake water. In fact, we can by and large simulate their results by admitting only their set of variables in the regression equation and using fish status information from a mixture of brown trout and perch populations.

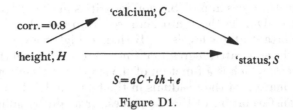

Figure D1.

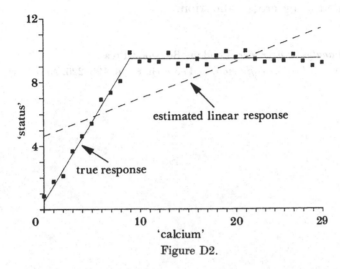

Figure D2.

A. ROSENBERG, (*RRAG, Imperial College, London, U.K.*). The stepwise regression analysis presented in this paper indicate that the variation in trout stock status in Norwegian lakes is best explained by three variables: pH, Al and height above sea level. No other chemical or physical variables contribute significantly to the fit. In particular, the authors suggest that these results imply no causal relationship between Ca ion concentration as measured in the survey and fish stock status.

Although the regression analysis *per se* is well founded and potential problems with error structure were allayed by using several different types of model, it is inappropriate to draw the particular conclusion above from this analysis. The reason lies in possible multicollinearity between Ca and the other explanatory variables in the final model. This could occur even if the correlation between Ca and, say, height above sea level is low, e.g. around 0.5. A general discussion of multicollinearity in problems of this kind can be found in Weisberg (1985, ch. 8).

To illustrate the problem, we simulated a three-variable data-set in which the response variable depends linearly on two correlated explanatory variables (figure D1). The explanatory variables were standardized to have the same variance. Stepwise regressions performed on sets of data generated this way indicate that, if the significance level is such that only one variable is selected, the actual choice depends on the relative strengths of the effects, as measured by the regression coefficients a and b in figure D1. If the effects are of similar strength, the choice is made randomly. Therefore it is difficult to rule out a causal effect on the basis of the stepwise analysis. A path analysis of the relationships between the variables may clarify matters (Li 1975).

One possible solution to this problem is to split the data-sets into 'less collinear' groups, perform stepwise analyses on each, and cross-validate the results.

An additional problem lies in possible nonlinearities in the relationship between Ca and the stock status. During the symposium several researchers noted that Ca may be much more important at low levels of pH than at higher levels. This may strongly affect the slope of the estimated regression 'line' (figure D2), and thus the probability of detecting an effect, which is a function of the slope. If this effect is large, it should be shown by examination of the residuals in the final model. It might be possible to resolve this by transformation of the Ca values, or again by splitting into low and high Ca régimes and using cross-validation.

References

Li, C. C. 1975 *Path analysis: a primer*. London: Boxwood Press.
Weisberg, S. 1985 *Applied linear regression*, 2nd edn, ch. 8, pp. 196–225. New York: John Wiley.

Factors affecting fish survival in Scottish catchments

By R. Harriman, E. Gillespie and B. R. S. Morrison

Freshwater Fisheries Laboratory, Pitlochry, Perthshire PH16 5LB, U.K.

Recent field and laboratory studies have shown that pH, calcium and ionic forms of aluminium are the key chemical parameters which influence the survival of juvenile salmonids. An experimental approach was used in the Allt a'Mharcaidh to evaluate the interactive effects of these parameters. This upland Scottish stream was selected because of its transitional nature, i.e. one which has a mean pH > 6 but is subject to acid episodes of pH < 5.5.

The raising of calcium levels by 30% during these episodes has significantly increased the survival of eggs to hatching. Population surveys at four sites, covering a range of habitats, indicated low numbers of 0+ fry at the end of the growing season compared with numbers in less acid streams. Also, the rate of density decline from the peak value showed between year variation which was correlated with the frequency of episodes of pH < 5.5. In choice chamber experiments an avoidance response to stream water of pH < 5.5 was exhibited by 0+ fry.

The relationship between mean pH and frequency of episodes of pH < 5.5 was evaluated using data from all the SWAP sites. An empirical relationship between brown trout densities and episodic frequency was obtained whereby the suitability of streams for salmonid fish could be assessed. The effects of long-term pH changes on trout densities in two transitional streams are discussed in terms of predicted and measured values.

Introduction

In most upland streams having adequate water quality, the capacity for holding juvenile fish is usually dependent on food availability and habitat characteristics. However, many streams and lakes in parts of Northern Europe and North America, with apparently excellent habitat characteristics have exhibited a decline or complete loss of salmonid fish as a result of surface water acidification (Beamish & Harvey 1972; Schofield 1976; Gunn 1986; Harriman *et al.* 1987). Although mortalities of adult Atlantic salmon *Salmo salar* L. and trout *Salmo trutta* L. have occasionally been observed (Jensen & Snekvik 1972; Leivestad & Muniz 1976), the major cause of the decline and eventual loss of fish stocks has been recruitment failure (Daye & Garside 1977; Lacroix 1985). Such failure has been attributed to disruption of normal egg development in spawning fish or to mortalities during early life stages (Peterson *et al.* 1980). Controlled exposure of adult dish, eggs and alevins to different chemical conditions in laboratory experiments has demonstrated that pH, calcium and some ionic forms of aluminium are the key factors affecting survival

(Baker & Schofield 1982; Brown 1983; Gunn *et al.* 1987). Lacroix (1985) suggested that the combined effects of elevated hydrogen and low calcium concentrations are the major influences on egg survival in organically rich waters. Salmon eggs and fry are vulnerable to the periods of depressed pH and calcium levels and elevated aluminium concentrations which characterize acid episodes in streams, especially during heavy rain and spring snowmelts (Johanessen & Henriksen 1978). The timing of these episodes may coincide with hatching and the emergence of the young salmon in April and May. If episodes occur continually on an annual basis at this crucial time in the salmon's life cycle then a stream may lose its juvenile salmon population during a period of four to five years; equivalent to the total mean length of their life cycle in Scotland.

As part of the Surface Waters Acidification Programme (SWAP) (Mason & Seip 1985) brown trout populations were measured at 13 sites over a three-year period (1986–88) and compared with a variety of water quality variables and habitats.

At one of the sites (Allt a'Mharcaidh) the salmon population dynamics were studied in relation to episodic acid events (*a*) to determine the hydrochemical conditions which influence egg survival, (*b*) to relate egg survival in natural and experimental conditions to the juvenile population in the stream.

Site description

The study sites are all located in Scotland and their identification numbers and general physical and chemical characteristics are described in detail by Harriman *et al.* (this symposium) and Ferrier & Harriman (this symposium). All the sites with mean pH < 5.0 are fishless but most of the experimental investigations were done in the Allt a'Mharcaidh (mean pH 6.5) which lies on the western edge of the Cairngorm range draining an area of approximately 10 km² into the River Feshie which is a tributary of the River Spey. Altitude ranges from 1111 m at the highest point to 250 m at the confluence with the River Feshie. Three fish species are present in the Allt a'Mharcaidh: Atlantic salmon, brown and sea trout and the brook lamprey *Lampetra planeri* (Bloch). Migratory fish have easy access from the confluence with the River Feshie with no physical obstructions as far as the stream flow gauging station (G1) upstream of which only brown trout are present (figure 1).

Methods and materials

(a) Egg survival studies

Two egg survival experiments were run at site 4 (Allt a'Mharcaidh), the first in 1985/86 and the second in 1986/87. The site for the experiments was upstream from electrofishing site D (figure 1) and chosen because it was an area of stream composed entirely of riffle which had previously been used by spawning salmon.

Eggs for both experiments were obtained from salmon caught in the River Avon, a tributary of the River Spey, which were fertilized and water hardened for 2 h before transportation to the experimental site. One hen salmon was used to supply eggs for the 1985/86 experiment and eggs from three hen salmon were used in 1986/87 and were arbitrarily labelled A, B and C for the purposes of the experiment.

In the 1985/86 experiment the trays (eight egg boxes/tray) were arranged in a 4 × 2 array with stream gravel in the upstream trays and limestone in the downstream trays. Gravel from the stream was used within the egg boxes.

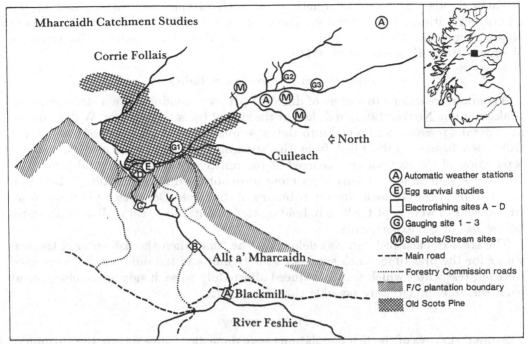

Figure 1. Location of fish sampling sites in relation to gauging and plot study sites in the Allt a'Mharcaidh. Location of Allt a'Mharcaidh is shown in inset.

The design was modified in the 1986/87 experiment to a double 3×3 latin square arrangement with limestone in both the boxes and the trays in the downstream square.

In both experiments the lifting of boxes was made at regular intervals during development and, in exceptional cases, immediately after an acid episode. At each sample lift, the dead and live eggs were counted in the field and in the laboratory two variables were measured:

(1) the ratio of the mass of the developing embryo to the total mass of the egg, less the mass of the outer membrane;

(2) the rate of embryo development as calculated using a 33-stage morphological score (Gorodilov 1983) so placing the embryos into one of 33 developmental categories by means of a series of diagnostic indices.

(b) Water chemistry/hydrology

At Allt a'Mharcaidh dip samples were taken twice weekly and average weekly stream chemistry samples were collected from a composite sampler. An autosampler was set to collect dip samples at pre-set regular intervals when a rain or snowmelt episode was expected. Stream level was recorded at 15 min intervals and converted to flow using a calibration equation.

During the 1986/87 experiment, two egg trays with modified egg boxes were used to allow the extraction of interstitial water (50 ml) for analysis.

At the other SWAP sites both dip and composite samples were taken on a weekly or bimonthly basis. Where continuous flow was measured, a relationship between mean pH and frequency of time below pH 5.5 was obtained. Comparable data for the rest of the sites were obtained using instantaneous flow/pH relationships.

The pH, alkalinity and conductivity of each sample were determined at the laboratory within 2 h of collection. The samples were filtered through a nucleopore 45 µm membrane filter and stored at 4 °C for subsequent analysis (Harriman & Morrison 1982; Harriman et al. 1990).

(c) Preference/avoidance studies

Avoidance reactions to waters of different pH were conducted in a steep-gradient tank (Gunn & Noakes 1986) in a shed at the side of the study stream. Water flowing at constant pressure was piped into the experimental tank at a rate of 0.46 l min⁻¹ from two header tanks filled from the stream. Dye tests confirmed that sharp separation of the test waters occurred at the centre of the experimental tank.

The 0+ salmon used in this experiment were obtained from the Allt a'Mharcaidh and also from the Girnock Burn, a tributary of the River Dee with a pH > 6.0 at all times. All fish were kept for 24 h in holding tanks containing Allt a'Mharcaidh water before use in the experiments.

The start of each test run was defined as the time when the fish entered the test water for the first time. Each run lasted for 10 min and ten different fish were used for each test water which was introduced alternately to each side of the chamber at each run to eliminate any possible chamber effects.

(d) Fish surveys

Annual surveys of the fish populations were done at all sites during the autumn of each year. In the Allt a'Mharcaidh, four sites with different habitat characteristics were fished every four to six weeks throughout the year. The fish populations were estimated by repeated electrofishing and removal of fish from within an area of stream (normally 100 m²) enclosed by stop-nets. This removal method allows an estimate of the population density to be calculated using a formula devised by Zippin (1958). Fork length and mass of the fish were recorded and a sample of scales for age determination was taken from each fish before its return to the stream.

Results

(a) Egg survival

Eggs from the boxes were grouped into three categories: live, dead, and infertile. The infertile eggs normally accounted for between 0% and 4% of the sample, although in one case in the second experiment, 16% of the eggs were infertile.

(i) 1985/86 experiment

In this preliminary experiment, the eggs within boxes which were packed with gravel for the duration of the experiment had a mean survival of 14% up to hatching. Eggs which were in boxes packed with limestone for the duration of the experiment had a mean survival of 56%. Survival in trays, from which limestone protection was removed after 11 weeks was only 1%, compared with 29% for those in which limestone protection was in place for 14 weeks.

(ii) 1986/87 experiment

No significant relationship (analysis of variance $P < 0.05$) was found between egg survival and the position of the boxes relative to the stream-bank or with the variations in stream flow either across or up and down the stream. There were no significant differences in the survival of eggs from the three parents within each substrate type. The only significant factor was the difference in mortality of the eggs between the gravel and limestone substrates.

Up to the time of hatch, mean survival of eggs in the limestone substrate was 90% compared to 63% for the eggs in the gravel. After hatching, the survival in the limestone dropped to 74% then to 54% in the subsequent two lifts. For the same sample lifts, the survival in the gravel substrates fell to 48% and 25% respectively. To investigate the effect of acid episodes a specific lift was made after an episode at the end of February 1987 during which pH fall to 5.88. There was no significant difference in mortality between this lift and one 5 d before the episode. A further episode at the end of March (min. pH 5.43), when eggs had hatched in both limestone and gravel, produced a significant decline in survival. After hatching, the next three lifts showed an increasing but similar decline in egg survival in both substrates. The survival at the point where swim-up would have occurred, had the alevins not been contained by the egg boxes, was 37% for the gravel and 54% for the limestone.

Comparison of the embryo/yolk ratios throughout the developmental period showed no significant differences in the ratios either between parents or between substrates. There were no significant differences in development within each batch of eggs/alevins or between the parents or the substrates.

(b) *Water chemistry/hydrology*

The flow data for the 1985/86 experiment showed that 21 episodes occurred, cumulatively lasting for approximately 13% of the experiment's duration; six of them occurred four weeks before hatching.

The frequency and duration of episodes during the second experiment were significantly lower than in the initial experiments. During the 1986/87 experiment nine episodes were recorded, five of these occurring in December 1986 at a time when the eggs were developing slowly in the water temperatures of less than 1.0 °C. Base flow conditions predominated during January and February with no episodes occurring. During March, April and May four episodes were recorded but for greater than 95% of the time the stream was at base flow. In the month before hatching, three episodes occurred, half the number of the previous year (table 1).

Concentrations of calcium in the surface-water showed a similar inverse relationship to flow as pH with a minimum of 26 μeq l^{-1} and a maximum of 48 μeq l^{-1}. Levels of labile aluminium in the stream water were consistently below 10 μg l^{-1} (mean 5 μg l^{-1}) in base flow conditions. The highest value measured during the period of the experiment was 40 μg l^{-1} on 13 April 1987 during the last recorded episode.

(c) *Interstitial water chemistry*

Analysis of the water from within the egg boxes containing granite gravel showed no significant chemical differences from that of the surface stream water. The pH and calcium concentration in the gravel showed a similar inverse relationship to flow as the stream water. There were, however, significant differences in pH and calcium concentrations between the interstitial water in the limestone-filled boxes and the granite gravel/stream water. The pH of the interstitial water in the limestone boxes was, on average, 0.2 pH units higher than that in the gravel boxes. Similarly, the mean calcium concentration in the limestone boxes was 57 μeq l^{-1} as opposed to 38 μeq l^{-1} for the gravel boxes; a difference of 33% (figure 2). Concentrations of labile

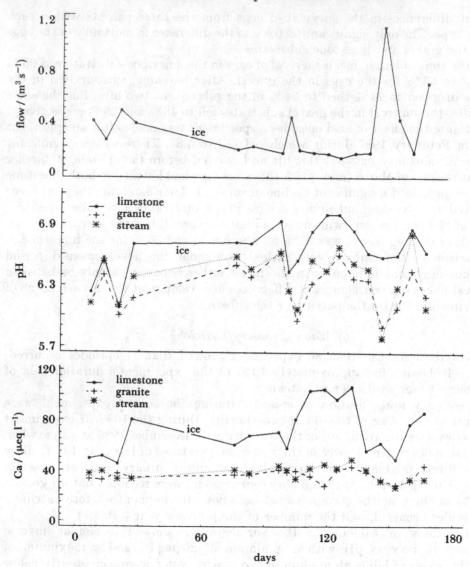

Figure 2. Temporal changes in stream flow, and pH and calcium in interstitial water extracted from boxes containing granite and limestone in Allt a'Mharcaidh.

Table 1. *Relationship between frequency and duration of episodes and survival of eggs up to the point of hatch*

	1985/86	1986/87
number of episodes, i.e. $Q > 0.7$ m^3 s^{-1}	21	9
duration of episodes (% of total duration of experiment)	13%	5%
number of episodes during month before hatching	6	3
survival to hatch: gravel	14	47
limestone	56	74

aluminium in the interstitial water of both the gravel- and limestone-packed boxes did not exceed 13 µg l^{-1}, with a mean of 3 µg l^{-1}.

(d) Preference/avoidance experiments

Salmon taken from both the Allt a'Mharcaidh and the chemically richer Girnock system exhibited strong avoidance reactions to test waters set to pH values of 5.5 and less. The value of pH 5.5 is clearly defined as the point at which avoidance reactions begin for the 0+ salmon used in this experiment. At pH 5.8 and above, fish exhibited no preference for either the test or the control water, and movements were random between the two sides of the test chamber (figure 3).

The two fish stocks differed significantly in the number of entries made into the test water. At pH > 5.5, fish from the Girnock on average showed twice the number of movements between each side of the chamber compared with the Allt a'Mharcaidh fish; 9–18 and 5–9 respectively. At pH < 5.5 the number of entries dropped to 5–8 for Girnock fish and 3–5 for Allt a'Mharcaidh fish.

Allt a'Mharcaidh fish were also exposed to different levels of the toxic labile fraction of aluminium at a similar pH, then to a combination of raised labile aluminium levels and lower pH. For runs 1 and 2, which simulated the range of labile aluminium levels present in the Allt a'Mharcaidh under most flow conditions, no avoidance reactions were observed. Increasing the labile aluminium concentration to 55 µg l^{-1} at a pH of 5.36 in run 3 produced a stronger avoidance reaction than that at pH 5.36 alone (see figure 3 for individual responses to aluminium).

(e) Fish densities

(i) Allt a'Mharcaidh

Trout and salmon fry were present at all four sites; the trout fry emerging from the redds up to four weeks earlier than the salmon.

Differences in densities between sections reflected the availability of suitable habitat for the salmon and trout. The mean 0+ salmon density for all sites was 57 per 100 m². The maximum densities of 0+ salmon found at site D varied from 105 per 100 m² in 1988 to 154 per 100 m² in 1987. The between year variations in maximum 0+ salmon densities are not significantly different from each other. Densities did not reach a maximum until at least four weeks after emergence from the gravel, which may reflect migration into site D from other sections of the stream which provided a less suitable habitat for salmon. In 1987, the first survey at site D to catch 0+ salmon was carried out on 22 June. However, the maximum density was not reached until 24 August; densities declined to winter levels by 21 December. From the time of peak densities (154 per 100 m²) at the end of August to the winter density of 39 per 100 m² on 21 December there were only four acid episodes ($Q >$ 0.7 m³ s^{-1}) having a total duration of 30 h. In 1988 from the period of peak 0+ densities in June to the autumn decline there were nine episodes of the same magnitude as 1987 having a total duration of 83 h. On 18 August 1988 the estimated density of 0+ salmon was only 7 per 100 m² compared with the density of 154 per 100 m² the previous year at the same time (figure 4).

(ii) Other sites

At the other SWAP sites only one fishing per year was undertaken in the same section during the study period. From the studies in Allt a'Mharcaidh there was evidence of a direct response of fish to frequency and duration of episodes of pH <

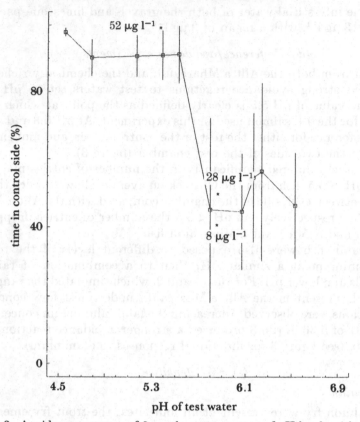

Figure 3. Avoidance response of 0 + salmon to a range of pH levels with and without added aluminium.

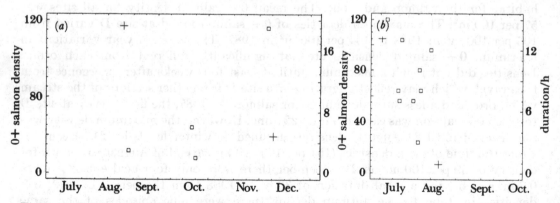

Figure 4. Relationship between frequency and duration of acid episodes (pH < 5.5) during the period between maximum and minimum salmon densities during (*a*) 1987 and (*b*) 1988 in the Allt a'Mharcaidh, where + denotes fish density; □ denotes acid episode. Density is the number per 100 m².

5.5. For convenience it would be useful to be able to convert an empirical mean pH/fish density relationship into one using frequency of episodes below a certain pH. The relationship between mean pH and frequency of pH below 5.5 is significant for waters with mean pH between 4 and 6 (figure 5*a*). For streams with mean pH > 6.0

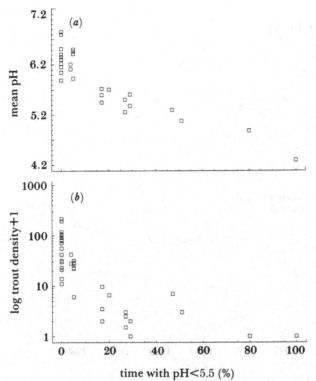

Figure 5. Relationship between (*a*) mean pH and frequency of acid episodes (pH < 5.5) and (*b*) log trout density (number per 100 m²) and frequency of acid episodes for all SWAP sites.

the estimated frequency of episodes of pH < 5.5 was less than 5% of the total cumulative time compared with streams with a mean pH between 5.5 and 6.0 which have a frequency between 5 and 30%. The critical point for trout viability is reached when the frequency of episodes of pH < 5.5 exceeds 30–40%; the population then falls almost to zero (figure 5*b*).

To assess the possible effects of reversibility of stream acidification on trout densities, we investigated the long-term changes in trout densities at two adjacent streams with a common access and mean pH in the 4.8–5.6 range (site 6) and 5.5–6.2 range (site 7). During the past decade, inputs of sulphate to these acidified catchments have declined with a similar reduction in stream sulphate levels (Harriman *et al.* 1990).

A plot of mean pH against densities for both streams (figure 6) indicates that mean pH is now improving to the extent that mean pH at site 6 is approaching the critical 5.5 level. Densities at site 6 have been very low during the past decade but the significantly higher density value in 1989 was twice that recorded in any previous year. This may be the first indication of a population response to an improving pH régime in the stream.

Mean total monomeric aluminium levels were 73 and 43 µg l⁻¹ respectively for sites 6 and 7 of which 15 and 8 µg l⁻¹ were in the labile form.

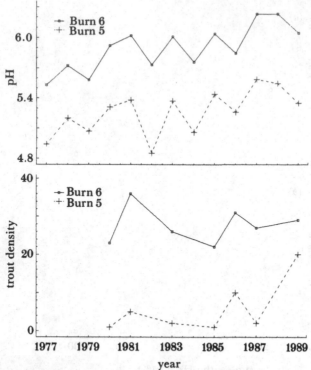

Figure 6. Long-term trends in pH and trout densities (number per 100 m²) in Burn 5 (Site 6) and Burn 6 (Site 7).

Discussion and conclusions

The studies of juvenile stages of Atlantic salmon in the acid-sensitive Allt a'Mharcaidh have identified an unusual pattern of survival and productivity during early life stages. The results indicate a significant egg mortality and low densities of $0+$ fry but with an ultimate production of $1+$ and $2+$ parr similar to that in non acid-stressed streams due perhaps to reduced sensitivity of later life stages to acid episodes and/or immigration of these older juveniles from outside the Mharcaidh system.

The expected egg survival to hatching stage in natural redds in streams having a pH > 6.0 at all flows is greater than 80% (Symons 1979), while that in egg containers planted into streams is 85–91% (Shearer 1961). The survival to hatch in egg boxes packed with local gravel in these experiments ranged from 14–47%, which is significantly lower than the survival values quoted above. For limestone substrate the range was 56–74% thus demonstrating the ameliorating effect of the 30% increase in calcium.

Brown (1983) showed that, in water of pH 5.4 and with total aluminium between 25 and 50 µg l^{-1}, there were significant brown trout alevin mortalities when calcium levels were less than 50 µeq l^{-1}. Mean calcium concentrations in egg boxes containing gravel during the current egg experiments were always less than 50 µeq l^{-1}.

During the period of the egg experiments, the average level of labile aluminium in the interstitial water was 3 µg l^{-1} with a maximum of 13 µg l^{-1} during episodes. Early studies by Baker & Schofield (1982) showed that at low pH values (pH < 4.6)

aluminium concentrations of 500 µg l^{-1} increased survival at the egg stage, but the same value was highly toxic at the alevin and fry stage. Studies by Muniz (1981) showed that 69 µg l^{-1} of total aluminium in water having a pH of 5.2 and calcium 15–30 µeq l^{-1}, produced 100 % mortality of alevins after 25 d treatment. However, studies in streams in England and Wales have shown that brown trout growth is affected by 20 µg l^{-1} of labile aluminium while streams with mean levels greater than 40 µg l^{-1} were fishless (Sadler & Turnpenny 1986). Lacroix (1987) showed that a level of labile aluminium in the range 10–35 µg l^{-1} was not the major limiting factor for Atlantic salmon survival in Nova Scotia streams. For the studies cited, a level of labile aluminium permanently higher than 40 µg l^{-1} appears to limit salmon and trout survival but if this level is only occasionally reached during episodes the effect on survival would appear to be minimal. The Allt a'Mharcaidh falls into the latter category indicating that aluminium is unlikely to be a significant influence upon egg and fry survival.

The maximum densities of 0 + fry in the study stream were less than those found in non acid-stressed streams with similar habitats such as the Shelligan Burn (around 700 per 100 m^2, Egglishaw & Shackley 1977) and a variety of streams in the North East Grampians (maximum densities around 250 per 100 m^2, W. R. Gardiner (personal communication). Differences in 0 + fry densities between the Allt a'Mharcaidh and the Shelligan would not appear to be due to habitat differences as both streams have similar physical and hydrological characteristics.

In those non-acid streams where egg deposition is not limiting, the decline in the fry densities throughout the summer is due primarily to density-dependent mortalities. Here the density at the end of the first year's growth is typically around 100 per 100 m^2, which is comparable to the maximum summer densities in the Allt a'Mharcaidh. Consequently, the decline in 0 + densities in the study stream throughout the growing season is probably not due to density-dependent factors.

The variation in 0 + fry density at the end of the growth season (between 8 and 39 per 100 m^2) appears to be influenced by the frequency and duration of acid episodes which in turn is dependent upon the frequency and duration of rain and snow events. The decline in the 0 + density to winter levels occurred between August and December in 1987 but in 1988 between June and August. Thus it is probable that the duration and also the temporal spacing of each episode is significant to the displacement or mortality of 0 + fry. Henriksen *et al.* (1984) showed that the first acid episode in a series may induce sub-lethal 'physiological stress' from which the fish may take several weeks to recover. Further episodes soon after the first may then cause mortalities. During the period of 0 + density decline in 1988, all of the episodes occurred within one week of each other, which may have given the fish little time to recover physiologically or to move back to their original territories if they had been displaced downstream.

Densities of 1 + parr in the Allt a'Mharcaidh are lower than those found in chemically richer streams with similar habitat but the difference is not as great as with the 0 + fry. The greatest density of 1 + parr found in the Allt a'Mharcaidh is similar to the lower range found in the Shelligan Burn (11 per 100 m^2) while the 2 + parr densities are both similar to that found in the Shelligan and close to the typical value of 5 per 100 m^2 reported by Symons (1979). It is probable that the ultimate densities of 2 + parr in the study stream are a function of habitat suitability rather than the direct effect of acidity.

Exposure to 0 + fry to different levels of pH in the test chamber has demonstrated that pH values of less than 5.5 produce a strong avoidance response. This ability by early life stages of salmonids to detect and avoid water having a low acidity is documented in several other studies (Gunn & Noakes 1986) but whether the 0+ fry die or move out of the system has yet to be established.

When the Allt a'Mharcaidh density data are fitted into the density range of the other SWAP sites its transitional nature can be appreciated (figure 5). All the sites with mean pH between 5.0 and 5.5 have very low trout densities with a population composed entirely of older fish, 0+ fry being absent. These adult trout are probably opportunistic, taking advantage of higher pH conditions and their greater resistance to lower pH episodes. Sites with a mean pH > 5.5 have a normal distribution of all the year classes including a good fry population. Where long-term pH increases have been identified (figure 6), the resulting change from a system with mean pH < 5.5 to one with mean pH > 5.5 should be accompanied by an increase in salmonid populations and the establishment of a juvenile stock.

The relationship between mean pH and frequency of episodes of pH < 5.5 will obviously vary depending on catchment size and soil/hydrological characteristics. In general, the range of pH around the mean value will tend to be smaller as catchment size increases.

Thus Lacroix (1989) showed a small pH range for two Nova Scotia rivers with mean pH of 4.9 (4.7–5.4) and 5.9 (5.6–6.3) with catchment areas of 142 and 193 km² respectively (i.e. 10–100 times that of the typical SWAP sites). He also concluded that the smolt production level of less than 0.8 per 100 m² in the more acid River Westfield was below the level where maintenance of salmon stocks could continue.

In general, the difference between mean pH and minimum pH is likely to increase as catchment size decreases. This is reflected at the 10 km² Allt a'Mharcaidh site where the mean pH (6.5) exceeds minimum pH (5.0) by 1.5 pH units, whereas the North River (Lacroix 1989) only exhibits a 0.3 pH difference (mean pH 5.9 down to 5.6).

We cannot say unequivocally whether salmon die during episodes or whether they respond by migrating out of the Allt a'Mharcaidh into the less acid environment of the River Feshie. However, Lacroix (1989) implied that mortality rather than fish movement was the main reason for low 0+ salmon densities in the Westfield river.

The authors thank colleagues at the Freshwater Fisheries Laboratory for their help and advice; in particular A. Watt, A. E. G. Christie, E. Taylor and D. King. We thank R. Henderson and S. R. Bonellie of Napier Polytechnic, Edinburgh for their help with data analysis and also J. Porter for collecting many of the water samples. The financial support for the Surface Waters Acidification Programme is gratefully acknowledged.

References

Baker, J. P. & Schofield, C. L. 1982 *Wat. Air Soil Pollut.* **18**, 289–309.
Beamish, R. J. & Harvey, H. H. 1972 *J. Fish. Res. Bd. Can.* **29**, 1131–1143.
Brown, D. J. A. 1983 *Bull. Environ. Contam. Toxicol.* **30**, 582–587.
Daye, P. G. & Garside, E. T. 1977 *Can. J. Zool.* **55**, 1504–1508.
Egglishaw, H. J. & Shackley, P. E. 1977 *J. Fish. Biol.* **11**, 647–672.
Gorodilov, Yu. N. 1983 Salmon, *Salmo salar* L. II. Scottish Fish. Res. Translation.
Gunn, J. M. 1986 *Environ. Biol. Fish.* **17**, 241–252.
Gunn, J. M. & Noakes, D. L. G. 1986 *Wat. Air Soil Pollut.* **30**, 497–503.
Gunn, J. M., Noakes, D. L. G. & Westlake, G. F. 1987 *Can. J. Zool.* **65**, 2786–2792.

Harriman, R., Morrison, B. R. S., Caines, L. A., Collen, P. & Watt, A. W. 1987 *Wat. Air Soil Pollut.* **32**, 89–112.

Harriman, R. & Morrison, B. R. S. 1982 *Hydrobiologia* **14**, 251–263.

Harriman, R., Gillespie, E., King, D., Watt, A. W., Christie, A. E. G., Cowan, A. A. & Edwards, T. 1990 *J. Hydrology* (In the press.)

Henriksen, A., Skogheim, O. K. & Rosseland, B. 1984 *Vatten* **40**, 255–260.

Jenson, K. W. & Snekvik, E. 1972 *Ambio* **1**, 223–225.

Johanessen, M. & Henriksen, A. 1978 *Wat. Res.* **14**, 615–619.

Lacroix, G. L. 1985 *Can. J. Fish. Aquat. Sci.* **42**, 292–299.

Lacroix, G. L. 1987 *Acid rain: scientific and technical advances* (ed. R. Perry, R. M. Harrison, J. N. B. Bell & J. N. Lester), pp. 516–521. London: Selper Ltd.

Lacroix, G. L. 1989 *Can. J. Fish. Aquat. Sci.* **46**, 2003–2018.

Leivestad, H. & Muniz, I. P. 1976 *Nature, Lond.* **259**, 391–392.

Mason, J. & Seip, H. M. 1985 *Ambio* **14**, 45–51.

Muniz, I. P. 1981 International Atlantic Salmon Foundation Special Pub. 10.

Peterson, R. H., Daye, P. G. & Metcalfe, J. L. 1980 *Can. J. Fish. Aquat. Sci.* **37**, 770–774.

Sadler, K. & Turnpenny, A. W. H. 1986 *Wat. Air Soil Pollut.* **30**, 593–599.

Schofield, C. L. 1976 *Ambio* **5**, 228–230.

Shearer, W. M. 1961 ICES/CM No. 98, 3 pp.

Symons, P. E. K. 1979 *J. Fish. Res. Bd Can.* **36**, 132–140.

Zippin, C. 1958 *J. Wildl. Mgmt* **22**, 82–90.

Discussion

B. Salbu (*Isotope Laboratories, Agricultural University of Norway, 1432-As-NLH, Norway*). In Dr Harriman's experiment, the distribution of Al species varies according to pH. In addition, temperature will influence the distribution. Unless the water is fractionated, the exposure cannot be defined, i.e. the source term. Thus, biotest experiments looking for bioavailability of elements require a high degree of definition on both sides of the membrane.

R. Harriman. The possible effect of temperature and size distribution was recognized during the biotest experiment. Samples were filtered *in situ* at the experimental site and the ion exchange separation was immediately completed at the water temperature of the stream. When the results of this technique were compared with ion exchange procedures in the laboratory the values were within 15 % of each other. It should also be emphasized that the measured levels of labile monomeric aluminium were on average around 15 μg l^{-1} and during short-term acid episodes only reached a maximum of 30–40 μg l^{-1}.

Even allowing for larger errors in the estimate of labile aluminium, these values are below the critical levels which have been reported to reduce the survival of juvenile salmonid species.

The major influence on survival was found to be pH and low levels of calcium.

The effects of controlled chemical episodes on the survival, sodium balance and respiration of brown trout, *Salmo trutta* L.

By R. Morris and J. P. Reader

Department of Zoology, University of Nottingham, Nottingham NG7 2RD, U.K.

Survival and recovery: yolk-sac fry, swim-up fry and 1–2 year juveniles of brown trout were exposed for up to 78 h to episodes of 12 μmol l^{-1} aluminium and pH 4.5 in an artificial soft-water medium ([Ca] 20 μmol l^{-1}; pH 5.6). Yolk-sac fry mortalities were negligible, even when tested at pH 4.0, although disturbances of sodium and potassium balance were evident at pH 4.5. A marked increase in susceptibility, with high mortalities, occurred when the yolk was fully absorbed. Mortalities continued for some time after the episode had finished. Juvenile mortalities were related to the length of the plateau phase (12 h gave 10% deaths; 30 h, 50% and 60 h, 75%). Exposure to two successive episodes gave lower mortalities than might have been expected from exposure to a single episode of equivalent length, and mortality declined as the interval between the two episodes increased. Surviving juveniles showed significant lowering of plasma sodium and chloride and increased haematocrit following an episode with a 30 h plateau. There was evidence of recovery of blood parameters towards normal values in this group five days after the episode had finished, although the gills showed signs of damage.

Sodium balance: juvenile brown trout were exposed to sub-lethal episodes in soft-water artificial media ([Ca] 20 μmol l^{-1}, pH 5.6). Sodium influx, outflux and net flux were measured during a series of short episodes consisting of acid alone (minimum pH 4.5), aluminium alone (pH 5.6 and 5.7 μmol l^{-1} Al; pH 5.4 and 28.6 μmol l^{-1} Al) and combined acid and aluminium (pH 4.5 and 2, 4.5, 11.5 or 15.6 μmol l^{-1} Al). Sodium balance showed similar characteristics regardless of type or length of episode. Sodium outflux increased during the rising phase of episodes and recovered to normal values by the end of the falling phase. The initial outflux increased with episodes of higher aluminium concentration. Sodium influx decreased during the early phases of episodes and partially recovered during the falling phase. Influx inhibition increased with increased aluminium concentrations. Calculations from net flux values and from terminal plasma sodium measurements show that sodium loss is low during acid episodes, but increases with the maximum total aluminium level attained during combined episodes. At similar aluminium concentrations, fish at pH 5.6 lose more sodium than at pH 4.5, emphasizing the importance of the ion balance effect of aluminium at the normal pH of many soft natural waters.

Respiration: fish showed no significant changes in oxygen consumption during most of the short episodes tested.

1. Introduction

The lethal effects of aluminium and low pH on salmonids in soft-waters have been demonstrated by both field (Schofield & Trojnar 1980) and experimental studies (Muniz & Leivestad 1980; Schofield & Trojnar 1980; Baker & Schofield 1982; Brown 1983). Most of the experimental studies have concentrated on the abrupt and continuous immersion of fish in waters of low pH and varying aluminium concentrations in attempts to define lethal conditions and the chain of physiological circumstances that can result in death. In some cases, the experiments have been conducted by using natural waters that may contain traces of toxic metals other than aluminium, and the water composition may even vary from one treatment to another. Fish-kills in soft acid waters occur during, or for a length of time after, episodic changes in water quality. They are typified by a fall in pH with an associated increase in concentrations of trace metals, e.g. aluminium, and a change in the concentration of major ions such as calcium, magnesium, sodium and potassium (Reader & Dempsey 1989). These sporadic fish-kills may have an important effect on declining fish populations.

Continuous exposure experiments are relevant to long-term episodes in the field but there has been no systematic study of the effects of short-term changes of water quality on the survival and physiology of fish under carefully controlled laboratory conditions. Our aim has been to design apparatus which can reproduce repeatable episodes, similar to those occurring naturally, with the capability of altering the episodic profile (Morris *et al.* (this symposium)). As the composition of natural waters and tap waters vary both with region and time, artificial media have been designed so that the ionic composition of the medium can be matched to simulate that of waters where episodes occur.

Two main physiological effects have been identified that are caused by the combined action of pH and aluminium. The first of these is respiratory difficulties (increased opercular movements and cough rates, excessive gill mucus production and hypoxia) that have been observed at aluminium levels at or above $2.8 \ \mu mol \ l^{-1}$ ($75 \ \mu g \ l^{-1}$) over a wide pH range (6.1 to 4.5) (Muniz & Leivestad 1980; Neville 1985). The second effect has been observed at aluminium levels above $2 \ \mu mol \ l^{-1}$ coupled with low pH (< 5.0). This combination can give rise to disturbances of ion balance, resulting in net loss of sodium and chloride from fish (Rosseland & Skogheim 1984; Neville 1985; Dalziel *et al.* 1986, 1987; Booth *et al.* 1988). Death may eventually take place as a result of the physiological consequences, which in the case of acid alone, are known to involve reduced blood volume and probably circulatory failure (Milligan & Wood 1982). Neville (1985) concluded that, in rainbow trout exposed to aluminium ($2.8 \ \mu mol \ l^{-1}$), the primary causes of death were ionoregulatory failure at low pH (< 5.0) and hypoxia at high pH (> 5.5).

This paper deals with two main aspects of the problems of fish deaths caused by acid and aluminium episodes. The first section deals with the survival and recovery of brown trout fry and juveniles during and after episodes, whereas the second describes the physiological effects of similar but sub-lethal episodes on the sodium balance and respiration of juvenile brown trout.

2. Materials and methods

Brown trout, *Salmo trutta* L., were obtained as 1–2 year juveniles (wet mass 26–76 g) or eyed ova. The fish were exposed to episodes of aluminium and low pH while in an aerated, soft-water artificial medium of analytical grade salts in deionized water. Survival studies on juveniles were performed at 10 °C in the apparatus designed for this purpose by Dempsey (1987), whereas the apparatus designed by Morris *et al.* (this symposium) was used for survival studies on fry (5 °C) and physiological experiments on juveniles (10 °C). The ionic constituents of the medium were chosen to simulate conditions recorded from acid waters with affected fish populations. Low-intensity lighting was continuous to allow samples to be taken without disturbing the fish unduly. Nominal composition of the medium was as follows: $CaCl_2$ 20, $MgCl_2$ 20, KNO_3 5 and NaCl 50 (μmol l^{-1}). Animals to be exposed to aluminium and low pH episodes were acclimated to the medium and the apparatus for 48 h before the onset of episodes, and recovery after episodes was under the same conditions as the acclimation. The pH of the medium was 5.6 without animals and rose to 5.7–6.0 under the influence of ammonia production by the animals in the absence of pH control. During episodes, aluminium was added as a solution of $AlCl_3 \cdot 6H_2O$ and pH was lowered by addition of diluted analytical grade H_2SO_4.

Episodes in survival studies consisted of a 6 h rising phase, a plateau or constant phase of variable duration and a 12 h falling phase. During physiological studies, the rising and falling phase was decreased by 90 minutes to allow time for sodium flux and respiratory measurements to be made. Flow was stopped for 30 minutes and influx calculated from the rate of uptake of added ^{22}Na, whereas net flux was determined from the change of sodium concentration during the same period. Outflux was obtained by difference. Flow was then restarted for 30 minutes to reduce isotopic levels, then stopped again to measure oxygen levels at the beginning and end of 30 minutes to enable oxygen consumption to be calculated. During the rising phase, the aim was to provide linear increase in $[H^+]$, from pH 5.6 to pH 4.5 accompanied by a simultaneous linear increase in total [Al] from zero to the desired plateau concentration (0–15 μmol l^{-1}). The aim during the plateau phase was to hold $[H^+]$ and total [Al] constant, and during the falling phase to provide simultaneous linear decrease in concentrations, returning to baseline conditions (pH 5.6 and zero [Al]). During the experiments, pH control was relatively accurate, being maintained at 4.48–4.55 (or better) at target pH 4.5 during an acid episode. Total aluminium concentrations in samples of medium were measured by the catechol violet method of Dougan & Wilson (1974), and monomeric aluminium by the same method, as modified by Seip *et al.* (1984).

3. Results and discussion

(a) *Survival and recovery studies on fry and juvenile fish*

During the following experiments, a standard peak or plateau concentration of 12 μmol l^{-1} Al was employed, usually accompanied by a plateau of pH 4.5.

(i) *Fry*

Early yolk-sac fry (spawned 1988–90 degree-days post-hatch) and later yolk-sac fry from the same spawning (425 degree-days post-hatch) experienced no mortalities in episodes of 6 h, 18 h or 30 h plateau duration at pH 4.5 or after a 6 h plateau of

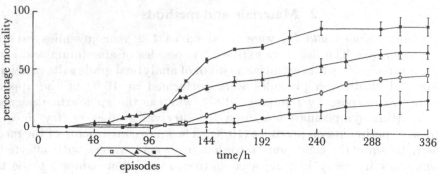

Figure 1. Cumulative mortality curves for brown trout swim-up fry subjected to episodes of varying plateau length. Episodic plateau values were pH 4.5 and 12 μmol l⁻¹ aluminium.

pH 4.0 at 5 °C. There were no mortalities during the recovery period of 120 h for earlier fry, and only three during the later series. Controls showed no mortalities ($n = 60$). Late yolk-sac fry (480 degree-days post-hatch) suffered no mortalities during a 30 h episode of pH 4.0 during the episode or a 120 h recovery period. Whole body digests of early and late yolk-sac fry samples in the recovery period, 168 h after the beginning of 18 h or 30 h plateau episodes showed significantly lower sodium and potassium levels relative to control fry, although there were no deaths after these episodes.

By contrast, mortalities were high among swim-up fry (700 degree-days post-hatch) during and after 18 h or 30 h plateau episodes (figure 1). Presence or absence of yolk appears to be the important factor in determining survival. A batch of fry at 425 degree-days in 1989 suffered 80–85 % mortalities during and after a 30 h plateau episode at pH 4.5. Whereas a similar group of 425 degree-day fry subjected to a similar episode in 1988 showed very few mortalities (see above). The susceptible group were reared at a somewhat higher temperature, were smaller, and had no evidence of yolk remaining and their survival was less than that of 700 degree-day fry subjected to the same type of episode in 1988.

(ii) Juveniles

One–two year juveniles, like swim-up fry, showed mortalities related to the plateau length of the episode to which they were subjected (figures 1 and 2). All juvenile deaths took place during the episode, whereas fry deaths continued during the post-episodic period. Plasma sodium and chloride concentrations of surviving juveniles sampled 24 h or 120 h after an episode of 30 h plateau duration were lower than those of controls sampled at the same time and the percentage haematocrit was higher in treated animals. Gradual post-episodic recovery of ion balance took place, evidenced by the fact that these parameters were closer to control values after 120 h of recovery than after 24 h. This is rather surprising because the gills are known to be the chief source of ion loss and uptake yet scanning electron microscope studies of the gills from animals subjected to a shorter episode (12 h plateau) but allowed to recover for 120 h, still showed signs of gill damage. This consisted of clubbing and damage to secondary lamellae so that the number of lamellae were reduced and debris was found in their place, in the few animals examined.

Juveniles subjected to two successive episodes at pH 4.5 and 10 μmol l⁻¹ Al, each of 21 h plateau duration, experienced mortalities comparable to those resulting from

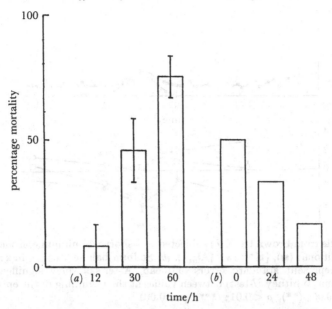

Figure 2. Mortality results for juvenile brown trout; (*a*) single episodes of varying plateau length; (*b*) double episodes, each of 21 h plateau length. Numbers give interval of time (hours) between episodes. Episodic plateau values were pH 4.5 and 12 μmol l⁻¹ aluminium.

a single episode of 30 h plateau, if the rising phase of the second episode commenced immediately after the falling phase of the first. Mortalities were lower as the interval between the two episodes increased (figure 2). This implies that fish must show some adaptation to the first episode and that the period of recovery between episodes also plays an important part in determining survival. In all cases of mortalities among juveniles, death was preceded by a period of loss of equilibrium, loss of colour control, and irritability. No animal surviving an episode was seen to behave in this manner.

(*b*) *Physiological responses in juvenile animals during episodes*

The experiments were done on groups of six animals at a time and were repeated or triplicated in most cases. They were made between the months of October–March to avoid seasonal variability (Stuart & Morris 1985). Two main groups of experiments have been done. These consist of short episodes (4.5 h rising phase, 7.5 h plateau, 10.5 h falling phase), which are reported in more detail, and a comparable series of longer episodes (4.5 h, 13.5 plateau, 10.5 h), which also received additional pretreatment. Occasional animals showed heavy negative sodium balance before the beginning of the episodes and the results from these animals have not been included in the present analyses as they are atypical and may have resulted from undetected skin damage or other unknown factors.

Some of the results from a series of experiments are summarized in figures 3 and 4. The series included the effects of the control situation at pH 5.6; an acid episode reaching a plateau maximum of 31.6 μmol H⁺ l⁻¹ (pH 4.5); an aluminium episode at pH 5.6 (2.5 μmol H⁺ l⁻¹) planned to give a nominal plateau of 10 μmol Al l⁻¹ (figure 3) and finally a combined episode of acid and aluminium intended to give plateau values of pH 4.5 and 10 μmol Al l⁻¹ (figure 4).

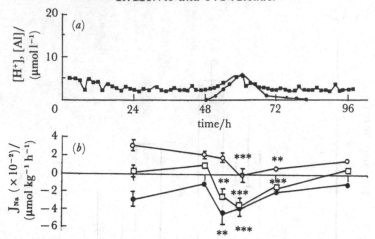

Figure 3. Mean results from brown trout (*b*) subjected to a short, aluminium episode at pH 5.6. (*a*) Environmental conditions: (■), [H⁺]; (●), [Al$_{total}$]. (*b*) Sodium balance: (○), influx; (□), net flux; (●), outflux. Bars represent standard errors of mean. Asterisks show significant differences (Student *t*-test or Mann–Whitney *U*-test) between values at the beginning of the episode and those so marked; (*), $p < 0.05$; (**), $p < 0.01$; (***), $p < 0.001$.

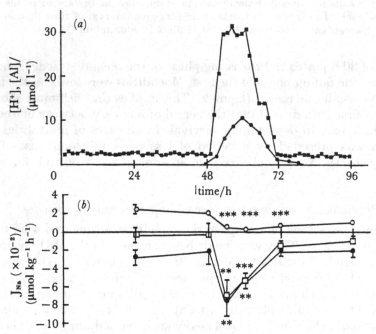

Figure 4. Mean results from brown trout (*b*) subjected to a short, combined acid (pH 4.5) and aluminium episode. Symbols and layout as in figure 3.

(i) *Environmental conditions*

The environmental conditions illustrated in figures 3*a* and 4*a* are the mean results from all six fish chambers taken during one experiment. The pH readings need little comment since they reflect the target values required quite accurately.

The effect of pH on the solubility and speciation of aluminium can be seen by referring to figures 3*a* and 4*a*. The rate of change of aluminium concentration is slower than that required by the target values during both rising and falling phases of the episode and reaches a peak rather than a plateau during short episodes of this type. The reason for this probably lies in the relative insolubility of aluminium at higher pH values such as those shown at the beginning and end of the pH 4.5 episode (figure 4a). At peak values, total [Al] amounted to 11.36 ± 0.32 µmol l^{-1} ($n = 12$) at pH 4.5 and corresponding monomeric values were 10.55 ± 0.11 µmol l^{-1} ($n = 12$) i.e. 92.8% of the aluminium was present in the form of soluble species.

During experiments done at pH 5.6 (figure 3a), the total [Al] peaked at 5.76 ± 0.42 µmol l^{-1} ($n = 12$) and monomeric [Al] was 3.87 ± 0.51 µmol l^{-1} ($n = 12$) so that the intended 10 µmol l^{-1} was never attained and the proportion of soluble monomeric species decreased to 67.3%. Thus during these experiments large quantities of insoluble aluminium must have been precipitated within the apparatus and the fish were subjected to a lower concentration than was intended.

(ii) *The effect on sodium balance*

Sodium balance effects from aluminium episodes are summarized in figures 3b and 4b. Results are the means of duplicated experiments (12 animals) unless otherwise stated.

The pattern of response of sodium balance reported here is similar in all the sub-lethal episodes tested. Fish respond by decreasing sodium influx and increasing outflux during the initial stages of the episode and in most cases this coincides with the rising phase of the episode. There is then a gradual recovery of outflux so that it may attain its original value by the end of the episode. Influx also recovers, usually during the falling phase of the episode, though at high aluminium levels it may not attain its original level, even during the recovery period.

Although flux values return to normal levels by the end of episodes, net flux values indicate that episodes cause a net loss of body sodium and it is possible to assess this by calculating the percentage change in total body sodium. Measurements of total body sodium from a similar group of untreated animals gave values of 37.34 ± 1.72 mmol Na kg^{-1} wet mass ($n = 10$) in December. This figure has been used in all subsequent calculations as characteristic of winter animals. The assessment involves calculating the area enclosed by the net flux curve during the episode and recovery period. It amounts to a calculated loss of -2.46% of the total body sodium as a result of acid exposure alone in these experiments. The relative contributions of influx and outflux to net flux can also be calculated by the same method.

The effects of a series of different aluminium concentrations, designed to give nominal peak concentrations of 0, 2, 5, 10 and 15 µmol l^{-1} of total [Al] on sodium balance are illustrated in figure 5. Here, the calculated percentage change in body sodium 24 h after the end of the episode is plotted against measured values of total [Al] after a pH 4.5 episode. The change can be seen from values derived from net flux measurements and the relative contributions of influx and outflux to these values are also shown. The figure shows that episodes of increasing aluminium concentration cause increased loss of body sodium and that these effects are still apparent after a recovery period of 24 h. Increased outflux is responsible for part of the sodium loss and is twice the level at 15 µmol l^{-1} of total [Al] than that of a pH 4.5 episode alone. Inhibition of sodium influx also influences sodium loss and its contribution is reduced to half of that shown during episodes involving lower aluminium concentrations (below 4.4 µmol Al l^{-1}).

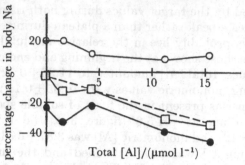

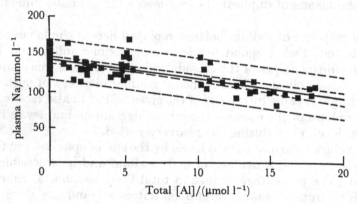

Figure 5. Calculated values of the mean percentage change in body sodium of brown trout 24 h after short, acid episodes (pH 4.5) involving different concentrations of aluminium; (□), mean values calculated from net flux values and thus represent the terminal state of sodium balance for the group of fish; (----), the same variable, but calculated from terminal plasma-sodium regression against aluminium for the same group of animals; (○), the calculated contribution of influx to sodium balance and (●) the outflux contribution. (For further explanation see text.)

Figure 6. The regression of brown trout plasma sodium on measured environmental aluminium 24 h after short, pH 4.5 episodes containing different amounts of total aluminium. Inner dotted lines show standard errors of the mean, outer dotted lines 95 % confidence limits. The regression equation is given in the text.

The results of plasma analyses done on blood samples obtained at the end of these experiments ($n = 68$) are recorded in figure 6. This shows the highly significant inverse linear relation ($p > 0.001$) between total [Al] and plasma sodium [Na_p], the regression equation is:

$$[Na_p] = -2.59 \, [Al_{total}] + 139.69.$$

Haematocrit values indicate a reduction in blood volume of ca. 10 % when animals are subjected to [Al_{total}] above 5 µmol l^{-1} so that there is little change in blood volume and probably in sodium space during these relatively short experiments. In this case, the plasma-sodium values may be said to be representative of total body sodium and it is possible to assess the percentage change of body sodium by comparing experimental and control values. The values obtained after arcsin conversion give a highly significant linear regression ($p > 0.001$) in which:

$$[\text{percentage Na}_{content}] = -1.88 \, [Al_{total}] + 95.8.$$

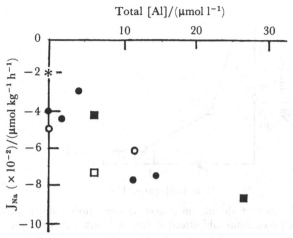

Figure 7. Maximum sodium outflux values obtained from brown trout as a result of different aluminium concentrations employed during episodes; (*), control values; (●), short, pH 4.5 episodes; (○), longer, pH 4.5 episodes; (■), short, pH 5.6 episodes; (□), longer, pH 5.6 episodes.

Values for percentage loss of body sodium calculated from this equation are plotted on figure 5 and show a reasonable correspondence with the values obtained by net flux calculations.

Sodium outflux. The behaviour of sodium outflux is reminiscent of shock or handling stress (Eddy & Bath 1979). It is brought about by acid alone, by aluminium alone, (figure 3b) or a combination of both (figure 4b), in which case the magnitude of the effect is dependent upon the concentration of aluminium (figure 7). Potts & McWilliams (1989) have calculated that the changes of trans-epithelial gill potentials (TEP) that take place when fish are adapted to acid conditions, would increase sodium outflux 1.7 times. However, outflux values exceed this considerably when control values are compared with higher concentrations of aluminium (figure 7). The highest outflux values we recorded were at low hydrogen ion concentration (pH 5.6 and 28 μmol total [Al] l^{-1}, figure 7), so that if the effects are direct effects on membrane permeability, aluminium itself must have a profound effect, perhaps displacing calcium from the membrane. There is also the possibility that animals may sense the presence of noxious substances, bringing about a train of events similar to handling shock. Whatever the explanation, it is clear that physiological compensation brings down the rate of efflux to near normal values within 18 h. This occurs presumably by mucus and prolactin secretion, which are known to reduce membrane permeability (Potts & Evans 1966; Wendalaar Bonga & Balm 1989).

Sodium influx. Sodium influx appears to be inhibited by acid, by aluminium and by combinations of both (figure 8). The relation between the maximum degree of inhibition of influx and aluminium level during short episodes at pH 4.5 indicates a hyperbolic curve. Reite & Staurnes (1987) have demonstrated reduced Na-K ATPase activity (−38%) from the gills of salmon held in water of pH 5 with 7.4 μmol l^{-1} aluminium added and the enzyme is known to be an important component of the sodium transport mechanism. Our findings support the hypothesis that aluminium acts as an enzyme inhibitor, as a relation of this type is typical of enzyme inhibition (Höfer 1981). The inhibition of sodium influx resulting from treatment in pH 4.5 alone and from pH 5.6 and aluminium, do not lie on the same curve as fish subjected

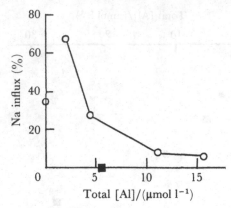

Figure 8. The effect of different aluminium concentrations during short, pH 4.5 episodes on sodium influx (maximum inhibition) in brown trout. (■), pH 5.6 episode.

to aluminium and pH 4.5. This suggests differences in the mechanism of inhibition related to aluminium speciation or in the rate at which aluminium reaches the enzyme. In the case of pH 5.6 and aluminium this gives rise to more effective inhibition.

Change in sodium content. Whereas flux measurements are good indicators of instantaneous changes of rates of sodium balance, they give no direct information on the way the sodium content of the fish changes with time. Because of this we have used flux rates to calculate the change of sodium content of the fish to draw attention to the accumulative changes that result in sodium loss. In the changing circumstances of episodes it is clear that limited sampling may distort real events. Because of this we have attempted an independent assessment of body sodium loss by using plasma-sodium samples taken 24 h after the end of the short episodes. The agreement between the two methods is shown in figure 5 and this justifies to some extent the sampling times chosen for flux measurements and the calculations of total body sodium.

The main findings from the different types of episodes are that at the relatively low calcium values (20 µmol l^{-1}) used in this study to simulate affected waters, the loss of body-sodium from fish increases with the peak or plateau concentration of aluminium at both pH 5.6 and 4.5. The main loss of sodium takes place during the rising phase of the episode at both pH values (figures 3b and 4b). This is largely a result of the increased rate of sodium outflux that may continue into the plateau phase at pH 5.6 (figure 3b). Thereafter, adaptation processes limiting ion loss take over, starting in the plateau phase in pH 4.5 episodes. The rate of outflux decreases and influx may make a slight recovery, particularly in the falling phase of the episode. It is important to realize that although the decrease in outflux limits sodium loss, the loss still continues, and in the absence of influx, the accumulative loss of sodium remains high. Thus the calculated change in body sodium at the end of a pH 4.5, 4.5 µmol l^{-1} total Al episode is -9.5% whereas at pH 5.6, 5.7 µmol l^{-1} total Al (figure 3b), the change is -14.5%. The effects of lengthening an episode are thus less drastic at acid pH values where some sodium influx takes place, than at pH 5.6 where influx is completely inhibited.

The effect of pretreatment. There was no statistical evidence of differences in sodium balance during longer-term episodes after fish had been pretreated for a month: (i) in tap water; (ii) in synthetic soft water of the same composition as that used for episodic experiments (pH 5.6); or (iii) in the same synthetic water with 1 μmol l⁻¹ aluminium added.

(iii) *Effects on respiration*

Several parameters have been used by other workers to assess the effects of pH and aluminium on fish respiration. These include respiratory rate, ventilation rate, cough rate, gape and blood respiratory characteristics. We were only able to measure oxygen consumption during these experiments because other measurements would have interferred with ion balance assessments. There are no well-defined, statistically significant trends in oxygen consumption in response to most of the episodic treatments. Examination of individual responses show that about a half of an experimental batch of six fish show an increase in oxygen consumption during the rising phase in response to the introduction of acid or acid plus aluminium, whereas the others remain relatively quiescent. During episodic deaths, oxygen consumption gradually reduced to zero over a period of *ca.* 12 h. The only well defined response given during short or longer episodes was that at pH 5.4 and 28.5 μmol total [Al] l⁻¹. Here oxygen consumption rose during and following episodic treatment, probably as a result of the very high aluminium levels involved. The episode was planned to give peak aluminium concentrations of 37 μmol l⁻¹ so that, as in previous experiments at pH 5.6, a great deal of aluminium was lost by adsorption and precipitation. Thus we conclude that short acid or combined acid and aluminium episodes have little deleterious effect on fish respiration other than an irritant effect on some individuals. The slight effects noted at pH 5.4 and very high supersaturated aluminium concentrations confirm Rosseland's findings (1980). These findings could be of importance in more prolonged episodes such as those caused by adding lime slurry in attempts to alleviate acid conditions (Skogheim *et al.* 1986).

(iv) *Solubility, speciation and the toxic effect of aluminium*

These studies confirm that the main effect of acid, aluminium and combinations of acid and aluminium, owe their toxicity to a disturbance of monovalent ion balance brought about by significant effects on both sodium influx and outflux. It is also clear that the magnitude of the effect is dependent upon aluminium concentration and that values above 2 μmol l⁻¹ at pH 4.5 have significant effects. There is also no doubt that in the absence of aluminium, acid pH values cause similar though diminished effects. It is not so certain that pH has a marked effect on loss of body sodium for a given measured aluminium concentration. The nearest comparable cases are those quoted in the previous section (pH 4.5, 4.5 μmol Al l⁻¹ and pH 5.6, 5.7 μmol Al l⁻¹), which show that the pH 5.6 combination causes most ion loss. This comes about because sodium influx is completely inhibited at this pH (figures 3b and 8) and outflux is also greater (figure 7). The main difference between aluminium speciation at the two pH values would seem to be an increase in the proportion of Al^{3+} (85 % of total Al) to $Al(OH)_2^+$ (5 % of total Al) at pH 4.5. This relationship reverses so that $Al(OH)_2^+$ (60 % of total) becomes the dominant species of pH 5.6, Al^{3+} declining to 5 % of the total Al (Johnson *et al.* 1981; Potts *et al.* 1985). This implies that the divalent hydroxal ion could be the more active ion in both ATPase inhibition and in stimulating ion efflux.

References

Baker, J. P. & Schofield, C. L. 1982 *Wat. Air Soil Pollut.* **18**, 289–309

Booth, C. E., McDonald, B. P., Simons, B. P. & Wood, C. M. 1988 *Can. J. Fish. aquat. Sci.* **45**, 1563–1574.

Brown, D. J. A. 1983 *Bull. environ. Contam. Toxicol.* **30**, 582–587.

Brown, D. J. A. & Sadler, K. 1989 In *Acid toxicity and aquatic animals* (ed. R. Morris, E. W. Taylor, D. A. Brown & J. A. Brown), pp. 31–44. SEB *seminar ser.* No. 34. Cambridge University Press.

Dalziel, T. R. K., Morris, R. & Brown, D. J. A. 1986 *Wat. Air Soil Pollut.* **30**, 569–577.

Dalziel, T. R. K., Morris, R. & Brown, D. J. A. 1987 *Annls Soc. r. zool. malac. Belg.* **117**, Suppl. 1, 421–434.

Dempsey, C. H. 1987 In *Acid rain: scientific and technical advances* (ed. R. Perry, R. M. Harrison, J. N. B. Bell & J. N. Lester), pp. 508–515. London: Selper.

Dougan, W. K. & Wilson, A. C. 1974 *Analyst* **99**, 413–430.

Eddy, F. B. & Bath, R. N. 1979 *J. comp. Physiol.* **129**, 145–149.

Höfer, M. 1981 *Transport across biological membranes* London: Pitman Publishing Ltd.

Johnson, N. M., Driscoll, C. T., Eaton, J. S., Likens, G. E. & McDowell, W. H. 1981 *Geochim. cosmochim. Acta* **45**, 1421–1438.

Milligan, C. L. & Wood, C. M. 1982 *J. exp. Biol.* **99**, 397–415.

Muniz, I. P. & Leivestad, H. 1980 In *Proceedings of the international conference on the ecological impact of acid precipitation* (ed. D. Drabløs, & A. Tollan), pp. 320–321. Oslo-Ås: SNSF project.

Neville, C. M. 1985 *Can. J. fish aquat. Sci.* **42**, 2004–2019.

Potts, W. T. W. & Evans, D. 1966 *Biol. Bull. mar. biol. Lab., Woods Hole* **131**, 362–368.

Potts, W. T. W. & McWilliams, P. G. 1989 In *Acid toxicity and aquatic animals* (ed. R. Morris, E. W. Taylor, D. A. Brown & J. A. Brown), pp. 201–220. SEB *seminar ser.* No. 34. Cambridge University Press.

Potts, W. T. W., Hunt, D. T. E., Blake, S. & French, P. 1985 London: SWAP.

Reader, J. P. & Dempsey, C. H. 1989 In *Acid toxicity and aquatic animals* (ed. R. Morris, E. W. Taylor, D. A. Brown & J. A. Brown), pp. 67–83. SEB *seminar ser.* No. 34. Cambridge University Press.

Reite, O. B. & Staurnes, M. 1987 In *Proceedings of the SWAP mid-term conference, Bergen, Norway* (ed. B. J. Mason), pp. 298–304. London: SWAP.

Rosseland, B. O. 1980 In *Proceedings of the international conference on the ecological impact of acid precipitation* (ed. D. Drabløs & A. Tollan), pp. 348–349. Oslo-Ås: SNSF project.

Rosseland, B. O. & Skogheim, O. K. 1984 *Rep. Inst. Freshwat. Res. Drottningholm* **61**, 186–194.

Seip, H. M., Muller, L. & Naas, A. 1984 *Wat. Air Soil Pollut.* **23**, 81–96.

Schofield, C. L. & Trojnar, J. R. 1980 In *Polluted rain* (ed. T. Y. Toribara, M. W. Miller & P. E. Morrow), pp. 341–366. New York: Plenum Press.

Skogheim, O. K., Rosseland, B. O., Hoell, E. & Kroglund, F. 1986 *Wat. Air Soil Pollut.* **30**, 587–592.

Stuart, S. & Morris, R. 1985 *Can. J. Zool.* **63**, 1078–1083.

Wendelaar Bonga, S. E. & Balm, P. H. M. 1989 In *Acid toxicity and aquatic animals* (ed. R. Morris, E. W. Taylor, D. A. Brown & J. A. Brown), pp. 243–263. SEB *seminar ser.* No. 34. Cambridge University Press.

Sodium balance in adult Atlantic salmon (*Salmo salar* L.) during migration from seawater to freshwater and to acid freshwater

By W. T. W. Potts,[1] C. Talbot,[2] F. B. Eddy[3] and M. Williams[3]

[1] *Department of Biological Sciences, University of Lancaster, Lancaster LA1 4YQ, U.K.*
[2] *Freshwater Fisheries Laboratory, Pitlochry, Perthshire PH16 5LB, U.K.*
[3] *Department of Biological Sciences, University of Dundee, Dundee DD1 4HN, U.K.*

Returning adult salmon could tolerate instantaneous transfer from seawater to freshwater, although in nature they usually choose to slow the transition. Sodium uptake was fully activated on entry to freshwater but sodium loss remained high for 6 h, resulting in a fall in blood concentration. Kidney function did not adapt for 24 h causing an increase in mass and a further fall in blood concentration.

In water of pH 5, sodium uptake was inhibited by 90%, resulting in a further decline in blood concentration, but sodium uptake returned to normal immediately in neutral water. After 6 h in water of pH 5, containing 20 μM Al, the probable conditions in the Duddon at the time of a recorded fish kill, sodium uptake was severely inhibited for 24 h after return to neutral water. Two short acid, high-aluminium episodes 24 h apart, as observed in the Duddon, would almost eliminate sodium uptake for 48 h, reducing the blood concentration to lethal levels.

1. Introduction

The transitions from seawater to freshwater and back again are physiologically so demanding that most species of fish are confined to one or the other. A well-known exception is the Atlantic salmon (*Salmo salar*) but even the salmon can survive the transition only at certain stages of its life cycle.

The change from seawater to freshwater is stressful and is accompanied by a fall in blood ion concentration and changes in the levels of several hormones. It also requires the replacement of an epithelium in the gills that excrete salt with one that takes up salt, the whole adaptation taking at least a week.

If a homing salmon were to encounter acidic water (i.e. below pH 5.5) on entering the river, the dual stress might make it more vulnerable than a fish entering freshwater at a higher pH. Seawater is normally alkaline, ca. pH 8.2. The Atlantic salmon *S. salar* is one of the more sensitive of the salmonid species and is more susceptible to acid waters than either the brown trout, *Salmo trutta* or the American brook trout *Salvelinus fontinalis*. Kills of homing salmon have been reported on several occasions in the rivers Esk and Duddon in Cumberland and the continuous pH monitoring programme, instituted by the NWWA on these rivers (Crawshaw 1984) has provided contemporary data on river flow and pH. During a fish kill in the river Duddon in 1983 the pH at Cropple How twice fell to 4.5 during two spates on 16 and 18 September and remained below pH 5.0 for about 8 h on each occasion (Crawshaw 1984, fig. 26).

Contemporary analyses of aluminium are not available but the correlation between Al and pH on the Duddon is very high (Crawshaw 1984). A pH of 5.0 corresponds to about 250 μg Al l^{-1}, *ca.* 10 μmol l^{-1} in the river. At the peak of acidity the Al might have reached 20 μmol l^{-1}.

Little work has been done on salt and water balance in the adult Atlantic salmon, in contrast with many papers on the physiology of its eggs, parr and smolts. The physiology of migration from seawater to freshwater is of both fundamental and practical importance. From the practical point of view, the changes in the gills and kidneys may affect the time spent by homing salmon in estuaries, where, in Britain, they are vulnerable to netting.

Salmon and trout can withstand instantaneous transfer between seawater and freshwater at appropriate times but some mortality may occur, and their behaviour is programmed to avoid sudden transitions whenever possible. There are no accessible mortality data comparing the effects of instantaneous with gradual transfer for adult salmon or sea trout but practising scientists will almost invariably transfer fish through a range of intermediate salinities to reduce mortality. In our own experiments (Potts *et al.* 1989) about 25 % of the salmon failed to survive direct transfer from seawater to freshwater but some of the deaths may have been caused by the stressful laboratory conditions. It was necessary to use relatively small tanks and to recirculate the water for considerable periods to monitor salt fluxes, and several of the fish had been prepared with dorsal aorta and urinary catheters.

Most salmon returning to their home rivers spend considerable periods in the estuary. To what extent they are waiting for suitable conditions of river flow and to what extent they are beginning to adapt to the lower salinities is uncertain, but it seems likely that they prefer to avoid an abrupt transition. Studies in the Ribble (Priede *et al.* 1988) in this country and in a Nova Scotian river (Stasko 1975) show that the fish move upstream on the flood and drift down or hold station on the ebb. As Stasko pointed out, this method of movement is energetically inefficient and the fish spend far longer in the estuary than their swimming speed would require. This suggests that there is some advantage in the slow progression. Once they have entered freshwater, progress is much more rapid, often averaging 10 or 20 km per day. Little work has been done on migration into small rivers where there is effectively no brackish zone, but the evidence available suggests that these fish make several excursions into freshwater and then drop back before finally ascending the river. With only a limited number of fish available, our experiments were confined to the effects of a rapid change from seawater to freshwater. This is the most stressful experiment and comparative work is required on the effects of more gradual transfer.

2. Methods

Adult salmon, *Salmo salar* L., were obtained from coastal nets in seawater at Carnoustie, Angus and were kept in seawater in the DAFS laboratory at Almondbank, Perth. Some comparative experiments were done with large four-year-old parr, reared in Almond river water and with post-smolts that had been adapted to seawater for three months. Salmon were prepared with an indwelling catheter in the dorsal aorta. Blood samples removed were replaced with equal volumes of saline to maintain blood volume. In some fish the urinary bladder was also catheterized. Blood samples were taken from the large parr by caudal puncture after the fish had been killed by a blow on the head. Net sodium flux in freshwater was determined from the rate of change of the sodium content of the experimental tank. Freshwater experiments were done, as far as possible, in running water from the River Almond. Closed circuit experiments were done in 40 l tanks in which the water was circulated by a rotary pump at 4 l min⁻¹, so that there was a flow of water from head to tail of the fish. The urinary catheter was passed through the wall of the tank to a fraction collector. The end of the catheter was held level with the water in the tank to avoid pressure differences.

Sodium influxes in freshwater were measured by adding 10–40 µC of ^{24}Na to the tank and taking blood samples after suitable periods of time. After centrifugation to remove cells, a plasma sample was counted in a well-type Panax scintillation counter and, after the activity had decayed, the sodium content was determined. In the freshwater experiments, water samples were taken at the beginning and the end of the experiment for sodium determinations. Sodium loss from the fish raised the sodium content of the tanks during some experiments so that the specific activity of the bath declined. The influx was calculated as by McWilliams (1980) on the assumption that all the sodium in the fish was exchangeable as a single compartment and that the fish contained 31.4 mmol Na kg⁻¹ in seawater and 25 mmol Na kg⁻¹ when adapted to freshwater (see below).

When the specific activity of the bath changed during the experiment it was assumed that the change was linear. Sodium efflux was calculated from the influx and the net flux. Where possible the fish were allowed to settle down in the tank for an hour or two before the isotope was added. To expedite mixing, the isotope was distributed widely over the surface of the tank.

The medium was changed from seawater to freshwater, whenever possible, by flushing the tank containing the fish with freshwater but this procedure required 5–10 minutes to reduce the sodium content of the tank to below 0.5 mmol Na l⁻¹. Studies on the rate of sodium uptake and loss immediately after transfer from seawater to freshwater, were done by lifting the fish from the seawater tank, in a net, first to a large tank of freshwater for one minute to wash, then to the experimental tank into which the isotope had already been mixed.

Sodium efflux in seawater was measured by injecting a known quantity of ^{24}Na into the peritoneum of a fish, whilst in the experimental tank, and measuring the activity of 250 ml samples of the bath over the following 6 h. If the quantity of ^{24}Na remaining in the fish at times, T_1 and T_2 is C_1 and C_2 then:

$$K = 1/(T_1 - T_2) \ln C_1/C_2.$$

After the first hour or so, during which the sodium became distributed throughout the sodium space of the fish, K became almost constant. Most estimates were based on the interval between 2 and 5 h after injection.

Sodium taken up at the gills appeared rapidly in the blood but equilibrated more slowly with the intracellular sodium in the white muscle, which has a very limited blood supply. Estimates of fluxes based on changes in the specific activity of the blood will therefore not be exactly comparable with fluxes based on changes in the specific activity of the sodium in the whole body (Potts *et al.* 1970). To assess the magnitude of this discrepancy, samples of white muscle were taken from six fish 1 h after the beginning of the experiments. The mean specific activity of the whole muscle sodium, after 1 h was 67 ± 13 % that of the plasma. The mean sodium content of the whole muscle was 12.6 mmol kg⁻¹. The muscle tissue amounted to 80 % of the mass of the fish and held 10.1 mᴍ kg⁻¹ of the total body sodium, assumed to be

31.4 mM kg⁻¹. After 1 h, the specific activity of the blood would be 10 % higher than that of the total body sodium. Estimated fluxes, calculated from rate constants and an assumed value of 31.4 mM Na kg⁻¹ total body sodium, have been reduced by this amount for comparison with fluxes calculated from changes in the sodium content of the baths.

Sodium determinations were made by using an EEL flame photometer and chloride determinations were made by using a Radiometer CM10 Chloride meter with 20 µl samples of plasma suitably diluted.

Water was acidified by the addition of dilute sulphuric acid, aluminium was added as $AlCl_3$ and monomeric aluminium was assayed by the Catechol violet method (Dougan & Wilson 1974). When fish were present the concentration of monomeric aluminium initially fell very rapidly. Where practicable, experimental tanks were pre-treated with aluminium. At the beginning of an experiment the quantity of aluminium required to reach the experimental concentration was added together with an additional 100 µmol Al kg⁻¹ per fish. After 15 min the aluminium concentration was measured and additional aluminium was added to bring the concentration back to the required level if necessary. This process was repeated about every half hour.

The individual variation between fish is considerable and so as far as possible fish were used as their own controls. Individual fish were treated on the cycle: freshwater, acid water, acid and Al water. In a few cases the cycle was reversed.

Urine was collected from fish in seawater for 24 h before the water was changed to freshwater. Blood samples were taken at the beginning of collection and just before the water was changed and at appropriate intervals afterwards. Experiments were done at 12 °C ± 1 °C.

3. Results

Salt balance in seawater

Salmon, like other marine bony fishes, maintain a blood concentration of about 40 % that of seawater. They are fairly permeable to salt and drink seawater to replace the water lost by exosmosis. Salt balance is maintained by excreting salt across the gills. The rate of exchange of sodium in four salmon from Carnoustie was 12.6 % h⁻¹. The mean plasma sodium concentration of adult salmon in seawater was 183 ± 3 mmol Na l⁻¹ plasma ($N = 12$).

Salt balance in freshwater

On entering freshwater, the problems of salt and water balance are reversed, water enters by osmosis and has to be removed in the urine, whereas salt is lost by diffusion and the urine and must be replaced by active uptake from the river water to the blood. An unexpected discovery in the course of this work was that sodium uptake attained normal freshwater levels immediately after transfer (Potts *et al.* 1985, 1989).

Sodium uptake in freshwater

In the first hour after transfer to Almond river water the rate of uptake was 0.95 mmol Na l⁻¹ ECF h⁻¹ equivalent to 0.5 % of the plasma sodium content. In the shortest experiment, in which one fish was removed from seawater and washed for only two minutes, the rate of uptake during succeeding five minutes in freshwater was equivalent to 1.0 % of the plasma sodium but the mean sodium concentration of the bath during this experiment was high, which would increase the rate slightly.

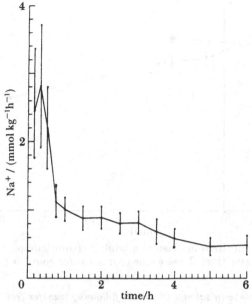

Figure 1. Rate of net loss of sodium from adult Atlantic salmon
following transfer from seawater to freshwater.

Sodium loss in freshwater

During the first few minutes after transfer from seawater to freshwater, sodium losses remained high (figure 1). The rate of loss declined rapidly during the first hour and the fish were close to equilibrium after 24 h, the mean rate of loss in three fish at this time being only 0.2% body salt h^{-1}. The cumulative salt loss reached 18% after 6 h and 20% of total body salt after 24 h.

Blood concentrations were also measured in salmon kept in seawater but subjected to the same cannulation and handling as the salmon that were transferred from seawater to freshwater. These fish showed only a slight elevation of blood concentration (figure 2). Plasma sodium fell by only 6% during the first six hours, from 182 to 171 mmol Na l^{-1}.

Sodium uptake in post-smolts

The unusual ability of the adult salmon to take up sodium from freshwater, immediately after transfer from seawater, at the same rate as in fish fully adapted to freshwater, raised the question of whether all salmon adapted to seawater maintain the ability to take up salt from freshwater or whether it is a transitory feature of migrating salmon that develops only as they approach their home rivers. To distinguish between these possibilities, rates of sodium uptake in freshwater were also measured in salmon that had smolted in the spring and had subsequently been confined to seawater for five months. These fish would not normally return to freshwater until the following year at the earliest. The rate of sodium uptake was initially much lower in these fish but rose slowly over several days following transfer. They were unable to maintain blood concentrations as high as those of the adult fish (table 1).

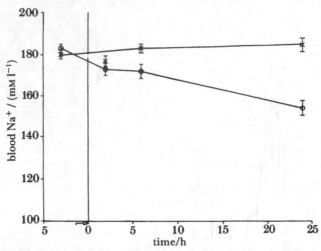

Figure 2. Sodium concentration in plasma in adult Atlantic salmon following transfer to
freshwater (FW); Freshwater (○); seawater controls (×).

Table 1. *Sodium uptake in salmon (S. salar) following transfer from seawater (SW) to
freshwater (FW) by post smolts*

	uptake (mmol l ECF h⁻¹)	uptake (% Na h⁻¹)	Blood concn (mmol Na l⁻¹)	N
1st hour	0.30 ± 0.06	0.23 ± 0.05	130 ± 2	6
42 hours FW	0.99 ± 0.08	0.96 ± 0.12	106 ± 4	6
72 hours FW	0.82 ± 0.06	1.02 ± 0.07	81 ± 4	3

Urine flow rates

The urinary flow rate of seawater-adapted salmon was low, only
0.72 ± 0.18 ml kg⁻¹ h⁻¹, and it did not change significantly for at least 25 h after
transfer to freshwater, but a marked increase occurred some 60 h after transfer with
the rate increasing 6.3-fold but then declining again (figure 3). The urinary flow rate
of fish adapted to freshwater for eight days and three months were similar at 0.99 and
1.19 ml kg⁻¹ h⁻¹, respectively. The glomerular filtration rate showed similar changes:
the seawater rate of 1.49 ml kg⁻¹ h⁻¹ remained more-or-less constant for the first
25 h, increasing 6.5 times by 60 h post-transfer before falling to 2.61 ml kg⁻¹ h⁻¹ after
eight days (Talbot *et al.* 1989).

Plasma and urine ions

Sodium

The sodium concentration of the plasma of seawater-adapted salmon, which had
been cannulated for urine collection and transferred directly to freshwater, fell from
the equivalent of about 44 % seawater (equivalent to 460 mmol/l SW) to only 27 %
seawater 25 h after transfer, but increased to 29 % after eight days. In the long-term
freshwater-adapted fish, the concentration was equivalent to 33 % (figure 4a). The
sodium concentration in the urine of seawater-adapted salmon was much lower than
in plasma and after transfer to freshwater the concentrations fell further (figure 4).

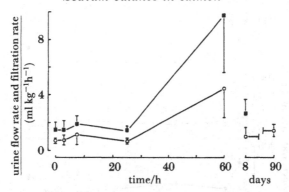

Figure 3. Urine flow rates (○) and filtration rates (■) from Atlantic salmon in seawater and following transfer to freshwater.

Table 2. *Sodium uptake by 4+ parr and by adult salmon (S. salar) at various pH at 12 °C*

	Sodium uptake/(mmol Na l ECF h⁻¹)		
	pH 6.5–7.0	pH 5.0	pH 4.0
Adult	0.903 ± 0.083	0.114 ± 0.026	0.131 ± 0.37
	$N = 3$	$N = 6$	$N = 10$
4 parr	1.38 ± 0.24	0.123 ± 0.029	0.075 ± 0.020
	$N = 4$	$N = 7$	$N = 10$

Chloride

Similar changes were observed in plasma chloride (figure 4b). Urine chloride in seawater-adapted fish (figure 4b), was some 4.6-times more concentrated than urine sodium but it fell rapidly after transfer to freshwater.

The effects of low pH on sodium balance

Sodium uptake was strongly inhibited in acid water. In water of pH 5.0 the rate of sodium uptake was less than 0.1 % of total body sodium per hour, only one fifth normal (table 2). Four-year-old parr showed a similar fall in sodium uptake in acid water (table 2).

Not only was uptake inhibited, but the net loss was increased from 0.5 % total body Na h⁻¹ at, pH 6.5 to 0.9 % at pH 5.0 and to 1.05 % h⁻¹ at pH 4.4 (table 3).

The effects of aluminium ions on sodium balance

The rate of uptake was low in acid conditions so that any further reduction in uptake following the addition of aluminium ions would be small (table 4). Even in 50 μmol Al l⁻¹ some sodium uptake was detected, although this may have taken place during the earlier part of the two hours of treatment.

Sodium uptake remained low for at least 24 h after treatment with aluminium, even after the fish had been well washed in running river water. After four hours of treatment with 20 μmol Al l⁻¹ at pH 5.0 the rate of uptake by 4+parr in freshwater was only 0.02 % body sodium h⁻¹ ($N = 6$) and even 24 h later it was still only 0.35 % ($N = 6$), about half the rate in the control parr. The rate of uptake in two salmon 24 h after similar treatment was around 0.3 % total body sodium h⁻¹, again about half the

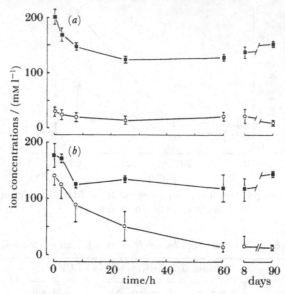

Figure 4. Ion concentrations in plasma (■) and urine (○) of Atlantic salmon
following transfer to freshwater; (a) sodium; (b) chloride.

Table 3. *Net sodium loss from Atlantic salmon S. salar adults*

Freshwater adaptation	treatment pH 6 h	Al treatment 2 h	Na loss/ (mmol kg^{-1} h^{-1})	N
1 day at pH 6.5–7.0	6.5–7.0	0	0.075±0.04	3
1 day at pH 6.5–7.0	5.0	0	0.136±0.018	4
3–5 days	5.0	0	0.237±0.067	4
3 days	5.0	20 µM Al	0.373±0.116	7
1 day	4.4	0	0.294±0.11	4

normal rate. The rates of uptake appeared to recover on return to freshwater but did
not reach normal levels, but the salmon that survived longest were probably those
with the highest rates of uptake. None of the salmon or parr survived for more than
three days after Al treatment. Sodium loss from salmon in the presence of
20 µmol Al l^{-1} averaged 1.6 % total body Na h^{-1}, ($N = 7$) higher than the rate in
aluminium-free water at the same pH but the levels were very variable and the
difference was not significant at the 5 % level (table 3). However, in three fish
transferred from seawater to freshwater containing 20 µmol Al at pH 5.0, after a
10 min wash in freshwater, the blood sodium dropped by 23 % in 24 h compared with
17 % in freshwater transferred to aluminium-free water (figure 2) so it is likely that
aluminium does increase sodium loss.

As the salmon become restless in small tanks in adverse conditions experiments in
acid and aluminium ions were usually limited to 4–6 h, during which the blood
concentrations fell by only a few per cent. Following treatment with aluminium at
10, 20 and 50 µmol l^{-1} the concentrations of plasma sodium continued to fall after
return to neutral freshwater. Blood concentrations of 75 % and 85 % of normal
freshwater levels were recorded in two salmon, two days after 4 h in 20 µmol Al l^{-1}
but again the survivors would have been those that had the highest concentrations.

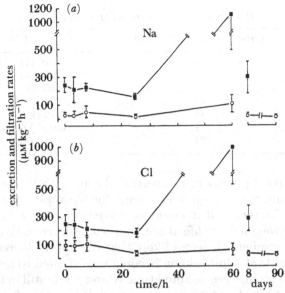

Figure 5. Excretion rate (○) and filtration rate (■) of (*a*) sodium and (*b*) chloride from Atlantic salmon in seawater and after transfer to freshwater.

Mass

Water influx must begin immediately after transfer to freshwater, although urine production remained low for a day or more (Talbot *et al.* 1989). The adult salmon showed a marked tendency to gain mass during the first 24 h after transfer, the net gain being *ca.* 6% after 8 h and 12% after 24 h, roughly equivalent to the urine deficit. Later, the mass returned towards the original level.

4. Discussion

There are technical difficulties in working with fish as large as mature salmon. Reasonably high specific activities of radio-isotopes could be obtained only by confining the fish in relatively small tanks and the stress this induced may have increased the effects of low pH and Al. A long series of urine samples could only be obtained from fish confined for several days and it is noteworthy that these fish had higher blood ion concentrations in seawater and lower concentrations in freshwater than fish that had more freedom. Attempts to collect urine from fish in acid and aluminium waters were never successful for more than a few hours as the fish became restless.

Salmon do not usually move directly from seawater to freshwater but either move up an estuary over a period of time (Hawkins & Smith 1986) or oscillate several times between the two waters. An infinite variety of programmes of transfer might be devised but all would make measurements of salt uptake and loss, and their interpretation, more difficult. For this reason direct transfer was used.

Sodium balance in seawater

The ability of adult salmon to take up sodium at the full rate immediately following entry into freshwater must make a useful contribution to maintaining sodium balance, reducing the net fall of plasma sodium during the first day by

Table 4. *Sodium uptake by Atlantic salmon (S. salar) after 4 hours in 20* μM *Al* l⁻¹ *pH 5.0*

	h recovery at pH 6.5–7.0	uptake/(mM Na l ECF h⁻¹)
adult salmon	1	0.010 (4)
	2	0.026 (4)
	18	0.45 (2)
	24	0.52 (2)
	43	0.72 (1)
4+ parr	1	0.026±0.010 (6)
	24	0.52±0.07 (6)

21 mmol l⁻¹ or 12 % of total plasma concentration. Evans (1982) has shown that in marine fish a little sodium is taken up in exchange for hydrogen ions as part of the regulation of acid–base balance, but it would be wasteful for the salmon to take up sodium at a high rate throughout its life at sea as it would have to be excreted again against a concentration gradient. It seems likely that the uptake increases as the fish approach freshwater and that adult salmon become pre-adapted to freshwater while still at sea, as the smolt becomes pre-adapted to seawater while still in the river. Post-smolts transferred back to freshwater had a very much lower rate of uptake and only reach the adult level after three days if they survived.

Because the pump saturates at a relatively low external concentration, the rate of sodium uptake, even in seawater, would reach only 1.2 % of total plasma sodium per h, only one tenth of the total sodium exchange in seawater.

The rate of exchange of sodium in seawater is 12.6 % h⁻¹. The initial rate of loss in freshwater is about 9 % h⁻¹ (figure 1) but this value may be inflated by loss from the skin and possibly from the gut and bladder following the shock of transfer.

When the blood concentration declines the tissues swell by osmosis. This swelling may be counteracted in two ways, by the transfer of ions from the cells to the plasma and by reduction in the organic solutes present in the cells, either by metabolism or by loss to the plasma, but any swelling will cause a shift of water from plasma to cells. The resulting reduction of plasma volume may increase viscosity and raise the blood pressure but will help to maintain the sodium and chloride concentration. Limitations of time and salmon did not allow us to follow the changes in extracellular and intracellular water but during the first six hours in neutral freshwater the net sodium loss was 18 % of total body sodium. However, after six hours the plasma sodium had fallen by only 6 %. Even if some of the total sodium loss was derived from the skin the small fall in plasma sodium suggests that the plasma volume had been reduced. After 24 h, and the loss of a further 30 % of the body sodium, the concentration of the plasma had fallen another 15 %. This suggests that plasma volume had increased again, probably assisted by the 8 % increase in body mass. During this time sodium uptake had been proceeding at 0.14 mmol kg⁻¹ h⁻¹ (about 0.5 % total body Na h⁻¹) so the total loss of sodium during the first 24 h was 9.8 mmol kg⁻¹, or 31 % of the original total (table 4).

One day after transfer from seawater to neutral freshwater the salmon has regained sodium balance, gaining and losing 0.14 mmol Na kg⁻¹ h⁻¹ or 0.5 % of total body sodium per hour. However, if the pH dropped to 5.0, sodium uptake was reduced by the equivalent of 10 % of total body sodium per day. In addition, sodium efflux is increased at low pH from 0.5 % h⁻¹ to 0.76 % at pH 5.0 and to 0.92 % at pH 4.0.

The most marked effect of aluminium ions was to prolong the inhibition of sodium uptake for about 24 h after the fish had been returned to neutral water. In the first 24 h after aluminium treatment uptake averaged one quarter of normal. Two successive short acid, aluminium episodes 24 h apart, would effectively block uptake for 48 h, leading to an additional net loss of 18 % of initial body sodium, with fatal results.

It is not certain from these results whether or not a salmon that encounters an acid episode as it enters freshwater is at a disadvantage compared with a salmon already adapted to freshwater. A salmon entering a river from the sea contains about 20 % more sodium chloride than a fish adapted to freshwater, which it loses during the first 24 h before reaching equilibrium. If during this period (or later) it encounters an acid episode, the inhibition of uptake will lose the equivalent of a further 10 % per day. However, if the low pH also increased sodium loss during the adaptation period then the situation would be even worse. As the sodium loss immediately after entering freshwater is so high, we did some experiments transferring fish directly from seawater to acid freshwater but the results were too irregular to draw firm conclusions although losses were on average higher. Further work is required on this point.

The results reported here suggested that the deaths of adult migrating salmon reported in the Duddon and Esk in 1983 were mainly due to the inhibition of the sodium uptake system by the low pH and aluminium, the inhibition caused by the aluminium continuing to inhibit uptake for at least 24 h after the end of the episode.

There is some inconsistency in the reported effects of low pH and aluminium ions on sodium uptake in fish. Many authors have found that low pH, even in the absence of aluminium ions, will inhibit sodium uptake, e.g. brook trout, *S. salvelinus* (Parker & Dunson 1970), sailfin molly, *Poecilia latipinna* (Evans 1975), rainbow trout, *S. gairdneri* (Kerstetter *et al.* 1970), brown trout, *S. trutta* (McWilliams & Potts 1978), goldfish, *Carassius auratus* (Maetz 1973) as well as several invertebrates such as the crayfish, *Austropotamobius* (Shaw 1960). In contrast, Dalziel (1986) found no effect of pH on sodium influx in the brown trout except in the presence of aluminium ions, where as little as 1 and 2 μmol Al l^{-1} had an appreciable effect. Some of the discrepancy between Dalziel's results and those of other workers may have been caused by the unsuspected presence of traces of aluminium in the water in some other experiments. The volumes of water required with experiments with adult salmon are so large that it is impracticable to prepare synthetic solutions from deionized water but no aluminium was detected (less than 1.0 μmol l^{-1}) in the Almond river water; nevertheless a marked reduction of sodium uptake was observed at low pH. Some of the discrepancy between different experiments may also arise from the various adaptation times of the fish to acid conditions. In long-term experiments the fall in blood concentration will stimulate uptake and tend to counteract the inhibition caused by H$^+$ ions.

It is possible to construct a sodium balance sheet for adult salmon migrating into an acid river directly from the sea or meeting an acid spate when already adapted to freshwater (table 5). In neutral river water, the gross loss will be about 9.8 mmol Na kg^{-1} or 31 % of total body sodium in seawater during the first 24 h after arrival from the sea, but the fish will replace about 3.4 mmol Na kg^{-1} or 10 % of the total sodium. As a result, the total body sodium will fall by 20 % from 31.4 to 25 mmol Na kg^{-1}, whereas the blood concentration will have fallen by about 16 % from 183 to about 154 mmol Na l^{-1} plasma (table 5).

Table 5. *Estimated sodium balance in adult S. salar following transfer from seawater to neutral freshwater, acidic freshwater and acidic freshwater containing* 20 μM Al l⁻¹

	neutral water for 24 h	water (pH 5) for 24 h	water (pH 5) containing 20 μmol Al l⁻¹ for 24 h	neutral water for 24 h after acid and acid and aluminium treatment
total body Na at beginning of period	31.4	31.4	31.4	21.9
loss in 24 h/(mmol kg⁻¹)	9.8	9.8	9.8	3.4
uptake in 24 h/(mM kg⁻¹)	3.4	0.3	0.3	0.8
total body Na at end of period/(mmol kg⁻¹)	25.0	21.9	21.9	19.3
blood concn at beginning of period/(mmol Na l⁻¹)	183	183	183	140
blood concn if fall ∝ to initial fall (mmol Na l⁻¹)	154	140	140	128

If the pH of the river was low on entry and remained below 5.5 for 24 h the net loss would increase to 9.5 mmol kg⁻¹ or 31%, even if the rate of gross loss were unchanged. The blood sodium is not directly proportional to total body sodium but if a fall of 6.4 mmol Na kg⁻¹ total body sodium corresponds to a fall of 29 mmol l⁻¹ blood (183 − 154 mmol Na l⁻¹) then the blood sodium would fall to only 140 mmol l⁻¹. Leivestad & Muniz (1976) found that some brown trout, *S. trutta*, began to die when the blood concentrations fell below 150 mmol Na l⁻¹ and mortality was great at 120 mmol Na l⁻¹ although some individuals were still alive with blood sodium as low as 90 mm Na l⁻¹. If low pH does increase gross loss then the situation would be even worse.

After only 4 h in water with Al concentration 20 μmol l⁻¹, the probable level reached in the Duddon in 1983, sodium uptake in the salmon would still be only half the normal rate 24 h after the pH and aluminium levels in the river returned to normal. If the recovery of sodium uptake is linear then the uptake during the 24 h following return to neutral water would be only 0.8 mm kg⁻¹ or 2.5% of the initial total sodium. Although the fall in blood concentration might reduce sodium loss to some extent, aluminium also appears to increase loss (table 3) and Dalziel (1986) found that in the brown trout efflux rose at pH 5.4 as the aluminium concentration was increased. It is therefore likely that the rate of loss would at least be maintained (table 5). If the pH returned to neutrality after 24 h the sodium loss during the following day would be 2.6 mmol kg⁻¹. The total loss would therefore be 9.5 + 2.6 or 12.1 mmol kg⁻¹ corresponding to a plasma concentration of 128 mmol Na l⁻¹ (table 4).

Several of the fish that were treated with aluminium died over-night. The plasma concentrations that were recorded in survivors were consistent with these calculations; 112 and 128 mmol Na l⁻¹ in two salmon and 120, 124, 117, 125 and 143 in five 4+ parr.

The effect of two short 8 h acid episodes separated by two days on a fish already adapted to freshwater might be as follows. During the total of 16 h at low pH and high aluminium, sodium uptake would be very small and the net loss would be 2.4 mmol. During each of the subsequent 24 h net loss would be about 2.4 mmol kg⁻¹

making a total net loss of 7 mmol kg^{-1}. A net loss of 7 mmol would lower the blood concentration about 32 mmol l^{-1} below normal, taking it well into the lethal range. It seems likely that two successive episodes would be more damaging than a single episode of the same total length if the salmon do not have time to recover in between (table 4).

Even lower concentrations of aluminium are effective in suppressing uptake in the trout (Dalziel 1986) but more work is required on the rate of recovery of sodium uptake from lower concentrations.

The decline in the ionic concentration of the plasma caused by low pH and even more by low pH and aluminium combined, is sufficient to bring the fish into the lethal range but the exact mechanism is less clear. Aluminium blocks the Na–K–ATPase that drives sodium uptake. Aluminium hydroxides also accumulate on the gills and either stimulate mucus production or prevent its dispersal. The resultant accumulation of mucus may interfere with respiration but four of the aluminium-treated fishes died during the 24 h after the return to neutral water and none of the fish died in Al solution of 10 or 20 μM l^{-1}. The reduction in plasma volume increases the protein concentration of the plasma and consequently the viscosity of the blood. This may be aggravated by swelling of the red cells and by the releaser of red cells from the spleen following hypoxic stress (Milligan & Wood 1982). This may be a secondary cause but the time interval between aluminium treatment and death suggests that it is a minor factor.

The authors express their gratitude to Mr M. Miles, Mr S. Keay and Mr J. Muir of the Almondbank Hatchery, without whose help this work could not have been done. Also, we thank Mr E. Duncan of Carnoustie and Mr I. Mitchell of Tay Salmon Fisheries, Perth for their help in procuring fish. This work was founded in part by the Surface Waters Acidification Programme of the Royal Society, the Royal Swedish Academy of Sciences, the Norwegian Academy of Science and Letters and by Natural Environment Research Council, grant No. GR3/6218.

References

Crawshaw, D. 1984 North West Water Rep. T.S.N. 84/3.

Dalziel, T. 1986 Ph.D. thesis, University of Nottingham.

Dougan, W. K. & Wilson, A. L. 1974 *Analyst* **99**, 413–430.

Evans, D. H. 1975 *J. comp. Physiol.* **96**, 111–115.

Evans, D. H. 1982 *J. exp. Biol.* **97**, 289–299.

Hawkins, A. D. & Smith, G. W. 1989 *Scottish Fisheries Res. Rep.* no. 36, pp. 1–24.

Kerstetter, T. H., Kirschner, L. B. & Rafuse, D. D. 1970 *J. Physiol., Lond.* **56**, 342–349.

Leivestad, H. & Muniz, I. P. 1976 *Nature, Lond.* **259**, 391–392.

Maetz, J. 1973 *J. exp. Biol.* **58**, 255–275.

McWilliams, P. G. 1980 *J. exp. Biol.* **88**, 269–280.

McWilliams, P. G. & Potts, W. T. W. 1978 *J. comp. Physiol.* **126**, 277–286.

Milligan, C. I. & Wood, C. M. 1982 *J. exp. Biol.* **99**, 397–415.

Packer, R. K. & Dunson, W. A. 1970 *J. exp. Zool.* **174**, 65–72.

Potts, W. T. W., Foster, M. A. & Strather, J. W. 1970 *J. exp. Biol.* **52**, 553–564.

Potts, W. T. W., Talbot, C. & Eddy, F. B. 1985 *J. exp. Biol.* **118**, 455–460.

Potts, W. T. W., Talbot, C., Eddy, F. B., Primett, D., Prunet, P. & Williams, M. 1989 *Comp. Biochem. Physiol.* **92**A, 247–253.

Priede, I. G., Solbe, J. F. De L. G., Not, J. E., O'Grady, K. T. & Cragg-Hine, D. 1989 *J. Fish Biol.* **33**A, 133–139.

Shaw, J. 1960 *J. exp. Biol.* **37**, 548–556.

Soivio, A., Nynolm, K. & Westman, K. 1975 *J. exp. Biol.* **62**, 207–217.

Stasko, A. B. 1975 *J. Fish Biol.* **7**, 329–338.

Talbot, C., Preston, T. & East, B. W. 1986 *Comp. Biochem. Physiol.* **85**A, 445–450.

Talbot, C., Eddy, F. B., Potts, W. T. W. & Primmett, D. R. N. 1989 *Comp. Biochem. Physiol.* **92**A, 241–245.

Physiological, foodchain and ecological effects among benthic invertebrates exposed to low pH and associated metal concentrations

By Jan Herrmann†

University of Lund, Department of Ecology, Animal Ecology, Ecology Building, Helgonavägen 5, S-223 62 Lund, Sweden

Acidification of surface waters affects their benthic invertebrates in various ways, many of which can be classified as either physiological or foodchain effects. Both may be manifest as changes in species community structure. Most crustaceans, molluscs and mayfly nymphs are vulnerable to acid conditions, whereas hemipteran waterbugs, dragonfly, stonefly and chironomid larvae seem to be less affected or even favoured.

Physiologically an animal can be harmed both by the lowered pH or by increased metal concentrations, notably aluminium (Al), but few good studies try to differentiate the importance of these factors in realistic experiments, using preferably sublethal parameters. At a low pH different metals can affect animals in different ways. A normal regulation of ions is vital for mayflies, and other organisms, and is adversely affected by low pH, or increased levels of Al or heavy metals, or combinations of these, causing net losses of sodium and other ions. Calcium (Ca) metabolism is often heavily disturbed at low pH, and the addition of Ca often aids an animal's survival.

However, in some cases Al seems to be able to mitigate the detrimental impact of hydrogen ion stress on Na ion regulation at pH 4. The same pattern of amelioration has also been found in mortality experiments. This beneficial effect of Al is probably only temporary and because of membrane permeability. A normal ion flux and content seems essential for successful moults, but ion regulation is also more disturbed during moulting. These physiological stress effects may be the cause of the increasing respiration rates of some mayfly nymphs, after Al exposure, and could also explain behavioural changes such as increased drift rates under conditions of low pH.

There is some evidence of bioaccumulation of Al and cadmium (Cd), i.e. increasing content with time and such compared to circumneutral water (or both). Biomagnification, i.e. higher levels of metals at higher trophic levels, of Al may occur, but this is more probable for Cd. Most of the Al on or in a mayfly is lost with the exuvium at moulting, but smaller amounts may be transferred to terrestrial foodchains, depending on species of Al and animals. The same pattern seems to hold for Cd. All metal effects can be mitigated by the presence of humus and other organic matter.

However, although hydrogen ion or metal levels may not be sufficiently high to be physiologically detrimental to invertebrates, the animals may still suffer in a slightly acid water. This may be because biotic interactions are affected. For example, certain

---

† Present address: Swedish Environmental Protection Agency, S-171 85 Solna, Sweden.

mayfly species show increased survival rates on, and higher preference for, algae from neutral streams. This is probably due to differences in algal species composition and their nutritive value.

Thus grazers are generally vulnerable to conditions in acidified surface waters. Both increases and decreases in the number of shredders have been observed with decreased pH, and these changes have been suggested to be related to reduced fish predation and increasing occurrence of detritus (or both). The number of predators do not generally change with altered pH. The change of one benthic community into another following acidification may also be due to the alteration of the bottom vegetation and thereby the habitat structure, as well as chemical or biological impact via some key species.

Observations indicate that some invertebrate populations can withstand a long natural exposure to low pH, but whether animals have become genetically adapted to the relatively rapid pH change is not clear.

Conclusions are also that (i) often it seems that pH itself is a more negative factor than Al; (ii) species resistant to low pH also seem to tolerate elevated Al concentrations; (iii) sublethal parameters should be used to discover and describe anthropogenic threats of acidification; (iv) the ultimate effect is on growth and reproduction; (v) non-significant tendencies must not be ignored; and (vi) the impact patterns are a mixture of abiotic and biotic, of physiological and foodchain, of short-term and long-term, of direct and indirect factors.

---

1. Introduction

We often regard the technological achievements in our modern society as a blessing, but they have also appeared to be a threat. The developments have been accompanied by unwanted side effects, from small-scale disturbances or even a total collapse of ecosystems. This is because we interfere too drastically with nature when we transform and redistribute energy and other natural resources. Acidification is a good and now classical example; its origin and chemical effect mechanisms are relatively well known, despite the discussion on how much of this can also be ascribed to the changes in land use during the last decades or even centuries. The lowered pH levels of surface waters in many parts of Europe and North America have caused a variety of problems for the biota. Knowledge of the character and functioning of these problems is far from satisfactory.

The negative effects associated with acidification can be detected at the individual, population, species, community or even ecosystem level, but the physiological and foodchain mechanisms behind the observed effects are preferably analysed at the individual level.

Traditionally, and with good reasons, the problems for the fish fauna have received most attention. Mainly physiological constraints for their survival have been presented, but also drastic changes in their food supply can occur, i.e. many prey items in the benthic invertebrate communities have disappeared. This is also clearly pronounced in the background document for SWAP (Mason & Seip 1985). Moreover, the invertebrates are important also in the processing of organic matter and energy.

If no functional analogue for a disappearing species is available, the fish and foodweb are severely affected. But even if the fish can survive on an alternative food, it might very well be less favourable because it is less common, less easy to catch, or less good in its nutritive value. In all these cases, less net energy gain is yielded by the fish to be allocated to growth and reproduction. Also then, the condition of the fish declines, and it becomes more sensitive to diseases and predation.

Acidification of surface waters affects their benthic invertebrates in various direct and indirect ways, which can be classified as either physiological or foodchain effects, approximately due to abiotic and biotic factors, respectively. Behavioural disturbances can probably also be traced back to these main groups of mechanisms. The concept 'foodchain' in the title of this paper does not only refer to biomagnification and the like, but also to the food, feeding, predation and other biotic relations along the foodchain.

The literature on the effects of lowered pH on benthic invertebrates was reviewed by Økland & Økland (1986). The specific Al effects at low pH were treated by Herrmann (1987a), whereas Campbell & Stokes (1985) reviewed the effects of mainly heavy metals under low pH conditions. The purpose of the present review is to discuss the different aspects within the area of its title, and present appropriate examples of our present knowledge, with emphasis on sublethal effects and the mechanisms involved, and less in a descriptive and taxonomic approach.

2. Distribution patterns

The natural species diversity of the bottom communities in lakes, at least the littoral parts, and especially streams, is often quite high, as the conditions show a large spatial and temporal variation. The benthic invertebrates in these waters are dominated by many kinds of insect larvae, snails, mussels, some crustaceans, leeches and other worms. Freshwaters show a large variation in their 'natural' pH (from below pH 3 up to 11–12), but the lakes and streams likely to be affected by acidification in their original state normally are within the range 5.5 to 7. At 1–2 pH units lower one inevitably finds a much more impoverished community, i.e. much fewer species, which includes loss of fish prey items, and often, but not necessarily, also a decreased abundance and biomass. Many species of crustaceans, molluscs and mayfly nymphs suffer most from the effects of low pH and disappear (Mossberg & Nyberg 1979; Otto & Svensson 1983; Sutcliffe 1983; Engblom & Lingdell 1984; Økland & Økland 1986).

Mussels are often relatively resistant to acidity (Mackie & Flippance 1983), and it has even been suggested that they can raise the alkalinity somewhat in their near vicinity (Rooke & Mackie 1984). Odonates (dragonflies and damselflies) are often quite abundant at acidic sites, perhaps partly due to the favourable structure of the *Sphagnum* mats that sometimes dominate the bottom in acidified lakes. In addition, their eggs seem unaffected by low pH (Hudson & Berrill 1986). This also accounts for some less sensitive mayfly nymphs, whereas the hatching rate was retarded for a sensitive species (Rowe *et al.* 1988a). Most water bugs (Hemiptera) are, compared with other macroinvertebrates, favoured by low pH, because of the absence of fish and, being air breathing, are less dependent on the conditions within the water. These and other biotic reasons are discussed by Stenson & Eriksson (1989) and in §7.

Chironomids often do well in acidic lakes (Mossberg & Nyberg 1979), possibly favoured by the increased amounts of non-decomposed organic matter. Also, increased amounts of bottom substrate rich in organic matter and calcium (or both) can act as a refuge and protect benthic animals from episodes of acidic water and associated metals (Raddum 1980; Weatherley & Thomas 1989; J. Herrmann, unpublished results).

For a large group of naturally acid brown-water streams in New Zealand, neither species richness nor the numbers of mayflies or other traditionally acid sensitive invertebrates was found to be correlated with pH (Winterbourn & Collier 1987). Benthic invertebrate communities of relatively high diversity have been found in small lakes with naturally very low pH, i.e. 3 or below (Havas & Hutchinson 1982; Rydgård *et al.* 1985).

Thus a variety of often interrelated abiotic and biotic factors, can go someway to explain the distribution of some invertebrate taxa. However, a problem with field data is that it is often not possible to separate different potential detrimental factors, which may co-vary, e.g. the lowered pH itself and associated metal concentrations such as aluminium or cadmium (Nilsson & Johansson 1985; Herrmann 1987a). Analyses of streams in southern Wales indicated Al to be an important factor determining the invertebrate community (Weatherley & Ormerod 1987; Henriksson 1988), whereas Canadian researchers found pH itself more important for the benthic faunal assemblages (Allard & Moreau 1987). Certainly food availability, competition, and predator conditions are also important (Havas & Hutchinson 1985a; Stenson & Eriksson 1989).

3. Mortality

Distribution patterns are in fact a function of where a species can survive, depending upon an array of lethal and sublethal factors (see following sections). Information from mortality tests can be of great value in indicating possible impact mechanisms. Descriptive studies relating mortality to only H^+ ions are mainly omitted in this paper. Rather metals, and generally all types of studies of physiological or food chain mechanisms will be emphasized. These can involve interactions between increased hydrogen ions and raised metal levels, as well as other ions, humic and other organic substances, food, predators, etc.

It has repeatedly been shown that Al can be lethal to freshwater invertebrates such as mayflies and crustaceans (see, for example, Havas 1985; Havas & Hutchinson 1982; Herrmann 1986). However, the mortality of the mayfly *B. rhodani* as well as chironomids and phantom midges mainly seems caused by low pH itself. A very interesting phenomenon is the fact that Al can, at least temporarily, increase the survival of some animals such as daphnids and mayflies exposed to low pH (*ca.* 4), that would otherwise be detrimental.

Campbell & Stokes (1985) critically reviewed published studies and predicted how some ten metals would act at a lowered pH, with respect to their speciation, their adsorption on biological surfaces and their uptake by and toxicity to aquatic biota. The biotic response and availability (or both) of cadmium (Cd), zinc (Zn) and copper (Cu) was suggested to decrease when pH is lowered from 7 to 4 (type I), whereas for aluminium (Al), and perhaps also mercury (Hg) and lead (Pb), the reverse seems to be the case, i.e. being more detrimental at low pH (type II). Mackie (1989) found that a metal can show a type I response for one species and type II for another. He also suggested a third group, with greatest biological response around neutrality. The large variation of metal effects on biota was discussed by Lithner (1989), who concluded that most elements can be hazardous at a concentration close to background levels, implying small margins for organisms to tolerate deviations from these levels.

Younger or smaller animals are, with respect to both pH or metal concentration, often more sensitive than older or larger ones of the same species, e.g. daphnids (Ravera 1984), chironomids (Williams *et al.* 1986), ephemeropterans (Allard & Moreau 1987; J. Herrmann, unpublished results) and amphipods (McCahon *et al.* 1989). Nilsson (1976) found female amphipods more sensitive to low pH than were the males.

Mackie (1989) found cadmium (Cd) more toxic than lead (Pb) and aluminium (Al) for all tested functional groups. He found shredders most sensitive to the metals, whereas predators were most tolerant. It was however suggested that the animals' response to a specific metal was more similar within a taxonomic group than within a functional group.

Amphipods are highly acidity sensitive, even at pH 6, but this impact was substantially decreased by moderate addition of humic substances, that also seemed to benefit food conversion (Hargeby & Petersen 1988). For blackfly larvae both low pH (*ca.* 6) as well as increased concentrations of Al or humus are detrimental, but the toxicity of either substance is counteracted by the other (Petersen *et al.* 1986). Humic acid also mitigated zinc toxicity of daphnids, although in this case pH was between 8 and 9 (Paulauskis & Winner 1988). An extensive review on the combined effects of pH, metals and humus on freshwater invertebrates was presented by Petersen *et al.* (1987).

Acid-stream bottoms can become covered with a reddish-brownish sludge, often a mixture of humic substances and iron (Fe) precipitates. Such conditions are certainly disastrous for those animals that require a relatively hard bottom to find a proper food, as for example mayflies and stoneflies (Rasmussen & Lindegaard 1988), but it might also cause oxygen supply problems. The food problem can be due to the adverse effect on the periphyton community (Sode 1983). The toxicity of Fe ions *per se* should also be elucidated. Maltby *et al.* (1987*a*) showed that the isopod *Asellus aquaticus* was fairly tolerant, as it is to pollution in general.

4. Ion regulation and respiration

Among the first scientists to work with these problems specifically for invertebrates were Sutcliffe and his co-workers (Sutcliffe 1978, 1983). This work focused mainly on ion regulation in crustaceans and how this could explain their distribution in streams of different pH in the Lake District. The importance of studying several ions, and the influence of temperature was emphasized. The importance of what type of food the animals could find (see below) was also indicated.

The basic knowledge of physiological mechanisms of insects is extensive, but views on how acidified water and associated changes in metal concentrations affect freshwater invertebrates are often influenced by fish physiologists. Thus Havas (1981) suggested a conceptual framework containing four major functions of greatest importance when pH affects freshwater animals, including invertebrates, namely sodium regulation, calcium regulation, acid-base balance and respiration, and how these are related.

In osmoregulation two processes, namely influx and outflux of sodium (Na) ions, are important. They can both be severely affected at low pH, depending on the animal and condition studied. In most cases an increased net loss has been shown, e.g. in daphnids (Havas *et al.* 1984), mussels (Malley *et al.* 1988), stoneflies (Twitchen, this symposium) and mayflies (Rowe *et al.* 1988*b*; Frick & Herrmann, this symposium). However, Berrill *et al.* (1987) noticed increased mortality and loss of sodium ions when exposing mayfly nymphs to pH 3.5, but not at 4.5, when compared with pH 6.5.

Increased Al levels, often associated with low pH, have also been shown to decrease Na levels in corixids (Witters *et al.* 1984), crayfish (Appelberg 1985), daphnids (Havas 1985; Havas & Likens 1985b), stoneflies (Twitchen, this symposium) and mayflies (Herrmann 1987b; Frick & Herrmann, this symposium). The fact that Al affects the ion regulation in several insect larvae has also been indicated by the strong Al-specific staining on structures known to mediate cation exchange for these animals (Havas 1986, 1987).

However, in the mentioned studies on daphnids, stoneflies and mayflies, Al was also shown to ameliorate the adverse impact of acidity at *ca.* pH 4 (Havas & Likens 1985b; Frick & Herrmann, this symposium; I. D. Twitchen, unpublished results), a phenomenon also seen in several mortality studies. The reason for this might be that Ca ions are rare at low pH, and Al ions can replace them as barriers towards hydrogen ions, thus stabilizing the cell membrane (Havas 1985). Other possibilities were presented by Herrmann (1987a). This can of course be of temporary advantage during a short episodic acidification event, but it must be emphasized that in the long run it might be a strain on the organism. Moreover Havas & Likens (1985a) indicated that maximum Al toxicity occurs when the water is oversaturated, i.e. at pH around 6.5, when 98% of the Al is theoretically insoluble. Also increased Cd levels, sometimes associated with lowered pH, have shown a clear tendency to diminish the Na loss, whereas Fe ions did not cause any notable change (K. G. Frick, unpublished results). Further, a normal ion flux and content seems essential for successful moults, but ion regulation also becomes more disturbed during moulting in acid waters.

The ability to incorporate calcium (Ca) into the exoskeleton, a physiological process, is essential for crayfish and has reportedly been seen to be disturbed at low pH (France 1983; Appelberg 1984). This also seems valid for molluscs, especially gastropods. The distribution patterns of these two groups, both in the microhabitat and the more regional scale, also seem to corroborate such a difference (Mackie & Flippance 1983; Økland & Økland 1986).

It has been shown several times that gastropods increase their tolerance to low pH with increasing Ca concentration (Økland & Økland 1986). The reason for this may be that the increased Ca levels in the blood act as a buffer against the pH, as suggested for mussels (Malley *et al.* 1988), or that Ca reduces membrane permeability of hydrogen ions. However, neither Willoughby & Mappin (1988) nor I. D. Twitchen (unpublished results) did find Ca to ameliorate the detrimental effect of low pH on sodium imbalances in mayflies and stoneflies, respectively. Jernelöv *et al.* (1981) suggested that haemoglobin was responsible for the acid neutralizing capacity found in red chironomid larvae.

Thus several physiological functions, e.g. sodium and calcium regulation, can be adversely affected by increased hydrogen or metal ion concentrations. Animals tend to compensate such 'physiological imbalances', which is a stressful situation, by a higher metabolic activity. This requires more energy, for which a larger respiration rate is needed, and consequently a higher oxygen consumption follows. Herrmann & Andersson (1986) demonstrated such a reaction for three species of mayflies exposed to increased Al levels, whereas decreasing respiration rates were found when caddisfly larvae were exposed to Al at low pH (Correa *et al.* 1986). The same type of response, i.e. increased oxygen consumption, was also found in a stonefly experiencing a pH deviating 2–3 units up and down from 6 (R. Rupprecht, unpublished results).

This reaction can be even more accentuated by Al hydroxide precipitations on body surfaces, as reported for fish (Muniz & Leivestad 1980). Whether this specific problem affects mayfly nymphs is not known, but at least high concentrations of Al have been observed on some parts of the body (J. Herrmann, unpublished results). Also increased mucus production at gill surfaces, potentially detrimental to their functioning, has been reported for fish, but McCahon *et al.* (1987) did not observe this on mayfly nymphs after exposure to pH 5 and 0.4 mg Al l^{-1}.

5. Drifting behaviour

It is reasonable to assume that an animal facing hostile conditions, e.g. high concentrations of hydrogen, aluminium or other metal ions, it will try to avoid this threat. In streams, the normal mechanism of escape is to drift away, i.e. to let the water-flow passively move the organism to more favourable conditions downstream. However, this fatalistic behaviour can take place also for more 'natural' reasons, being a part of the animals' life cycle or feeding behaviour in an appropriate strategy of resource exploitation, and triggered by light, water velocity or temperature. Thus relating the drift to an anthropogenic factor proposed to affect an animal population in the field, can pose some problems in the analysis.

Nevertheless, some invertebrate species, especially the mayfly *B. rhodani*, increase their drift intensity at a lowered pH (Nilsson 1976; Hall *et al.* 1982; Ormerod *et al.* 1987; Raddum & Fjellheim 1987; Herrmann *et al.*, unpublished results), thus partially depleting the population at the affected locality (Hall *et al.* 1980). An indirect observation of the influence of pH on the drift was made by Kullberg & Petersen (1987), who noticed a decreased drift of macroinvertebrates when an acid stream was limed. Furthermore, elevated Al concentrations have been proved to cause an increased drift behaviour (Ormerod *et al.* 1987).

Exceptions also exist. McCahon *et al.* (1989) indicated that single short-term episodes did not cause any changed drift. Some mayfly species, e.g. *Heptagenia sulphurea*, and netspinning caddis larvae of the genus *Hydropsyche*, reacted by halting the drift totally, thus taking the risk of perishing (Nilsson 1976; J. Herrmann, unpublished results).

6. Bioaccumulation and biomagnification of metals

Adopting the definitions of Dallinger *et al.* (1987), the term bioaccumulation should be restricted to when a foreign substance occurs in an organism at a concentration higher than in the surrounding water. Unfortunately, this concept is too often confused with the phenomenon when an organism at a higher trophic level shows a higher concentration than that at a lower one. This food-chain accumulation should, however, be referred to as biomagnification.

The word bioavailability indicates the amount and type of a substance that can potentially be taken up by an organism. The substance may either be 'disarmed' by some internal or external compound or process (e.g. at acidic and high metal conditions calcium can act so), be incorporated into the organism without any harm, or just pass through the organism, and eventually reach a predator (Ravera 1984). The actual pathway depends on involved osmotic processes, extent of metabolic uptake, type and amount of the substance, although the type of organism, its feeding habits, developmental stage and exposure time are also important. A measure of bioavailability, as attained from methods not using organisms, e.g. a dialysis bag, does not necessarily give all the information about the biological potential or real activity of a substance.

Some investigators have demonstrated that the Al content of benthic invertebrates in acidic waters are generally higher than in more circumneutral conditions (Wickham *et al.* 1987; Bendell Young & Harvey 1988; Herrmann 1990), whereas Hall *et al.* (1988) and McCahon & Pascoe (1989) did not observe such a pattern.

In moderately acid waters accumulation of Al with time has been shown for daphnids, mayflies, amphipods and stoneflies (Havas 1985; McCahon *et al.* 1989; Frick & Herrmann 1990). There is also evidence for Cd (A. Gerhardt, unpublished results), and Cu (Wickham *et al.* 1987) accumulation. Servos *et al.* (1987) did not find increased Al, Cd or Zn levels during an artificial low pH episode. However, most of these studies could not discriminate between metals on the body surface and those within the animal.

This separation is crucial to solve the controversy over whether insect nymphs can convey Al to the terrestrial ecosystem on their emergence. This was suggested by Nyholm (1981), but regarded as less probable by Otto & Svensson (1983). This could be a link in a type of biomagnification process, in this case from stoneflies to birds, where they might disturb reproduction in different ways. To test the general applicability of this hypothesis, mayfly nymphs were exposed to different Al conditions, exposure times and moulting sequences (Frick & Herrmann 1990). Only a smaller fraction of Al was accumulated internally, and most was deposited on or in the exuviae. It would then be left behind when the insect emerged from the water, thus in general not supporting the idea of biomagnification of Al in the foodchain. However, under special conditions, some externally accumulated Al might also be transferred to terrestrial foodchains. Confirming this, Bendell Young & Harvey (1988) found that less than 10 % of larval Al was transferred to the adults. This result was also confirmed by Malley *et al.* (1987), who found that neither pH nor external Al substantially affected the internal Al in crayfish.

Hall *et al.* (1988) found that after field experiments with mayfly nymphs, most Al was found in their guts, but only small amounts were adsorbed to the outer body surface. However, Krantzberg & Stokes (1988) concluded that adsorption was important, but somewhat suppressed by lowered pH. McCahon *et al.* (1987) demonstrated that Al could be found on all parts of mayfly nymphs. Obviously, the controversy is still there, and different observations are somewhat contradictory. If substantial amounts of Al, under some conditions, are conveyed to e.g. birds, the chemical form in which Al exists and in what form it can be 'used' by the next trophic level is still not known.

In several studies it has been shown that acidified lakes and streams exhibit drastically increased mercury levels in water, sediment, and invertebrates (Lindqvist *et al.* 1990). Elevated levels of mercury and cadmium have been reported for fish, loons and ducklings in acidified waters (Stenson & Eriksson 1989), but only relatively weak evidence for biomagnification for these metals were presented.

If biomagnification takes place within the benthic foodchain, predators would have high Al levels. This is obviously not the case, either at pH 6 or pH 4 (Herrmann 1990). On the contrary, they found the highest Al concentrations in shredders, perhaps due to increased Al levels associated to bottom material.

7. Food, feeding and other biotic relations

Most effects discussed so far are of a rather immediate type, involving physiological responses. When found, tolerance levels also for sublethal parameters, are not satisfyingly reflected in distribution patterns for animals. This can be because acidification can also have several indirect biotic effects, via impaired growth and reproduction, resulting in altered structures of populations, communities and ecosystems (cf. Appelberg 1987). The most obvious is that chemical conditions can in some way affect the food quality and quantity (or both), as shown for *Gammarus pulex* by Petersen *et al.* (1986).

The concept 'foodchain' in the title of this paper does not only refer to biomagnification and the like, but also to food, feeding, predation and other biotic relations along the foodchain, dealt with below.

Decomposition of organic matter, especially coarse detritus, has repeatedly been shown to be slower in acid waters (Otto & Svensson 1983; Andersson 1985; H. Oscarsson, unpublished results). This can be a result of reduced microbial activity (Leivestad *et al.* 1976; Garden & Davis 1988), or fewer shredding detritivores (Andersson 1985), or both. The microbial explanation was, however, disputed by Gahnström & Fleischer (1985) and Townsend & Hildrew (1988). The decline of shredders might be due to the fact that some invertebrate predators increase when fish become scarce. The feeding activity of the animals can also be reduced (McCahon & Pascoe 1989). The importance of the shredding detritivores is also indicated by observations of increased numbers of shredders (*A. aquaticus*) after the liming of a lake (H. Oscarsson, unpublished results). This could also be due to the introduction of fish, parallel to the liming, suppressing influential predators on *A. aquaticus*.

However, there are also observations in acidic streams that shredders can be relatively more abundant than in more neutral streams (Friberg *et al.* 1980; Nilsson & Johansson 1985; J. Herrmann, unpublished results). This may be because the adverse effect of low pH might be compensated by an increased supply of non-degraded organic matter, although of poorer quality (Otto & Svensson 1983). As a consequence, shredders grow larger, but also slower, and are less efficient in decomposing the organic matter. That caddisfly larvae can grow larger in acid lakes can, however, also be caused by the absence of fish in such waters (Raddum 1980).

Invertebrates utilizing fine detritus, i.e. deposit and filter feeders, are often disfavoured in acid waters. But the most affected group of animals are the scrapers or grazers (Otto & Svensson 1983; Sutcliffe 1983; Nilsson & Johansson 1985). Among these, the dominating organisms are nymphs of Ephemeroptera (mayflies). There certainly are physiological reasons for this (see above), but for some species at least, it also seems clear that their food, mainly the microflora attached to the bottom material in streams, has become less abundant in acidified waters, less favourable in its composition, or occurs at the wrong time. In extensive laboratory experiments, Sjöström (this symposium) found that *Baetis rhodani* had a reduced survival on algae from an acid stream compared to algae from a neutral one, irrespective of water type used. In fact, in neutral water, the survival of nymphs reared on algae from an acid stream was as low as for starving animals. Willoughby & Mappin (1988) indicated that the variation in accessibility of algae in time, rather than their quality as food *per se*, excluded the mayfly *Ephemerella ignita* from some acid streams.

Predators quite often seem to be favoured, or at least unaffected, in acid lakes and streams. This may be because of their relatively opportunistic feeding habits (Otto

& Svensson 1983), the absence of foodchain accumulation (biomagnification) of Al (Herrmann 1990), or 'an ecological release' when an acidified water loses its fish stock.

Fish predation can be regulatory in freshwaters, but when fish are lost, invertebrates such as odonates, phantom midges and water boatmen often become dominant in the benthic community. This may strongly affect both the pelagic and littoral food webs and the increased abundance of odonates are suspected to be partly responsible for the subdued shredder guild in lakes (Stenson & Eriksson 1989).

Increased retention of phosphorus in soils and bottom sediments caused by retarded decomposition and mineralization, and precipitation with Al, causes a reduced primary productivity in acidified lakes and streams. This nutrient limitation may eliminate less competitive algal species and changes in algal communities may affect the bottom invertebrates. This 'oligotrophication' (Grahn *et al.* 1979) following acidification can be enhanced by immobilization of phosphorus in the fecal pellets of odonates and the nutrient sink in the *Sphagnum* mat, often a conspicuous feature of acidified lakes (Nyman, this symposium). This mat also harbours some invertebrates, by acting as a shelter against predators but also acting as a foraging area (B.-I. Henriksson, unpublished results). Further, there are several observations of increased occurrence of filamentous green-algae in acidified lakes and streams (Grahn *et al.* 1974; Müller 1980; J. Herrmann, unpublished results). This may be caused by increased deposition of nitrogen, i.e. partly the same agent that causes the acidification, but also because maximal photosynthesis of some algal species is just below pH 5 (A. Rosmarin, personal communication).

Finally, increased transparency of water often prevails in acidified lakes, and two mechanisms contribute to this. One is the decreased primary productivity, the other the precipitation of humic substances by Al. Increased transparency compensates to some extent for the lowered phosphorus levels (Stenson & Eriksson 1989). Better light can also allow predators to be more efficient in their hunting. Pelagic insects as well as birds might benefit from this and their increased feeding success affects the bottom fauna. Conversely, the prey might better become aware of the predator.

8. Adaptation?

Hopefully acidification caused by air pollution and other activities will be reduced in the future. The rate of recovery sometimes proceed only slowly, remedial activities are only temporary complements, and the research effort and effectiveness of restoration methods are still too low. Therefore it is also worthwhile to discuss whether organisms can acclimatize, or even adapt, to the man-induced increased acidity and metal levels.

Fish can, by experimental treatment, attain significant but not very long-lasting improvements for survival in acid conditions. This is achieved, to some extent, by acclimatization, i.e. a non-genetical 'response', but perhaps also by adaptation, which may give rise to a genetically altered population (Gjedrem 1980; Rosseland & Skogheim 1987). But it has also been suggested that pH may exert selection pressure on benthic invertebrates, with the success of more acid-tolerant genotypes (Maltby *et al.* 1987 *b*).

The red chironomid larvae that have existed for at least several hundred years in extremely acid conditions (pH < 3) in NW Canada may provide an example of adaptation. They withstand low pH conditions much better than conspecific animals from a 'normal' lake in Scandinavia, probably because of their haemolymph that can buffer the acidity better (Jernelöv *et al.* 1981). Freshwater isopods, which have been exposed to mine effluents for several decades, appeared to be more tolerant to low pH compared to those exposed during shorter periods of time (Maltby *et al.* 1987 *a*). Observations that amphipods, isopods and mussels that were pre-adapted to a low-alkalinity or low pH locality, proved to be more able to tolerate a low pH than those from high-alkaline conditions (Mackie 1989) also support this possibility. If an ability like this has the function as a mechanism of genetic divergence in adjacent populations, at different pH régimes, natural selection can be advantageous when facing environmental stress, as well as in the time to recover from it (France 1987). However, McCahon & Pascoe (1989) reported absence of any preadaptive value for a mayfly nymph.

Klercks & Weis (1987) reviewing the topic, concluded that in most cases a pre-adaptation effect is possible, giving more tolerance. This seems more common among 'lower' animals, i.e. invertebrates. Krantzberg & Stokes (1988) pointed out that pre-exposure can give rise to lowered sensitivity for some metals, but not for others.

9. Concluding remarks

Two general tendencies seem to hold for Al, being the most studied metal in the connection with low pH and invertebrates: (i) pH itself quite often seems more important than Al in explaining the reactions and distribution patterns of benthic invertebrates; and (ii) low pH resistant species also seem to tolerate elevated Al concentrations.

Although a good deal of knowledge on distribution and mortality of organisms in relation to pH-metal exposure in the field was presented above, this does not contribute much to a real understanding of the problems we are addressing, even if they may indicate what they are. Most of this paper was therefore devoted to the variety of sublethal parameters, reflecting the various problems that organisms are exposed to. For several reasons such 'early warning signals' must dominate when discussing and analysing environmental disturbance patterns. Sublethal parameters should make it possible to detect and understand an environmental problem earlier and better, in the sense of a more correct interpretation.

When exposed to either adverse environmental conditions or constrained by intra- or interspecific competition (or both), organisms have less energy available for growth and reproduction, and ultimately the existence of an organism on a locality depends on its reproductive success.

However, an adverse effect may be partly compensated for and might not be manifest until later. To reveal such subtle and truly sublethal effects, one must also remember that negative results, i.e. non-significant relationships, can hide an emerging effect. They can have poor statistical significance, but still be biologically significant and become serious with time. This was also pointed out by Hayes (1987), recommending more rigorous experiments and larger numbers of samples, in order not to lose information and early indications of unknown processes.

Even if this paper mainly deals with basically abiotic factors, i.e. more chemical and physiological ones, often implying short-term and direct explanations and phenomena, the large variety of biotic and thus more long-term and indirect factors (or both) are as important. At least in their ultimate and crucial phase, ecological and thus many environmental problems are a highly interrelated mixture of abiotic and biotic; of physiological and foodchain; of short-term and long-term; of direct and indirect factors and phenomena that determine the response of nature and what we observe. This is why we still comprehend too little and why we need more really integrated studies to understand our relation to and impact on nature.

For critical reading and many useful comments on an early draft I sincerely thank Dr K. Frick, Dr S. Rundle and Dr B. Svensson. The work for this review has been financed by the Surface Waters Acidification Programme, the Royal Swedish Academy of Sciences and the Swedish Environmental Protection Agency.

References

Allard, M. & Moreau, G. 1987 *Hydrobiologia* **144**, 37–49.

Andersson. G. 1985 *Ecol. Bull.* **37**, 293–299.

Appelberg, M. 1984 *Rep. Inst. Freshw. Res. Drottningholm* **61**, 48–59.

Appelberg, M. 1985 *Hydrobiologia* **121**, 19–25.

Appelberg, M. 1987 In *Ecophysiology of acid stress in aquatic organisms* (ed. H. Witters & O. Vanderborght) (*Annls Soc. r. zool. malac. Belg.* **117**), pp. 167–179.

Bendell Young, L. & Harvey, H. H. 1988 *Verh. Internat. Verein. Limnol.* **23**, 246–251.

Berrill, M., Rowe, L., Hollett, L. & Hudson, J. 1987 In *Ecophysiology of acid stress in aquatic organisms* (ed. H. Witters & O. Vanderborght) (*Annls Soc. r. zool. malac. Belg.* **117**), pp. 117–128.

Campbell, P. G. C. & Stokes, P. M. 1985 *Can. J. Fish. Aquat. Sci.* **42**, 2034–2049.

Correa, M., Coler, R., Yin, C.-M. & Kaufman, E. 1986 *Hydrobiologia* **140**, 237–241.

Dallinger, F., Prosi, F., Segner, A. & Back, H. 1987 *Oecologia* **73**, 91–98.

Engblom, E. & Lingdell, P.-E. 1984 *Rep. Inst. Freshwat. Res., Drottningholm* **61**, 60–68.

France, R. L. 1983 In *Freshwater crayfish V* (ed. C. R. Goldman), pp. 98–111. West Port: Avi Publishing Company Inc.

France, R. L. 1987 In *Ecophysiology of acid stress in aquatic organisms* (ed. H. Witters & O. Vanderborght) (*Annls Soc. r. zool. malac. Belg.* **117**), pp. 129–137.

Friberg, F., Otto, C. & Svensson, B. S. 1980 In *Proceedings of the international conference on the ecological impact of acid precipitation* (ed. D. Drabløs & A. Tollan), pp. 304–305. Norway: SNSF project.

Frick, K. G. & Herrmann, J. 1990 *Ecotox. Environ. Safety* **19**, 81–88.

Gahnström, G. & Fleischer, S. 1985 *Ecol. Bull.* **37**, 287–292.

Garden, A. & Davies, R. W. 1988 *Environ. Pollut.* **52**, 303–313.

Gjedrem, T. 1980 In *Proceedings of the international conference on the ecological impact of acid precipitation* (ed. D. Drabløs & A. Tollan), p. 308. Norway: SNSF project.

Grahn, O., Hultberg, H. & Landner, L. 1974 *Ambio* **3**, 93–94.

Hall, R. J., Likens, G. E., Fiance, S. B. & Hendrey, G. R. 1980 *Ecology* **61**, 976–989.

Hall, R. J., Pratt, J. M. & Likens, G. E. 1982 *Wat. Air Soil Pollut.* **18**, 273–287.

Hall, R. J., Bailey, R. C. & Findeis, J. 1988 *Can. J. Fish Aquat. Sci.* **45**, 2123–2132.

Hargeby, A. & Petersen, R. C. Jr 1988 *Freshwat. Biol.* **19**, 235–247.

Havas, M. 1981 In *Effects of acidic precipitation on benthos* (ed. R. Singer), pp. 49–65. Springfield: North American Benthological Society.

Havas, M. 1985 *Can. J. Fish Aquat. Sci.* **42**, 1741–1748.

Havas, M. 1986 *Wat. Air Soil Pollut.* **30**, 735–741.

Havas, M. 1987 In *Ecophysiology of acid stress in aquatic organisms* (ed. H. Witters & O. Vanderborght) (*Annls Soc. r. zool. malac. Belg.* **117**), pp. 151–164.

Havas, M. & Hutchinson, T. C. 1982 *Can. J. Fish Aquat. Sci.* **39**, 890–903.

Havas, M., Hutchinson, T. C. & Likens, G. E. 1984 *Can. J. Zool.* **62**, 1965–1970.

Havas, M. & Likens, G. E. 1985*a* *Can. J. Zool.* **63**, 1114–1119.

Havas, M. & Likens, G. E. 1985*b* *Proc. natn. Acad. Sci. U.S.A.* **82**, 7345–7349.

Hayes, J. P. 1987 *Ecotox. Environ. Safety* **14**, 73–77.

Henriksson, L. 1988 In *Liming of Lake Gårdsjön, an acidified lake in SW Sweden* (ed. W. Dickson), National Swedish Environmental Protection Board Report No. 3426, pp. 309–327.

Herrmann, J. 1986 In *Proceedings of the 3rd European Congress of Entomology* (ed. H. H. W. Velthuis), p. 176. Amsterdam.

Herrmann, J. 1987a In *Speciation of metals in water, sediment and soil systems* (ed. L. Landner) (Lecture notes in earth sciences) vol. 11, pp. 157–175. Berlin: Springer-Verlag.

Herrmann, J. 1987b In *Ecophysiology of acid stress in aquatic organisms* (ed. H. Witters & O. Vanderborght) (*Annls Soc. r. zool. malac. Belg.* 117), pp. 181–188.

Herrmann, J. 1990 (In preparation.)

Herrmann, J. & Andersson, K. G. 1986 *Wat. Air Soil Pollut.* 30, 703–709.

Hudson, J. & Berrill, M. 1986 *Hydrobiologia* 140, 21–25.

Jernelöv, A., Nagell, B. & Svensson, A. 1981 *Holarc. Ecol.* 4, 116–119.

Klerks, P. L. & Weis, J. S. 1987 *Environ. pollut.* 45, 173–205.

Krantzberg, G. & Stokes, P. M. 1988 *Environ. Tox. Chem.* 7, 653–670.

Kullberg, A. & Petersen, R. C. Jr 1987 *Freshwat. Biol.* 17, 553–564.

Leivestad, H., Hendrey, G., Muniz, I. P. & Snekvik, E. 1976 In *Impact of acid precipitation on forest and freshwater ecosystems in Norway* (ed. F. H. Braekke), pp. 87–111. Norway: SNSF Project, FR 6/76.

Lindqvist, O., Johansson, K., Aastrup, M., Andersson, A., Bringmark, L., Hovsenius, G., Håkanson, L., Iverfeldt, Å., Meili, M. & Timm, B. 1990 *Wat. Air Soil Pollut.* (Submitted.)

Lithner, G. 1989 *Sci. tot. Environ.* 87/88, 365–380.

Mackie, G. L. 1989 *Arch. Environ. Contam. Toxicol.* 18, 215–223.

Mackie, G. L. & Flippance, L. A. 1983 *J. mollusc. Stud.* 49, 204–212.

Malley, D. F., Chang, P. S. S., Moore, C. M. & Lawrence, S. G. 1987 *Can. Tech. Rep. Fish Aquat. Sci.* 1480, 54–68.

Malley, D. F., Huebner, J. D. & Donkersloot, K. 1988 *Arch. Environ. Contam. Toxicol.* 17, 479–491.

Maltby, L., Calow, P., Cosgrove, M. & Pindar, L. 1987b In *Ecophysiology of acid stress in aquatic organisms* (ed. H. Witters & O. Vanderborght) (*Annls Soc. r. zool. malac. Belg.* 117), pp. 105–115.

Maltby, L., Snart, J. O. H. & Calow, P. 1987a *Environ. pollut.* 43, 271–279.

Mason, J. & Seip, H. M. 1985 *Ambio* 14, 45–51.

McCahon, C. P. & Pascoe, D. 1989 *Arch. Environ. Contam. Toxicol.* 18.233–242.

McCahon, C. P., Brown, A. F., Poulton, M. J. & Pascoe, D. 1989 *Wat. Air Soil Pollut.* 45, 345–359.

McCahon, C. P., Pascoe, D. & Kavanagh, C. Mc. 1987 *Hydrobiologia* 153, 3–12.

Mossberg, P. & Nyberg, P. 1979 *Rep. Inst. Freshwat. Res., Drottringholm* 58, 77–87.

Muniz, I. P. & Leivestad, H. 1980 In *Proceedings of the international conference on the ecological impact of acid precipitation* (ed. D. Drabløs & A. Tollan), pp. 84–92. Norway: SNSF project.

Müller, P. 1980 *Can. J. Fish Aquat. Sci.* 37, 355–363.

Nilsson, A. N. & Johansson, A. 1985 *Inf. Sötv.-lab. Drottningholm*, 1985/11, 56pp. (In Swedish, with English summary.)

Nilsson, L. M. 1976 *Report to the Swedish Environmental Protection Agency.* 14pp.

Nyholm, E. 1981 *Environ. Res.* 26, 363–371.

Økland, J. & Økland, K. A. 1986 *Experientia* 42, 471–486.

Ormerod, S. J., Boole, P., McCahon, C. P., Weatherley, N. S., Pascoe, D. & Edwards, R. W. 1987 *Freshwat. Biol.* 17, 341–356.

Otto, C. & Svensson, B. S. 1983 *Arch. Hydrobiol.* 99, 15–36.

Paulauskis, J. D. & Winner, R. W. 1988 *Aquat. Toxicol.* 12, 273–290.

Petersen, R. C. Jr, Hargeby, A. & Kullberg, A. 1987 *SNV Rep. no.* 3388, pp. 1–147.

Petersen, R. C. Jr, Petersen, L. B.-M., Persson, U., Kullberg, A., Hargeby, A. & Paarlberg, A. 1986 *Wat. Qual. Bull.* 11, 44–49.

Raddum, G. G. 1980 In *Proceedings of the international conference on the ecological impact of acid precipitation* (ed. D. Drabløs & A. Tollan), pp. 330–331. Norway: SNSF project.

Raddum, G. G. & Fjellheim, A. 1987 In *Ecophysiology of acid stress in aquatic organisms* (ed. H. Witters & O. Vanderborght), (*Annls Soc. r. zool. malac. Belg.* **117**), pp. 77–87.

Rasmussen, K. & Lindegaard, C. 1988 *Wat. Res.* **22**, 1101–1108.

Ravera, O. 1984 *Experientia* **40**, 2–14.

Rooke, J. B. & Mackie, G. L. 1984 *Can. J. Zool.* **62**, 793–797.

Rosseland, B. O. & Skogheim, O. 1987 In *Ecophysiology of acid stress in aquatic organisms* (ed. H. Witters & O. Vanderborght) (*Annls Soc. r. zool. malac. Belg.* **117**), pp. 255–264.

Rowe, L., Hudson, J. & Berrill, M. 1988*a* *Can. J. Fish Aquat. Sci.* **45**, 1649–1652.

Rowe, L., Berrill, M. & Hollett, L. 1988*b* *Comp. Biochem. Physiol.* **90** A, 405–408.

Rydgård, M., Brodin, Y., Nilsson, A. N. & Olsson, T. I. 1985 *Ent. Tidskr.* **106**, 133–138. (In Swedish.)

Servos, M. S., Malley, D. F., Mackie, G. L. & La Zerte, B. D. 1987 *Bull. Environ. Contam.* **38**, 762–768.

Sode, A. 1983 *Arch. Hydrobiol. Suppl.* **65**, 134–162.

Stenson, J. A. E. & Eriksson, M. O. G. 1989 *Arch. Environ. Contam. Toxicol.* **18**, 201–206.

Sutcliffe, D. W. 1978 *Rep. Freshwat. Biol. Ass.* **46**, 57–69.

Sutcliffe, D. W. 1983 *Rep. Freshwat. Biol. Ass.* **51**, 30–62.

Townsend, C. R. & Hildrew, A. G. 1988 *Verh. Internat. Verein. Limnol.* **23**, 1267–1271.

Weatherley, N. S. & Ormerod, S. J. 1987 *Environ. Pollut.* **46**, 223–240.

Weatherley, N. S. & Thomas, S. P. 1989 *Environ. Pollut.* **56**, 283–297.

Wickham, P., van de Walle, E. & Planas, D. 1987 *Env. Poll.* **44**, 83–99.

Williams, K. A., Green, D. W. J., Pascoe, D. & Gower, D. E. 1986 *Oecologia* **70**, 362–366.

Willoughby, L. G. & Mappin, R. G. 1988 *Freshwat. Biol.* **19**, 145–155.

Winterbourn, M. J. & Collier, K. J. 1987 *Hydrobiologia* **153**, 277–286.

Witters, H., Vangenechten, J. H. D., Van Puymbroeck, S. & Vanderborght, O. L. J. 1984 *Bull. Environ. Contam. Toxicol.* **32**, 575–579.

Apparatus for simulating episodes

By R. Morris[1], J. P. Reader[1], T. R. K. Dalziel[2]
and A. H. W. Turnpenny[3]

[1] *Department of Zoology, University of Nottingham, Nottingham NG7 2RD, U.K.*
[2] *Powergen Technology Centre, Ratcliffe-upon-Soar, Nottingham NG11 0EE, U.K.*
[3] *National Power Technology and Environment Centre, Fawley,
Southampton SO4 1TW, U.K.*

Details are given of the design and construction of a computer controlled apparatus for simulating episodes of high acidity and raised aluminium concentrations. The apparatus uses artificial soft-water media and has facilities for measuring sodium fluxes and the respiratory rate of individual fish. It can also be used for mortality and recovery studies on larger numbers of animals. Attention has been given to the problems of pH measurement in low conductivity waters and to the reduction of stress in fish caused by handling and sampling during experiments. Examples of the performance of the apparatus are given elsewhere in this symposium.

---

1. Apparatus

An apparatus for simulating episodic changes in water quality for physiological investigations on freshwater fish in which deionized water and salts are pumped into an aerated mixing tank to give artificial water of the desired composition (figure 1). Water from the mixing tank is then pumped through each of the litre fish bottles at a rate of *ca.* 700 ml h^{-1} for each bottle.

Details of the fish bottle design are shown in figure 1. Each is a one litre screw-capped polyethylene bottle provided with its own air supply and has the necessary equipment for controlling pH. Each bottle has a closed circulatory system attached holding an oxygen electrode.

Aluminium concentration of the medium is controlled by the rate of delivery of concentrated $AlCl_3$ into the mixing tank. Its rate of flow is determined by a series of six hour target values fed into a computer program at the beginning of an experiment. An algorithm within the program determines pump speed at given times to give the required time against concentration sequence (figure 2).

The pH against time sequence of an episode is controlled by two systems by using feedback from the pH measuring systems.

1. The initial system controls pH in the mixing tank. Like aluminium, the rate of flow of the acid pump is determined from a series of six hour target values fed into the computer initially. The computer calculates five minute targets from these and feedback from incoming information from the tank about its actual pH is then used to modify pump speed.

2. The same five minute targets are also used to control the pH of each fish bottle using feedback from pH values measured in each bottle. This is done because the fish

[397]

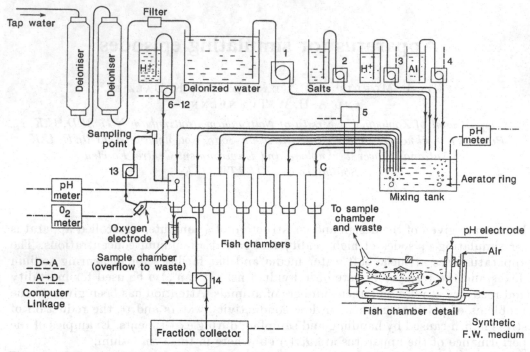

Figure 1. Diagram of the episodic rig. The pumped loop system used for the oxygen electrode and for remote sampling, the pH electrode and pH meter and the acid input to the fish chamber together with the automatic sampling mechanism are shown for one of the six fish chambers only. All chambers have identical systems. The seventh chamber (C) has no fish and provides control samples. Key to types of peristaltic pumps: (1, 2) Watson Marlow 502S/501R (single channel, variable speed, run at constant speed and set by hand); (3, 4) Watson Marlow 101 U/R (single channel, variable speed, computer controlled); (5) Watson Marlow 502S/501M (seven channel, variable speed, run at constant speed and set by hand); (6–12) Scientific Industries International (single channel, constant speed: 2 r.p.m., computer duration controlled); (13) Watson Marlow 502S/501R (six channel, variable speed, run at constant speed and set by hand); (14) Scientific Industries International (seven channel, variable speed, run at constant speed and set by hand; on, off, and direction of flow computer controlled).

modify the pH by producing ammonia that can alter the pH significantly in the small volume of water within the bottles.

The computer controls a sampler and fraction collector so that samples for aluminium analysis can be collected at one and a half hour intervals.

The computer program also provides static periods for the measurement of sodium balance and oxygen consumption at the beginning of each six hour period (figure 2). ^{22}Na is used to measure sodium influx over the first half-hour static period and changes of sodium concentration to measure net flux. Outflux is calculated from these two parameters. This period is followed by a half-hour through-flow to flush out the isotope. The second static period is used to measure oxygen consumption. Five minutes before this occurs, the computer switches off the air to the bottles and the flow displaces the air. Measurements of dissolved oxygen, by the oxygen electrodes, are then used to calculate oxygen consumption.

The apparatus is controlled by an Apple Europlus computer containing a timing card and a 16-channel, 8-bit A/D: D/A card. A continuous review of conditions in

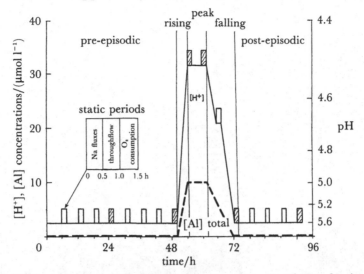

Figure 2. Schematic diagram to show the episodic phases and target values to be simulated during a pH 4.5, 10 μmol l⁻¹ concentration of aluminium. The episode illustrated is a short one, which rises for 4.5 h, peaks for 7.5 h and falls over 10.5 h with static periods occupying 1.5 h every 6 h so that the total duration of the episode is 22.5 h. Static periods are represented by blocks that are shaded to indicate those periods during which sodium flux measurements were to be done. Unshaded blocks show where oxygen consumption measurements only were done.

the apparatus can be obtained through a screen monitor and print-out. Typical results are shown in Morris & Reader (this symposium).

Some biological mechanisms reducing the availability of phosphorus in acid lakes

By Hans G. Nyman

Department of Ecological Zoology, University of Göteborg, Box 250 59, S-400 31 Göteborg, Sweden

Acid deposition on the aquatic ecosystem has changed the biological structure in acidified lakes. Some changes lead to the immobilization of phosphorus and hence to more oligotrophic conditions in the lakes.

Some mechanisms behind the phosphorus immobilization are discussed based on a series of field and laboratory experiments.

In many lakes in Sweden acidification leads to a dominance of calanoid copepods in zooplankton, whereas cladocerans are reduced. A field experiment using enclosures with calanoids and cladocerans in lake Gårdsjön, Sweden, showed that this shift could lead to an increased loss of phosphorus from the pelagic system.

Benthic invertebrates such as odonats, common in acid lakes, can act as oligotrophicating agents, trapping nutrients in large faecal pellets surrounded by a peritrophic membrane. A field experiment showed that the contents of litterbags containing faeces from odonata decompose more slowly in acid lakes compared with normal lakes. It also showed that the occurrence of a *Sphagnum* mat further decreased decomposition. The *Sphagnum* mat can also act as 'pellet hider' reducing fragmentation and cophrophagy by larger detrivores and hence the release of nutrients trapped in faecal pellets. However, other experiments showed that chironomids in the *Sphagnum* mat can fragment odonate faecal pellets. Another experiment showed that excretion (which could be done by chironomids) or any other release of phosphorus inside the *Sphagnum* mat is countered by the ability of *Sphagnum* to rapidly trap phosphorus and assimilate. The experiments indicate that the *Sphagnum* mat may play a very important role and act as a nutrient sink in the nutrient dynamics of a lake.

Introduction

The impact of acid deposition on water quality and the aquatic ecosystem has dramatically changed the biological structure in acidified lakes (see, for example, Haines 1981). Several groups, the most obvious being fish, disappear or are reduced in abundance, whereas other species of animals and plants appear or increase in abundance. When the acidification process leads to the elimination of fish (Overrein et al. 1980; Sevaldrup & Skogheim 1986), it can also lead to a series of changes in the flora and fauna of a lake (see Eriksson et al. 1980; Henrikson et al. 1980; Eriksson 1989). The absence of fish leads to new predator–prey relations and one of the results is an increase of Odonata and other large invertebrates in the littoral areas. Low pH and clearer water makes it possible for the moss *Sphagnum* to expand and colonize lake beds where it would earlier have been too dark for them (Hultberg & Grahn 1975). The development of the *Sphagnum* mat is one of the most dramatic changes that occur in many acidified lakes.

Numerous lakes in western Sweden have shown a decreasing phosphorus concentration (Hultberg & Andersson 1982) during the acidification process. Investigations of acid and non-acid oligotrophic lakes in Sweden and Canada show significantly lower phosphorus concentrations in acidified lakes (Hörnström *et al.* 1973; Dillon *et al.* 1979). Phosphorus is often the limiting nutrient for algal growth in acidified lakes (see, for example, Dillon *et al.* 1979; Persson & Broberg 1985) and is thus crucial for lake productivity. Any restriction in availability or loading of phosphorus due to acidification will therefore be of utmost importance for lake metabolism.

This oligotrophication of acidified lakes could be the result of a lower input of phosphorus from the acidified drainage area as an effect of more efficient trapping in the soil (Broberg & Persson 1984). Increased leaching of aluminium from the drainage area and then precipitation of aluminium–phosphorus complexes in the lake water has been suggested by Almer *et al.* (1978) and by Dickson (1980). However, this process is questioned by Broberg (1984) regarding the concentrations and pH levels occurring in already acidified lakes.

A reduced phosphorus level could also be an effect of the biotic changes following fish removal in the lakes (a top-down regulation). In a lake experiment by Henrikson *et al.* (1980) biotic changes after fish removal in a non-acidified lake, resulted in a development towards oligotrophic condition as shown by a drop in limnetic primary production, pH, total phosphorus, total nitrogen and increased transparency. The works of Carpenter *et al.* (1985), Carpenter & Kitchell (1988), Kitchell & Carpenter (1987) and Shapiro & Wright (1984) also indicate that biological regulation alone can influence phosphorus levels in lakes.

In this paper my aim is to identify some of the possible biological mechanisms for phosphorus reduction in acidified lakes based upon a series of field and laboratory experiments. As the results are to be published in full elsewhere, this paper will only briefly summarize the methods and only emphasize the most important results.

Materials and methods

The study area for all the experiments has been lakes in the Gårdsjö area of western Sweden.

The experiment regarding the influence of different zooplankton populations on the sedimentation rate was done in 15 large polyethylene bags (300 l) in limed lake Gårdsjön from 10 June to 4 July, 1988. The bags were filled with filtered (100 µm) lake water and left at the site for three days before the start of the experiment. A zooplankton population dominated by cladocerans was added to five of the bags and one dominated by calanoid copepods (*Eudiaptomus gracilis*) was added to five other bags. The remaining five served as a control group. Sediment traps were placed in all bags are a depth of 1.5 m and water samples were taken for phosphorus analysis at the start of the experiment, three times during the experiment and again at the end. Zooplankton and phytoplankton populations were sampled at the start and end of the experiment in all the bags.

The experiment regarding the durability of defecated material from different zooplankton was performed at the lakeside with zooplankton, freshly collected and 'cleaned' through several baths of filtered water (0.5 μm). Samples were examined under a microscope before and after a 2 min stirring of the defecated material in a few millilitres of water.

The experiments for phosphorus excretion from Odonata were done at the laboratory at 20 °C for 24 h with well-fed animals.

The experiments on decomposition of faecal pellets from Odonata in the littoral in different lakes were done with faeces in litterbags (two or three per lake) in 10 'normal' lakes and 11 acid lakes. In the acid lakes the litterbags were placed at a *Sphagnum* locality (*ca.* 10 cm deep) and at a *Sphagnum*-free locality (if available). The contents of the litterbags were photographed at the beginning and end of the experiment (lasting 16 days). The 'faeces area' projected from the photographs were used to estimate the degree of decomposition.

The experiments on the *Sphagnum* role as 'a faecal pellet hider' were done in the laboratory with odonats and faecal pellets in small aquaria where one series had a bottom of *Sphagnum* whereas one series served as a control.

The experiment on fragmentation of odonate faecal pellets by chironomids from the *Sphagnum* mat was done in the laboratory.

The experiments for the ability of *Sphagnum* to trap phosphorus from water were done in the laboratory with *Sphagnum* collected from the 'top layer' at the lake bottom and cleaned through several baths of filtered (5 μm) lakewater. Six 3.5 l aquaria were filled with filtered (5 μm) lakewater and a portion of *Sphagnum* (dryweight *ca.* 5.2 g) whereas six served as a control. Phosphorus was added to all the aquaria (an addition of 35 μg l^{-1}) and the phosphorus levels were monitored at intervals until the end of the experiment which lasted 120 h.

Results and discussion

(a) *Zooplankton dominance of cladocerans vs calanoid copepods: contrasting effects on the sedimentation rate*

The results of the phosphorus analysis from the sedimentation traps in the experiment are presented in figure 1. The Kruskal–Wallis one-way analysis of variance by ranks applied to the results gave a significant treatment effect ($p < 0.009$). The calanoida group tested with the Mann–Whitney U-test against the control group and the cladoceran group showed that the calanoids differed significantly (two-tailed, $p < 0.012$ and $p < 0.012$, respectively). These differences in sedimentation rates were further strengthened with the results from the water analyses (see Nyman 1990 *a*).

One explanation for the increased sedimentation rate for the calanoid-dominated group could be a difference in defecation behaviour between Calanoida and cladocerans. Calanoid copepods produce dense, compact faecal pellets surrounded by a peritropic membrane (Gould 1957; Ferrante & Parker 1977), whereas cladocerans such as *Daphnia* have been noted to produce a slurry that dissipates rapidly (Rigler 1971). To determine if this defecation behaviour resembled that of the cladocerans and calanoid copepods in the Gårdsjö experiment, newly defecated material was collected from dominant isolated species of zooplankton and examined under a microscope. To determine if the defecated material easily dissipates the faeces samples collected were stirred for 2 min and examined again. The faecal pellets from calanoids were easily recognizable before and after the stirring and did not dissipate, whereas faeces from the cladocerans had mostly already dissipated before the stirring

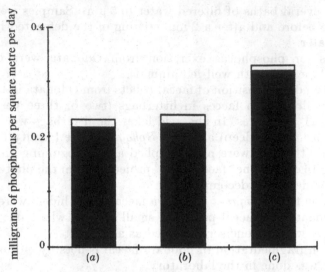

Figure 1. Sedimentation rate in milligrams of phosphorus per square metre per day (with s.e. values) in (a) control group ($n = 5$), (b) cladoceran group ($n = 5$) and (c) calanoida group ($n = 5$) from field experiment in lake Gårdsjön, 1988.

and had almost totally dissipated into small fragments after stirring (see Nyman 1990a). Faecal pellets from calanoid copepods also have a high sinking rate in water and decompose slowly (Ferrante *et al.* 1977) and could probably reach the sediments before breakdown in most lakes.

The bottom-up:top-down hypothesis on the regulation of trophic level biomass proposed by McQueen *et al.* (1986) suggests that a change of top predators (as is the usual case during the acidification of a lake) will have the strongest impact on primary producers in oligotrophic environments and that the effects are more easily 'damped out' in more eutrophic environments. Most lakes endangered by acidification were more or less oligotrophic originally thus making them the most sensitive group to top-down regulation.

My basic hypothesis, strengthened by these experiments, is that acidification-induced change of the zooplankton community towards dominance of calanoid copepods could alter the vertical flux of nutrients in the lake, making it faster and thus lowering the nutrient content of the water.

(b) Littoral in acid and non-acid lakes: contrasting roles as phosphorus sinks

Attention has previously been given to the role of benthic invertebrates in processing detritus material in the sediments by fragmentation and recirculating nutrients by bioturbation (see Anderson & Sedell 1979; Anderson 1985). However, many of the benthic invertebrates that are common in acidified lakes, partly due to reduced predation from fish, produce large compact faecal pellets surrounded by a peritropic membrane (Andersson & Nyman 1990; Ladle & Griffiths 1979). If these faecal pellets are not fragmented by animals, this 'pelletization' of organic material into large packets may immobilize nutrients for a considerable time as breakdown is partly a function of particle size (see, for example, Hargrave 1975) and is also slowed by the peritropic membrane (Anderson & Nyman 1990). Large benthic inverte-brates, common in acid lakes, can thus act as oligotrophicating agents, trapping

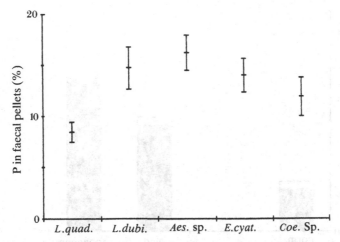

Figure 2. Amount of phosphorus in faecal pellets as a percentage of the total amount excreted by different odonatan species during 24 h in laboratory experiments (with ±s.e.): *Libellula quadrimaculata* ($n = 14$), *Leucorrhinia dubia* ($n = 9$), *Aeshna*, sp. ($n = 47$), *Enallagma cyathigerum* ($n = 10$), *Coenagrion* sp. ($n = 16$) (from Anderson & Nyman 1990).

nutrients into durable faecal pellets instead of letting them circulate among the compartments of the foodweb. A large proportion of compact faecal pellets with peritropic membranes could act as a phosphorus sink and retard the nutrient cycling of an entire community, as has been demonstrated by Neill (1975) in microcosmos studies under normal pH conditions.

One of the groups that produce faecal pellets with peritropic membranes and are favoured by acidification is Odonata (see, for example, Eriksson *et al.* 1980; Henrikson 1988) and species from this group were selected for a closer study of the phosphorus excretion. Some of the results from these experiments are presented in figure 2. The amount of phosphorus confined to faecal pellets was between 8 and 16 % of the total amount of excreted phosphorus (see Anderson & Nyman 1990).

The litterbag experiment with Odonata faeces was done to determine the influence of acid water and acid water + *Sphagnum* exposure, compared with 'normal' conditions, on decomposition of the faecal pellets. The results are presented in figure 3. The Kruskal–Wallis test applied to the results indicated a significant variation among the three groups ($p < 0.00002$). The 'normal' pH group tested with the Mann–Whitney U-test against the 'acid' group showed that the breakdown in an acid environment was slower ($p < 0.0009$, two-tailed). The 'acid' group tested against 'acid + *Sphagnum*' group (measurements of both groups from eight lakes) with the Wilcoxon paired-ranks test showed that breakdown was slower on the *Sphagnum* location in the lakes ($p < 0.03$, two-tailed). The experiment indicates that the breakdown of organic matter is much slower in acid lakes compared with 'normal' lakes. The experiment also indicates that *Sphagnum* retards the breakdown of organic matter.

The *Sphagnum* mat in acid lakes gives the littoral a new three-dimensional environment compared to an unaffected area. An experiment with faecal pellets on a *Sphagnum* surface compared with an uncovered surface, showed that the *Sphagnum* mat could act as a 'faecal pellet trap' and hide pellets from cophrophagy and fragmentation by large invertebrates (see Nyman 1990*b*) and thus slow the breakdown of the detritus.

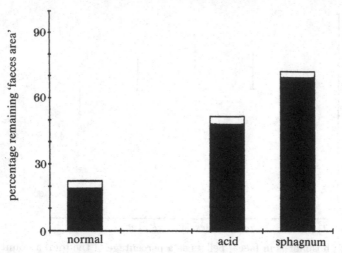

Figure 3. Decomposition of Odonata faecal pellets in litterbags in the littoral in different lakes. In normal lakes ($n = 10$), acid lakes ($n = 8$) and *Sphagnum* in acid lakes ($n = 11$). The 'faeces area' projected from photographs (with s.e. values) were used to estimate the degree of decomposition.

The three-dimensional environment of the *Sphagnum* mat could also be hypothesized to damp out water movements down in the *Sphagnum*, making this environment a large sedimentation trap and preventing redistribution and retaining sedimented material. The samples collected from this environment were also full of detritus. In contrast, the bottoms of the littoral areas in oligotrophic, non-acidified lakes, consist mostly of erosion or transport bottoms. This type of environment does not accumulate detritus to any great extent. Sedimented material is instead redistributed into deeper parts of the lakes. The increased accumulation of detritus is probably one of the reasons (together with changed predator–prey relations) for the unexpected multitude of invertebrates that inhabit the *Sphagnum* mat in acidified lakes (see Henrikson 1990).

To test if the small chironomids in the *Sphagnum* could fragment and use large faecal pellets with peritropic membranes, odonat pellets were fed to them. In the experiment it was determined that the chironomids could handle and fragment even large pellets (see Nyman 1990*b*). However, a quantitative invertebrate sample from the *Sphagnum* mat (see Henrikson 1990) revealed that many faecal pellets (658) could be found intact in the mat (see Nyman 1990*b*) indicating that many are not detected by the fragmentors.

There is, however, an increased input of organic detritus and hence nutrients into the *Sphagnum* mat due to the 'retaining of sedimented material' function and a large population of invertebrates living in it and excreting phosphorus.

The experiment on the ability of *Sphagnum* to trap and retain phosphorus revealed that uptake was fast (see figure 4) and that the *Sphagnum* retained the phosphorus during the 120 h of the experiment (see figure 5). It is quite possible that the *Sphagnum* in the lakes is in a very good position to compete with other organisms for the nutrient released from animals living down into the mat and then trap the nutrients into more and more *Sphagnum* biomass. The animals living down in the mat would then have the function of a 'shunt' pumping nutrients from the trapped detritus into the *Sphagnum* biomass rather than nutrient-release agents to other parts of the foodweb.

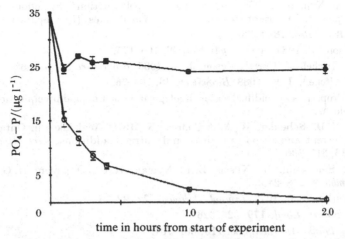

Figure 4. Uptake by *Sphagnum* of PO$_4$-P from water in a laboratory experiment. Measurements of the remaining PO$_4$-P content of water ($\pm$s.e.) during the first 2 h (o) compared with a control group without *Sphagnum* ($n = 6$) (●).

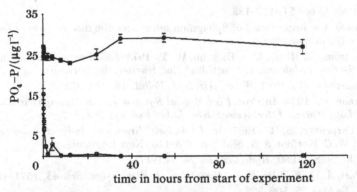

Figure 5. Uptake by *Sphagnum* of PO$_4$-P from water in a laboratory experiment. Measurements of the remaining PO$_4$-P content of water ($\pm$s.e.) during the whole experiment (120 h) (o) compared with a control group without *Sphagnum* ($n = 6$) (●).

My basic hypothesis, based upon the outcome of the series of experiments above, is that the *Sphagnum*-covered littorals in acidified lakes, in contrast to the littorals in 'normal' lakes, have become large phosphorus sinks trapping detritus in an environment that retards the breakdown and release of nutrients. If this is the case, the *Sphagnum* environment is actively preventing nutrients from recirculating in the foodweb thus making the acidified lakes more oligotrophic.

This work was supported by grants from the Surface Waters Acidification Programme. I thank Ulf Andersson for his contributions and I also thank my students for careful technical assistance in the *Sphagnum* experiment.

References

Almer, B., Dickson, W., Ekström, C. & Hörnström, E. 1978 In *Sulfur in the environment, part II.* (ed. Nriagu), pp. 273–311. New York: John Wiley.

Anderson, G. 1985 *Ecol. Bull.* **37**, 293–299.

Anderson, N. H. & Sedell, J. R. 1979 *A. Rev. Entomol.* **24**, 351–377.

Andersson, U. P. & Nyman, H. G. 1990a Faecal 'pelletization' by odonata and other invertebrates – a potential nutrient sink mechanism in acid lakes. (In preparation.)

Broberg, O. 1984 *Water Res.* **18**, 1273–1278.

Broberg, O. & Persson, G. 1984 *Arch. Hydrobiol.* **99**, 160–175.

Carpenter, S. R., Kitchell, J. F. & Hodgson, J. R. 1985 *BioScience* **35**, 634–639.

Carpenter, S. R. & Kitchell, J. F. 1988 *BioScience* **38**, 764–769.

Dickson, W. 1980 Properties of acidified water. Ecological impact of acid precipitation, pp. 75–83. *SNSF projekt.* Oslo-Ås.

Dillon, P. J., Yan, N. D., Scheider, W. A. & Conroy, N. 1979 Acid lakes in Ontario, Canada: Characterization, extent and resposes to base and nutrient additions. *Arch. Hydrobiol. Beih. Ergebn. Limnol.* **13**, 317–336.

Eriksson, M. O. G., Henrikson, L., Nilson, B. I., Nyman, H. G., Oscarson, H. G. & Stenson, J. A. E. 1980 *Ambio* **9**, 248–249.

Ferrante, J. G. & Parker, J. I. 1977 *Limnol. Oceanogr.* **22**, 92–98.

Gauld, D. T. 1957 *Nature, Lond.* **179**, 325–326.

Haines, T. A. 1981 *Trans. Am. fish. Soc.* **110**, 669–707.

Hargrave, B. T. 1976 In *The role of terrestrial and aquatic organisms in decomposition processes* (ed. J. M. Anderson & A. Macfadyen), pp. 301–321. Blackwell.

Henrikson, L., Nyman, H. G., Oscarson, H. G. & Stenson, J. A. E. 1980 *Hydrobiologia* **68**, 257–263.

Henrikson, B. I. 1988 *Oikos* **51**, 179–183.

Henrikson, B. I. 1990 The importance of Sphagnum mosses for the distribution of benthic fauna in acidified lakes. (In preparation.)

Hörnström, E., Ekström, C., Miller, U. & Dickson, W. T. 1973 Acidification impact on west-coast lakes. *Inf. from the freshw. lab.* no. 4, Drottningholm, Sweden (in Swedish).

Hultberg, H. & Andersson, T. 1982 *Water, Air, Soil. Pollut.* **18**, 311–331.

Hultberg, H. & Grahn, O. 1975 In *Proc. First Special Symposium on Atmospheric Contribution to the Chemistry of Lake Waters. Inter. Assoc. Great Lakes Res*, pp. 208–217.

Kitchell, J. F. & Carpenter, S. R. 1987 In *Predation: direct and indirect impacts on aquatic communities* (ed. W. C. Kerfoot & A. Sih), pp. 136–146. New England.

Ladle, M. & Griffiths, B. S. 1980 *Hydrobiologia* **74**, 161–171.

McQueen, D. J., Post, J. R. & Mills, E. L. 1986 *Can. J. Fish. aquat. Sci.* **43**, 1571–1581.

Neill, W. E. 1975 *Ecology* **56**, 809–826.

Nyman, H. G. 1990a Zooplankton dominance of cladocerans vs calanoid copepods – contrasting effects on the sedimentation rate in a field experiment. (In preparation.)

Nyman, H. G. 1990b Littoral in acid and non-acid lakes: contrasting roles as phosphorus sinks. (In preparation.)

Overrein, L., Seip, H. & Tollan, A. 1980 Acid precipitation – effects on forest and fish. *Acid Precipitation – effect on Forest and Fish Projekt, Res. Rep.* no. 19. Ås Norway.

Persson, G. & Broberg, O. 1985 *Ecol. Bull.* **37**, 158–175.

Rigler, F. H. 1971 In *A manual for the assessment of secondary productivity in freshwaters* (ed. W. T. Edmondson & G. G. Winberg), pp. 264–269. Oxford: Blackwell.

Sevaldrup, I. H. & Skogheim, O. K. 1986 *Water, Air, Soil Pollut.* **30**, 381–386.

Shapiro, J. & Wright, D. I. 1984 *Freshwat. Biol.* **14**, 371–383.

Stenson, J. A. E. & Eriksson, M. O. G. 1989 *Arch. environ. Contam. Toxicol.* **18**, 201–206.

Aluminium and pH effects on sodium-ion regulation in mayflies

By Kjell G. Frick and Jan Herrmann†

Department of Ecology, Animal Ecology, University of Lund, Helgonavägen 5,
S-223 62 Lund, Sweden

The toxicity of hydrogen ions to aquatic animals is generally believed to depend on respiratory problems and difficulties to maintain the proper ion balance (see review in Havas (1981)). Recent studies on stoneflies (Plecoptera) suggest that different kinds of sodium ion transporting cells have different capacities to discriminate between hydrogen and sodium ions (Twitchen 1987). Aluminium is thought to interfere with the function of ion transporting cells in salmonid fish by binding to important enzyme sites (Staurnes *et al.* 1984) and fish mortalities at moderately low pH (5.5) have been explained by the toxic effect of aluminium causing ion regulation problems combined with deficient respiration (Muniz & Leivestad 1980). Aluminium has also been demonstrated to affect sodium ion regulation in some aquatic invertebrate taxa, e.g. daphnids (Havas & Likens 1985), corixids (Witters *et al.* 1984) and mayflies (Herrmann 1987).

Sodium is one of the major osmo-effectors in mayflies and earlier work has documented decreasing whole-body sodium content in mayfly nymphs as a result of increasing Al concentrations at low pH (Herrmann 1987). This study aimed at a more detailed understanding of sodium ion fluxes in two mayfly species of different acid sensitivity, *Heptagenia sulphurea* Müll. and *Ephemera danica* Müll. (Ephemeroptera), at different Al concentrations (0.0, 0.5 and 2.0 mg l^{-1}) and pH levels (4.0, 5.0 and 6.0). The nymphs were taken from circumneutral streams in southern Sweden. By using an isotope (^{22}Na) tracer technique both influx and outflux rates of sodium ions for individual nymphs were established. From these values the net flux of sodium was calculated. Influx was measured by changes in ^{22}Na initially added to the external medium, whereas outflux was measured by the loss of sodium isotope from nymphs that had been loaded with ^{22}Na for 24 hours.

The sodium ion regulation in freshwater insects consists of an influx and an outflux component. The outflux of ions is mainly a result of renal activity and cotransport of ions as excess water is expelled from the body (Shaw & Stobbart 1963). An active ion reabsorption occurring in the rectum has been documented in several aquatic insects (Shaw & Stobbart 1963), thus minimizing ion loss. This reabsorption would constitute part of the influx component. Influx sites are also found on body surfaces thus defining another part of the influx component.

In view of the short duration of the experiments and the flux rates (less than 10% per hour) indicated here, outflux, as measured in these experiments, would mainly consist of the difference between the renal losses and the reabsorption of ions in the rectum, whereas the influx experiment probably measured mainly the sodium ion uptake at body surfaces.

† Present address: SEPA, S-171 85 Solna, Sweden.

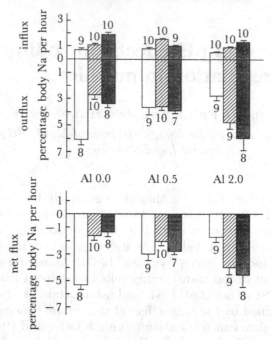

Figure 1. Sodium ion regulation rates (influx, outflux and net flux) for *Heptagenia sulphurea* at different hydrogen ion (pH 4.0, pH 5.0 and pH 6.0) and aluminium (0.0, 0.5 and 2.0 mg l⁻¹) concentrations. Bars show mean flux rate $\pm$ s.e. Numerals above and below bars signify number of samples in each treatment; ($\square$), pH 4.0; ($\blacksquare$), pH 5.0; ($\boxtimes$), pH 6.0.

Both pH and Al had an effect on net sodium regulation for both species. The effect of low pH was an increase of net loss of sodium mainly because of increased outflux. For *E. danica* at pH 4 this loss reached approximately 8 % of the total body sodium content per hour (figure 2), and for *H. sulphurea* the loss was 4 % per hour (figure 1). This difference between the species seems to reflect their relative acid sensitivities.

Aluminium had different effects on the sodium regulation at different hydrogen concentrations. At low pH (4.0) net sodium losses were cut by *ca.* 50 % for both species when Al concentrations were 2.0 mg l⁻¹ compared with 0.0 mg l⁻¹ (figures 1 and 2). This was accomplished by a significant reduction of outflux rates and with *E. danica*, a smaller but significant reduction of influx rates as well. The greater change in the outflux rate measurements than in the influx component could be explained by the added effect of a decreased excretion of ions from the malpighian tubule. This renal activity is responsible for the main share of ion loss in aquatic freshwater insects. Ameliorating effects of Al on ion regulation at low pH have also been observed in daphnids (Havas & Likens 1985) and fish (Muniz & Leivestad 1980).

The net effect of Al at low pH thus seems to be a blocking of ordinary sodium ion regulation giving temporary ameliorating effects at episodic acidification events. However, this stasis could, in the long term, put a strain on the organism. Respiration studies on the effect of pH has showed decreased respiration rates at pH levels both lower and higher than the preferred for *Nemurella pictetii* (Plecoptera) (Rupprecht & Frisch 1990) and Al additions gave further increases in respiration rates for *H. sulphurea* and *E. danica* at low pH, which was interpreted as a compensatory mechanism for stress (Herrmann & Anderson 1986). When pH was high (pH 6.0) both species suffered from a decrease in influx rates at high Al

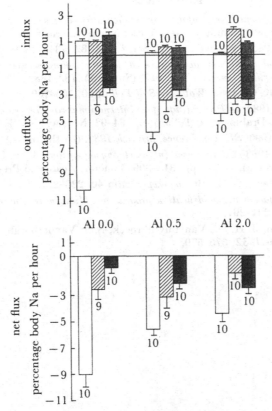

Figure 2. Sodium ion regulation rates (influx, outflux and net flux) for *Ephemera danica* at different hydrogen ion (pH 4.0, pH 5.0 and pH 6.0) and aluminium (0.0, 0.5 and 2.0 mg l^{-1}) concentrations. Bars show mean flux rate ±s.e. Numerals above and below bars signify number of samples in each treatment; (□), pH 4.0; (■), pH 5.0; (▨), pH 6.0.

concentrations. In *H. sulphurea* a simultaneous outflux increase resulted in an increased net sodium loss (figure 1) but the outflux rates and net balance of *E. danica* remained the same (figure 2).

Moulting nymphs of *H. sulphurea* showed substantially increased rates of sodium influx and outflux rates. When compared with the mean flux rate of each respective treatment, influx rates doubled and outflux rates increased almost three times. Considering the restraints on influx rates by high hydrogen ion and Al concentrations and the probability that high influx rates and low outflux rate (or both) are essential for successful moults, it can well be that moulting nymphs are especially sensitive to low pH and elevated Al. However, the ability to moult successfully in such conditions, and possibly also to alter the system of ion transporting cells to a more favourable one after the moult, could constitute an adaptive bottleneck resulting in a less acid sensitive population.

In conclusion, the main difference between *H. sulphurea* and *E. danica* was expressed in their response to elevated Al concentrations at high pH where sodium loss increased for *H. sulphurea* but not for *E. danica*. The two species also diverged in the magnitude of their response to low pH at which *E. danica* showed a greater net loss of sodium.

References

Havas, M. 1981 (ed.) *Effects of acidic precipitation on benthos*. Procedures of a symposium on acidic precipitation on benthos, 1980. Hamilton, New York: North American Benthological Society.

Havas, M. & Likens, G. E. 1985 *Proc. natn. Acad. Sci. U.S.A.* **82**, 7345–7349.

Herrmann, J. 1987 In *Ecophysiology of acid stress in aquatic organisms* (ed. H. Witters & O. Vanderborght), *Annls Soc. r. zool. malac. Belg.* **117**, (Suppl. 1), 182–188.

Herrmann, J. & Andersson, K. G. 1986 *Wat. Air Soil Pollut.* **30**, 703–709.

Muniz, I. P. & Leivestad, H. 1980 In *Proceedings of an international conference on ecological impact of acid precipitation* (ed. D. Drabløs & A. Tollan), pp. 84–92. Norway: SNSF Project.

Rupprecht, R. & Frisch, S. 1990 *SIL Conference, Munich 1989*. (In the press.)

Shaw, J. & Stobbart, R. H. 1963 In *Advances in insect physiology* (ed. J. W. L. Beament, J. E. Treherne & V. B. Wigglesworth), vol. 1, pp. 315–399. London: Academic Press.

Staurnes, M., Sigholt, T. & Reite, O. B. 1984*a* *Experientia* **40**, 226–227.

Twitchen, I. D. 1987 In *Surface water acidification programme. Midterm review conference, Bergen, Norway, 22-26/6 1987*, pp. 291–297.

Witters, H. E., Vangenechten, J. H. D., Van Puymbroeck, S. & Vanderborght, O. L. J. 1984*b* *Bull. environ. Contam. Toxicol.* **32**, 575–579.

The physiological bases of resistance to low pH among aquatic insect larvae

By IAN D. TWITCHEN†

Department of Biological Sciences, University of Lancaster, Lancaster LA1 4YQ, U.K.

The results presented demonstrate that the direct physiological effect of low pH and elevated aluminium concentrations (or both) is the key factor limiting the distribution of acid-sensitive species of stonefly nymphs. Acid-tolerant species survive in acid waters because their apical carriers have a much greater discrimination between sodium and hydrogen ions on both sides of the apical membrane. They also have a reduced permeability to hydrogen ions compared to acid-sensitive species.

---

1. The impoverished invertebrate communities of acid waters

Acid waters (pH < 5.7), such as Gaitscale Gill (pH 4.8) a tributary of the River Duddon in Cumbria, have an impoverished invertebrate community in which stoneflies, Plecoptera, predominate amongst the larger animals. There are two main theories to account for the impoverished fauna in acid streams; the direct physiological effect of low pH and elevated aluminium concentrations (or both), and the indirect effect of reduced food availability (Sutcliffe & Hildrew 1989).

The ability of acid-tolerant species of stonefly nymph to survive in both acid and neutral waters, whereas acid-sensitive species are restricted to neutral waters (pH > 5.7), made stonefly nymphs the ideal choice as the study insect in this investigation. The object of the study was to determine the physiological bases of resistance to low pH in acid-tolerant species.

2. Sodium fluxes in neutral and acid waters

The acid-tolerant species, *Amphinemura sulcicollis*, *Protonemura meyeri* and *Leuctra moselyi*, were able to maintain sodium balance in Gaitscale Gill water (pH 4.8), Oxnop Gill water (pH 7.6), and in the artificial river water medium (ARWM) at both pH 7.0 and 4.0 (e.g. figure 1 *a–e*). In contrast, the acid-sensitive species, *Perla bipunctata*, *Dinocras cephalotes* and *Isoperla grammatica*, were only able to maintain sodium balance in Oxnop Gill water (pH 7.6) and the ARWM at pH 7.0 (e.g. figure 2 *a*, *d*). The net sodium loss in Gaitscale Gill water (pH 4.8) was due to an inhibition of sodium uptake, while in the ARWM at pH 4.0 sodium uptake was inhibited and efflux stimulated, resulting in a severe sodium imbalance (e.g. figure 2 *b* and *e*).

---

† Present address: Department of Biological Sciences, University of Dundee, Dundee DD1 4HN.

[413]

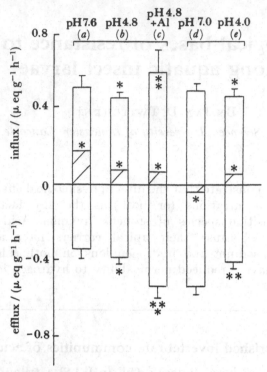

Figure 1. Short-term (24 h influx, 24 h efflux) sodium fluxes for the acid-tolerant *A. sulcicollis* in neutral and acid waters. The net flux is indicated by the hatched column. Mean values ±s.e. Net flux (*), NSD between influx and efflux; (**), SD ($P < 0.05$). Unidirectional fluxes (*), NSD compared with control values; (**), SD. (**) SD between (*a*) and (*c*) but NSD between (*b*) and (*c*). (**) SD between (*a*) and (*c*) and (*b*) and (*c*). (*) NSD between (*a*) and (*c*) or (*b*) and (*c*). (*a*) Oxnop Gill water (pH 7.6, 0.25 meq Na^+ l^{-1}, 1.3 meq Ca^{2+} l^{-1}) control for natural waters (*a*), (*b*) and (*c*); $n = 8$. (*b*) Gaitscale Gill water (pH 4.8, 0.13 meq Na^+ l^{-1}, 0.05 meq Ca^{2+} l^{-1}) $n = 15$. (*c*) Gaitscale Gill water containing aluminium (pH 4.8, 0.13 meq Na^+ l^{-1}, 0.05 meq Ca^{2+} l^{-1}, 20 μmol Al^{n+} l^{-1}) $n = 7$. (*d*) The ARWM at pH 7.0 (0.25 meq Na^+ l^{-1}, 0.7 meq Ca^{2+} l^{-1}) control for ARWM (*d*) and (*e*); influx $n = 8$, efflux $n = 26$. (*e*). The ARWM at pH 4.0 (0.25 meq Na^+ l^{-1}, 0.7 meq Ca^{2+} l^{-1}) influx $n = 8$, efflux $n = 12$.

3. The effect of aluminium on sodium fluxes in natural acid waters

The acid-tolerant *A. sulcicollis* could maintain sodium balance in Gaitscale Gill water (pH 4.8) containing 20 μmol l^{-1} aluminium (figure 1*c*), a concentration that could be encountered in Gaitscale Gill (Crawshaw 1984). In contrast, aluminium significantly contributed to the inhibition of sodium uptake and the stimulation of sodium efflux in the acid-sensitive *P. bipunctata* (figure 2*c*), identifying aluminium as a key additional factor excluding acid-sensitive species from acid waters.

4. The mechanism of sodium transport in stonefly nymphs

Sodium uptake in stonefly nymphs (figures 3 and 4), is analogous to enzyme-substrate saturation kinetics at all pHs investigated, and incorporates a 'carrier-mediated' component of sodium efflux (Stobbart 1974), which is linked to the sodium uptake mechanism (I. D. Twitchen, unpublished data). This is incorporated into the

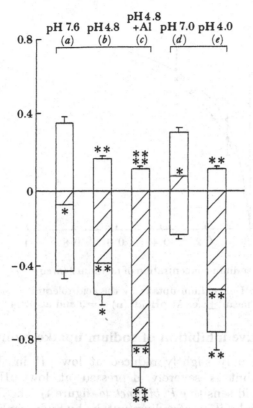

Figure 2. Short-term (24 h influx, 24 h efflux) sodium fluxes for the acid-sensitive *P. bipunctata* in neutral and acid waters. For details see legend to figure 1; (a) $n = 8$; (b) $n = 8$; (c) $n = 9$; (d) influx $n = 21$, efflux $n = 34$; (e) influx $n = 13$, efflux $n = 21$.

carrier mechanism by making all the components for the movement of sodium reversible:

$$Na^+ + C \underset{k_{-1}}{\overset{k_{+1}}{\rightleftharpoons}} CNa \underset{k_{-2}}{\overset{k_{+2}}{\rightleftharpoons}} C + Na^+,$$

where C is the carrier, Na^+ the substrate, CNa the carrier–sodium ion complex and the transported sodium ions the product of the reaction. Sodium transport is adequately accounted for by the Briggs & Haldane (1925) model, in which the rate of sodium uptake, v, is given by:

$$v = \frac{k_{+2} \cdot [E_0] \cdot [S]}{((k_{-1} + k_{+2})/k_{+1}) + [S]} = \frac{k_{+2} \cdot [E_0] \cdot [S]}{K_m + [S]}$$

Where $[E_0]$ is the total concentration of the carrier and $[S]$ the concentration of the substrate, Na^+.

The reciprocal of K_s ($= k_{-1}/k_{+1}$) is the affinity constant of the carrier for sodium ions. V_{max} ($= k_{+2} \cdot [E_0]$) is the maximum rate of sodium uptake when the sodium transport system is saturated.

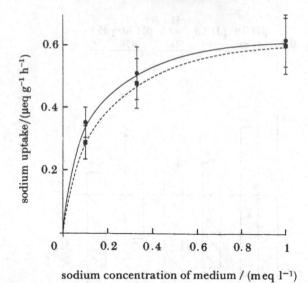

Figure 3. The effect of pH on sodium uptake by the acid-tolerant *A. sulcicollis*. Each point represents the mean $\pm$ s.e. At pH 7.0 ($\bullet$) $n = 9$ and at pH 4.0 ($\blacksquare$) $n = 8$.

5. The competitive inhibition of sodium uptake by hydrogen ions

Sodium uptake was only slightly reduced at low pH in the acid-tolerant *A. sulcicollis* (figure 3), but is severely depressed at low pH and low sodium concentrations in the acid-sensitive *P. bipunctata* (figure 4). The Dixon plot (figure 5), demonstrates that the inhibition of sodium uptake by hydrogen ions in *P. bipunctata* is competitive in nature (see Dixon & Webb 1979). The effect of competitive inhibition of sodium uptake by hydrogen ions is formally similar to the effect of inhibitors on enzyme kinetics (Dixon & Webb 1979). The inhibitor (H^+), reversibly binds to the active site of an enzyme (carrier) and reduces the binding of the substrate (Na^+):

$$
\begin{array}{c}
H^+ \\
+ \\
Na^+ + C \xrightleftharpoons[k_{-1}]{k_{+1}} CNa \xrightleftharpoons[k_{-2}]{k_{+2}} C + Na^+ \text{ transported} \\
\Big\downarrow {\scriptstyle k_{-1i}} \Big\uparrow {\scriptstyle k_{+1i}} \\
CH \xrightleftharpoons[k_{-2i}]{k_{+2i}} C + H^+ \text{ transported.}
\end{array}
$$

The components for the movement of hydrogen ions are also reversible because hydrogen ions are considered to be the counter ion in the mechanism of sodium transport in aquatic insects (Stobbart 1974; Wright 1975).

When the external substrate concentration [Na^+] is raised to saturating levels the effects of the inhibitor (H^+) become negligible.

The apparent K_m (calculated from Wilkinson (1961)) increases by a factor of

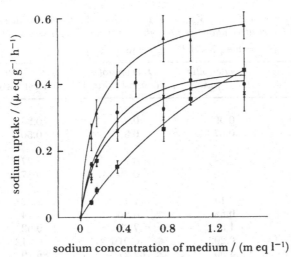

Figure 4. The effect of pH on sodium uptake by the acid-sensitive *P. bipunctata*. Each point represents the mean ±s.e. At pH 8.0 (▲), 7.0 (●), 5.0 (×) and 4.0 (■), all data points $n = 6$.

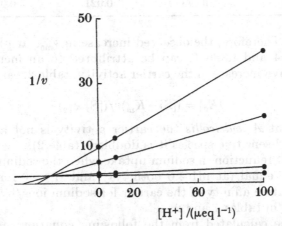

Figure 5. A Dixon plot of the data in figure 4, to demonstrate the competitive nature of sodium uptake inhibition by hydrogen ions in the acid-sensitive *P. bipunctata*.

$(1 + [I]/K_i)$ (Dixon & Webb 1979). Therefore sodium uptake at low pH can be described by the following relation:

$$v = \frac{k_{+2} \cdot [E_0] \cdot [S]}{((k_{-1} + k_{+2})/k_{+1}) \cdot (1 + [I]/K_i) + [S]},$$

where $[I]$ is the concentration of the inhibitor, K_i ($= k_{-1i}/k_{+1i}$) is the dissociation constant for the enzyme–inhibitor complex (carrier–hydrogen ion complex), and $1/K_i$ is the affinity constant of the carrier for hydrogen ions.

Competitive inhibition affects only K_m and not V_{max} as infinitely high concentrations of sodium would displace the inhibitor from the enzyme (Dixon & Webb 1979). However, V_{max} varies considerably in individuals depending on their moulting cycle. Before and after moulting V_{max} increases, decreasing again during the inter-moult period (I. D. Twitchen, unpublished data). In response to ion depletion at low pH, for example figure 2b, c, e, freshwater animals increase V_{max} by increasing $[E_0]$

Table 1. *The observed increase in* K_m *and* V_{max} *with decreasing* pH, *the dissociation constants* K_i *and* K_s *and the relative affinities for sodium ions and hydrogen ions*

$$(K_m = \text{meq Na}^+ \text{ l}^{-1}; \ V_{max} = \mu\text{eq Na}^+ \text{ g}^{-1} \text{ h}^{-1})$$

	A. sulcicollis (acid water)	D. cephalotes (neutral water)	P. bipunctata (neutral water)
pH 7			
K_m	0.09	0.20	0.23
V_{max}	0.67	0.37	0.50
pH 5			
K_m	—	0.25	0.27
V_{max}	—	0.29	0.47
pH 4			
K_m	0.14	1.4	2.22
V_{max}	0.68	0.80	1.14
$(1 + [I]/K_i)$	1.5	7.23	9.62
K_i	0.202	0.016	0.012
Aff.H$^+$1/K_i	4.95	62.27	86.19
Aff.Na$^+$/Aff.H$^+$	4.70	0.75	0.49
Aff.Na$^+$1/K_s	23.25	46.68	42.16
K_s	0.043	0.021	0.024

and k_{+2} (or both). Therefore, the observed increase in V_{max} at pH 8.0 and 4.0 for *P. bipunctata*, figure 4 and table 1, can be attributed to an increase in the carrier activity. The relative increase in the carrier activity, table 2, can be calculated from the relation:

$$[E_0] = ([S] + K_m)v/([S] \ k_{+2}).$$

In the acid-tolerant *A. sulcicollis* the carrier activity is not affected at low pH, whereas in the acid-sensitive species it is doubled (table 2).

The proportional reduction in sodium uptake when the sodium and hydrogen ion concentrations are equal (at pH 4.0 both Na$^+$ and H$^+$ = 0.1 meq l^{-1}), provides a measure of the relative affinity of the carrier for sodium ions/hydrogen ions, shown as Aff.Na$^+$/Aff.H$^+$ in tables 1 and 2.

K_i values can be calculated from the following equation modified from Wong (1975):

$$K_i = [\text{H}^+]_{\text{pH}=4.0}/[(K_m(\text{pH } 4)/K_m(\text{pH } 7)) - 1].$$

As $1/K_i$ is equal to the affinity of the carrier for hydrogen ions multiplying $1/K_i$ by Aff. Na$^+$/Aff.H$^+$ gives $1/K_s$, the affinity of the carrier for sodium ions (tables 1 and 2).

6. The physiological bases of resistance to low pH

The observed affinity constants $1/K_s$ and $1/K_i$ indicate that the ability of the acid-tolerant *A. sulcicollis* to survive in acid waters is due to the much greater affinity of its carrier for sodium ions compared with hydrogen ions, Aff.Na$^+$/Aff.H$^+$ = 4.7 (table 1). Therefore, even where the carrier activity is not increased at low pH, sodium uptake would be reduced by *ca.* 18% at pH 4.0, where [Na$^+$] = [H$^+$], whereas at pH 5.0 the reduction would be less than 2%. For *A. sulcicollis* sodium efflux rates were also reduced by *ca.* 20% in the ARWM at pH 4.0 (figure 1*e*).

In the neutral water species *D. cephalotes* and *P. bipunctata* the carrier has a higher affinity for hydrogen ions than sodium ions, Aff.Na$^+$/Aff.H$^+$ = 0.75 and 0.49,

Table 2. *The relative increase in the number of carrier sites at low* pH, *the expected* K_t *and* K_s *and the relative affinities for* Na+ *and* H+ *where the number of carrier sites is not increased*

	relative $\dfrac{\text{pH } 4[E_0]}{\text{pH } 7[E_0]}$	$\dfrac{\text{Aff.Na+}}{\text{Aff.H+}}$	Aff.Na+ $1/K_s$	K_s
A. sulcicollis (acid water)	1.021	4.20	20.77	0.048
D. cephalotes (neutral water)	2.19	0.24	15.15	0.066
P. bipunctata (neutral water)	2.30	0.17	14.33	0.070

respectively (table 1). Therefore, even when the carrier activity was doubled at pH 4.0, sodium uptake was reduced by 47% and 72%, respectively. Sodium uptake would have been reduced even further at low pH, if the carrier activity had not increased, because the Aff.Na+/Aff.H+ ratios would have been only 0.24 for *D. cephalotes* and 0.17 for *P. bipunctata* (table 2).

The striking differences in the effects of low pH on sodium efflux rates between acid-tolerant and sensitive species (figures 1*b*, *c*, *e* and 2*b*, *c*, *e*), are caused by differences in their permeability to hydrogen ions and the 'carrier-mediated' component of sodium efflux. The increased carrier activity at low pH for the acid-sensitive species, leads to a proportional increase in the 'carrier-mediated' component of sodium efflux. At low pH the apical carriers of *A. sulcicollis* have a greater discrimination between sodium and hydrogen ions on both sides of the apical membrane, than those of the acid-sensitive species.

The results provide a more detailed model for the mechanism of sodium transport and the inhibition of sodium uptake by hydrogen ions in aquatic animals than had been available before. In particular it has demonstrated that the essential differences between acid-tolerant and sensitive species, lies in quantifiable differences in the relative affinities of the carrier for sodium and hydrogen ions. A re-examination of a few critical species, in the light of the theories presented, should clarify our understanding of the differences in sodium uptake inhibition and efflux stimulation in acid waters, between acid-tolerant and sensitive species of fish.

References

Briggs, G. E. & Haldane, J. B. W. 1925 *Biochem. J.* **19**, 338.

Crawshaw, D. H. 1984 *North West Water Rivers Division, report TSN 84/83*, pp. 1–35.

Dixon, M. & Webb, E. C. 1979 In *Enzymes*, 3rd edn. London and New York: Longmans Green and Company, Academic Press.

Michaelis, L. & Menten, M. L. 1913 *Biochem. Z.* **49**, 333.

Stobbart, R. H. 1974 *J. exp. Biol.* **60**, 493–533.

Sutcliffe, D. W. & Hildrew, A. G. 1989 In *Acid toxicity and aquatic animals* (ed. R. Morris, E. W. Taylor, D. J. A. Brown & J. A. Brown), pp. 13–29. (*Society for Experimental Biology. Seminar series* **34**.) Cambridge University Press.

Wilkinson, G. N. 1961 *Biochem. J.* **80**, 324–332.

Wong, J. T.-F. 1975 *Kinetics of enzyme mechanisms*. London, New York and San Francisco: Academic Press.

Wright, D. A. 1975 *J. exp. Biol.* **62**, 141–155.

Trophic relationships in acid waters

By Elizabeth Y. Haworth

Institute of Freshwater Ecology, Ambleside, Cumbria, U.K.

Studies concerned with streams and lakes of varied ranges of acidity and alkalinity clearly show that there are great changes of fauna and flora as acidity increases. This is compounded by a 'knock-on' effect where there is a primary response to water chemistry and nutrient levels are enhanced by a secondary response caused by an absence of food source.

1. Introduction

The suggestion, over a decade ago, that depositions from the atmosphere were becoming increasingly acidic due to industrial pollution and that this was significantly affecting the aquatic ecosystem, encouraged a series of studies to verify the problem and estimate the extent of the damage (Sutcliffe 1983; Sutcliffe & Carrick 1988).

The main objectives of the series of studies on acidification made by the Institute of Freshwater Ecology have concerned the variations that exist within waters of low conductivity and the conditions that support the fauna and flora found there. To this end, comparative studies have mainly concentrated on circumneutral and acid-water streams within the catchment of the River Duddon in S.W. Cumbria, U.K., although wider ranging studies were also included.

2. Results

Models to show the pattern and variation in the chemistry of acid waters include DAM (the Duddon Acidification Model) which illustrates the process-based interactions between atmospheric deposition and the catchment (Tipping 1990). Such a model is compared with plots of the aluminium concentration and pH of the Lake District tarns that lie on the base-poor Borrowdale volcanic bedrock. The model also provides a means by which the ionic concentration of lake water can be back-calculated in terms of historical pH and the trend has been compared with the historical pH inferred from palaeolimnological diatom stratigraphies (Haworth *et al.* 1988). It was found to be in reasonable agreement for the upland tarns of Wasdale in west Cumbria.

The variation in water chemistry, even within small areas, such as on a single catchment, is caused by differences in geology, sometimes modified by soils and vegetation. Not only is the ionic composition important but also the food and nutrients in the chain that are the basis of the aquatic ecosystem. The source of these is the detritus that comes in from the terrestrial vegetation and soils. The area of poor, acid, somewhat peaty soils in the Upper Duddon valley supports a mainly *Nardus stricta* grassland (with some *Molinia*, *Juncus*, bracken and *Sphagnum*), hence basic provision for the aquatic life-system is poor, even when compared with the more varied leaves of trees coming into the lower streams.

The rates of detrital decomposition by the aquatic microflora (fungal hypho-mycetes and bacteria) in acid and circumneutral waters were assessed by comparative experiments in such waters. Submerged bags containing leaves of alder, oak or mat-grass in streams of different pH revealed the differing rates of decay as measured by weight loss (table 1). Chamier (1987) found that alder leaves degraded faster than oak or grass, especially in circumneutral waters. Bacterial counts were higher from the material in circumneutral water than that in acid water samples, while fungal colonization on grass in streams of pH ≤ 5.5 was the slowest, with increasing colonization further downstream and in more neutral water. The numbers of fungal species present demonstrated the lower diversity (0–4 spp.) that is present in the more acid waters of upland grasslands compared with that on the leaf detritus (*ca*. 14), especially oak, in wooded areas with waters of *ca*. pH 6.8. Microbial studies have shown that not only is the type of detritus found in acid waters poor (being mostly *Nardus stricta*) but that the slow rate of microbial decomposition much restricts the amount of food and nutrients available in the acid streams.

The poor rate of decomposition that has been found results from the striking difference in the microflora of the acid and circumneutral waters. There is a significant total absence of the yellow bacteria ascribed to the genus *Cytophaga* that are the important leaf litter decomposers of the aquatic ecosystem capable of degrading pectin or cellulose, the essential ingredients of plant polymers. Results were based on the comparison of plate counts of bacteria from identical types of detritus (stems of *Nardus stricta*) put out in 25 streams and the presence and development of these species thus tested (Jones *et al*. 1987). This demonstrated their absence from acid waters of pH < 5.5, and the steady decline in their numbers when detritus was transferred from circumneutral waters to acid ones. Subsequent recolonization was observed upon their return to neutral waters and further confirmed the lack of acid-tolerance of these forms.

Marker & Willoughby (1988) have studied the biomass of epilithic and epithytic algae in the acid and circumneutral streams of the Duddon catchment. The different species and biomass in the two types are due to both water chemistry and substrate (especially the type of bryophytes). Algal productivity in part of one stream, Crosby Gill, was further affected by the shading from the tree canopy. The different bryophytes found in the upper, acid and lower, circumneutral waters (*Nardia compressa* (the liverwort) and *Hygrohypnum luridum* (a moss) have determined the type of epiphytes present; these were mainly diatoms (*Eunotia rhomboidea* and *Peronia fibula* on the former and *Achnanthes minutissima* and *Cocconeis placentula* on the latter); in streams in southern England the major aquatic macrophyte is *Ranunculus aquatilis*, which supports mainly *Diatoma vulgare* and *Cocconeis placentula*. Four algae dominated the upper, more acid streams *Hormidium*, *Zygogonium*, *Homoeothrix* and *Stigonema*, whereas the less acid streams encouraged the growth of *Ulothrix*, *Spirogyra*, *Rhodochorton*, *Batrachospermum* and *Lemanea*. The southern streams supported mainly diatoms, encrusted Chlorophyceae and Cyano-phyceae, as well as *Vaucheriae* and *Cladophora*. Measurements of chlorophyll *a* (table 1) show that the hard-water, nutrient-rich southern streams support larger populations than the northern, soft-water and acid streams. Aspect, bedrock and shading can be as important as acidity in determining biomass levels in soft-water streams.

Diatom assemblages have been identified from a large number of unproductive waterbodies in the Lake District and the statistical ordination of these, by using the

Table 1. *The differences in algal productivity, as mean annual chlorophyll a, and in decomposition rates of three types of leaf litter in streams of the Duddon catchment, compared with some in southern England*

location	pH	altitude/m	mean annual chlorophyll a mg m^{-2}	days to 50% mass loss (leaves)		
				grass	oak	alder
(a) Duddon catchment						
Wrynose Bottom	4.8–6.3	ca. 250	34.2	231	282	233
Gaitscale Gill	4.6–5.0	ca. 250	18.1	179	245	215
Mosedale Feck	4.6–5.2	ca. 250	33.9	222	284	233
Hardknott Gill	6.1–6.9	ca. 250	13.3	176	217	69
Crosby Gill	6.0–7.3	<100	135.0	114	159	42
(b) southern England						
Dockens Water (Hants.)	6.7	<50	49.1	—	—	—
Ober Water (Hants.)	6.7	<50	24.1	—	—	—
Bere Stream (Dorset)	7.9–8.5	<50	116.6	—	—	—
Bere Stream (Dorset)	7.9–8.5	<50	123.4	—	—	—

program CANOCO (canonical community ordination program), has revealed an over-riding relation between diatom distribution, and pH and alkalinity (Haworth, *et al.* 1990). Certain groups of taxa typify the differing groups of waters. *Achanthes minutissima* is a typical dominant in most circumneutral waters and subgroups are identifiable by the presence of *Fragilaria*, *Cyclotella* or *Brachysira* spp.; diatom assemblages of the most acidic waterbodies (excluding upland pools less than 1 m deep) can be identified by assemblages including some or all of the following: *Frustulia rhomboides* var. *saxonica*, *Peronia fibula*, *Eunotia tenella*, *E. naegeli*, *E. incisa*, *Achnanthes marginulata*, *Brachysira brebissonii*, *Tabellaria binalis*, *T. quadriseptata* or *Aulacoseira tethera*. Such an ordination of sites can also be used to show within-site variation over time, including that between different time horizons in the sediment profile that can emphasize the shift in taxa in acidified sites, by relating them to present analogues.

The question of, how the variation of food availability in sites of differing water chemistry determine the distribution of certain aquatic animals, has been addressed in these studies. Experiments have been done to examine the survival and feeding requirements of several invertebrates (Willoughby 1988). *Baetis rhodani* and *B. muticus* (mayflies) could survive unfed for less than ten days in acid waters whereas *B. rhodani* could survive longer in acid water only if the ionic concentration was high. When *B. muticus* were fed on detritus from acid (pH 5) and neutral (pH 7) streams, they survived far better in the latter. Neither animal can tolerate water of pH < 5.0 but feeding experiments have shown that these two species are also limited by different food requirements; *B. rhodani* is an algal grazer (mainly on the diatom *Cocconeis placentula* growing on moss and stones in circumneutral streams), whereas *B. muticus* feeds on higher-plant detritus.

Not only is there a lack of tolerance to acid-water chemistry but the lack of detrital food, due to poor vegetation type and the lack of microbial decomposers limits the growth of *B. muticus*. The lack of the right substrate and water quality for the diatom *Cocconeis* limits *B. rhodani* from populating the upland acid streams of the Duddon valley. In the lower, more circumneutral streams the supply of these foods is better because there is a more varied vegetation on more base-rich soils.

Early studies by Sutcliffe and Carrick showed the relation between pH and the distribution of invertebrates throughout the River Duddon. They found that mayflies, certain species of stoneflies, caddisflies and beetles, the freshwater shrimp and limpets only occurred where pH remained above 5.7 throughout the year. But, wherever the pH either fell below 5.7 in the autumn–winter, or remained permanently acid, fewer species were found in an impoverished fauna dominated by stonefly larvae. It was the herbivorous taxa that appeared sensitive to low pH, whereas most carnivorous taxa are not (Sutcliffe 1983), and this also applied to several other catchments in the Lake District, all on base-poor rocks. The same impoverished fauna also occurs elsewhere in Northern Britain and on the continent of Europe, where streams have pH values below 5.7.

3. Conclusions

Given the clear indications that the acid waters have a severely restricted fauna and flora, and that this is compounded throughout the food-chain by the food deficiencies, there is an obvious requirement that the amount of atmospheric depositions be reduced to prevent further decline of other sensitive areas that are base-poor.

There are also cases where some managed adjustments may be necessary or beneficial. Modelling studies on the strategies for neutralizing acid waters have shown that by stimulating productivity through the addition of nutrients, base may be generated within lakes (Davison 1987). Although the pH rise will only be slight (less than 1.0), the resultant biologically driven buffering capacity will provide a resistance to external influences, such as further acidic loading. Where liming is necessary, sodium salts may be more effective than calcium carbonate or hydroxide as they dissolve more readily and can be used to achieve higher alkalinities (Davison & House 1988). A lake thus treated would therefore be more able to tolerate further acidic depositions for a longer period after treatment.

Grateful thanks to all the authors who gave their most helpful assistance in the preparation of this review. Thanks also to all the establishments that provided additional funding for various projects.

References

Chamier, A.-C. 1987 *Oecologia* **71**, 491–500.

Davison, W. 1987 *Schweiz. Z. Hydrol.* **49**, 186–201.

Davison, W. & House, W. A. 1988 *Wat. Res.* **22**, 577–583.

Haworth, E. Y., Atkinson, K. M. & Carrick, T. R. In *Proceedings of the 10th International Diatom Symposium.* (In the press.)

Haworth, E. Y., Lishman, J. P. & Tallantire, P. 1988 *Report to the Department of the Environment,* (32 pp.)

Jones, J. G., Sutcliffe, D. W., Simon, B. M. & Carrick, T. R. 1987 In *Surface water acidification programme; mid-term review conference, Bergen, Norway 1987,* pp. 325–330.

Marker, A. F. & Willoughby, L. G. 1988 In *Algae and the aquatic environment* (ed. F. E. Round), pp. 312–325. Bristol: Biopress.

Sutcliffe, D. W. 1983 *A. Rep. Freshwat. Biol. Ass.* **51**, 30–62.

Sutcliffe, D. W. & Carrick, T. R. 1988 *Freshwat. Biol.* **19**, 179–189.

Tipping, E. 1990 *Freshwat. Biol.* **23**, 7–23.

Willoughby, L. G. 1988 *Int. Revue ges. Hydrobiol. Hydrogr.* **73**, 259–273.

Food and feeding of *Baetis rhodani* (Ephemeroptera) in acid environments

By Per Sjöström†

Department of Ecology, Animal Ecology, University of Lund,
Helgonav. 5. S-223 62 Lund, Sweden

One of the organisms most vulnerable to acidification is *Baetis rhodani* (Ephemeroptera). The nymphs feed predominantly on the organic layer, consisting of algae and detritus present on surfaces in streams.

1. Introduction

This study focuses on differences in food quality of the organic layer to *B. rhodani* from acid and neutral streams and whether the nymphs show different feeding behaviour when encountering food of these different types.

2. Methods

Nymphs were individually introduced into test chambers: one set of 160 test chambers with acid water (pH 4.5) and one with neutral water (pH 6.5). The experiments were done in spring and autumn. The nymph in each test chamber was provided with equal amounts of food collected from either the acid or the neutral stream, or no food at all. After six days feeding ceased and the nymphs were starved. Mortality, moulting and emergence were registered daily.

3. Results

Nymphs in both acid and neutral water showed a lower mortality rate when fed with food from the neutral stream. The mortality rate was the same when comparing neutral and acid water for this food régime. In neutral water there was no difference in the mortality rate between nymphs provided with food from the acid stream and those provided with no food at all. In acid water those nymphs with no food showed a slightly higher mortality rate than those provided with acid food.

The results were consistent in both spring and autumn, though less pronounced in the autumn experiment. Because mortality occurred predominantly when nymphs were moulting and the moulting frequency was higher during the spring experiment, this could account for the higher spring mortality rate rather than any seasonal differences in food quality.

The food derived from the acid stream had a significantly higher organic content than the food from the neutral stream. During spring the organic layer from the acid stream consisted primarily of a green alga (*Stigeoclonium* sp.) whereas detritus contributed less than 10 %. During autumn the layer consisted of more than 90 %

† Present address: Swedish State Power Bd, S-162 87 Vällingby, Sweden.

flocculated humic material and few algae. In the neutral stream during spring, diatoms of several species dominated the organic layer and detritus contributed less than 50 %, whereas in autumn it contributed more than 95 %, with few diatoms present.

In a preference experiment the position of nymphs on either food from acid or neutral streams was recorded every hour. One experiment was run in acid water and one in neutral water. In both acid and neutral water the nymphs were found more frequently on the neutral food. The number of changes of position was higher in acid water than neutral water.

From continuous recordings during day and night with an infrared sensitive video camera it was confirmed that the nymphs spent more time on food from the neutral stream, though this was more pronounced in acid water. These recordings also confirmed that the activity was much higher in acid water than neutral water during both day and night. In the acid water the nymphs did not graze on the food from either the acid nor the neutral stream. In the neutral water, time spent grazing was about 20 % on food from the neutral stream and less than 5 % on food from the acid stream.

4. Discussion

The decreased survival of B. rhodani when provided with food from an acid stream and the low preference for it exhibited by the nymphs indicate that the ability to utilize this food source is low; in fact so low that it is comparable with no access to food. In addition, physiological stress as indicated by the increased activity and the lack of grazing in acid water is probably another factor that contributes to the vulnerability of B. rhodani to acidification.

General discussion

C. EXLEY AND J. D. BIRCHALL (*Institute of Aquaculture, University of Stirling, Scotland*; *ICI p.l.c. P.O. Box 11, The Heath, Runcorn, U.K.*). Aluminium is toxic to fish in acid water and a most significant recent finding is the elimination of the acute toxicity of *ca.* 7 μM l^{-1} aluminium to salmon fry in the presence of *ca.* 100 μM l^{-1} silicic acid, Si(OH)$_4$, (Birchall, J. D., Exley, C. *et al.* 1989 *Nature, Lond.* **338**, 146–148). It has been shown that aqueous aluminium species react with silicic acid from below pH 5 to form hydroxyaluminosilicate species with the Si:Al ratio approaching 0.5 (Chappell, J. S. & Birchall, J. D. 1988 *Inorg. chim. Acta* **153**, 1–4). These species, formed when the silicic acid concentration exceeded 100 μM l^{-1}, greatly reduce the bioavailability of aluminium. The strength of binding of silicate to hydroxyaluminium exceeds that of sulphate above pH 5, phosphate above pH 6 and even that of strong chelators such as citrate at pH 7. Thus in the presence of silicic acid, aluminium is prevented from binding at gill epithelial sites, an effect possibility enhanced by a gill boundary layer more alkaline than the bulk water. An important conclusion of this work was that the Si:Al balance is a key, hitherto neglected, factor affecting the bio-availability and toxicity of aluminium.

H. ANDERSON (*Macaulay Land Use Research Institute, Aberdeen, Scotland*). Chris Exley has proposed that Al may be detoxified by formation of soluble Al-silicates.

At MLURI, especially in view of our involvement in research into the involvement of (proto-) imogolite in podzol formation, we have been interested in any association of Al and Si in the SWAP plot studies. Si is routinely determined in all input–throughput and output samples.

All attempts to show Al–Si association in filtered samples have failed. Whereas Si has always appeared in small molecular-size fractions (less than 1000 daltons), Al is associated with the larger molecules, mainly as organic complexes. In certain samples with large contents of humic substances, we would not expect proto-imogolite to form.

However, as Dr Harriman has shown, fish survival at Mharcaidh is threatened by H$^+$, not Al toxicity. Imogolite has been found in B horizons of Mharcaidh soils.

C. EXLEY (*Institute of Aquaculture, University of Stirling, Stirling FK9 4LA, Scotland*). Aluminium is the principal toxicant of dilute acid water, (pH *ca.* 4.5–6.0). In fish, acute toxicity is the result of aluminium interactions at epithelial membranes, (mainly the gill epithelium) altering their barrier properties. Symptoms include ionregulatory dysfunction and respiratory disturbances. The toxic mechanism is unknown; however, the mitigative effects of Ca^{2+}, Na$^+$ and H$^+$ and protective effects of citrate and silicic acid on acute aluminium toxicity suggests that aluminium binds to ligand(s) integral to gill and membrane biochemistry (or both).

Research has concentrated on the symptoms of toxicity and has ignored the fundamental toxic mechanism. To study mechanisms the chemical rationale for aluminium toxicity must be considered and a great deal more work is needed. Studies on the amelioration of aluminium toxicity have proved very useful in elucidating the toxic mechanism. For example, the abolition by silicic acid of acute aluminium toxicity in bulk water of pH 5.0 is conclusive evidence that the gill boundary layer pH is near-neutral under such conditions. The importance of pH in acute aluminium toxicity is not in the proportions of aluminium hydrolysis species present but in the stability of Al^{3+} interactions with gill biochemistry.

GWYNETH HOWELLS (*Department of Zoology, University of Cambridge, Downing Street, Cambridge, U.K.*). Mr Exley (and Professor Birchall, in the same research group) have claimed that silicate can be an effective detoxicant of dissolved inorganic aluminium. Although I can accept the evidence that silicate in laboratory bioassays can operate in this way (like citrate), it seem unlikely to be effective in reality. The ratio of Si:Al needed is 13, but this is, in my experience, not achieved either as mean values or during short-term chemical changes in acid lakes or streams, although it may be of possible significance in soil waters.

Modelling long-term trends in surface water acidification

By P. G. Whitehead[1], A. Jenkins[1] and B. J. Cosby[2]

[1] *Institute of Hydrology, Wallingford, Oxfordshire OX10 8BB, U.K.*
[2] *Department of Forestry, Duke University, North Carolina, U.S.A.*

To assess long-term acidification trends models are required that can characterize the principal mechanisms operating, account for the changing levels of deposition inputs and provide good estimates of past, present and future soil and water chemistry. At the same time the model should be transferable so that it can be readily applied to a wide range of catchments in differing pollution climates with differing land-use regimes and differing soils and parent geology. In this paper the application of MAGIC (model of acidification of groundwaters in catchments) is described for moorland and afforested catchment sites in Scotland and Wales. In addition, MAGIC has been applied in a regional analysis to predict distributions of water quality across Wales and the Galloway region of Scotland. The sensitivity of the model to parameter variations between sites is explored and the model used in a predictive mode to assess effects of land-use change such as afforestation and the likely changes in future atmospheric pollutant deposition levels.

The model results support the findings of palaeoecological studies that acidification has occurred in many U.K. catchments and demonstrates a clear link between deposition of atmospheric pollutants and acidification.

1. Introduction

Acidification may be regarded as essentially a problem over two very different timescales. Short-term fluctuations in acidification are generally driven by meteorological factors and hydrological processes operating in the catchment. The timescale of these events are in the order of hours, or at most days, and the level of acidity will be largely controlled by the ability of the catchment to buffer incoming acidity within the catchment's hydrological response time. On the other hand, long-term changes in soil and water chemistry can occur over years or decades causing chronic acidification.

Approaches to short-term response modelling are described elsewhere (Christophersen *et al.* 1982; Whitehead *et al.* 1986 *a*, *b*; Wheater *et al.*, this symposium). Modelling long-term changes in acidification has been approached in two ways. The first is an empirical approach whereby extrapolations from present conditions are made by using empirical relations between rainfall chemistry and surface water quality (Henriksen 1979). The second approach utilizes mechanistic, process-orientated, numerical models of hydrology and geochemistry to make the quantitative linkage between deposition and water quality (Schnoor *et al.* 1984; Seip & Rustad 1983; Cosby *et al.* 1985 *a*, *b*). It is essential that such models take into account the long-term interactions occurring between physical and chemical

[431]

characteristics within a catchment and can account for the dominant processes operating. In this paper we describe one such model, MAGIC (model of acidification of groundwaters in catchments) and illustrate its application to Wales and Scotland. Results from specific sites are presented in addition to a regional analysis. The model is used in a predictive manner to assess the effect of different future pollutant deposition patterns and the impacts of land-use change such as afforestation. An analysis of the calibrated model parameters from a variety of sites is considered.

2. Conceptual basis of MAGIC

Extensive details of the background theory and equations used in MAGIC have been given by Cosby *et al.* (1985*a*). However, the dominant processes incorporated include.

1. Anion retention by catchment soils (e.g. sulphate adsorption).

2. Adsorption and exchange of base cations and aluminium by soils.

3. Alkalinity generation by dissociation of carbonic acid (at high carbon dioxide partial pressures in the soil) with subsequent exchange of hydrogen ions for base cations.

4. Weathering of minerals in the soil to provide a source of base cations.

5. Control of Al^{3+} concentrations by an assumed equilibrium with a solid phase of $Al(OH)_3$.

MAGIC simulates these processes by using the following.

1. A set of equilibrium equations which quantitatively describe the equilibrium soil processes and the chemical changes that occur as soil water enters the stream channel.

2. A set of mass balance equations which quantitatively describe the catchment input–output relationships for base cations and strong acid anions in precipitation and stream water.

3. A set of definitions which relate the variables in the equilibrium equations to the variables in the mass balance equations.

3. Application to the Allt a'Mharcaidh – a moorland transitional site

The Allt a'Mharcaidh is a transitional site located in the Cairngorm Mountains of NE Scotland. Full details of catchment characteristics, instrumentation and sampling methodology are given elsewhere (Jenkins *et al.* 1988; Ferrier & Harriman, this symposium).

MAGIC has been applied to the catchment by using a two-stage optimization procedure. First, the nitrate and ammonia uptake rates are determined together with the soil-sulphate adsorption capacity. These parameters are all independent and therefore can be optimized to give a unique value to match the output stream chemistry. The second stage of the optimization considers the parameters controlling cation behaviour, namely weathering rates, which control cation supply from bedrock, and selectivity coefficients, which control ion exchange in the soils. The output stream chemistry and measured base saturation are used to drive the Rosenbrock optimization procedure, a robust and generally reliable technique. Full details of the optimization of model parameters are given by Jenkins *et al.* (1988) and a comparison of model parameters in relation to other site applications of MAGIC are given later in this paper (see table 4).

Table 1. *Observed and simulated stream chemistry for the Allt a'Mharcaidh*
(in microequivalents per litre)

	1846 model simulated	1986 model simulated	1986 observed	2126 model simulated
Ca	21.4	37.5	37.1	43.8
Mg	28.6	29.6	29.9	29.7
Na	102.1	117.0	116.1	115.1
K	7.9	8.9	8.4	10.2
NH_4	0.0	2.0	—	2.2
SO_4	15.2	50.3	50.1	70.0
NO_3	0.0	2.2	2.1	2.2
Cl	111.3	111.3	111.3	111.3
Alk	33.5	31.6	33.0	17.2
H	1.6	1.7	2.0	2.8
pH	5.8	5.8	5.7	5.6

MAGIC produces a close match between observed and simulated stream chemistry as shown in table 1. Simulated soil base saturations are also good estimates of observed values at 8%. Historical reconstruction and future response, assuming constant deposition chemistry at current levels to 2126, are also shown in table 1.

In the Allt a'Mharcaidh catchment, soils have retained a high buffering capacity since 1846 with almost unchanged alkalinity and pH in streamwater. By 2126, however, assuming constant deposition into the future, a decrease in alkalinity to *ca.* 30% of the present value is forecast with a drop in mean pH of only 0.1 unit. Stream sulphate levels increase steadily from 1846 to the present day and continue to rise to 2126. Soil base saturation remains almost constant to the present, despite increased output of base cations, due to high weathering rates. Base saturation deteriorates beyond 1986, however, as soil exchange sites become saturated with hydrogen and strong acid anions.

The simulation results differ markedly from the pattern of change demonstrated by the MAGIC simulation of Dargall Lane, a heavily acidified catchment in SW Scotland (Cosby *et al.* 1986). In particular, sulphate concentrations in streamwater at the Allt a'Mharcaidh do not reflect changes in the deposition sequence and the pH response is smooth and damped at this site. This is a direct consequence of the high value of E_{mx}, the maximum sulphate adsorption rate, that allows a high degree of sulphate adsorption and a long time-lag between any change of input chemistry and its resulting effect on output chemistry. Conversely, at Dargall Lane little sulphate adsorption has occurred so that any change in sulphate deposition is reflected almost immediately in the run-off chemistry. By keeping all optimized parameters and deposition factors constant, and running the hindcast and forecast simulation for different values of E_{mx}, a variety of responses can be produced and these are shown in figure 1*a*. As E_{mx} is decreased, response time decreases and input and output chemistry become similar, whereas increasing sulphate adsorption causes an attenuated sulphate response. A key parameter controlling acid inputs to the catchment is the sulphate dry-deposition factor. This increases the sulphate loading to the catchment to account for aerosol and dry deposition of anthropogenically derived sulphate. Estimating the value of this parameter is particularly difficult and the sensitivity of the Allt a'Mharcaidh to the parameter is illustrated in figure 1*b*.

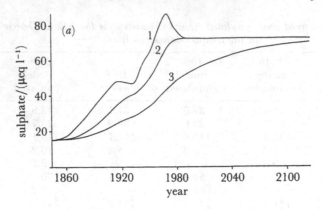

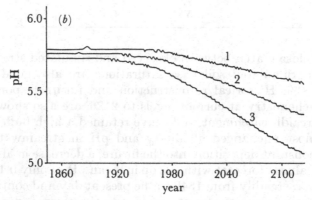

Figure 1. (a) Simulated stream sulphate concentrations assuming E_{mx}, the maximum sulphate adsorption rate, is 1.0, 10.0 and 28.6 for 1, 2 and 3, respectively. (b) Simulated pH showing the effect of enhanced dry deposition of sulphate. Dry deposition is 20%, 40% and 60% of wet deposition for 1, 2 and 3, respectively.

The higher the dry-deposition factor, the larger the pH decline. The major change in the pH trend occurs between 1980–1990 suggesting that the Allt a'Mharcaidh is a truly transitional site that will undergo faster acidification in the future if deposition remains at present-day levels. The rate of decline is highly dependent on aerosol deposition rates, which can be significant at the higher altitudes.

4. Application to Chon and Kelty – acidified catchments in central Scotland

MAGIC has been applied to the Chon and Kelty catchments, two forested sites, located in an area of high deposition 40 km north of Glasgow. Final optimized values of weathering rates and soil exchange selectivity coefficients are compared with other catchments later in this paper (see table 4). Values of weathering of calcium and magnesium are higher in Chon and this accords well with field observation of a dolerite dyke within the catchment that affects the outflow concentrations of these ions. Simulated stream chemistry (table 2) matches observed chemistry closely at both Chon and Kelty. The model also successfully simulates present-day soil

Table 2. *Observed and predicted present day stream chemistry at Chon and Kelty*
(in microequivalents per litre)

	Chon		Kelty	
	observed	predicted	observed	predicted
Ca	43.3	45.7	19.0	20.6
Mg	48.3	46.2	36.9	36.0
Na	181.3	184.1	200.9	199.8
K	7.1	8.2	7.8	8.2
NH_4	7.2	6.9	13.2	12.9
SO_4	93.3	102.7	100.0	105.2
Cl	224.5	217.6	216.5	214.7
NO_3	3.0	2.9	10.2	10.1
Al_3	24.4	19.5	48.8	48.2
H	24.5	24.2	95.5	88.9
pH	4.6	4.6	4.0	4.1

chemistry as soil base exchange fractions at the two sites are well matched with measured data (Jenkins *et al.* 1990).

Stream pH, hydrogen ion, alkalinity, calcium and sulphate reconstructions for the two sites from 1847–1987 are shown in figure 2. Chon shows a very low background hydrogen ion concentration with an increasing trend which accelerates in the period 1950–1960 to give a rapid increase in hydrogen ion concentrations. Kelty shows a similar accelerated increase during that period but has a very high background concentration. This is because of the high level of organics in the catchment that are assumed to be at a constant level throughout the simulation. The period of rapidly increasing hydrogen corresponds to the planting and growth of the forest. At the time of canopy closure (1965), both catchments are subject to the most severe acidifying processes: (i) total load of anthropogenic wet and dry deposition is at a high level as the assumed deposition curve peaks at this time; (ii) input from canopy filtering is also at a maximum because canopy closure (and thus maximum filtering by the trees) and maximum deposition coincide; (iii) maximum cation uptake coincides with canopy closure; and (iv) the effect of increased evapotranspiration is also at a maximum. Around 1970 stream concentrations level off and by 1980 have started to decrease. This is in response to the falling deposition levels in recent years and to the decrease in uptake of base cations as the forest matures. This apparent recovery is in accord with reconstructions from diatom evidence (Battarbee 1988).

The base saturation reconstruction (figure 2*e*) indicates a progressive soil acidification through time as base cations are leached in response to the incoming acidity. High weathering rates at Chon produce a high initial base saturation although this falls steadily until 1950 and then accelerates downwards at the onset of afforestation. At Kelty, although the apparent initial base saturation is not as high as at Chon. The model indicates that cation losses from the soil will result in slightly higher percentage base saturation. No recovery of base saturation is seen at either site in response to decreased emissions since 1970, although the rate of decrease slows, and this accords well with the expected slower recovery of soils as they continue to desorb sulphate.

The effect of forest growth at the two sites is to accelerate acidification of the surface water as the result of a gradual increase in anthropogenic deposition. Similar

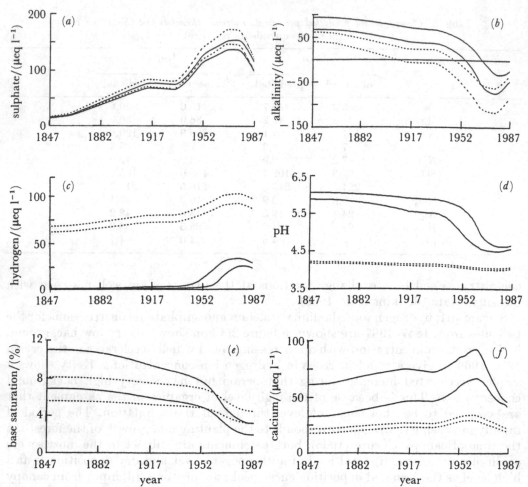

Figure 2. Simulated stream chemistry and soil base saturation for the Chon and Kelty sites in central Scotland; (----------) represent Kelty; (————) represents Chon, double lines represent the upper and lower ranges of expected behaviour.

results have been obtained previously by Whitehead *et al.* (1988*a, b*) for forested catchments in Wales at Plynlimon and Llyn Brianne. It should be emphasized that forests growing in pristine areas will not have such an acidifying effect as demonstrated by Neal *et al.* (1986). The model clearly demonstrates that as well as the effects of canopy interception, evapotranspiration and cation uptake cause a significant acidification of the soil. This can in turn cause enhanced surface water acidification depending on the activity of the mobile anion and the base cation status of the soil. At Kelty the increased soil acidification appears to lead to water acidification because hydrogen is removed from the soil associated with sulphate, whereas at Chon, base cations are exchanged thereby affording some buffer to the stream acidity.

5. Regional application of MAGIC to Wales

MAGIC has been applied in a regional context to assess water quality changes across a range of streams or lakes. In the regional approach, the MAGIC is run repeatedly

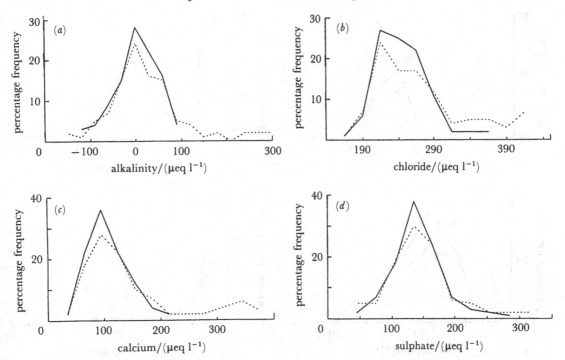

Figure 3. Observed (- - -) and simulated (——) stream chemistry distributions for Wales using the full observed regional data set; (*a*) alkalinity, (*b*) chloride, (*c*) calcium and (*d*) sulphate.

with different parameter values chosen randomly from given distributions. The ensemble of model runs is then evaluated and compared with the observed distributions of water quality across the region obtained from analysis of survey data. This Monte Carlo approach allows a simulation of the water quality changes across the region as a whole. The approach was developed by Cosby *et al.* (1988) and Hornberger *et al.* (1989) in an analysis of 700 Norwegian lakes. A similar approach has been used by an analysis of surface water in the Galloway region of Scotland (Musgrove *et al.* 1990) and in Wales by Jenkins *et al.* (1990).

Figure 3 shows the simulated distribution of calcium, magnesium, alkalinity and sulphate against the observed distributions for Wales obtained from 130 streams. In all cases the simulated distribution is close to the observed distribution suggesting that the model has captured the principal features of water quality across the region. Of particular interest is the question of how the water quality distributions in the region have changed over time. As shown in figure 4, the pre-industrial 1844 distributions for sulphate and alkalinity are very different from the present day. Sulphate levels are much lower and alkalinity is significantly higher. The temporal changes in the entire regional distributions have also been investigated under an assumed 30% deposition reduction in excess sulphate deposition linearly between present day and the year 2000. Although sulphate chemistry in the streams is reduced, there is no major shift in alkalinity. This is probably because the base saturation levels of the soils are low and even a 30% reduction in sulphate is insufficient to achieve a significant recovery on the base-poor Welsh soils.

The portion of the simulated region that has alkalinity less than zero was subjected to various hypothetical future deposition patterns in an attempt to look at the likely

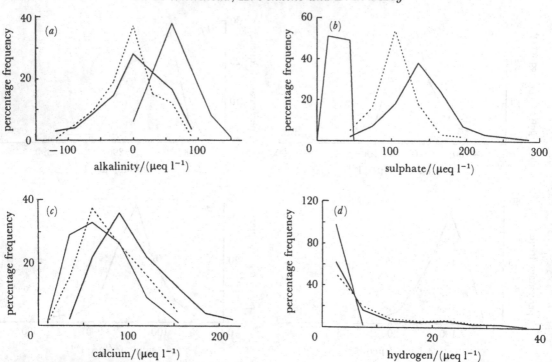

Figure 4. Simulated background (—), present day (——) and future (---) stream chemistry for Wales assuming a 30% decrease in acid deposition; (*a*) alkalinity, (*b*) sulphate, (*c*) calcium and (*d*) hydrogen.

response of the worst affected sites. Three future deposition scenarios were used, linear reduction in non-marine sulphate deposition over a 20 year period to 30%, 60% and 90% of the 1984 value, thenceforth remaining constant. Table 3 describes the predicted chemistry equlibria for each deposition scenario. The major feature of the results is the trend in increased pH and alkalinity and decreased sulphate in the streams as the deposition is decreased. Reductions in deposition of 30% and 90% lead to mean stream sulphate concentrations of 115 and 43.7 μequiv l⁻¹, respectively. This reduction of sulphate concentrations raises alkalinity, from a mean of 9.4 to 75.2 μequiv l⁻¹. Coupled with the change in alkalinity, the pH rises from a mean of 5.5–6.2. Ormerod (1989) assessed changes from a biological viewpoint and suggested that trout survival in Welsh streams would be significantly improved with a 60% reduction in sulphate deposition.

6. Regional application of MAGIC to southwest Scotland

The Galloway Region of SW Scotland contains many lochs and streams that drain moorland, forest and pasture catchments. The bedrock consists mainly of lower palaeozoic rocks of Ordovician and Silurian age with a few intrusions of granite of the Old Red Sandstone age. The Galloway region differs significantly from Wales in that it is an area of high deposition and has many acidified lakes. MAGIC has been applied to Galloway using the same procedure as for Wales. As is the case for Wales a close fit is obtained between the observed and simulated water quality distributions for Galloway. The changes over time in the distributions of simulated chemistry were

Table 3. *Welsh stream concentrations in equilibrium with reduced deposition for*
10 % most sensitive sites
(in microequivalents per litre)

variable	mean	standard deviation	minimum	maximum
(a) effect of a 90 % reduction in deposition				
pH	6.2	0.2	5.6	6.5
Na	230.4	43.2	146.8	322.0
Ca	68.7	27.8	25.8	149.0
Mg	62.3	14.0	36.86	106.3
SO$_4$	43.7	6.4	34.5	58.5
Cl	253.8	47.2	168.9	347.1
Alk	75.2	32.0	14.5	150.7
(b) effect of a 60 % reduction in deposition				
pH	5.9	0.3	5.1	6.3
Na	231.5	42.8	154.5	322.0
Ca	26.8	26.8	28.0	149.0
Mg	64.8	13.7	40.3	107.1
SO$_4$	79.8	12.9	58.4	106.9
Cl	253.8	47.2	168.9	347.1
Alk	43.9	25.8	−6.9	96.6
(c) effect of a 30 % reduction in deposition				
pH	5.5	0.4	4.7	6.0
Na	231.9	42.4	159.2	322.0
Ca	71.8	26.4	28.8	149.0
Mg	65.8	13.6	41.4	107.2
SO$_4$	115.8	21.3	80.1	161.3
Cl	253.8	47.2	168.9	347.1
Alk	9.4	23.7	−59.9	44.6
(d) current chemistry				
pH	5.1	0.4	4.9	5.3
Na	236.6	41.3	167.0	333.1
Ca	79.2	23.4	41.9	147.6
Mg	76.4	19.1	47.0	138.7
SO$_4$	153.2	30.2	98.4	216.8
Cl	253.8	47.2	168.9	347.1
Alk	−4.9	3.6	−11.9	0.4

ascertained by investigating the output for 1844, 1982 and 2060. The future deposition was modelled as a linear decrease to 30 % of the 1982 level by the year 2001 and constant thereafter. The results are presented in detail by Musgrove *et al.* (1990) and model reconstruction shows a large drop in both pH and alkalinity over the past 140 years. The pH level falls by 0.8 pH unit and alkalinity falls by *ca.* 70 µequiv l^{-1}. This is in accord with the findings of Battarbee & Flower (1985), who report changes in pH level up to 1.0 pH unit during the same period, for those lochs in the granitic region of Galloway. Very little recovery is seen during the future scenario. The small size of this recovery reflects the depletion in the soil of base cations with the low rate of soil weathering in the region, that enable only a slow recovery rate. Alkalinity levels indicate a major shift in distribution from 1840 levels to current values. Even with a further 30 % reduction in sulphate deposition alkalinity shows no major shift and indicates the scale of the problem across a sensitive region such as Galloway.

Table 4. Parameter values for a range of MAGIC applications

Site – location	E_{mx}[a]	weathering rates[b]				selectivity coefficients[c]				reference
		Ca	Mg	Na	K	Ca	Mg	Na	K	
Allt a'Mharcaidh – Scotland	28.9	29.2	12.1	11.1	9.9	3.02	3.18	−0.28	−3.64	Jenkins et al. 1988
White Oak Run – U.S.A.	8.0	0.0	10.0	4.0	12.0	4.1	4.1	−0.5	−1.01	Cosby et al. 1985
Dargall Lane – Scotland	0.1	38.0	35.0	2.0	8.0	2.49	2.91	−0.22	−3.92	Cosby et al. 1986
Lake Gardsjon – Sweden	1.0	10.0	12.0	13.0	2.0	1.9	4.1	0.7	−1.0	Wright et al. 1986
Lake Hovvatn – Norway	1.0	3.0	1.0	0.5	0.5	1.0	2.2	−0.5	−4.0	Wright et al. 1986
Llyn Brianne – Wales	0.01	25.0	15.0	10.0	1.0	1.94	1.67	−2.1	−5.33	Whitehead et al. 1988
Plynlimon – Wales	3.38	116.9	88.2	66.4	0.0	2.70	3.27	−0.65	−4.70	Whitehead et al. 1988
Loch Chon – Scotland	7.0	53.0	58.0	8.0	10.0	1.16	2.22	0.28	−3.42	Jenkins et al. 1990
Loch Tinker – Scotland	4.4	110.0	22.0	5.0	1.0	0.06	−0.69	−0.69	−3.91	Jenkins et al. 1990
Yli Knuatila – Finland	1.5	17.0	10.5	1.8	1.8	2.4	4.21	0.60	−3.63	Lepisto et al. 1988
Kelty Water – Scotland	3.5	1.0	17.0	27.0	14.0	−0.82	0.77	0.07	−3.62	Jenkins et al. 1990

[a] In milliequivalents per kilogramme.
[b] In milliequivalents per square metre per annum.
[c] $\log S$ (Al cation).

7. Transferability of MAGIC

MAGIC has now been applied to 26 individual sites, two plot experiments (Wright 1987; Skeffington, personal communication) and in several regional analyses in the U.K. (Scotland and Wales), in Norway and extensively in the U.S.A. Table 4 shows a typical range of applications of MAGIC together with key parameters such as E_{mx}, (the maximum sulphate adsorption capacity of the soils) weathering rates and selectivity coefficients. As might be expected with the highly heterogeneous nature of geology, soils, vegetation and hydrochemical flowpaths in catchments, parameters vary from site to site. For example E_{mx} is particularly high at the Allt a'Mharcaidh and White Oak Run reflecting the sulphur adsorbing properties of soils in these catchments. Also weathering rates vary from site to site and reflect solid and drift geology. Loch Chon shows high weathering rates and this arises from a doleritic dyke in the catchment providing a significant source of base cations from weathering reactions. Similarly, Plynlimon shows high weathering rates in the moorland catchment reflecting liming that occurred over 40 years ago. Selectivity coefficients also vary from site to site but provide a consistent set of values.

8. Conclusions

MAGIC has been applied to a range of sites in Scotland and Wales and comparison with palaeocological results suggest that it gives a good representation of the long-term behaviour of catchments (Jenkins *et al.* 1990). Its wide application to many sites in Scandinavia, North America and the U.K. support this view as does the regional application in Wales and southwest Scotland.

The program confirms that acidification of surface water is a serious problem in some parts of Britain. Acidification levels are high in areas with thin base-poor soils on granitic type geology. Unfortunately, afforestation in high deposition areas tends to enhance the acidity by scavenging or filtering acidic particles and mist and this effect can be very significant often doubling the loads of acidic deposition entering catchments. It should be emphasized that afforestation in low deposition regions appears to have minimal effect (Neal *et al.* 1988). However, a strategy for forestry management to minimize acidification effects is still required for Scotland.

Although acidification appears to be at least partly reversible the MAGIC model indicates that significant levels of emission reductions are required for there to be a sustained recovery. Different catchments will show different reversibility responses dependent on factors such as soil-base, saturation levels, sulphate adsorption and release mechanisms, weathering rates, hydrological flow paths and deposition rates. Although there are many uncertainties in the model, it provides the only means at present of making site specific or regional predictions of long-term future behaviour of stream and lake acidity.

The authors are indebted to staff of the Institute of Hydrology for their technical support throughout SWAP and to the SWAP Committee for providing funding and support over the past four years.

References

Battarbee, R. W., Flower, R. J., Stevenson, A. C., Jones, V. J., Harriman, R. & Appleby, P. G. 1988 *Nature, Lond.* **332**, 530–532.

Battarbee, R. W., Flower, R. J., Stevenson, A. C. & Rippey, B. 1985 *Nature, Lond.* **314**, 350–352.

Christophersen, N., Seip, H. M. & Wright, R. F. 1982 *Wat. Resour. Res.* **18**, 977–997.

Cosby, B. J., Hornberger, G. M. & Wright, R. F. 1988 In *Uncertainty and regional modelling* (ed. J. Kamari), Reidel. (In the press.)

Cosby, B. J., Whitehead, P. G. & Neale, R. 1986 *J. Hydrol.* **84**, 381–401.

Cosby, B. J., Wright, R. F., Hornberger, G. M. & Galloway, J. N. 1985*a Wat. Resour. Res.* **21**, 51–63.

Cosby, B. J., Wright, R. F., Hornberger, G. M. & Galloway, J. N. 1985*b Wat. Resour. Res.* **21**, 1591–1601.

Henriksen, A. 1979 *Nature, Lond.* **278**, 542–545.

Hornberger, G. M., Cosby, B. J. & Wright, R. F. 1989 *Wat. Resour. Res.* **25**, 2009–2018.

Jenkins, A., Ferrier, R. C., Walker, T. A. B. & Whitehead, P. G. 1988 *Wat. Air Soil Pollut.* **40**, 275–291.

Jenkins, A., Whitehead, P. G., Musgrove, T. J. & Cosby, B. J. 1990 *J. Hydrol.* (In the press.)

Lepisto, A., Whitehead, P. G., Neal, C. & Cosby, B. J. 1988 *Nordic. Hydrol.* **19**, 99–120.

Musgrove, T. J., Whitehead, P. G. & Cosby, B. J. 1990 In *Impact models to assess regional acidification* (ed. J. Kamari), pp. 131–135. Amsterdam: Kluwer.

Neal, C., Whitehead, P. G. & Cosby, B. J. 1986 *J. Hydrol.* **84**, 381–401.

Ormerod, S. J., Weatherley, N. S., Varallo, P. V. & Whitehead, P. G. 1988 *Freshwat. Biol.* **20**, 127–140.

Schnoor, J. L., Palmer, W. D. & Glass, G. E. 1984 In *Modelling of total acid precipitation impacts* (ed. J. L. Schnoor), vol. 9, pp. 155–173. Boston: Butterworth.

Seip, H. M. & Rustad, S. 1983 *Wat. Air Soil Pollut.* **21**, 217–223.

Whitehead, P. G., Neal, C., Seden-Perriton, S., Christophersen, N. & Langan, S. 1986*a J. Hydrol.* **85**, 281–304.

Whitehead, P. G., Neal, C. & Neale, R. 1986*b J. Hydrol.* **84**, 353–364.

Whitehead, P. G., Bird, S., Hornung, M., Cosby, B. J., Neal, C. & Paricos, P. 1988*a J. Hydrol.* **101**, 191–212.

Whitehead, P. G., Reynolds, B., Hornung, M., Neal, C., Cosby, B. J. & Paricos, P. 1988*b Hydrol. Process.* **2**, 357–368.

Wright, R. F., Cosby, B. J., Hornberger, G. M. & Galloway, J. N. 1986 *Wat. Air Soil Pollut.* **30**, 367–380.

Wright, R. F. & Cosby, B. J. 1987 *Atmos. Environ.* **21**, 727–730.

Discussion

M. CRESSER (*Aberdeen University, Department of Plant and Soil Science, Old Aberdeen, Scotland, U.K.*). I would like to return to the very important point raised by Sir John Mason about the reliability of MAGIC for making long-term predictions. It is often not clear precisely where the numbers inserted in some of the numerous black boxes of the model come from. Could Dr Whitehead confirm that often such numbers are selected from within a wide range of possible values to make the predicted water solute composition fit the observed data? If this is so, is it not possible that some key processes (especially biological processes) may be being ignored completely? Could this not substantially limit the validity of some of the conclusions reached about long-term effects?

P. G. WHITEHEAD. First, MAGIC is a process-based model rather than a black-box model. Key processes are included in the model are ion exchange, weathering, CO_2 degassing, sulphate adsorption, etc. Most of these are chemical processes. Biological processes are not included explicitly because there is still considerable debate as to what biological controls are operating and to what extent they influence stream and soil chemistry. Organic acid is included in the model as are temperature effects. I do not feel that any major processes are missing in MAGIC or else this model could not reproduce the chemistry in so many different sites or regions. Hence I do not believe the major conclusions concerning long-term effects are invalid. Indeed all the field evidence is that reversibility predicted by MAGIC is already occurring.

Hydrochemical models for simulation of present and future short-term changes in stream chemistry: development and status

By Nils Christophersen[1], Colin Neal[2], Hans M. Seip[3]
and Alex Stone[3]

[1] Center for Industrial Research, P.O. Box 124 Blindern N-0314 Oslo 3, Norway
[2] Institute of Hydrology, Maclean Building, Crowmarsh Gifford,
Wallingford OX10 8BB, U.K.
[3] Chemistry Department, University of Oslo, P.B. 1033 Blindern N-0315
Oslo 3, Norway

The historical development and the present version of the Birkenes model in relation to the available field data, including conservative tracers, are summarized. An outline for further model development is presented, stressing streamwater as a mixture of different soil waters. Predictions of future short-term changes in streamwater chemistry at Birkenes, for different deposition scenarios, are then addressed along two lines. The first approach comprises a two-step procedure, involving predictions of long-term changes in soil chemistry using the MAGIC model, and simulations of short-term changes in streamwater chemistry, for various sulphate deposition levels, using the Birkenes model. Major improvements in streamwater chemistry are predicted for reductions corresponding to 60% of the present excess loading. The second approach utilizes a charge balance argument to estimate the minimum reductions in sulphate that are required to bring the present H^+ and inorganic monomeric Al concentrations at highflow down to values acceptable for fish survival. The calculations are uncertain but considering both approaches and neglecting intervention by liming, even a reduction in the excess sulphate deposition of 90% would not guarantee successful restocking at Birkenes. However, it is essential to decrease the deposition and also smaller reductions may limit further soil acidification in southern Norway and make fish restocking possible in less acidified areas.

1. Introduction

Starting from a situation dominated by conflicting interpretations concerning the major processes controlling streamwater chemistry (Rosenqvist 1977, 1978), the Birkenes model (Christophersen et al. 1982) was developed. The philosophy has since been to test and improve the model in an iterative way. This has been accomplished, largely through SWAP funding, by applying it to other catchments and by actively designing interdisciplinary field programmes leading to new data on hydrology, soil chemistry and geochemistry (cf. Christophersen et al., this symposium). Close cooperation has also evolved with other studies outside SWAP; in particular, with the work conducted by Plynlimon in mid-Wales (Neal et al. 1986) and at the Panola Mountain in Georgia, U.S.A. (Hooper et al. 1990).

[445]

After a summary of the historical background and discussion of the interaction between modelling and field work, emphasis will be placed on predictions of future changes in streamwater chemistry for given sulphate deposition scenarios.

2. Background

A commonly held view explaining acid surges occurring in streamwater during storm and snowmelt events was that the H^+ ions in rainfall and meltwater were simply transferred to the stream, without much chemical interaction with the terrestrial environment. This is in agreement with the classical hydrological picture of stormflow as predominantly consisting of rain or meltwater, e.g. Hortonian overland flow or saturated overland flow.

Rosenqvist (1977, 1978) proposed an opposing view disputing the role of acid rain as a main cause of freshwater acidification. He argued that land use changes causing forest regrowth were dominant factors and that terrestrial controls, in particular cation exchange, were the major factors regulating streamwater chemistry. He postulated, for instance, that stormflow waters at the Birkenes catchment in southern Norway were acidic irrespectively of the rainstorm acidity; this was confirmed by Nordö (1977) and by Christophersen *et al.* (1982). His views, right or wrong, produced a scientific discord that forced a more rigorous scientific assessment of catchment behaviour and persists to this day.

The Birkenes study was initiated in 1971–72. (See Seip *et al.* and Christophersen *et al.* (this symposium) for an updated site description.) Predominantly based on input–output data from this catchment, a simple and mainly hydrologic model (the Birkenes model) was developed in the mid-seventies (Lundquist 1976, 1977), with the chemical parts being developed in the early 1980s (Christophersen & Wright 1981; Christophersen *et al.* 1982). Inspired by the work of the soil scientist John Reuss (1980) and stimulated by the Rosenqvist controversy, the model was derived assuming dominance of terrestrial controls in determining streamwater chemistry. Based on the mobile anion concept (Seip 1980), the model (figure 1) contained two distinct soil environments controlling runoff chemistry. It was hypothesized that stormflow, which is acidic and aluminium-rich, was generated by rapid transfer of water in the upper soil horizons, where the solution composition was controlled by cation exchange and aluminium hydroxide (gibbsite) mineral equilibrium. The less acidic and more calcium-rich baseflow, stemming from the lower reservoir in the model, was hypothesized to be controlled by the same gibbsite equilibrium but also by mineral weathering reactions in deeper soil horizons.

The Birkenes model was later applied to several catchments and was able to reproduce major variations in streamwater chemistry. These catchments include tributaries to Harp Lake (Seip *et al.* 1985; Rustad *et al.* 1986) and Turkey Lake (Lamb *et al.* 1988) both in Ontario, Canada, and streams at Kloten (Grip 1982) and Gårdsjön in Sweden (Grip *et al.* 1985). Among the SWAP sites in addition to Birkenes, the model has been applied to the Allt a'Mharcaidh site in Scotland (Stone *et al*, this symposium) and work is also progressing for the Svartberget catchment in northern Sweden (Taugböl & Bishop, this symposium) and Atna in mid-Norway (S. Samdal 1990, personal communication).

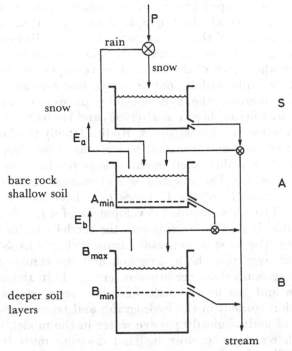

Figure 1. The Birkenes model comprises two reservoirs where highflow originates from the upper reservoir and the lower reservoir produces baseflow. Each reservoir has a threshold value (A_{min}, B_{min}), and the lower reservoir also has a maximum storage, B_{max}; (P), precipitation; (E), evapotranspiration; (Q), water flux. See Stone & Seip (this symposium) for more detailed information on the model structure.

3. Hydrological, hydrogeochemical and model developments

(a) Observations and modelling of chemically inert substances

The Birkenes model, as of 1982, maintained the classical hydrological picture of rain or meltwater entering the catchment being rapidly transported to the stream during highflow. The new feature was that, during water transport, a series of reactions occurred within the soil and groundwater areas, that changed the chemistry. However, the inability of the model to simulate the behaviour of chemically inert (conservative) substances, i.e. chloride and the stable oxygen isotope O^{18}, called for modifications of the model's hydrological structure (Christophersen *et al.* 1985; Neal *et al.* 1988; Hooper *et al.* 1988; Stone & Seip 1989). Basically, the inert substances, although very variable in rainfall, are highly damped in the runoff. This had been observed in earlier hydrological studies using O^{18} and H^2 (Sklash & Farvolden 1979; Rodhe 1981) where it was concluded that most of the runoff during an event (typically 70%) was comprised of pre-event water residing in the catchment before the rain or snowmelt.

The model that had satisfactorily reproduced the observed aluminium, hydrogen, base cations, and sulphate concentrations in streamwater could not, as it stood, reproduce the degree of damping observed for conservative substances. This posed the conceptual problem of how to reproduce both the damped conservative species and the very responsive and flow related behaviour of the chemically reactive solutes (Hooper *et al.* 1988).

To accommodate the damped behaviour of conservative species in the stream, while still reproducing a 'spiky' hydrograph, Seip *et al.* (1985) and Stone & Seip (1989, this symposium) showed that an improved fit for the Birkenes model could be obtained by ensuring an enhanced mixing within the model. Specifically, they included piston flow where part of the runoff from the upper reservoir is transferred to the lower reservoir while pushing out an equivalent amount of water from the latter. One can also increase the hydrologically passive stores of water in the reservoirs as shown by Christophersen *et al.* (1985) and Hooper *et al.* (1988). A similar modification was applied by Lindstrøm & Rodhe (1986) to the PULSE model to reproduce observed O^{18} concentrations in the streams of Swedish catchments.

Although intuitively sensible, such modifications to the hydrologic sub-models could be classified as *ad hoc*. This question was elucidated as part of investigations into parameter identifiability of the Birkenes model (cf. de Grosbois *et al.* 1988; Hooper *et al.* 1988). The work included development of a multisignal optimization procedure, that makes it possible to calibrate the model simultaneously to several measured time series. The most recent results from Birkenes (A. Stone, unpublished results) showed that by fitting the hydrograph and the conservative tracers, two parameter sets giving equivalent performance emerged. In the one set, the upper reservoir was large and the lower small, and vice versa in the other case. The combined information content in the hydrograph and tracer signals is sufficient to define the amount of hydrologically passive water in the model, but insufficient to determine in which reservoir to store it. That decision must be taken based on judgement or by calibrating on additional signals.

(b) *Observations of chemically reactive species*

The results from the field work at Birkenes are summarized by Christophersen *et al.* (this symposium). Briefly, the observations show that to a first approximation the concentrations of H^+, inorganic monomeric aluminium (Al_i), and Ca^{2+} in streamwater can be explained as a variable mixture of three types of soil-waters (endmembers); these being water from the O/H and B horizons on the slopes, and the peat deposits in the valley bottom (VB). At highflow the contributions from the O/H and B horizons dominate whereas at baseflow much of the streamflow originates from VB. These mixing ideas are explored under the heading of end-member mixing analysis (EMMA) (cf. Christophersen *et al.* 1990*a*; Hooper *et al.* 1990).

Chemically, the available field data support the importance of cation exchange processes in the organic horizons, although quantification of these processes is not straightforward. These soil-waters are strongly undersaturated with respect to any gibbsite equilibrium. For the B and VB horizons the pAl^{3+}–pH relations for the soil solutions approximately fulfil the necessary (but not sufficient) condition for gibbsite equilibrium (cf. Neal & Christophersen 1989).

(c) *Model status and development*

The picture emerging at this stage is that the original Birkenes model captured an essential feature, namely the importance of water flow paths in the catchment depending on hydrological conditions. Also, as long as the upper reservoir in the model is taken to be the O/H horizons, the model correctly included cation exchange as a controlling mechanism, and, in later versions, discarded gibbsite for the upper reservoir (Stone & Seip 1989). However, in the lower reservoir, the baseflow in the model, with the higher pH and Ca^{2+} concentration, is produced by a kinetic weathering reaction (half time of 11.4 days). Such a temporal change in composition has not been observed in the field for the B or VB layers. It would seem more appropriate, to accommodate both vertical and lateral chemical gradients, to introduce three reservoirs as a minimum, each with a certain endmember composition, and then produce runoff by mixing. However, caution must be exercised to avoid an overparametrized model with unidentifiable parameters.

To achieve this aim, work on the hydrological side is presently directed at relating the endmember contributions predicted by EMMA to hydrological and physical variables such as soil-moisture and hydraulic conductivity. This effort has been initiated at the Panola (Georgia, U.S.A.) catchment since the EMMA approach has so far been most successful for this site; six solutes in streamwater were explained well by mixing three soil-water types (Hooper *et al.* 1990). Concerning the mechanisms controlling the soil-water chemistry, a promising approach, both for the O/H and B horizons, seems to be the application of the CHAOS model of Tipping & Hurley (1988). This model includes (de)complexation reactions from soil organics with a charge dependent equilibrium constant (Andersen *et al.*, this symposium).

Other models like MAGIC (Cosby *et al.* 1985) and ILWAS (Gherini *et al.* 1985) have several processes in common with the Birkenes model, although the quantitative descriptions could be different (cf. Reuss *et al.* 1986 and Bergström & Lindström 1989). In our view, all acidification models should be tested more against observations and be developed iteratively. MAGIC, for instance, gives plausible simulations of long-term changes in streamwater chemistry, but is still based on the assumption of gibbsite mineral equilibria.

4. Predicting changes in streamwater chemistry in response to reductions in acid deposition

Predictions of future changes in streamwater chemistry must be made with recognition of the picture outlined above. One possible line of argument might be that both the Birkenes and MAGIC models capture essential features of acidified catchments and that, given the lack of second generation models, they are still among our best tools for making predictions. Alternatively, one can argue that as the process descriptions are known to be uncertain, a process oriented approach is not profitable, and other methods should be sought. Different scientists will seek different solutions to this dilemma, and an approach is chosen here which includes both the use of the process orientated mathematical models and a simple charge balance model. The focus is on the Birkenes catchment. See Henriksen *et al.* (1988) and Hornberger *et al.* (1989) for discussions of reversibility of acidification on a regional scale.

Regarding the yardstick one can use to measure the reversal of stream acidification, the criterion used here will be that restocking of brown trout can take place. Approximate limits of water quality tolerable to fish are indicated in figure 2*a*, which also includes a scatter plot of measured streamwater concentrations of Al_i and H^+ from the episodic studies. High-flow samples are found on the right whereas baseflow samples are located on the left.

The limits in figure 2*a* correspond to 4 μM Al_i ($\approx$ 100 μg l^{-1}) and 20 μeq l^{-1} of H^+ (pH 4.7). These thresholds can only be taken as indicative as brown trout survival depends on many additional factors. However, comparing these limits with the

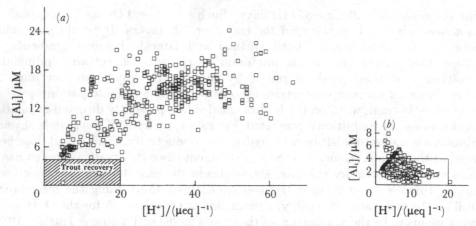

Figure 2. (a) Plot of observed [Al$_i$] against [H$^+$] for the episodic studies at Birkenes covering the period 1984–87. Rough water-quality limits for brown trout are indicated by the rectangle. (See text for details.) (b) Simulated concentrations of Al$_i$ and H$^+$ in 2125 for the sulphate deposition scenario corresponding to a 90% reduction in excess sulphate from the present value. The total fluoride level was assumed insignificant in estimating Al$_i$.

regional picture of lake water chemistry and fish status found by Henriksen *et al.* (1989), it would seem that water chemistry at the boundary of the rectangle would imply a stressed but possibly surviving population.

(a) The modelling approach

The Birkenes model has been used to predict changes in episodic variations in streamwater chemistry for different deposition scenarios (Rustad *et al.* 1986; Seip *et al.* 1987; Stone & Seip 1989). The technique used here (cf. Stone *et al.*, this symposium) is different in that long-term predictions of the soil chemistry were first performed using the two-box version of MAGIC (Jenkins & Cosby 1989). The two boxes were defined in the same manner as in the Birkenes model: the amounts of water passing annually through each box in MAGIC were estimated by using the Birkenes model. At selected points in the future (years 2025 and 2125), the Birkenes model was fed with the resulting soil chemistry parameter values provided by MAGIC and simulations of the short-term episodic chemistry were done (Stone *et al.*, this symposium). The deposition scenarios corresponded to linear declines over 40 years in the excess sulphate deposition of 30%, 60%, and 90% of the present value. The deposition was then held constant at its reduced value for another 100 years. The hydrologic conditions in the Birkenes model were assumed similar to those in 1983 to 1986.

The MAGIC simulations indicate that a reduction equal to 30% of present values is not sufficient to halt the soil acidification. At 60% reduction, significant improvements in the streamwater chemistry occur. A plot of the estimated Al$_i$ and H$^+$ concentrations produced by the Birkenes model for 90% sulphate reduction is shown in figure 2b. It is seen that the streamwater chemistry under the most critical episodes generally lies on the border of the rectangle. Note that in estimating [Al$_i$], the total fluoride concentration was assumed negligible. With fluoride present, higher Al$_i$ values are obtained. The fact that [Al$_i$] peaks at intermediate H$^+$ concentrations in figure 2b, is because of the mixing of the two different water types with subsequent degassing of CO$_2$.

(b) *The charge balance approach*

Using the diagram in figure 2a, the question can be posed as to how large a reduction in sulphate is required to bring the sum of [H$^+$] and [Al$_i$] from the present level at high-flow (*ca.* 110 µeq l^{-1}) to within the rectangle. A sum of 30 µeq l^{-1} for these species will be considered sufficient as that is the approximate value represented by the upper right-hand corner of the rectangle (Christophersen *et al.* 1990b). Given the EMMA picture, the response of the streamwater chemistry will depend on how the soil-water endmembers move in the diagram (cf. Christophersen *et al.*, this symposium). This can only be addressed if the soil processes can be quantified, but it is still possible to estimate the required reductions in sulphate based only on the present and desired future streamwater chemistry (Christophersen *et al.* 1990b). The present average sulphate concentration in streamwater, in excess of the seasalt contribution, is 114 µeq l^{-1} (SFT 1989). To reach the rectangle in figure 2a, a simple calculation, based on ANC, gives a lower estimate of the required reduction in sulphate deposition of 85% of the present level. (Here 10 µeq l^{-1} of sulphate is assumed for the 'natural excess background level'.) A constant level of total organic carbon (TOC) is assumed and the calculation takes into account the effect of organic anions. But if the Ca^{2+} concentration in streamwater is also reduced because of lower sulphate levels, or nitrate or TOC increase, larger declines in sulphate are required.

(c) *Conclusion*

Most European countries are presently committed to reduce sulphur emissions by 30%–80% of the 1980 values. It is concluded that this will be insufficient, without intervention by liming, to revitalize fish stocks in catchments that are as acidified as Birkenes. In such areas even a 90% reduction would not appear to guarantee successful fish restocking. The proposed reductions in emissions are, however, essential in limiting further soil acidification in southern Norway and could also allow reintroduction of fish in less acidified streams and lakes.

This work was funded by the British–Scandinavian Surface Waters Acidification Programme (SWAP), the National Environmental Research Council (NERC), U.K., the Norwegian Department of the Environment, and the Norwegian Council for Scientific and Industrial Research (NTNF).

References

Bergström, S. & Lindström, G. 1989 *National Swedish Environmental Board, Stockholm*, Report No. 3601, 37 pp.

Christophersen, N. & Wright, R. F. 1981 *Wat. Resour. Res.* **17**, 377–389.

Christophersen, N., Seip, H. M. & Wright, R. F. 1982 *Wat. Resour. Res.* **18**, 977–996.

Christophersen, N., Kjaernsrød, S. & Rodhe, A. 1985 Report No. 10, pp. 29–40. Oslo: Nord. Hydrol. Programme.

Christophersen, N., Neal, C., Hooper, R. P., Vogt, R. D. & Andersen, S. 1990a *J. Hydrol.* (In the press.)

Christophersen, N., Neal, C. & Mulder, J. 1990b *J. Hydrol.* (In the press.)

Cosby, B. J., Hornberger, G. M., Galloway, J. N. & Wright, R. F. 1985 *Wat. Resour. Res.* **21**, 1591–1601.

de Grosbois, E., Hooper, R. P. & Christophersen, N. 1988 *Wat. Resour. Res.* **24**, 1299–1307.

Gherini, S. A., Mok, L., Hudson, R. J. M., Davis, G. F., Chen, C. W. & Goldstein, R. A. 1985 *Wat. Air Soil Pollut.* **26**, 425–460.

Grip, H. 1982 UNGI Report No. 58, University of Uppsala, Sweden, 144 pp.

Grip, H., Jansson, P. E., Johnson, H. & Nilsson, S. I. 1985 *Ecol. Bull.* **37**, 176–192.

Henriksen, A., Lien, L., Rosseland, B. O., Traaen, T. S. & Sevaldrud, I. S. 1989 *Ambio* **18**, 314–321.

Henriksen, A., Lien, L., Traaen, T. S., Sevaldrud, I. S. & Brakke, D. F. 1988 *Ambio* **17**, 259–266.

Hooper, R. P., Stone, A., Christophersen, N., de Grosbois, E. & Seip, H. M. 1988 *Wat. Resour. Res.* **24**, 1308–1316.

Hooper, R. P., Christophersen, N. & Peters, N. R. 1990 *J. Hydrol.* (In the press.)

Hornburger, G. M., Cosby, B. J. & Wright, R. F. 1989 *Wat. Resour. Res.* **25**, 2009–2018.

Jenkins, A. & Cosby, B. J. 1989 In *Regional acidification models* (ed. J. Kamari, D. Brakhe, A. Jenkins, S. Norton & R. F. Wright), pp. 253–266. Heidelberg: Springer-Verlag.

Lam, D. C. L., Bobba, A. G., Jeffries, D. S. & Craig, D. 1988 *Can. J. Aquat. Sci.* **45**. Suppl. 1, 72–80.

Lindstrøm, G. & Rodhe, A. 1987 *Nord. Hydrol.* **17**, 325–334.

Lundquist, D. 1976 *SNSF-project, IR 23/76*, 28 pp. Oslo: Norwegian Institute for Water Research. (In Norwegian.)

Lundquist, D. 1977 *SNSF-project, IR 31/77*, 32 pp. Oslo: Norwegian Institute for Water Research. (In Norwegian.)

Neal, C., Smith, C. J., Walls, J. & Dunn, C. S. *J. geol. Soc. Lond.* **143**, 635–648.

Neal, C., Christophersen, N., Neale, R., Smith, C. J., Whitehead, P. G. & Reynolds, B. 1988 *Hydrol. Process.* **2**, 155–165.

Neal, C. & Christophersen, N. 1989 *Sci. tot. Envir.* **80**, 195–203.

Neal, C., Robson, A. & Smith, C. J. 1990 *J. Hydrol.* (In the press.)

Nordö, J. 1977 In *Acid soil – acid water* (ed. I. Th. Rosenqvist), pp. 106–110. Oslo: Ingeniørforlaget. (In Norwegian.)

Reuss, J. O. 1980 *Ecol. Model.* **11**, 15–38.

Reuss, J. O., Christophersen, N. & Seip, H. M. 1986 *Wat. Air Soil Pollut.* **30**, 909–931.

Rodhe, A. 1981 *Nord. Hydrol.* **12**, 21–30.

Rosenqvist, I. Th. 1977 *Acid soil – acid water*, 87 pp. Oslo: Ingeniørforlaget. (In Norwegian.)

Rosenqvist, I. Th. 1978 *Sci. tot. Envir.* **10**, 39–49.

Rustad, S., Christophersen, N., Seip, H. M. & Dillon, P. D. 1986 *Can. J. Fish. Aquat. Sci.* **53**, 625–633.

Seip, H. M. 1980 In *Ecological impact of acid deposition* (ed. D. Drabløs & A. Tollan), pp. 358–365. Oslo: SNSF project, Norwegian Institute for Water Research.

Seip, H. M., Seip, R., Dillon, P. J. & de Grosbois, E. 1985 *Can. J. Aquat. Sci.* **42**, 927–937.

Seip, H. M., Christophersen, N. & Rustad, S. 1987 In *Reversibility of acidification* (ed. H. Barth), pp. 149–155. New York: Elsevier.

Sklash, M. G. & Farvolden, R. N. 1979 *J. Hydrol.* **43**, 45–65.

Stone, A. & Seip, H. M. 1989 *Ambio* **18**, 192–199.

Tipping, E. & Hurley, M. A. 1988 *J. Soil Sci.* **39**, 505–519.

Vogt, R. D., Seip, H. M., Christophersen, N. & Andersen, S. 1990 *Sci. tot. Envir.* **96**, 139–158.

Discussion

D. W. JOHNSON (*Desert Research Institute, Reno, Nevada, U.S.A.*). Four modeling studies were presented that included the use of two existing dynamic models (MAGIC and Birkenes), existing population dynamics and statistical models (for the effects upon fish), and the development of a third new model for investigating short-term fluctuations in stream chemistry. I would like to restrict most of my comments to the chemistry models, an area with which I am most familiar.

Conspicuous by its absence is the ILWAS model; mention was made of it once (in the paper by Christophersen *et al.* (this symposium), but it does not seem to have been employed by any of the investigators either for forecasting or for scientific reasons. I must ask why this is so, given its scientific stature and given that it is by far the most rigorously and realistically constructed of all the dynamic models. One could guess as to why it was not used (too complicated, too many knobs to turn, too expensive), but I think that I would like to leave the point open for discussion.

It is interesting to note that each of the dynamic models of soil and water acidification has a similar set of equations for cation exchange, anion adsorption, and carbonic acid leaching, reflecting a rather advanced state of knowledge (or at least confidence) as to the nature of these processes. Because of these similarities, differences in short-term (e.g., episodic) predictions between the models should not be significant. Mysteries remain, however; the episodic behavior of stream chemistry in the Allt a'Mharcaidh catchment is thought to be best described by a simple, semi-empirical mixing model. Indeed, the results look encouraging, but caution must be exercised in extending simplicity beyond its limits, especially with respect to the carbonate system. Mixing two solutions mathematically can produce a third solution that cannot exist in nature. For instance, mix 3 l of water pH 6, bicarbonate of 200 μequiv. l^{-1} with 5 l of water with pH 4, bicarbonate of pH 0, and you obtain a theoretical water of pH 4.2, bicarbonate of 75 μequiv. l^{-1}. No such water exists, of course, because the hydrogen ion in water number 2 the titrates the bicarbonate in water number 1 (causing a release of CO_2). The authors do not report pH results in their paper, but given the large differences in alkalinity and pH of the waters that they are mixing, I strongly suspect that a comparison of pH and alkalinity values for the mixed waters would reveal some serious problems. The Birkenes model also uses mixing; does it allow for titration or other precipitation reactions?

Whereas most of the dynamic models have basically the same formulation for exchange, adsorption, and precipitation reactions, no two models have the same formulation for soil mineral-weathering reactions. This is not because of a lack of theoretical base in soil mineral-weathering reactions, but because of a lack of accurate quantitative information. Mineral dissolution and formation reactions can be represented with thermodynamic equations, as is the case in the ILWAS model, but large uncertainties remain because of the large disparities between laboratory-derived mineral dissolution rates and field-estimated weathering rates (the latter usually being much slower). MAGIC attempts to get at a field weathering rate empirically, employing educated guesswork and a critical assumption: a steady-state soil condition at some point in the pre-industrial era (nominally, 1846). The latter assumption is questionable for North America, even in pristine forests, and is probably completely invalid for the U.K. and Scandinavia where human activities such as forest harvesting have surely kept the soil well away from a steady-state condition (if indeed it was ever in one) for a long period before the industrial era.

Given the inherent uncertainties in weathering rates alone (not to mention uncertainties as to the hysteresis and kinetics of sulphate adsorption, vegetation uptake, and a host of other important processes), long-term fore- or hindcasts of the dynamic models, although they are of considerable scientific interest, must be regarded with considerable scepticism when it comes to use in forming policy. There is a strong tendency among all of us in the scientific field to avoid rigorous examination of our pet hypotheses, so we typically assign others to play Devil's advocate as I am doing here. In the case of these models, however, whose predictions may well be used to help formulate policy, it is imperative that their makers and users give them a very critical, even hostile examination in search of their probable real value as predictive tools. Along the same vein, these models need a rigorous test, involving predictions of future events with no post-mortem 'calibrations'. I noted with satisfaction that Bravington *et al.* boldly proposed such a test of their models by a re-examination of the SNSF lakes. One hopes that this will be done, and that similar proposals for testing the dynamic models will surface. Opportunities for such tests have been largely missed with the RAIN Project, and another opportunity is about to be missed with the U.S. EPA's Watershed Manipulation Project. Other opportunities also exist in form of long-term stream chemistry records that have not yet been calibrated.

With a few notable exceptions in the papers by Bravington *et al.* and Christophersen *et al.* at this symposium there is little evidence of self-examination and self-criticism in the acid deposition modeling literature, but rather a hard sell for using the models as predictive tools for exactly the purpose of guiding policy. If we are allowed the same standards of accuracy set by economic forecasters, this will present no problems; but expectations seem to be much higher than that.

Modelling short-term flow and chemical response in the Allt a'Mharcaidh catchment

H. S. Wheater[1], F. M. Kleissen[1], M. B. Beck[1], S. Tuck[2], A. Jenkins[2]
AND R. Harriman[3]

[1]*Imperial College of Science, Technology and Medicine, London SW7 2AZ, U.K.*
[2]*Institute of Hydrology, Wallingford, Oxfordshire OX10 8BB, U.K.*
[3]*Freshwater Fisheries Laboratory, Pitlochry, Perthshire PH16 5LB, U.K.*

Like many upland streams, the 10 km² Allt a'Mharcaidh experimental catchment in Scotland has a rapid response to precipitation and snowmelt events; the stream hydrograph characteristically rises and recesses in a few hours. Accordingly models must have a comparable time resolution to capture the transient pulses of, e.g. acidity and aluminium, which may have a significant impact on aquatic ecosystems.

Detailed analysis of plot and catchment data at the Allt a'Mharcaidh catchment, described by Wheater *et al.* (this symposium), has led to a hypothesis of catchment response which forms the basis of a new hydrological catchment model operating in time steps of 1 h. The structure of the model is described and results of fitting stream flow observations for a calibration period in 1988 and split-sample tests from 1986 and 1987 are presented. The characteristic time-dependent features of the stream flow are generally well reproduced, and the model is broadly consistent with plot-scale observations.

Comprehensive data on changes in stream chemistry after major rainfall events are available from six storms. These changes cannot be simply explained in terms of plot-scale observations, which are inconsistent with low flow chemistry. However, fluctuations in stream flow relationships are correlated with variations in stream discharge as defined by bi-weekly spot samples. The implications of this relationship are explored in terms of the mixing of constant concentration waters and a consistent base-rich groundwater component is indicated. The hydrological model is extended to simulate alkalinity and is seen to capture the principal features of the stream response to rainfall events.

---

1. Introduction

In most catchments subject to acidification, occurrence of high stream flow acidity is associated with transient episodes due to snowmelt or high rainfall. Pulses of acidity and aluminium have been shown to have serious adverse effects on aquatic ecosystems (Henriksen *et al.* 1984; Harriman *et al.*, this symposium); episodic stream flow chemistry is therefore of major importance. It has been widely recognized that stream flow chemistry is dependent on hydrological response, which for upland streams is highly dynamic. The hydrograph record may rise to a peak in just a few hours. It is therefore evident that the observations must be represented on a relatively fine time scale to characterize the stream response. If this is not done, the information content of the data for process identification will be substantially reduced.

Despite these considerations, almost all hydrochemical modelling has been based on daily or longer time steps. In modelling long-term changes, an annual time step seems reasonable, although even on this time scale it is important to represent correctly contributions from different flow paths (Jenkins & Cosby 1989). For modelling of short-timescale events, the daily time step has primarily been determined by the frequency of available chemical data. Even with automatic samplers, there are serious logistical difficulties in capturing the data necessary for full chemical analysis, and the associated analytical effort is substantial.

The aim of the experimental programme for the U.K. SWAP sites has been to characterize mean catchment chemistry through routine sampling of stream and soil waters at many locations, and to capture a representative range of stream flow events for full chemical analysis. This paper is concerned with the analysis and model simulation of episodic changes in stream flow on an hourly timescale for the Allt a' Mharcaidh catchment. The hydrological model is based on interpretation of plot and catchment observations. Based on the analysis of stream water chemistry during six representative chemical events in that catchment, the hydrological model has been extended to simulate changes of alkalinity in the stream.

2. A hydrological model for the Allt a'Mharcaidh

(a) Catchment hydrological observations

In a companion paper in this symposium (Wheater *et al.*), a detailed interpretation of the hydrological response of the Allt a'Mharcaidh catchment is presented, based on observations from experimental plots on representative soils and hill slope sequences. In summary, the catchment has an area of 10 km² and an elevation range from 1111 m to 330 m at the outfall gauge. At the higher elevation plateaux, alpine podzols predominate, giving way to peaty podzol soils on the steeper flanks of the catchment. In the lower regions, at reduced slope, soils are mainly peat, with extensive erosion features.

The alpine soils have little through flow or overland flow and soil water conditions are generally unsaturated. Free drainage to lower horizons is inferred. The mid-slope peaty podzol soils have a complex response, due to local heterogeneity and spatially discontinuous indurated layers. Through flow can occur rapidly in response to precipitation events and the upper organic soil horizon shows a more prolonged response than the underlying (Bh) layer. There is evidence of preferred flow paths in mid-profile, associated with rapid downslope flow, but water–table conditions at the base of the soil profile produce a damped response indicative of slow downslope drainage. The peat soils are relatively poorly drained and remain wet for much of the year, although highly transient flows are observed at the base of the organic soil.

Inspection of the stream flow hydrograph shows that two principal modes of response occur following precipitation events. The hydrograph shows sharp, short duration 'spikes'; underlying these is a much more damped response 'hump', which may extend over several days. These two could be considered to be superimposed on a base flow component.

The various responses are illustrated in Wheater *et al.* (this symposium). The rapid response observed at the base of the peaty podzol slope lags rainfall by, typically,

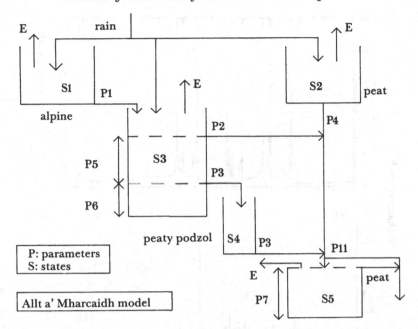

Figure 1. Structure of the hydrological model.

2 h, but occurs simultaneously with the rapid response observed at the base of the peat and with the hydrograph spike. It is hypothesized that flows through preferred paths are rapidly transmitted to the stream once the peat soils are wet, and are responsible for the hydrograph spike. The water table feeding the podzol seep shows a very similar response to the hydrograph hump, which is attributed to downslope soil water drainage. Base flow may be generated by underlying groundwater or soil water drainage.

(b) Model formulation

Some problems associated with model formulation are discussed by Beck *et al.* (1990). There is an important dilemma for the modeller concerning the level of complexity to be adopted. Most hydrological models, even of modest complexity, are strictly non-identifiable (Wheater *et al.* 1986) in the sense that a unique set of best fit parameter values cannot be defined. The search for a fully identifiable model will lead to a model structure that is highly simplistic; it is unlikely to give a particularly good fit to observed data or be readily associated with a physical understanding of the controlling processes (as demonstrated for Loch Ard in Beck *et al.* (1990).

The approach adopted here has been to formulate a model that is a conceptualization of the response modes hypothesized above, and hence retains physical significance, but requires as few calibration parameters as possible. Since ambiguity in parameter identification is inevitable, the model cannot be regarded as a formal test of those hypotheses. However, a failure of the model can be regarded as a failure of the hypotheses; success denotes some conditional support for the physical interpretation.

Initial stages of model development are described by Webster (1989). The structure finally adopted is shown in figure 1. The model is spatially distributed in the sense that precipitation input is assigned separately to the alpine, peaty podzol

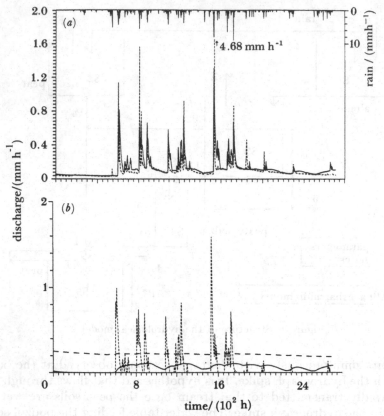

Figure 2. Hydrological simulation, 1 June to 30 September 1988. (a) Stream flow results: ——,
simulated discharge; – – – –, observed discharge. (b) Component flows: ——, from lower peaty
podzol; – – –, from upper peat; ·····, from upper peaty podzol.

and peat areas. The peaty podzol soils are represented by a single conceptual store
with two linear outlets. Parameter P2 defines a transient response (the 'spike'), P3
represents the 'hump' and to give appropriate damping is routed through a two
reservoir system (see also Beck *et al.*, this symposium). In the catchment, some peaty
podzol slopes drain directly to the stream, others as described above are upslope of
extensive peat areas. Hence peaty podzol discharges are partitioned by parameter
P11.

Initially, the peat soils were represented by a single store, which when full allowed
transmission of upslope inputs to the stream. This proved an over-simplification,
since rainfall on the peat areas mainly serves to maintain high moisture storage
within the peat soils, hence the peat was partitioned as shown, with S5 representing
transmission pathways (e.g. gully base) and S2 the bulk of the unsaturated peat soil.

As will be evident from the previous discussion, the routeing of alpine waters is
ambiguous. It is assumed here that the alpine soils drain downslope to the peaty
podzol areas. It is possible that a proportion will move as groundwater, bypassing
the observed soil system, but such response cannot be identified from the hydrological
data alone.

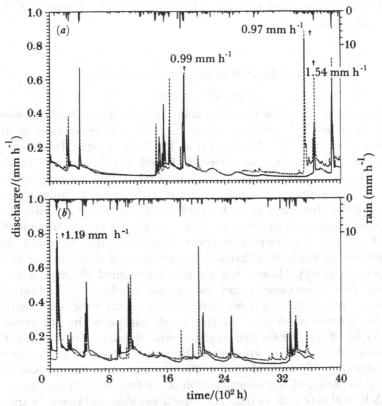

Figure 3. Hydrological simulation, verification periods. (*a*) 1 June–15 November 1986.
(*b*) 3 June–31 October 1987. Simulated discharge, ———; observed discharge, ----.

(*c*) *Model performance*

The model was calibrated by fitting the observations for the period 1 June to 30
September 1988, using catchment average precipitation and evaporation calculated
from Penman potential evaporation estimates, subject to linear reduction as a
function of the soil water storage states.

The result for stream flow is shown in figure 2*a*, with the component flows given
in figure 2*b*. The R^2 goodness of fit criterion has a value of 0.46. Overall, this
performance is considered to be satisfactory as the major features of the hydrograph
are fairly well reproduced. The relatively low R^2 value is mainly due to a failure to
match the observed peak discharges precisely. This is not surprising, given the
relatively large catchment area, the highly dynamic stream flow response and
uncertainties of flow data in the high flow régime. It will be shown below that for
chemical response, accurate representation of low and intermediate flows is more
important.

The component flows are consistent with the hypothesized behaviour, although it
can be noted that base flow is largely due to the peat storage and, as discussed above,
is an ambiguous component. Split-sample tests for 1986 and 1987 summer data
(using the same parameter set) are shown in figure 3. As would be expected, there is
some deterioration in performance (the triggering of hydrograph spikes is dependent
on correct reproduction of peaty podzol soil water storage and some over-estimation

of evaporation is evident) but in general the characteristic features of the data are reasonably well reproduced.

3. Modelling of event chemistry

(a) Methodology

The problems of model identifiability also underlie hydrochemical modelling. In practice, two approaches have been taken. In the first generation of hydrochemical models (Christophersen et al. 1982; Chen et al. 1982; Cosby et al. 1985) a hypothesized set of chemical interactions was adopted, based largely on the Reuss–Johnson formulation (Reuss & Johnson 1985), and associated with an (explicit or implicit) hypothesized hydrological response. These models are not uniquely identifiable (Kleissen et al. 1989), but have provided useful hypotheses for acidification research, and a method for long-term prediction. However, there have been model failures, for example with respect to conservative tracer results (Neal et al. 1988), and this has generated interest in modelling through the analysis of data, rather than through fitting a synthesis of hypotheses. For models determined through data analysis, a simple formulation is necessary, and the approach adopted has been to consider conservative mixing of two or more components of the flow, each with a defined chemistry. An interesting feature of this work has been that in some cases (for example Hooper et al. 1990) the components required to explain the stream chemistry have all been from known catchment sources. In other cases, however (Christophersen et al. 1990; Neal et al. 1990), the observed stream flow data cannot be interpreted in terms of the superposition of identified components.

For the Allt a'Mharcaidh catchment, the available database is insufficient to pursue complex chemical modelling. Furthermore, there is a major difficulty in trying to represent the basic processes. As noted by Ferrier et al. (1990), the range of observed soil water chemistries for the alpine, peaty podzol and peaty soils is relatively small and does not encompass stream water chemistry. The approach adopted has therefore been in the first instance to use simple conservative mixing concepts to analyse the stream response to precipitation events. On the basis of these results, limited synthesis has been possible, with the aid of a simple model of alkalinity superimposed on the hydrological representation.

(b) Mixing model analysis

Data from the bi-weekly spot samples and the much more frequent chemical sampling during high flow events are reviewed by Harriman et al. (1990). A striking result is a well-defined relationship between alkalinity and stream discharge rate. For flows in excess of 0.6–$0.9 \, \mathrm{m^3 \, s^{-1}}$ alkalinity values are practically independent of discharge rate (ranging between -5 and $10 \, \mu\mathrm{eq} \, \mathrm{l^{-1}}$). For smaller flows a consistent inverse relationship is observed; under low flow conditions alkalinities of 90–$120 \, \mu\mathrm{eq} \, \mathrm{l^{-1}}$ have been observed. As noted above, this immediately indicates a major difficulty in producing a process-based synthesis of catchment response since soil water alkalinities are generally negative (Ferrier et al. 1990). It is apparent that a major source of alkalinity, which dominates low flow conditions, has not been identified. Some groundwater spot samples, collected from shallow boreholes adjacent to the main stream, indicated very high alkalinity, but results were inconsistent and varied over a wide range (200–$800 \, \mu\mathrm{eq} \, \mathrm{l^{-1}}$).

Table 1. *Hydrochemical event data* (*1988*)

date	8 Mar.	18 Mar.	1 Jul.	22 Jul.	12 Aug.	12 Sept.
peak discharge/ $(m^3 \ s^{-1})$	0.45	1.8	0.18	1.2	10.5	0.42

To explore the dynamic response of the more conservative chemical species, a simple two-component mixing model was applied to the available data (six events in 1988, table 1), for alkalinity as defined by Harriman *et al.* (1990), total organic carbon (TOC), Ca, Mg, Na, Cl and SO_4 (not all species were available for all events). It should be noted that the approach assumes conservative mixing of invariant components. It is considered that these assumptions are reasonable first approximations; the important point (Neal *et al.* 1990) is that they allow the interaction of hydrological and chemical processes to be studied.

To represent the mixing of variable flows q_1 and q_2 from two sources of constant concentration, the concentrations must be specified (c_1 and c_2 respectively), then

$$q_1/q_2 = (c_s - c_2)/(c_1 - c_s),$$

where c_s is the stream flow concentration. (The response pattern of q_1 and q_2 is, in practice, not sensitive to the prior assumption of c_1, c_2 unless $c_s \approx c_1$ or c_2.) For this analysis the component concentrations were generally selected from the extremes of observed stream flow.

The alkalinity analysis was carried out for two sets of components of 0 and 90 µeq l^{-1}, and -5 and 120 µeq l^{-1}, with similar results (120 µeq l^{-1} represented the highest observed stream alkalinity over the study period; soil water chemistries were generally associated with negative alkalinity). The snowmelt events of 8 March and 18 March 1988 and the small summer storm of 1 July 1988 (after a dry June) showed almost identical response, illustrated in figure 4*a*. A constant, high alkalinity, base flow (*ca.* 0.1 $m^3 \ s^{-1}$) was identified, with the hydrograph response entirely represented by low alkalinity water. The remaining three storms were similar, but with an important variation, as illustrated in figure 4*b* for 22 July 1988. As before, the hydrograph response is predominantly generated by low alkalinity water, but a short duration high alkalinity response is also observed, superimposed on the constant alkaline-rich base flow.

This simple analysis has therefore been quite revealing. It suggests that a consistent source of high alkalinity water is generating base flow, that the principal source of event water is of low alkalinity (equivalent to that observed in soil waters), and that a third component contributes alkaline water during significant rainfall events. This could be postulated for example as a spatially discrete source, or a flushing of alkaline groundwater.

A similar analysis has been undertaken for the other species. Results for Ca are essentially the same as for alkalinity, as are those for SO_4, but those for Mg are more complex. The events show variable responses for both high and low Mg sources, the events of 18 March and 12 August giving overall dilution, the remainder showing overall concentration.

Results for Na and Cl are generally similar. Both high and low concentration components have a rapid response; for the larger rainfall events the low concentration component predominates, giving overall dilution, whereas the converse occurs for

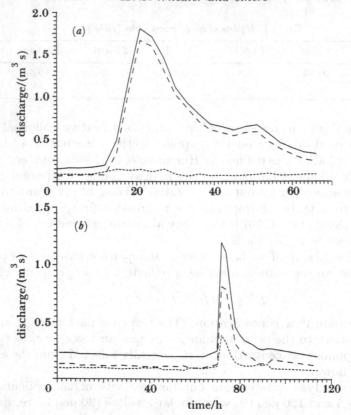

Figure 4. Mixing model analysis of alkalinity. (a) 18 March 1988. (b) 22 July 1988. Observed flow, ——; flow, alkalinity 90 µeq l⁻¹, ⋯⋯; flow, alkalinity 0 µeq l⁻¹, – – –.

the smaller events. Marked differences occur in different snowmelt events. The 8 March event is dominated by a flush of high Cl, which is less marked for Na. A small flush of high Cl occurs in the 18 March event, but there is overall dilution of Na.

TOC data are available for three of the events. As would be expected from organic soil contributions, the rainfall events are dominated by high TOC response, but interestingly for the large 12 August storm there is a response of zero TOC water which matches that of the high alkalinity water noted above. The snow event of 18 March shows a preponderance of zero TOC water.

This simple analysis has given interesting insights into catchment response. A consistent, three-component response is apparent for Alk, Ca and SO$_4$. A more variable response is evident for Na and Cl, with dependence on event magnitude, and for Mg. TOC in general reflects expected soil water inputs, but also a zero TOC response was noted. The analysis is of course based on a limited data set, but clearly indicates a diversity of responses which could not readily be incorporated into a multi-component end-member analysis. This can also be seen from component interrelationships, for example in figure 5. In figure 5a sulphate is plotted against alkalinity. Observed soil water chemistries have low alkalinity and a range of mean SO$_4$ concentrations from 43–51 µeq l⁻¹ (Ferrier *et al.* 1990). Assuming a base flow component of high alkalinity, at least two further components would be required to explain the data. The calcium–alkalinity data is shown in figure 5b. Mean soil water concentrations range from 10–29 µeq l⁻¹ (Ferrier *et al.* 1990). Again, assuming a high

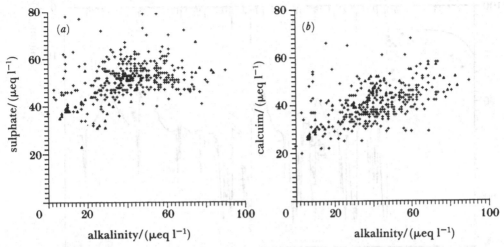

Figure 5. (*a*) Sulphate–alkalinity data. (*b*) Calcium–alkalinity data. Spot samples, + + +; event data, △△△.

alkalinity component, a source of low alkalinity, high Ca water is required to explain the scatter in the data. However, the range of observed chemistries is small in comparison with other published data.

(*c*) *Synthesis of stream alkalinity*

The mixing model is essentially a tool for analysis. The hydrological model presented above provides a tool for the simulation of stream flow discharge. The chemical data in total showed complex behaviour, but for alkalinity a relatively clear set of responses has been identified. As alkalinity is a useful index of acidification status, generally considered to be conservative, the development of a hydrochemical simulation model was therefore based on the hydrological representation allied to alkalinity sources inferred from the mixing model analysis.

The parametrization of the hydrological model had given three primary modes of response, i.e. base flow (nominally generated from the peat, but in reality ambiguous), the hydrograph spikes, generated from the upper peaty podzol (and associated with preferred flow paths) and the intermediate hump, generated in the lower peaty podzol reservoir (and associated with downslope matrix drainage). Two of these components therefore closely mirrored the mixing model analysis. The base flow could reasonably be associated with the high alkalinity source and the spiked response with the low alkalinity source. Alkalinity of the humped response was ambiguous but was assigned to lie in the range 0–50 µeq l^{-1}. The alkalinity model therefore consisted of the hydrological model with prescribed alkalinity sources: 90 µeq l^{-1} associated with P4, 0 µeq l^{-1} associated with P2 and 0–50 µeq l^{-1} with P3.

Parameter identification of the model has been carried out by manual optimization (automatic optimization is currently in hand). The alkalinity results are particularly sensitive to low and intermediate response, hence some minor modification of the original hydrological parametrization was required. The best fit for hump alkalinity was 25 µeq l^{-1}. The resulting simulations are shown in figure 6 (other results are shown in the companion paper by Beck *et al.* (this symposium). By hydrochemical modelling standards, the results are most encouraging. Discrepancies clearly show the importance of correct hydrological representation, however, for example in the

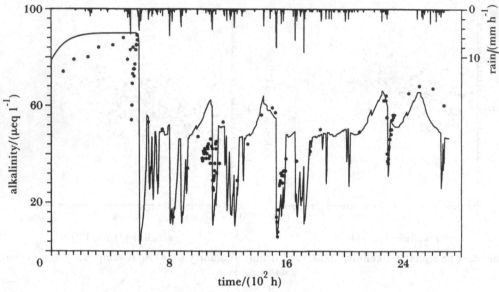

Figure 6. Alkalinity model results, 1988.

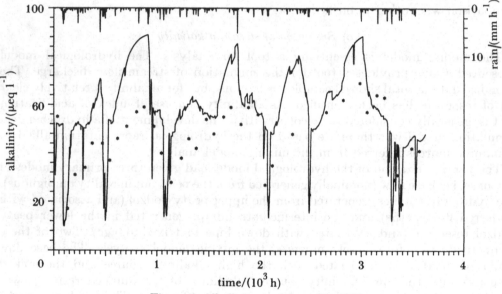

Figure 7. Alkalinity model results, 1987.

failure to simulate the event at 550 h and in the final overestimate of stream flow (following time 2600 h). It should be noted that this plot excludes a handful of anomalous and extremely high alkalinity samples. For example, the first sample of the event of 12 August 1988 had an alkalinity of 109 µeq l⁻¹. The results lie outside the scope of conceptualized process.

Maintaining the same parameters and alkalinity source concentrations, split sample testing was carried out for 1986 and 1987 data. The results show some divergence with respect to low flow performance (figure 7). The overall pattern is well reproduced, but simulated low and intermediate flows are too low at times.

4. Conclusions

To interpret the processes responsible for the observed rapid changes in stream flow and stream chemistry, fine temporal resolution is required. Hence a hydrological model has been developed which embodies the processes inferred from field experimental data at different scales and reproduces the principal features of the observed stream flow hydrograph.

Two component analysis, assuming conservative mixing, identified diverse responses to precipitation events for different chemical species. Consistent behaviour was observed for alkalinity, calcium and sulphate in which the transient hydrograph response was primarily associated with low concentration water, superimposed on a uniform, high concentration base flow. For large rainfall events, a short pulse of high concentration water was also noted.

Based on the hydrological model, a simple hydrochemical simulation model was developed in which low alkalinity water was associated with the transient response of the peaty podzol soils, intermediate alkalinity with downslope drainage water and high alkalinity with base flow. This successfully reproduced the major features of the 1988 data and, without further calibration, gave reasonable representation of the 1986 and 1987 observations.

Of the three components of alkalinity response, only that associated with rapid changes in the hydrograph could be identified with observed soil water chemistries. Hence a major ambiguity exists with respect to the representation of processes involved in catchment hydrochemistry. This is particularly significant for the Mharcaidh catchment since it is high alkalinity water from unknown sources which is maintaining stream pH levels. However, this result is mirrored in work elsewhere. The results of Neal *et al.* (1990), for Plynlimon are similar in several respects. It can be concluded that examination of event response has given some insight into hydrological and stream chemistry behaviour, and allowed a quantification of the dynamic response under current conditions. It has also, however, revealed important gaps in knowledge and thus pointed to the need for further research if we are to define the future response of the high alkalinity waters to acidification impact.

The financial support of the Surface Waters Acidification Programme is gratefully acknowledged.

References

Beck, M. B., Kleissen, F. M. & Wheater, H. S. 1990 *Rev. Geophys.* (In the press).

Chen, C. W., Dean, J. D., Cherini, S. A. & Goldstein, R. A. 1982 *Proc. ASCE* EE3, 455–472.

Christophersen, N., Seip, H. M. & Wright, R. F. 1982 *Water Resour. Res.* **18** (4), 977–996.

Christophersen, N., Neal, C., Hooper, R. P., Vogt, R. D. & Andersen, S. 1990 *J. Hydrol.* (In the press.)

Cosby, B. J., Wright, R. F., Hornberger, G. M. & Galloway, J. N. 1985 *Water Resour. Res.* **21**, 51–63.

Ferrier, R. C., Walker, T. A. B., Harriman, R., Miller, J. D. & Anderson, H. A. 1990 *J. Hydrol.* (In the press.)

Harriman, R., Gillespie, E., King, D., Watt, A. W., Christie, A. E. G., Cowan, A. A. & Edwards, T. 1990 *J. Hydrol.* (In the press.)

Henriksen, A., Skogheim, O. K. & Rosseland, B. O. 1984 *Vatten*, **40**, 255–260.

Hooper, R. P., Christophersen, N. & Peters, N. E. 1990 *J. Hydrol.* (In the press.)

Jenkins, A. & Cosby, B. 1989 In *Regional acidification models* (ch. 19) (ed. D. Brakke, A. Jenkins, J. Kaemari, S. Norton & R. F. Wright), Springer Verlag.

Kleissen, F. M., Beck, M. B. & Wheater, H. S. 1990 *Water Resour. Res.* (Submitted.)

Neal, C., Christophersen, N., Neal, R., Smith, C. J., Whitehead, P. G. & Reynolds, B. 1988 *J. Hydrol. Processes* **2**, 155–165.

Neal, C., Smith, C. J., Walls, J., Billingham, P., Hill, S. & Heal, M. 1990 *J. Hydrol.* (In the press.)

Reuss, J. O. & Johnson, D. W. 1985 *J. Environ. Qual.* **14**, 26–31.

Webster, R. 1989 M.Sc. dissertation, University of London, U.K.

Wheater, H. S., Bishop, K. & Beck, M. B. 1986 *J. Hydrol. Processes* **1**, 89–109.

Discussion

B. SALBU (Isotope Laboratories, Agricultural University of Norway, 1432-As-NLH, Norway). When mixing water models are considered, the kinetics of processes taking place in the mixing zones are essential. In non-equilibrium systems fluctuations influence for instance the distribution of Fe species should be expected; for example rapid polymerization and slow dissolution kinetics. As this aspect of the models has not been discussed, comments on the importance of kinetics should be given.

H. S. WHEATER. Severe problems exist with respect to the identification of chemical processes at catchment scale. Complex hypotheses cannot be readily tested at this scale, hence simple mixing models have been used to analyse stream water quality dynamics. It is, however, an essential prerequisite that mixing model analysis is applied only to chemical species which are, or may reasonably be approximated as, conservative. The analysis cannot be readily extended to non-equilibrium systems.

Modelling and quantitative analysis of the impact of water quality on the dynamics of fish populations

By M. V. Bravington[1], A. A. Rosenberg[1], R. Andersen[3], I. P. Muniz[2] and J. R. Beddington[1]

[1] Renewable Resources Assessment Group, Centre for Environmental Technology, Imperial College of Science, Technology and Medicine, London SW7 1NA, U.K.
[2] Institute of Zoophysiology, University of Oslo, Oslo 2, Norway
[3] Lafjord Aquaproducts A/S Fjellse, 4400 Flekkefjord, Norway

In general, the effects of altered stream and lake acidity on fish populations are reflected in changes in mortality, recruitment and growth. Data from two Norwegian sources are used. The first is a long-term study of a single lake (Selura), before and after the start of a continuous liming programme. The second is a cross-sectional study of 85 representative lakes across the country. Both studies contain information on age compositions of brown trout populations undergoing responses to differing physical and chemical conditions. There are two main responses: senescence, when partial or complete recruitment failures lead to a population with a preponderance of large, old fish, and juvenilization, when increased post-spawning mortality leads to a population having a large proportion of immature fish with large year and class variability. These responses have been related to physical and chemical characteristics of the environment.

Apparent changes in mortality over time are used to determine the degree of juvenilization or senescence experienced by each population. The populations are then projected forward in time to predict prospects for survival.

The effects of long-term liming on the population structure in Selura are clearly shown by changes in growth and recruitment. Prospects for other lakes under liming and restocking are considered.

1. Introduction

One of the clearest and most noticeable effects of surface-water acidification in both Europe and North America is the reduced abundance or local extinction of fish populations. Whereas the analysis of presence or absence data may indicate the basic pattern of extinctions (Henriksen et al. 1989; Sevaldrud et al. 1980), it may take 20 years or more for a fish population to become extinct after the first signs of acidification are noted. An analysis of the population dynamics that incorporates the current health of the stock may potentially be a more informative diagnostic tool than a presence or absence survey.

In this paper we present analyses of the population dynamics of brown trout (Salmo trutta) in Norwegian lakes, focusing on the information contained in population age structure. The effects of acidification can be localized to specific periods in the life history of the fish that are then reflected in the age structure of the populations in lakes of different chemical and physical characteristics. We then consider the likely trajectory of a population in future by estimating the pre-acidification and current population parameters. These parameters implicitly describe a population projection model (Leslie 1945; Sadler 1983) and enable us to estimate the probability that the stock will be lost over various timescales. The viability of restocking lakes with fish is then evaluated based on these probability calculations.

For this research two principal sets of data containing appropriate age structure information were available. The regional survey data from the Norwegian programme 'Acid precipitation – effects on forests and fish', the SNSF project, sampled fish in 85 lakes, largely in southern Norway (Rosseland et al. 1979). These data are used to describe the basic pattern of acidification effects and to estimate the probability of population extinction over time for each type of lake.

The second data set is from Lake Selura in southwestern Norway. Selura has been sampled for abundance and age structure in most years since 1976. Addition of lime to the lake began in 1983 so the recovery of the population from a pre-liming, reduced level can be documented as an indication of the reversibility of acidification effects.

2. The SNSF regional survey

The SNSF survey strategy and database are described in detail in Rosseland et al. (1979). The survey was conducted from 1976 to 1979 and the basic pattern of the results is described in Sevaldrud et al. (1980). The survey sampled 85 lakes, measuring physical and chemical as well as biological characteristics. Two of these lakes were resampled a second time and four others were resampled on three other occasions giving a total sample size of 105. Only 52 of the samples (49.5%) were from populations large enough to yield a sufficient sample size for the estimation of relative abundance at different ages.

(a) Grouping lakes by physical and chemical characteristics

The survey measured a wide range of characteristics (Rosseland et al. 1979). To reduce the dimensionality of the problem, factor analysis (Harman 1970) was used to extract four variables which describe most of the variance (80%) in those lake characteristics thought to be relevant to fish status (Rosseland et al. 1979). Factor 1 is closely related to the cations calcium, sodium and magnesium and also to the total sulphate content, including the marine component, and is some measure of buffering capacity. Factor 2 is strongly positively correlated with pH and negatively correlated with aluminium ion concentration. Factor 3 is negatively related to the level of organic material and positively related to the clarity of the water. Organic compounds are thought to mitigate the toxic effects of aluminium, so a high value for factor 3 indicates higher toxicity of aluminium (Baker & Schofield 1980). Finally, factor 4 is a measure of the area of the lake and its catchment area. Larger lakes and catchments are more likely to provide stable conditions and refuge areas.

Using these factors, a quick cluster analysis was applied to group the lakes into nine clusters (SPSS 1986). A brief description of the clusters follow. Cluster 1 is inland in northern Aust and Vest Agder and is particularly low on factor 1 (buffering). Cluster 2 is made up of southeasterly lakes in Akershus and Telemark, which are high in organics (factor 3). Clusters 3 and 5 are coastal lakes in the southwest (Vest Agder and Rogaland) which have high buffering (factor 1). Cluster 3 is high on factor 2 (pH), however, unlike cluster 5. Clusters 4 and 8 are made up of larger, more northerly or westerly lakes. Cluster 6 is low on both factor 1 and 2 and is located in Aust and Vest Agder near the coast. Clusters 7 and 9 are northerly in Hedmark and More and Romsdal counties, have low buffering, but moderate to high pH. The lakes in cluster 8 are in northern Aust Agder with low pH and buffering.

(b) *Patterns in the age structures*

There are two main processes reducing the abundance in fish in acidified lakes: increased mortality and decreased reproduction. The evidence for the occurrence of these processes was reviewed by Harvey (1981) and Rosseland (1986), and they were modelled by using laboratory data by Van Winkle *et al.* (1986).

Healthy populations are assumed to have constant mortality at age and constant average recruitment. The mortality rate of such a population is given by the slope of the relation between log abundance and age. If mortality increases, particularly for the post spawners, the slope of this relation is steeper and the older age groups are lost. This is termed juvenilization of the population.

If mortality remains constant but recruitment declines over time, the above slope is flattened and the relative abundance of older fish is greater than for a healthy population. This is termed senescence of the stock.

Clearly, both increased mortality of older fish and reduced recruitment of young fish to the stock may be occurring in the same population. Any analysis that depends on changes in the above slope, such as that presented here, will tend to underestimate both effects if they occur together.

(i) *Estimating mortality and recruitment variability*

These parameters were estimated for each population having adequate sample sizes. Only those age classes fished unselectively were analysed, in accordance with the gear selectivity data in Rosseland *et al.* (1979). An outline of the procedure is given below.

For a non-schooling fish such as brown trout, the numbers caught in each age class should follow a Poisson distribution, with encounter rate proportional to abundance. If recruitment is constant, then the expected numbers decline exponentially with age, the rate of decline determining mortality. We can compute a 'sample variance' based on the goodness-of-fit of the data to this exponential curve. If the 'sample variance' exceeds its expectation, the excess can be used to estimate recruitment variability. The actual estimation is done by using maximum likelihood methods, assuming a Gamma distribution for recruitment. It is assumed that, compared with recruitment, mortality is stable over adult age classes and over time.

Estimates of pre-acidification mortality rates were obtained from data on about 800 lakes sampled between 1910 and 1940 (Dahl & Lund 1944). From the age of the oldest fish found and the number examined, we obtained an estimate of mortality for each of these lakes. These were then combined in a weighted average over neighbouring lakes to give an estimate of pre-acidification mortality for each SNSF lake.

The 'significance' of an observed change in mortality can be determined by considering the variances of the pre- and post-acidification estimates. The level of significance was chosen to give the same proportion of 'healthy' lakes as in

Table 1. *Distribution of population structures by lake group*

cluster	extinct	senescent	juvenile	healthy	total
1	11	1	10	2	24
2	8	0	1	1	10
3	2	0	4	1	7
4	0	0	2	0	2
5	12	4	5	2	23
6	8	0	6	1	15
7	2	4	3	1	10
8	5	0	0	0	5
9	6	1	3	0	10
total	54	10	34	8	106

Henriksen (1989), adjusted for time and area. Lakes showing a significant increase in mortality were classified as juvenilized, and those showing significant 'decreases' as senescent. Lakes for which an insufficient sample was obtained were classed as extinct.

(ii) *The pattern of population condition by lake group*

Several reports have detailed the pattern of fish population extinctions with water chemistry and region of Norway (Muniz *et al.* 1980; Henriksen *et al.* 1989). This pattern is evident in the SNSF data analysed here on examination of the number of lakes in each cluster in which the population is extinct, senescent, juvenilized or healthy (table 1). For example, cluster 8 in northern Aust Agder has low pH as well as low buffering capacity (figure 1) and all lakes have lost their fish populations. Similarly, cluster 2, whose lakes are in Telemark and Akershus, has a very high proportion of barren lakes. The healthy lake in cluster 2 has the highest pH and buffering capacity in the group as well as high organics and large size.

Barren lakes occur throughout the country with the exception, in this data set, of the west coast (Hordaland) and the northwest (Møre and Romsdal). In these two localities the sampled lakes were either healthy or senescent. The other healthy or senescent populations are generally in coastal lakes, with two exceptions.

(iii) *The pattern of population condition with water quality*

Although there is little question that there is a relation between water quality, in a general sense, and the loss of fish populations, it is far from clear how the patterns of stressed populations (juvenilized or senescent) relate to the water chemistry. From table 1 it is clear that juvenilization is a much more wide spread phenomenon than senescence of the populations.

The relation between water quality and the current status of the population was examined by a contingency table analysis by using the water-quality factors 1 and 2 described above (table 2). The factors' scores are normalized so that, for example, positive scores on factor 2 can be considered high pH and low aluminium, whereas negative scores are low pH and high aluminium. Chi-squared statistics for this table indicate that here is a clear effect of water quality on population condition ($\chi^2 = 20.96$; d.f. $= 9$). The effect is still significant if only the extant populations are considered ($\chi^2 = 14.01$; d.f. $= 6$).

The poorer water quality lakes have more extinct and juvenilized populations than

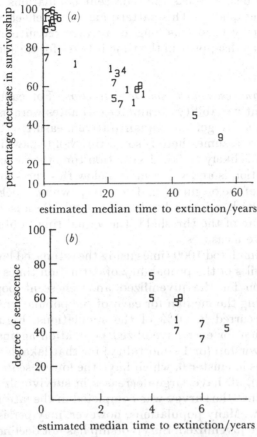

Figure 1. Times to extinction. (*a*) Juvenilized populations. Percentage decrease in survivorship from estimated pre-acidification levels against the median time to extinction for each lake from simulations projecting the population forward in time. The numbers indicate the type of lake by cluster number. (*b*) Senescent populations.

Table 2. *Contingency table of the effect of water quality on population condition. Expected values are shown in brackets*

	extinct	senescent	juvenile	healthy	total
high pH, low buffer	12 (17.49)	6 (3.24)	13 (10.69)	3 (2.59)	34
high pH, high buffer	5 (6.69)	0 (1.24)	5 (4.09)	3 (0.99)	13
low pH, high buffer	17 (13.89)	4 (2.57)	4 (8.49)	2 (2.06)	27
low pH, low buffer	20 (15.94)	0 (2.95)	11 (9.74)	0 (2.36)	31
total	54	10	33	8	105

expected from a null hypothesis of no effect on population condition. A similar, though weaker pattern is obtained by using factors 2 and 3 ($\chi^2 = 12.92$; d.f. = 9).

In general, it appears that the juvenilized populations are often found in poorer, possibly marginal conditions, whereas the senescent populations tend to appear in waters of somewhat better quality. This pattern may be a reflection of the fact that juvenilized populations may persist as long as average recruitment is maintained. Senescent populations may disappear in the time it takes for the older animals to die out, if recruitment fails.

Changes in mortality and estimated time to extinction. For each population, the mortality and recruitment variability parameter estimates were used to project the population forward in time. To get some quantitative measure of time to extinction for a given population, we assume there is some threshold spawning stock biomass below which recruitment is likely to fail. Extinction time then becomes the number of years until the population is and will remain below this threshold. The threshold level was chosen as 15% of the original (unaffected) spawning stock biomass, as very few extant populations with a lower level were found. Because of the fixed and somewhat arbitrary nature of the threshold, the actual times obtained should only be treated as comparative measures.

Each population was simulated 1000 times using the estimated level of recruitment variability to obtain profiles of the probability of extinction times. The distribution of the time to extinction for the juvenilized and senescent populations can be summarized by considering the median for each of the populations, i.e. the time by which extinction had occurred in 50% of the simulations. Figure 1*a* shows the median times to extinction for each juvenilized population against the percentage decrease in annual survivorship (or 1−mortality) for that lake, labelled by the lake cluster number. The lakes in cluster 6, which have the lowest scores on factor 2 (low pH and high aluminium), all have large decreases in survivorship and very short extinction times. Given that the survey was completed in the late 1970s, these lakes are very likely barren now. Many populations, however, may persist for quite a long time even though they are juvenilized. Survivorship has not declined so much in, for example, clusters 3, 7 and 9 where the level of factor 2 is higher (higher pH).

The median times to extinction for the senescent populations (figure 1*b*) are in general much shorter than for the juvenilized populations. The southern, coastal lakes in cluster 5 are more severely affected than the northern (cluster 7) lakes. These lakes were predicted to have extant populations for only five or six years after the survey and are probably barren now.

The suggestion that senescent populations are more transient than juvenilized populations is confirmed by this analysis. Even a relatively slow decline in recruitment means that the expected time to extinction is less than ten years. Juvenilized populations seem to be more persistent even in lakes of similar water quality. For example, lakes in clusters 5, 7 or 9 had extant senescent populations for only 4 to 8 years (figure 1*b*), whereas juvenilized populations in these same clusters, so subjected to similar water quality, are expected to survive for much longer in general (20–40 years; figure 1*a*). The underlying reason why a population juvenilizes or becomes senescent is still unclear, but the pattern of the observations may result in part from this difference in timescales.

(iv) *Restocking potential*

It is possible to evaluate the viability of restocking as a strategy for maintaining populations in acidified waters by calculating the probable time to extinction in restocked lakes for known population parameters. This calculation, as in the last

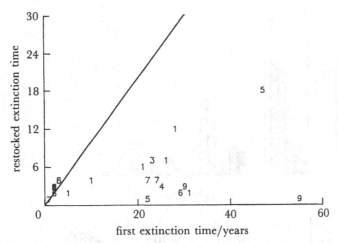

Figure 2. Juvenilized populations. Estimated extinction times for projected restocked lakes against the original extinction time. The scenario here is of a sudden episodic mortality removing the juvenilized stock, which is then restocked.

section, depends on the level of recruitment variability and mortality. Clearly, for senescent lake, with low reproduction, restocking is really only viable if it is done repeatedly. However, for a lake that had a juvenilized population, the new stock may persist for some time before additional restocking is required.

For the lakes that had juvenilized populations in the survey, we calculated the median time to extinction after restocking a single year class at the average recruitment level (figure 2). The results indicate as expected that the restocked populations have a shorter median extinction time than the original population. However, restocking would provide populations which may persist for substantial periods in several lakes, particularly in clusters 5, 7, and 1. Restocking may be an economical means of repopulating lakes with trout in these areas.

3. Lake Selura: effects of liming

Lake Selura, in coastal southwestern Norway, has the physical and chemical characteristics of a cluster 5 lake. Its large size and relatively small spring snowmelt have protected it from the very worst effects of acidification to date.

In 1977–78, and from 1982 onwards, two distinct populations were studied: one spawning in the lake and senescent, and one spawning in an inlet creek and juvenilized (Andersen *et al.* 1984). Liming of this inlet creek began in 1983, and has been augmented in subsequent years to cover the whole lake.

Using the same techniques as for the SNSF data, the pre-acidification annual survivorship was estimated at roughly 0.54. For the creek spawners, this figure implies a decrease of 32% in survivorship to 1978. The coefficient of variation (cv) of recruitment was three times higher than for the longer-lived lake spawners, for whom the spawning environment is more stable. Despite this degree of juvenilization, projections do not point to a significant risk of extinction, assuming no further decreases in water quality.

The rate of decline of recruitment to the senescent, lake spawning population was estimated at about 9% per year, giving a projected median time to extinction of 16

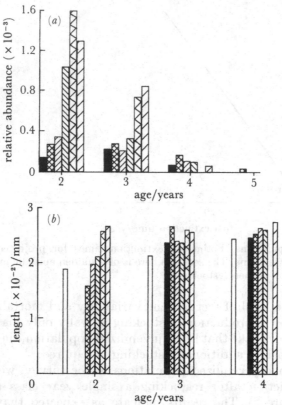

Figure 3. Selura lake spawners. (*a*) Relative abundance at age after liming began in 1983. Recruitment increased strongly. Survivorship first increased, then decreased again. (*b*) Average length at age for each year; (□), 1979; (■), 1983; (▨), 1984; (▨), 1985; (▨), 1986; (▨), 1987; (▨), 1988.

years. However, this figure would be reduced if acidification damage began before 1973. Evidence from a lake-spawning arctic charr population suggests that serious problems were already occurring by 1968.

After liming, survivorship of the creek spawners increased by 24%, and growth rates also improved. Recruitment also increased, and its cv around the new level had declined by 15% in 1987.

The effects of the liming on the lake spawning population were even more pronounced. Because of the relative volumes of lake and creek, two to three years were required for the effects to become fully apparent. Figure 3*a* shows the changes in recruitment, and figure 3*b* the changes in mean size at age. Because of gear selectivity problems, the estimation of post-liming survivorship is difficult. However, there would appear to have been an initial increase in survivorship in 1985, followed by a decline to 1988. This is probably because of density dependence induced by very high recruitment, and does not represent a threat to long-term health. For this population, the imminent extinction has been averted.

4. Conclusions

One of the most difficult problems in this work is to identify the critical combination of environmental variables that determine the changes in demography leading to the two basic phenomena of juvenilization and senescence. It seems clear that the two occur in rather different waters and environmental conditions. Analysis of the survey data shows that juvenilization is the more widespread phenomenon. Such populations have been identified in poorer environmental conditions than have senescent stocks, and are expected to persist longer under such conditions.

The basic analyses of the SNSF data do not permit any unequivocal statement to be made about the likely causal mechanisms involved. Since two populations in Lake Selura alone exhibited different responses, such mechanisms must operate on a very local scale. Nevertheless, with liming, there was a clear reversal of the demographic change in both senescent and juvenilized stocks in the Lake. Accordingly, it seems reasonable to conclude that the main causal factor determining the demographic changes is the acidification.

Given that the demographic changes are reversible under liming, it is possible to assess the implications of this treatment on the likelihood of continued existence of populations of brown trout. Similarly, if the acid precipitation and its effects on the water quality are not reversed, then the estimates of median time to extinction should give a reasonable quantitative guide to the behaviour of lakes of different types.

It is thus possible to make testable predictions about the behaviour of trout populations in the set of lakes studied. A re-examination of the lakes analysed under SNSF could be used to investigate: (i) whether extinction has indeed occurred in those lakes for which a short lifetime was predicted, and (ii) in lakes for which subsequent extinction has not occurred, what are the changes in age composition and relevant environmental variables since the lake was last surveyed.

In addition, restocking experiments in lakes whose trout stocks have collapsed could give clearer insights into the extent to which senescence and juvenilization occur together by tracing the fate of cohorts.

With these data it would be possible to properly quantify the underlying relation between changes in environmental variables (reflected in the factor scores for the SNSF data) and changes in the demography. Such studies would substantially improve the predictive power of the system, and should allow reasonably accurate assessments to be made of likely trends in similar lake populations.

We gratefully acknowledge Professor Lars Walløe, Sir Richard Southwood and Sir John Mason for supporting our research. This study was funded by a grant from the Royal Society Surface Waters Acidification Programme.

This paper is dedicated to the memory of Dr Chris Mills of the Freshwater Biological Association, who was instrumental in the early stages of developing this work.

References

Andersen, R., Muniz, I. P. & Skudal, J. 1984 Inst. of Freshwat. Res., Drottningholm report no. 61, pp. 5–15.

Baker, J. P. & Scofield, C. L. 1980 In *Ecological impact of acid precipitation* (ed. D. Drablos & A. Tollan). Proceedings of the international conference on the ecological impact of acid precipitation, pp. 292–293. Oslo: SNSF project.

Dahl, K. & Lund, H. M. K. 1944 Verstanalyser over Orret fra 383 Norske vatn og vassdrag. Samlet ved Statens Forsoksvirksomhet for ferskvannsfiskeri 1910–1943, pp. 27. Oslo: Utgitt ved Landbruksdepartementet.

Harman, H. H. 1970 *Modern factor analysis*, 487 pp. Chicago: University of Chicago Press.

Harvey, H. H. 1981 In *Proceedings of an international symposium on acid rain and fishery impacts* (ed. R. E. Johnson), pp. 227–242. American Fisheries Soc.

Henriksen, A., Lien, L., Rosseland, B. O., Traaen, T. S. & Sevaldrud, I. S. 1989 *Ambio* **18**, 314–321.

Leslie, P. H. 1945 *Biometrika* **33**, 183–212.

Muniz, I. P. & Leivestad, H. 1980 In *Ecological impact of acid precipitation* (ed. D. Drablos & A. Tolan). Proceedings of the international conference on the ecological impact of acid precipitation, pp. 84–92. Oslo: SNSF project.

Rosseland, B. O. 1986 *Wat. Air Soil Pollut.* **30**, 51–460.

Rosseland, B. O., Balstad, P., Mohn, E., Muniz, I. P., Sevaldrud, I. & Svalastog, D. 1979 Presentasjon av ukvalgskriterier, innsamlingsmetodikk og anvendelse av programmet ved SNSF – prosjektets provefiske 1 perioden 1976–79. Teknisk notat Bestandsundersokelser Datafisk – SNSF – 77, TN 45/79, p. 63.

Sadler, K. 1983 *Freshwat. Biol.* **13**, 453–463.

Sevaldrud, I. H., Muniz, I. P. & Kalvenes, I. 1980 In *Ecological impact of acid precipitation* (ed. D. Drablos & A. Tollan). Proceedings of the international conference on the ecological impact of acid precipitation, pp. 350–351. Oslo: SNSF project.

Van Winkle, W., Christensen, S. W. & Breck, J. E. 1986 *Wat. Air Soil Pollut.* **30**, 639–648.

Identification of hydrological processes of surface water acidification

By M. B. Beck, F. M. Kleissen and H. S. Wheater

Department of Civil Engineering, Imperial College of Science, Technology and Medicine, London SW7 2BU, U.K.

Establishing unambiguously the paths taken by water through the soils of a catchment is an essential part of the development of models for surface water acidification. In particular, field observations of inert chemical tracers have been recognized as an important source of information for identifying flow paths and moreover for assessing the adequacy of catchment acidification models. Therefore a major question is the amount and type of field information required to estimate unique values for the parameters of these models, and to identify the flow paths within the catchment. Case studies addressing this question are reported for two SWAP field sites in Scotland: Burn 10 at Loch Ard, and the Allt a'Mharcaidh catchment. The results emphasize the crucial role not only of the chemical tracer observations but also of the association between specific antecedent field conditions and the tracer fluxes of quite modest precipitation events.

1. Introduction

Identification of the movement of water through soils is of central importance to the characterization of surface water acidification. The chemical composition of the streamwater depends on the time of contact between the water and soils, and on the types of soils that the water has had contact before it reaches the stream. One currently popular view is that this chemical composition is a function of the mixing of varying volumes of water deriving from different parts of the catchment, each of these flows having its own characteristic chemical 'signature' (Neal *et al.* 1988).

The problem of determining the sources and pathways of water can be addressed by two complementary approaches, i.e. (i) by conducting specialized field experiments across the catchment, sub-catchment, hill-slope, and plot scales (as discussed in the companion papers of Wheater *et al.* (this symposium) and (ii) through the systematic study of parameter estimation in conceptual hydrological and hydrochemical models (the subject of this discussion). Interpretation of the way that the chemical signatures of the various water sources have been combined, i.e. hydrograph separation (or flow-path identification), can be formulated as a problem of estimating the coefficients of a mathematical model of the catchment. The solution of this problem is a direct function of the availability and number of observations of the conservative tracer substances that (in sum) define the signatures of the contributing parts of the catchment.

Within the framework of an extensive programme of research on predictability and model identifiability (Beck 1987, 1990; Beck *et al.* 1990) the paper describes an analysis of the role of tracer observations in reducing the uncertainty associated with the parameters of conceptual hydrological models, in other words, in reducing ambiguities in the constituent hypotheses about flow paths.

2. The significance of tracer observations

The prototypical model of a catchment's hydrochemical response to an acidic precipitation event is that of Christophersen & Wright (1981) and Christophersen *et al.* (1982). It comprises a chemical sub-model constructed on the basis of what was believed to be the firm foundations of a well identified hydrological sub-model. Subsequent re-examination of these foundations, prompted notably by the model's inability to characterize correctly the movement of an (assumed) inert tracer substance through the catchment, has revealed them to be inadequate (Bergström & Lindström 1987; Neal *et al.* 1988). Given daily observations of both flow and tracer (chloride) concentration in the stream, identification of the hydrological sub-model for the Birkenes catchment in southern Norway showed that a single-store representation was more appropriate than the previous conceptual distinction of two stores, one for the water of the upper soil horizons and one for that of the lower horizons (Hooper *et al.* 1988). Our own studies, on the identifiability of models both of the Birkenes catchment (Beck *et al.* 1987) and of one of the Loch Ard sub-catchments (Beck *et al.* 1990), show very much the same results. In the light of this experience, the crucial test of an acceptable model of stream acidification has become whether it is able to describe the response of a conservative substance during a precipitation or snowmelt event.

There has long been controversy over the relative contributions from 'old' and 'new' waters in the storm response of a catchment. It is an obvious and fundamental problem in hydrology, but one that has acquired practical urgency because of its central importance in determining the acidification of surface waters. That recent field-work in hydrology has drawn attention to the role of chemical species in the interpretation of hydrological pathways and residence times is thus of considerable interest (Hooper & Shoemaker 1986; Kennedy *et al.* 1986; Pearce *et al.* 1986). Much of what can now be concluded from the field studies of the Allt a'Mharcaidh catchment in Scotland has been distilled into a consistent relation between observed catchment outflow and its alkalinity, which is assumed to be a conservative chemical property (Harriman *et al.* 1990). These data have already been the subject of an analysis of hydrograph 'separation' wherein the composition of the streamwater is defined as a mixture of two sources of water within the catchment, each with its own chemical, and in this case alkalinity, 'signature' (Wheater *et al.*, this symposium).

A fundamental practical and theoretical question arises: to what extent can the flow paths, and hence residence times of water passing through a catchment, be established unambiguously from conceptual hydrological models identified with reference to field observations of precipitation, stream flow and chemical-tracer concentrations?

3. Burn 10, Loch Ard

Perhaps the most natural means of conceiving of the hydrology of a catchment is in terms of a system of inter-connected stores, or reservoirs of water. Flow paths and residence times are thus quantified in a model of this conceptual (analogue) system by features such as: the volumes of water present in each store; the recession constants relating outflow from a store to the volume of water retained in it; and the routing parameters determining the fractions of these outflows passing along the various inter-connections among the stores and receiving stream.

The conceptual stores can usually be associated with the different areas and types of soil of the catchment; this association, together with the manner in which the stores are connected, constitutes the (model) composition of hypotheses about flow paths. Corroboration, or refutation, of these constituent hypotheses can then be assessed as a matter of solving the equivalent problem of estimating the unknown coefficients in the model from a given set of input and output field observations. In general, the smaller the uncertainty (error variance) attaching to the resulting parameter estimate, the more crisply defined is the corresponding flow-path.

In the specific case of Burn 10, at Loch Ard in Scotland, a candidate model of the catchment is one of just two conceptual reservoirs, for the upper and lower soil horizons (as in the model of the Birkenes catchment). The model has therefore three unknown coefficients to be estimated by reference to the field observations. These are k_a, the recession constant of the outflow from the upper reservoir, k_b, the recession constant for the lower reservoir, and f, the routing parameter that divides the outflow from the upper reservoir into a portion flowing to the stream via the lower reservoir and the remainder to the stream directly. The question concerning the unambiguous identification of flow paths can now likewise be stated in more specific terms. Thus: how accurately (with what degree of uncertainty) can the coefficients k_a, k_b, and f be estimated given (i) observations of precipitation and stream flow, (ii) additional observations of a single tracer in the precipitation and stream flow, (iii) still further observations of a second tracer as well, and so on?

Figure 1 shows results for the estimation of the routing parameter f under precipitation conditions observed for Burn 10 during November and December 1986 (the data are sampled hourly). In the absence of any tracer observations (the actual conditions) the dashed curve of figure 1 illustrates how the identifiability of f is degraded during what are the dry periods of $t_{360}-t_{440}$, $t_{530}-t_{670}$, and $t_{780}-t_{960}$. As expected, it is only at times of significant perturbation of the catchment, notably during the wet weather of $t_{440}-t_{520}$ and $t_{670}-t_{730}$, that the standard deviation of the parameter estimation errors is narrowed to any great extent. Significant perturbation of the catchment's behaviour can, of course, arise in the form of pulses of tracer fluxes. The other two curves of figure 1 indicate that given additional (hypothetical) observations of a single tracer substance, and then of two tracer substances, the identifiability of the routing parameter is markedly improved, albeit only marginally so for the second tracer once the first tracer is available.

Of particular importance in these theoretical results is the significance of tracer events associated with rather modest, or even small, amounts of precipitation coincident with quite specific field conditions, i.e. specific states of water storage and reservoir tracer concentrations just before the start of precipitation. The instance of a well identified f at hours t_{585} and t_{610} are cases, respectively, where the passage from small precipitation events of the first tracer and then the second tracer (after exhaustion of the first) enable clear access to the way that the conceptual flow paths separate below the upper soil horizon. Further results for the catchment at Loch Ard, with special reference to the relation between experimental/field conditions and model parameter estimation, are given elsewhere (Wheater *et al.* 1986; Beck *et al.* 1990).

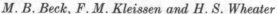

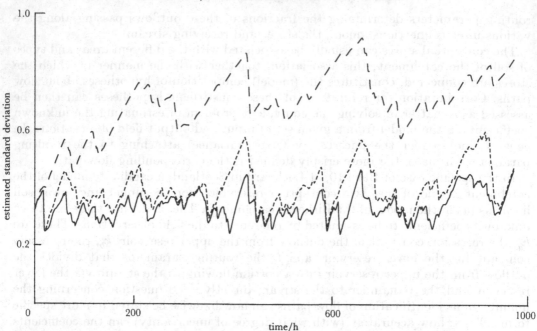

Figure 1. Variations with time of the standard deviations of the errors of estimation in the routing parameter, f, of a conceptual hydrological model of Loch Ard. The various curves show the standard deviation of the parameter estimation errors (the parameter has a nominal value of 0.15); (———), 2 tracers; (– – – –), 1 tracer; (— — —), no tracer signal.

4. Allt a'Mharcaidh catchment

The companion papers of Wheater *et al.* (this symposium), and Jenkins (this symposium) show clearly the difficulties in characterizing the complex behaviour of the relatively large Allt a'Mharcaidh catchment in Scotland. For the present discussion there are three salient points. First, from the complementary hydrological field studies across catchment, sub-catchment, hillslope and plot scales it is apparent that the catchment's hydrology exhibits three distinct dynamical features in its response: the almost constant base flow, an intermediate 'hump', and a fast transient flash flow or 'spike' (Wheater *et al.*, this symposium). Secondly, the consistent stream-flow–alkalinity relation of Harriman *et al.* (1990) allows the inference that if there were just two contributing sources of water in the catchment's event response, one would be time invariant and have a high alkalinity, while the other would be strongly dynamic with a low alkalinity (Wheater *et al.*, this symposium). There is evidence of a putative third source of water of intermediate alkalinity, although this does not suggest that three contributing sources must necessarily equate to a hydrological response with three dynamical components. Thirdly, the stream-flow chemistry has observed properties that cannot be generated from an assumption of simple, conservative mixing of the chemical 'signatures' observed in the pits located in the catchment's characteristic soil types (Ferrier *et al.* 1990).

The conceptual model composition of constituent hypotheses about flow paths is described elsewhere in Wheater *et al.* (this symposium). As in the foregoing studies of Burn 10 at Loch Ard, questions concerning the unambiguous identification of the catchment's flow paths (and thus the soils with which the water has had contact) can

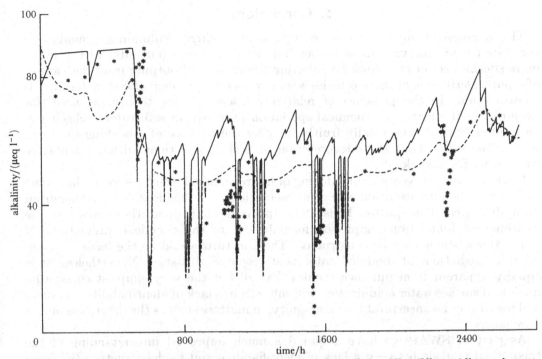

Figure 2. Simulated and observed alkalinity in the outflow of the Allt a'Mharcaidh catchment, using two alternative conceptual models. The period up to t_{520} represents very dry conditions. However, the alkalinity observations suggest that significant downslope drainage is not apparent until after the second precipitation event at t_{620}; (———), simulated alkalinity (one reservoir for lower peaty podzol); (-----), simulated alkalinity (two reservoirs for lower peaty podzol); (∗∗∗∗∗), observed alkalinity.

be put in the potentially more incisive and revealing form of questions about the identifiability of this conceptual model's parameters, for example: (i) does access to the observations of stream alkalinity alter the values for the model's recession constants and routing parameters as estimated on the basis of the stream flow observations alone in Wheater *et al.* (this symposium); (ii) to which flows in the interconnected conceptual system of stores can time-invariant alkalinities be assigned; (iii) are these signatures 'imprinted' in parts of the catchment far from the stream or immediately adjacent to it; and (iv) can they also be identified as unknown parameters from the input/output data instead of being specified *a priori* by assumption?

Figure 2 is indicative of the kind of results to be expected from analysis of these issues. It shows observations of alkalinity in the catchment outflow (see also Jenkins *et al.* (this symposium)) and the simulated alkalinity responses of two alternative hydrological model structures. These alternatives differ in respect of their hypotheses about the residence time of a downslope drainage path, ostensibly passing through predominantly peaty podzol soils. The two alkalinity events at t_{520} and t_{600} are critical in illuminating the condition necessary for the intermediate 'hump' flow to be simulated after a long dry period. Identification of the alternative models shows that the hypothesis of a slow (as opposed to a fast) downslope drainage path is the more successful hydrological interpretation of these observations.

5. Conclusions

The purpose of any model is to replace increasingly voluminous bundles of correlated field observations of causes (inputs) and effects (outputs) with a much more succinct set of principles for inferring the variety of outputs from the variety of inputs. Derivation of these principles is known as the problem of system (or model) identification. In the presence of relatively insecure prior theory, as here (for example, in the matter of chemical speciation processes in soil water; Schecher & Driscoll (1988)), it is especially fruitful to view the purpose of modelling as one of posing questions about the interpretation of field observations within a tentative conceptual framework.

Over the past ten years the modelling of hydrochemical event responses has come to focus on the interpretation of tracer observations and the identification therefrom of hydrological flow paths. It is thus apparent that upon these more secure hydrological foundations improved formulations of the chemical interactions of soil–water systems may be constructed. These in turn would be the basis for more reliable predictions of modified catchment response statistics. Nevertheless, it is equally apparent from our own studies that all but the very simplest conceptual models of surface water acidification will suffer from a lack of identifiability, i.e. there will inevitably be uncertainty, or ambiguity, in matters such as the determination of flow paths.

As part of SWAP we have acquired a much improved understanding of the theoretical relation between a lack of identifiability and model structure (Kleissen *et al.* 1990). Novel methods of quantifying model identifiability and of estimating parameters have been developed, and these have been applied extensively to practical investigation of those field (experimental) conditions most likely to promote model identifiability (Beck *et al.* 1990). The challenge now must be to generate predictions of future behaviour that are maximally insensitive to model uncertainty.

References

Beck, M. B. 1987 *Wat. Resour. Res.* **23** (8), 1393–1442.

Beck, M. B. 1990 *J. Forecast.* **9**. (In the press.)

Beck, M. B., Drummond, D., Kleissen, F. M., Langan, S. J., Wheater, H. S. & Whitehead, P. G. 1987 In *Systems analysis in water quality management* (ed. M. B. Beck), pp. 133–150. Advances in water pollution research. Oxford: Pergamon Press.

Beck, M. B., Kleissen, F. M. & Wheater, H. S. 1990 *Rev. Geophys.* (In the press.)

Bergström, S. & Lindström, G. 1987 In *Proceedings symposium on acidification and water pathways, Bolkesjö, Norway*, vol. 1, pp. 163–172.

Christophersen, N., Seip, H. M. & Wright, R. F. 1982 *Wat. Resour. Res.* **18** (4), 977–996.

Christophersen, N. & Wright, R. F. 1981 *Wat. Resour. Res.* **17** (2), 377–389.

Ferrier, R. C., Walker, T. A. B., Harriman, R., Miller, J. D. & Anderson, H. A. 1990 *J. Hydrol.* (In the press.)

Harriman, R., Gillespie, E., King, D., Watt, A. W., Christie, A. E. G., Cowan, A. A. & Edwards, T. 1990 *J. Hydrol.* (In the press.)

Hooper, R. P. & Shoemaker, C. A. 1986 *Wat. Resour. Res.* **22** (10), 1444–1454.

Hooper, R. P., Stone, A., Christophersen, N., de Grosbois, E. & Seip, H. M. 1988 *Wat. Resour. Res.* **24** (8), 1308–1316.

Kennedy, V. C., Kendall, C., Zellweger, G. W., Wyerman, T. A. & Avanzino, R. J. 1986 *J. Hydrol.* **84**, 107–140.

Kleissen, F. M., Beck, M. B. & Wheater, H. S. 1990 *Wat. Resour. Res.* (In the press.)

Neal, C., Christophersen, N., Neale, R., Smith, C. J., Whitehead, P. G. & Reynolds, B. 1988 *Hydrol. Process.* **2**, 155–165.

Pearce, A. J., Stewart, M. K. & Sklash, M. G. 1986 *Wat. Resour. Res.* **22** (8), 1263–1272.

Schecher, W. D. & Driscoll, C. T. 1988 *Wat. Resour. Res.* **24** (4), 533–540.

Wheater, H. S., Bishop, K. H. & Beck, M. B. 1986 *Hydrol. Process.* **1**, 89–109.

Hydrological and chemical modelling of the Allt a'Mharcaidh SWAP site in the highland mountains of Scotland

By Alex Stone[1], Hans Martin Seip[1], Steven Tuck[2], Alan Jenkins[2], Robert C. Ferrier[3] and Ron Harriman[4]

[1] Chemistry Department, University of Oslo, P.B. 1033 Blindern, N-0315 Oslo 3, Norway
[2] Institute of Hydrology, Maclean Building, Crowmarsh Gifford, Wallingford, Oxfordshire OX10 8BB, U.K.
[3] Macaulay Land Use Research Institute, Craigiebuckler, Aberdeen AB9 2QJ, U.K.
[4] Freshwater Fisheries Laboratory, Pitlochry, Perthshire PH16 5LP, U.K.

The Birkenes model (BIM) of stream acidification was adapted and used on a site that possesses very different physical and chemical characteristics from the Birkenes catchment on which the original model was based. The Allt a'Mharcaidh (AM) catchment in Scotland differs in several important parameters such as size, soil depth, rate of discharge and the concentrations of several important chemical species such as SO_4^{2-}, Al^{3+} and H^+. The model, with minimal alteration, was able to produce reasonable simulations for discharge and major chemical species. This substantiates and validates the premise that the most important processes in the catchment are incorporated. However, the model represents a highly simplified picture of the real processes and further model development is necessary.

1. Introduction

The results obtained from computer simulation models of acidification vary greatly and are very dependent upon the original intent and scope of the model. Types range from lumped models, which are used mostly for predictive purposes and deal with large timescales, to discrete models with shorter timescales that may reproduce rapid hydrological and chemical changes. The BIM is an example of the latter but has tentatively been used for predictions as described in Stone & Seip (1989, 1990, this symposium). However, regardless of their original intent, all models are valid only if they encompass processes that can be transferred between catchments. For more information concerning the current status of surface acidification models, see Christophersen et al. (this symposium), Stone & Seip (1989, 1990, this symposium).

2. Methodology

The structure and contents of the BIM have been well presented in the literature (Christophersen et al. 1982; Stone & Seip 1989). The BIM consists of three reservoirs, two of which represent the upper (organic) and lower (mineral) soil horizons and a third that is used when necessary for the accumulation and melting of snow. The hydrological and chemical sub-models were essentially the same as used in earlier work (Stone & Seip 1989, 1990). Observations of a conservative tracer, such as the heavy isotope of oxygen found in water, $H_2^{18}O$, have proved very useful for the determination of the hydrological sub-model. Organic acids are not included in the version used here. The model is combined with a mathematical optimization routine for the selection of parameter values (Hooper *et al.* 1988).

Several institutions in the U.K. share responsibility for the collection precipitation and stream data. Because of the practical difficulties related to catchment size and accessibility, the data for stream discharge and precipitation possessed greater uncertainty than data from the Birkenes catchment. A more complete account of the work is given elsewhere (Stone *et al.* 1990).

The AM catchment, located in the Cairngorm mountains of Scotland, differs from the Birkenes catchment in several important areas. AM is larger (10 km² versus 0.41 km² for Birkenes), possesses greater height variation (from 225 to 1111 m elevation range versus 205 to 300 m for Birkenes) and greater soil depth (approximate means are 0.83 versus 0.25 m). In addition, the concentrations of several important chemical species were very different for AM and Birkenes. Stream output of SO_4^{2-} at AM is considerably lower than input indicating the presence of important adsorption–reduction processes. The stream-sulphate budget at Birkenes is roughly in balance. In addition, the concentrations of Al^{3+} and H^+ at AM are very low (*ca.* 0.0 and 1.9 µeq l⁻¹, respectively) compared with the corresponding concentrations at Birkenes (34.9 and 26.3 µeq l⁻¹, respectively). For more information concerning the physical and chemical characteristics of the two catchments, see Stone *et al.* (1990).

3. Results and discussion

The observed and simulated stream discharge and Cl⁻ concentrations are presented in figure 1 *a, b*. Unlike the Birkenes catchment, data for $H_2^{18}O$ do not currently exist for AM. As an approximation of a conservative tracer, the chloride ion was used as in the Birkenes catchment (Hooper *et al.* 1988; Stone & Seip 1989; Stone *et al.* 1990). The high input of chloride from seaspray in the precipitation, mist and fog overwhelms chloride reactions in the soil and enables the species to function conservatively. Initially, parameter optimization was free to choose the amount of water to transfer between the two reservoirs and to the stream. A tendency was observed, however, to require all water from the upper reservoir to flow into the lower, thereby increasing damping of the conservative tracer and minimizing the amount of water from the upper reservoir directly to the stream. This was unacceptable for the chemical simulations and the model structure was changed so that a portion of the water from the upper reservoir was always forced to the stream regardless of the other parameter values used. This assumption did not adversely affect the simulation of chloride. The overall pattern and baseline trends were well represented by the current model structure.

The discharge simulation possesses some of the deficiencies noted in earlier Birkenes simulations (Hooper *et al.* 1988; Stone & Seip 1989; Stone *et al.* 1990). The model simulation fails to meet peak heights. These discrepancies are not necessarily caused by model deficiencies. No weir exists at AM and all values of discharge are taken from a simple stage–discharge relation calculated by using the salt-pulse detection method, which leads to considerable uncertainties in the measurements,

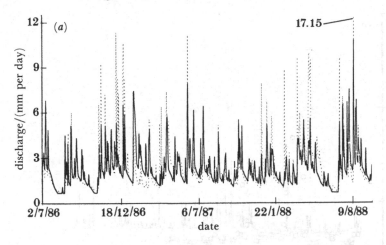

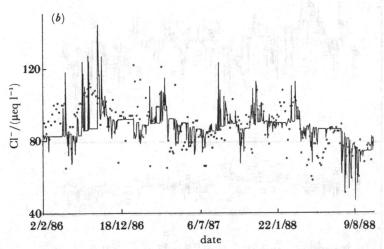

Figure 1. (*a*) observed (------) and simulated (———) discharge measured in millimetres per day and (*b*) observed (○) and simulated (———) Cl⁻ stream concentrations in microequivalents per litre for Allt a'Mharcaidh. The period is from 2 June 1986 to September 22 1988. Note that the single peak in observed discharge in August 1986 is off scale.

particularly for periods of high discharge. However, past experience indicates that better agreement is obtained for baseline flow than for peak discharge.

Because of its larger size and soil depth, the AM possesses the ability to store large amounts of water. During and after periods with high precipitation, a 'humped' effect can be seen in the discharge record particularly for the autumns of 1986 and 1988. The normal peak discharge is superimposed upon this feature. The model simulates this effect adequately. For more information on the discharge simulation, see Stone *et al.* (1990).

The model was then optimized to obtain the best parameter values for the sulphate sub-model and finally for major cations. The results for H⁺, Na⁺, SO₄²⁻ and Ca²⁺ are shown in figure 2*a*–*d*, respectively. The simulated and observed values for Al³⁺ are not included as the aluminium concentration in the stream was so low that insufficient information was available for the optimization routine to obtain

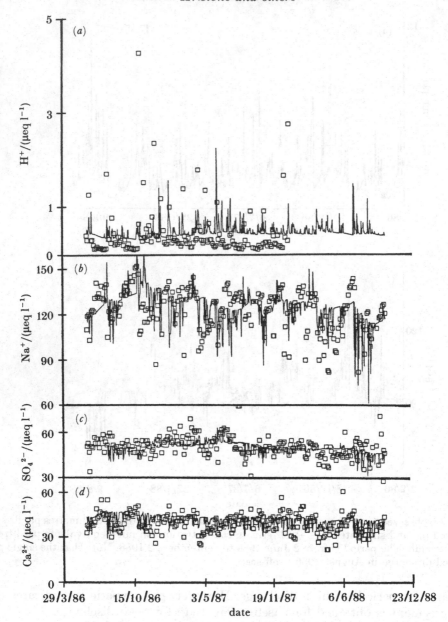

Figure 2. Observed (□), and simulated (——), stream concentrations for (a) H^+, (b) Na^+, (c) SO_4^{2-} and (d) Ca^{2+} for the Allt a'Mharcaidh catchment. All concentrations are in microequivalents per litre. The period shown is from 2 June 1986 to 22 September 1988. Note the different scale for H^+.

acceptable values for this species. The parameter values were therefore fixed at reasonable values for the soils in the two reservoirs (for more details see Stone *et al.* (1990)).

As can be seen from the graphs, the BIM simulates the various chemical species reasonably well. The sulphate simulation fails to demonstrate as much short-term variability as seen in the observed stream concentrations. It was able, however, to simulate the large absorption/reduction experienced in the AM catchment and produce baseline values that closely approximate stream concentrations. This suggests that, although improvement is possible with the sulphate sub-model, the gross processes are reasonably well included. The same is true for the hydrogen ion simulation where there exists a slight tendency to overestimate baseline values. The difference, however, is very small, approximately 0.3 μeq l^{-1}. In addition, the simulated hydrogen ion concentration does not show as great a short-term variability as observed in the stream. However, the general trends are in fairly good agreement with the observed values. The calcium and sodium simulations are indicative of the remaining ionic species that possess better agreement with observed stream concentrations. Both the baseline trends and short-term variability are well represented by the current model structure.

4. Conclusions

It is very important that acid precipitation simulation models undergo sufficient evaluation and validation. The hydrological sub-model should be tested for its ability to reproduce a conservative tracer in addition to discharge. It is also important that the model can be used for catchments with widely varying chemical and hydrological characteristics. The BIM has been evaluated by using both of these criteria and found to provide quite good simulations for discharge and the important chemical species. Improvements are required in particular for aluminium, but also for SO_4^{2-}. However, the ability to simulate the ionic levels observed in the stream for a catchment such as AM substantiates the premise that the general chemical and physical properties inherent in soils are reasonably well represented in the current model structure. Without such strenuous evaluation, model predictions and simulations are open to a high degree of uncertainty and one must be aware of and consider model limitations when making policy decisions.

References

Christophersen, N., Seip, H. M. & Wright, R. F. 1982 *Wat. Resour. Res.* **18**, 977–996.

Hooper, R. P., Stone, A., Christophersen, N., de Grosbois, E. & Seip, H. M. 1988 *Wat. Resour. Res.* **24**, 1308–1316.

Stone, A. & Seip, H. M. 1989 *Ambio* **18**, 192–199.

Stone, A. & Seip, H. M. 1990 *Sci. tot. Envir.* **96**, 159–174.

Stone, A., Seip, H. M., Tuck, S., Jenkins, A., Ferrier, R. C. & Harriman, R. 1990 *Wat. Air Soil Pollut.* (In the press.)

Hydrological and chemical modelling of the Birkenes SWAP site in southern Norway

By Alex Stone and Hans Martin Seip

Chemistry Department, University of Oslo, P.B. 1033 Blindern, N-0315 Oslo 3, Norway

The Birkenes model (BIM), created to simulate short-term changes in water quality for a small stream in southernmost Norway, was expanded to encompass a 3.5 year data record and combined with a mathematical routine that enabled the selection of optimal parameter values. The hydrological sub-model was subjected to validation through the use of conservative tracers (^{18}O and the chloride ion). The model structure was expanded to include chemical sub-models for the following species: Al^{3+}, Ca^{2+}, Cl^-, H^+, HCO_3^-, Mg^{2+}, Na^+ and SO_4^{2-}. The model produced adequate simulations for all species. The calibration parameter values were then used to simulate two additional periods with very different hydrological and chemical qualities. The results were in satisfactory agreement with the observations substantiating the premise that the major processes are included within the current model structures. However, some areas in need of improvement are identified, such as the sub-models for Al^{3+} and SO_4^{2-}.

1. Introduction

Computer models have long been used as a tool both for understanding the complexities of a natural system and for predictions. The various models that currently exist differ in intent and purpose. Some models simulate yearly trends for mainly predictive purposes, others produce daily trends for the evaluation of short-term chemical processes. The BIM is an example of the latter. However, it has tentatively been used for predictions either directly (Stone & Seip 1989) or in combination with the MAGIC model (Stone et al., this symposium). For more information on the types of simulation models and their current status, see the articles by Christophersen et al. (this symposium) and Stone & Seip (1989, 1990).

2. Methodology

The BIM is a stream acidification model based upon data collected for a small (0.41 km²), forested catchment at Birkenes in southernmost Norway (Christophersen et al. and Seip et al., this symposium). Data collection started in 1971. The model consists of three reservoirs, two of which are used to approximate lower and upper soil horizons. The third is a snow reservoir and is used only when needed. The BIM structure is well reported in the literature and has been used in other catchments. For more information on the BIM, see the articles by Christophersen et al. (this symposium), Stone & Seip (1989, 1990) and Stone et al. (1990).

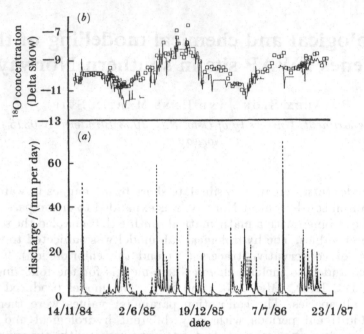

Figure 1. (a) Observed (------) and simulated (——) discharge in millimetres per day and (b) simulated (——) and observed (□) ^{18}O concentrations in Delta smow (standard mean ocean water) in per mille for the Birkenes catchment. The period shown is from 1 January 1985 to 31 December 1986 but was taken from a full simulation from 1 June 1983 to 31 December 1986.

3. Results and discussion

The quality of the simulations of chemical species is very dependent upon the ability to adequately reproduce discharge. Furthermore, it is vital that the water-flow pathways are realistically represented. Flow contributions from the various soil reservoirs are important to assure adequate simulation of chemical stream quality. The hydrological model was subjected to parameter optimization (Hooper *et al.* 1988) by using a data set of 3.5 years. In addition to discharge, two conservative tracers, water with the heavy isotope of oxygen ($H_2^{18}O$) and Cl^-, were included in the parameter determination procedure to provide information on water pathways. The chloride ion can be used at Birkenes as an approximation of a conservative tracer because of the large input from seaspray that overwhelms chemical interactions involving chloride. The results for discharge and ^{18}O are presented in figure 1 a, b. The graphs show only the last two years of the simulation period but were taken from the full 3.5 year period. The BIM provides an adequate simulation of discharge although there exists a problem in producing peak flows. The calibration period covers a wide range of variations in the discharge and the optimization procedure provided one set of parameter values that adequately represented these fluctuations. The overall patterns for the conservative tracers were also well reproduced.

The main parameters in the chemical sub-models of BIM were then subjected to an optimization procedure that led to simulation of the ionic species. The results for SO_4^{2-} and the most important species biologically, Al^{3+} and H^+, are presented in figure 2a–c, respectively. These concentrations are well reproduced by the current model structure although some deficiencies do exist. The sulphate concentration is

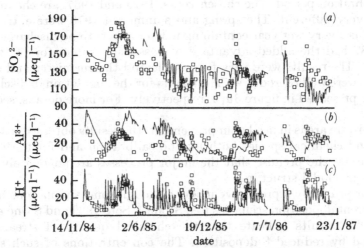

Figure 2. Observed ($\square$) and simulated (———) concentrations for (a) SO_4^{2-}, (b) Al^{3+} and (c) H^+ in µeq l^{-1} for the Birkenes catchment. The period shown is from 1 January 1985 to 31 December 1986 but was taken from a full simulation from 1 June 1983 to 31 December 1986.

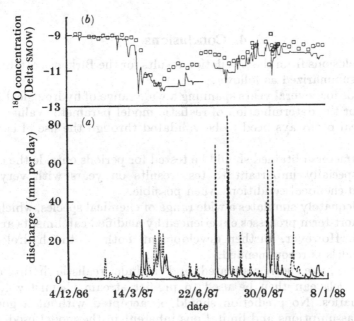

Figure 3. (a) Observed (------) and simulated (———) discharge in millimetres per day and (b) simulated (———) and observed ($\square$) ^{18}O concentrations in Delta SMOW (per mille) for the Birkenes catchment. The period shown is from 1 January to 31 December 1987 but was taken from a full simulation from 1 June 1986 to 31 December 1987.

underestimated at times particularly during snow melt and is overestimated in others. The aluminium and hydrogen species are occasionally underestimated during peak discharges and aluminium often lacks the short-term variation seen in the observations.

The parameter values obtained were tested against observations for two years

outside the calibration period. The chosen years, 1976 and 1987, are chemically and
hydrologically very different. The spring and summer of 1976 were extremely dry
whereas 1987 was a very wet year containing few periods with no discharge from the
catchment. 1987 had the added advantage of possessing values for ^{18}O that were
lacking in 1976. The results were very favourable and the chemistry and discharge
for both periods were well reproduced. The results for the simulation of discharge and
^{18}O for 1987 are presented in figure 3a, b, respectively. For more details, see Stone &
Seip (1990).

Although there are areas in need of improvement, the results obtained with the BIM
for discharge and chemical species both for the calibration and for the validation
periods substantiate the premise that the major processes are adequately included
within the current model structure.

The BIM is not primarily a predictive model although some simulations have been
made with different deposition scenarios (see Stone & Seip 1989 and Stone et $al.$ (this
symposium)). The results indicate that the chemical quality of streamwater is
strongly affected by reduced S-deposition. The concentrations of such species as
Al^{3+} and H^+ decline substantially with reduced sulphate deposition. The results
are, however, very dependent upon the limitations of the various models and the
assumptions made during their use.

4. Conclusions

The main conclusions from the simulation results for the Birkenes catchment using
the BIM can be summarized as follows:

1. Data records for several years spanning a wide range of hydrological conditions
are important for the determination of realistic model parameter values.

2. Hydrological pathways need to be validated through the use of conservative
tracers.

3. Model parameters obtained should be tested for periods outside the calibration
period. It is especially important to test results on years with very different
hydrological and chemical conditions when possible.

4. The BIM adequately simulates a wide range of chemical species, which indicates
that the basic short-term processes experienced by acidified catchments are included
reasonably well. However, further development both of the hydrological and
chemical sub-models is recommended.

5. Models that have been validated and include realistic hydrological and
chemical sub-models can then be used to predict stream chemistry for various
depositions scenarios. No prediction should be accepted without a good understanding of the assumptions and limitations inherent in the model used.

References

Hooper, R. P., Stone, A., Christophersen, N., de Grosbois, E. & Seip, H. M. 1988 $Wat.$ $Resour.$ $Res.$
    24, 1308–1316.
Stone, A. & Seip, H. M. 1989 $Ambio$ **18**, 192–199.
Stone, A. & Seip, H. M. 1990 $Sci.$ $tot.$ $Envir.$ **96**, 159–174.
Stone, A., Seip, H. M., Tuck, S., Jenkins, J., Ferrier, R. C. & Harriman, R. 1990 $Wat.$ Air $Soil$
    $Pollut.$ (In the press.)

Predictions of stream acidification by using the Birkenes and MAGIC models

BY ALEX STONE[1], NILS CHRISTOPHERSEN[2], HANS MARTIN SEIP[1] AND
RICHARD F. WRIGHT[3]

[1] Chemistry Department, University of Oslo, P.B. 1033 Blindern, N-0315 Oslo 3,
Norway
[2] Centre for Industrial Research, P.B. 124 Blindern, N-0314 Oslo 3, Norway
[3] Norwegian Institute for Water Research, P.B. 69 Korsvoll, N-0808 Oslo 8, Norway

The Birkenes model (BIM) of stream acidification was used to simulate several
different scenarios for reducing excess sulphate deposition. The changes in the soils
in 40 and 140 years predicted by the MAGIC model for sulphate deposition reductions
equal to 30%, 60% and 90% of present values were used in BIM simulations with
stream-sulphate concentrations reduced proportionately. The reductions in excess
sulphate indicate that sulphate plays an important role in stream acidification.
Major decreases in sulphate translate into marked decreases in toxic species such as
Al^{3+} and H^+. In all cases, changes within the soil had only a minor effect upon
streamwater concentrations of these specks. At 30% reduction, soil recovery was
minimal or non-existent and a tendency was observed for the stream Al^{3+} and H^+
concentrations to move toward current values. At 60% and 90% reductions, the
soils began to recover and the amount of recovery was substantially greater for the
90% than for the 60% levels.

1. Introduction

The ability of computer models to predict the effect of change in anthropogenic
deposition upon natural systems is dependent upon their ability to approximate
naturally occurring processes. Existing models vary widely in type and original
intent. Simulation models exist that are basically lumped models dealing with long-
term changes such as MAGIC (Model of Acidification of Groundwater In Catchments)
or with shorter-term variations as is the case with BIM. Comparison of the models
may be hampered by several factors such as differences in timescales and structure.
If, however, the models possess a valid approximation of natural processes, the
predictions should reach similar conclusions. For more information on the various
acidification models see Christophersen et al. (this symposium), Stone & Seip (1989,
1990) and Stone et al. (1990). In this work, MAGIC and BIM were combined to predict
future changes in water chemistry for several deposition scenarios.

2. Methodology

The structures of both MAGIC and BIM have been well reported in the literature
(MAGIC, see Cosby et al. (1985); BIM, see Christophersen et al. (1982) and Stone
& Seip (1989)). Both were classified as soil-oriented charge balance models in the

comparison of models by Reuss *et al.* (1986) and important soil processes are treated similarly. A recent version of MAGIC, which includes two soil reservoirs (Jenkins & Cosby 1989), was used and allowed for easier comparison between the two models. Although the aluminium concentration in the stream in BIM is computed from the concentration in the soil reservoirs, the version of MAGIC used assumes that this concentration corresponds to an $Al(OH)_3$ phase in the stream. The two reservoirs represent the upper (organic) and lower (mineral soil) layers. BIM possesses a third reservoir when needed for the accumulation and melting of snow (see Christophersen *et al.* (this symposium), fig. 1), a factor which is not required by MAGIC because of its yearly simulation period.

The Birkenes simulations were made by using results from earlier work (Stone & Seip 1989, 1990) that were validated through the use of conservative tracers. The data for the simulations were obtained from the Birkenes catchment, situated in southernmost Norway and encompassed a 3.5 year period from June 1983 to December 1986. Birkenes has a long history of acidification with sulphate inputs approximately equivalent to outputs. The parameters used for the BIM simulation were obtained from previous research. For more information on both the data and the model parameters, see Stone *et al.* (1990) and Stone & Seip (1989, 1990).

The MAGIC model was used to simulate a period from 1845 to 2125. Precipitation concentrations were assumed constant over time with the exception of anthropogenic generated species, i.e. SO_4^{2-}, NO_3^- and NH_4^+. The concentrations of these species were assumed in 1845 to be very low and were assigned values based on their concentrations in seaspray or what is believed to be their natural, background concentrations. The concentrations were increased slowly until the turn of the century when their concentration began to rise rapidly. The SO_4^{2-} deposition was reduced slightly during the past five years. For the predictions, SO_4^{2-} concentrations were reduced linearly until 2025 and were then held constant until 2125. NO_3^- and NH_4^+ reductions were assumed to be half that of SO_4^{2-}. The values for a 30%, 60% and 90% reduction of current excess SO_4^{2-} deposition were used to determine the effect of SO_4^{2-} reduction. Because of sea-spray contribution, these reductions correspond to reductions in total deposition of approximately 27%, 54% and 81% of the 1985 values.

The current values for cation exchange capacity (CEC) and base saturation (BS) were obtained from data collected by the Norwegian Forest Research Institute (NISK). Values of 100 and 25 meq kg^{-1} were used to approximate the upper and lower soil layers, respectively. These numbers are between values for the O/H and E horizons and close to the lower B horizon, respectively. The corresponding values for BS were 11.3% and 7.1%, respectively.

Weathering was set to values of 8.5 and 12.8 meq m^{-2} a^{-1} for the upper and lower reservoirs and about half of this was caused by calcium. Weathering was constant over time. The precipitation and discharge amounts were set equal to 1.51 and 1.183 m a^{-1}. The routing of water through the various soil layers was calculated from the averaged, yearly values obtained from the optimized simulation of the Birkenes catchment (Stone & Seip 1989, 1990). It was determined that 71% of the flow from the upper soil was transferred to the lower reservoir and that 29% flowed directly from the upper soil to the stream. CEC values in 1845 were then adjusted until observed stream concentrations and CEC in 1985 were approximately met.

The results from the MAGIC simulations were then evaluated and adapted to fit into the current BIM structure. Because of the difference in structure between the

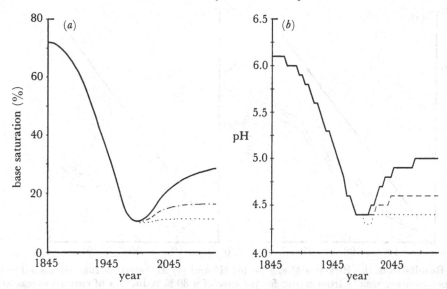

Figure 1. (a) Base saturation and (b) pH simulations results (upper soil horizon) from MAGIC for the 30% (⋯⋯), 60% (———) and 90% (——) reductions in current excess SO_4^{2-}. The model was run from 1845 to 2125. Base saturation is in percentage of CEC.

two models, it was not possible to use a direct transfer of parameter variations. Comparisons for the upper soil layer were straightforward. The ion exchange constants responsible for the $Al^{3+}:Ca^{2+}$ and $Ca^{2+}:Mg^{2+}$ ratios were evaluated to determine how they varied with time and deposition sequence. The relative variations were then transferred to similar parameters in BIM. For the lower soil layer, the changes in H^+ concentrations were evaluated. The corresponding fluctuations were used to vary a parameter in BIM that determines equilibrium $[H^+]$.

Each deposition scenario produced four simulations. The first, common to all scenarios, was a Birkenes simulation with current conditions and parameter values. The second was with current parameter values but reduced SO_4^{2-} deposition. The last two were simulations with both reduced SO_4^{2-} deposition levels and parameter values adjusted according to the results obtained from MAGIC simulations for the years 2025 and 2125, respectively. All results were compared with current conditions.

3. Results and discussion

The BIM is not primarily a predictive model but is intended to simulate short-term changes in soil and water quality. Unlike models such as MAGIC, no processes are included that follow the change in soil characteristics (i.e. BS) over extended periods of time. For example, with BIM it is not possible to predict the length of time needed to reach new equilibrium conditions. This information is best provided by long-term simulation models such as MAGIC.

The results from the MAGIC simulations for the various deposition sequences are presented in figure 1. The values for BS at a 30% reduction of the excess deposition for the upper soil horizon (figure 1a) demonstrate that there is little soil recovery at any time in the future. Considerable soil recovery occurs at a 90% reduction, although the soil is never able to return to earlier levels within the time span evaluated. The pH in streamwater increased to *ca.* 5.0, which is potentially sufficient

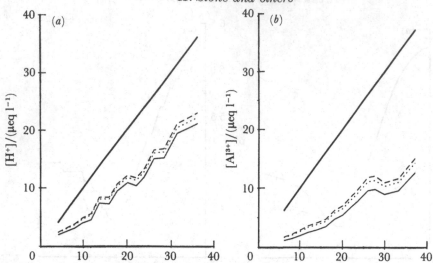

Figure 2. Results from the Birkenes Model for (*a*) H$^+$ and (*b*) Al^{3+} against 1985 simulated values for each species using results from MAGIC for the case of a 60 % reduction of current excess SO$_4^{2-}$. The results include the effect of reducing SO$_4^{2-}$ by 60 % with current model 1985 parameters ($\cdots\cdots$) and with soil parameters from MAGIC in 2025 (---) and in 2125 (——). All results were compared to current Birkenes simulation values (——). All concentrations are in micro equivalents per litre.

to support fish populations but indicates that, on a smaller timescale, there may be fluctuations around this value that may exclude complete recovery. In effect, the data indicates that given no additional treatment, streamwater concentrations may not reach a level that would support fish populations without a SO$_4^{2-}$ deposition decrease in the order of 90 %.

The BIM simulations based upon the MAGIC result agree with this conclusion. Figure 2*a*, *b* compares results for the 60 % reduction sequence for both Al^{3+} and H$^+$ to present-day values. Because of the many points, the results were grouped into distinct ranges with smaller ranges being used for the lower values of both aluminium and hydrogen where the greatest number of points occurred. The results demonstrate that SO$_4^{2-}$ is the most important factor in determining stream concentration. With a 60 % reduction in current deposition, the soils recover very slowly and the process has barely begun in 2125. The changes, however, are only a fraction of the change caused by the SO$_4^{2-}$ reduction. Both [Al^{3+}] and [H$^+$] remain high during high flow events that suggests that conditions suitable for fish survival even at 2125 could probably not be maintained. The results at 90 % reduction of current deposition levels contain substantially less aluminium and fewer high-flow events with high H$^+$ concentrations (see also Christopher *et al.*, this symposium).

4. Conclusions

Computer acidification models provide a powerful tool that can help both researchers and policy decision makers to better understand both the intricacies of and how man interacts with nature. Assuming that the models are a valid representation of naturally occurring processes, the benefits derived from their usage are twofold. First, they lead to an increase in understanding of natural processes and how humans interact and alter these delicate balances (or both). Secondly, they can be used to predict how changes in anthropogenically generated pollutants can potentially alter future conditions. Armed with this information, policy decision makers and the general public could be made more aware of the consequences and costs of their decisions that would hopefully lead to more intelligent and informed policies.

The results hopefully add to the understanding of future conditions given present day changes. A failure to decrease anthropogenic SO_4^{2-} would not greatly alter those catchments that are currently strongly acidified. Other areas, however, that are not yet strongly acidified would change substantially. Existing plant and animal species would be severely stressed and, potentially, be eradicated. Reductions in SO_4^{2-} would have a marked effect upon such areas and would halt the acidification process before it reached severe limits.

The results from MAGIC and BIM simulations indicate that, for those areas that are currently highly acidic, only substantial reductions of man-produced pollutants will produce any pronounced change in streamwater concentrations. Reductions of 30 % of current values would not appear to halt the acidification process and a decrease to 40 % of current values produces insufficient improvement. Such reductions may, of course, have a beneficial effect in less-affected areas. Reductions of the order of 90 % are needed before water quality conditions suitable for fish survival are met. Survival may not, however, be guaranteed as episodic events producing high [H$^+$] may still occur. These results should be tempered with the realization that the soil-regeneration process could be hastened by such practices as liming that help to replace depleted base cations. Therefore a 60 % reduction in current deposition may be sufficient if soil depletion, originally caused by anthropogenic deposition, is reversed by human intervention.

In summary, the effects of acidification, once effected, are difficult to reverse. Reductions of man-made SO_4^{2-} pollutants are needed to prevent those areas undergoing acidification from completing the process. Those areas that are already severely damaged may require some additional action in combination with SO_4^{2-} reduction to regenerate soil quality.

References

Christophersen, N., Seip, H. M. & Wright, R. F. 1982 *Wat. Resour. Res.* **18**, 977–996.

Cosby, B. J., Wright, R. F., Hornberger, G. M. & Galloway, J. N. 1985 *Wat. Resour. Res.* **21**, 1591–1601.

Jenkins, A. & Cosby, B. J. 1989 In *Regional acidification models* (ed. J. Kämari, D. Brakke, A. Jenkins, S. Norton & R. Wright), pp. 253–266. Heidelberg: Springer-Verlag.

Reuss, J. O., Christophersen, N. & Seip, H. M. 1986 *Wasp*, **30**, 909–930.

Stone, A. & Seip, H. M. 1989 *Ambio*, **18**, 192–199.

Stone, A. & Seip, H. M. 1990 *Sci. tot. Envir.* **96**, 159–174.

Stone, A., Seip, H. M., Tuck, S., Jenkins, J., Ferrier, R. C. & Harriman, R. 1990 *Wasp*. (In the press.)

Modelling runoff water chemistry in an organic-rich catchment by using the Birkenes model

By Geir Taugbøl[1], Kevin H. Bishop[2] and Harald Grip[2]

[1] Chemistry Department, University of Oslo, P.B. 1033 Blindern, N-0315 Oslo 3,
Norway
[2] Department of Forest Site Research, Swedish University of Agricultural Sciences,
S-901 93 Umeå, Sweden

In many acid waters, increases in stream discharge are accompanied by large decreases in pH and increases in potentially toxic aluminium. The critical factors determining water quality, however, differ substantially. Knowledge concerning the mechanisms that determine natural acidity and its response to acid deposition may be increased through the use of computer-simulation models.

Svartberget and Birkenes are both calibrated catchments with an overall area of *ca.* 0.5 km². Birkenes has the largest input of sulphate, whereas Svartberget has much higher levels of dissolved organic carbon in runoff. pH-values in runoff are similar *ca.* 4.5. During episodes of peak discharge, the pH tends to decrease at both sites. The accompanying anions, however, seem to be mainly organic ions at Svartberget whereas sulphate and chloride dominate at Birkenes.

Svartberget provides a suitable data set to study the basic mechanisms of organic acidity of natural waters. The Birkenes model (BIM) was developed by using Birkenes data and simulates short-term variations quite successfully. Simulations by using BIM on Svartberget data were in good agreement with observed discharge. The modelling of the conservative tracer, ^{18}O, was fairly successful. Simulations for other chemical species are, as yet, unsatisfactory. A sub-model that includes organic acid–aluminium interactions seems necessary to improve the performance of the model.

---

1. Introduction

The Birkenes model (BIM) (Christophersen *et al.* 1982, Stone & Seip and Christophersen *et al.*, this symposium) is a computer-based simulation model used for examination of hydrochemical processes determining the water quality of small natural catchments. BIM has been developed by using the data available from the Birkenes catchment in southernmost Norway. It has been calibrated with fair success on other catchments as well (Rustad *et al.* 1986; Lam *et al.* 1989; Stone & Seip, this symposium), indicating a general validity of the model.

An extended period of reliable monitoring data on hydrological, climatic and chemical parameters is required to apply BIM. SWAP has yielded some additional sites that may give valuable knowledge on certain aspects by using BIM as a means for interpretation of data and for testing theories (Seip *et al.*, this symposium).

Svartberget is a catchment particularly rich in organic content of runoff water. The geographical location far to the north results in low levels of acidic depositions compared with Birkenes, though the pH of runoff is in the same range, down to 4, during peak flow events. The mobility of SO_4^{2-} is known to be the primary factor governing the water chemistry in many catchments receiving high depositions of acidic input. Hence the sulphate dynamics are a main part of BIM. At Svartberget the low pH levels are not correlated to rising sulphate levels and other mechanisms should be considered.

Many authors (Brakke *et al.* 1987; Marmorek *et al.* 1987; Tipping *et al.* 1988) have pointed to the role played by natural organic acids determining the acidity of runoff water, though this role has often been overlooked. Bishop & Grip (this symposium) and Grip & Bishop (this symposium) show that leakage of organic compounds is the main source of acidity during storm events at Svartberget. BIM in its current version is, therefore, not expected to satisfactorily simulate the chemical processes when acid deposition impacts a naturally acidic catchment like Svartberget. A sub-model including dynamics of organics should be developed and calibrated on the Svartberget data to give additional information about possible mechanisms. Rustad *et al.* (1986) have worked out a simple sub-model, but a serious validation was never done.

The first step, reported here, is to verify the general hydrological structure of BIM, calibrating the hydrological parameters on Svartberget data. Chemical simulations are not meaningful until the hydrology is well characterized.

2. Data base

Birkenes is a catchment that has been continuously monitored since 1972. Extensive modelling work has been done and model structure is described in several papers (Christophersen *et al.* 1982; Hooper *et al.* 1988; Stone *et al.*, this symposium). The Svartberget catchment has been monitored since 1981, but chemical analyses were not included until 1985. The two catchments are comparable in many aspects: i.e. size *ca.* 50 ha†, forested with mature Scots pine in upper regions and Norway spruce in lower; glacial till covering gneissic-granitic bedrocks (though Svartberget has a much thicker till); podzolic soils; drainage of a peat bog area (substantially larger at Svartberget). Climatic data reflect the more northern and inland position of Svartberget, rainfall and runoff are 720/390 mm a^{-1} and 1450/1100 mm a^{-1}, at Svartberget and Birkenes, respectively. Svartberget normally has a snow season with temperatures consistently below zero from November through to late April, whereas Birkenes experiences milder winters with variable snow cover and several melting episodes. Svartberget experiences considerably less impact from long range transported air pollutants; sulphur wet deposition amounts to 0.5 g m^{-2} a^{-1} compared with 1.7 g m^{-2} a^{-1} at Birkenes. The contribution from sea salts is also more important at Birkenes.

Hydrological data are monitored daily at calibrated V-notch weirs located at the stream outlets. Meteorological data are taken daily from nearby stations. Chemical input and output data are normally provided from weekly samples, though daily analyses have been done during some periods, e.g. snow melt and storm events. The data series are much longer at Birkenes, giving fair periods for model calibration and validation. At Svartberget a period of less than 2.5 years is available for model simulations, which is, at most, only sufficient for calibration. The monitoring programme continues, however. This will yield data for further studies.

† 1 ha = 10^4 m^2.

3. Results and discussion

The BIM has been developed by using Birkenes data to provide a simulation of hydrology and chemical stream concentrations, which is in good agreement with observed values. The work reported here is an adaption of BIM to the Svartberget data using a parameter optimization procedure built into the computer program (Hooper *et al.* 1988). To date the hydrological parameters have been adjusted by using precipitation and flow amounts and observed values for the conservative tracer, $H_2^{18}O$. The best fit was obtained for snow-free seasons only (June–October).

The size of the B-box (lower reservoir, representing deeper soil horizons) is increased about three times at Svartberget compared with Birkenes, a feature that is consistent with the much deeper soils at Svartberget. The minimum amount of water in B-box is reduced six times from Birkenes to Svartberget. It is likely that this results from the much larger ^{18}O-fluctuations at Svartberget.

The database presently lacks several ^{18}O-observations in precipitation that have yet to be analysed. The missing data are corrected for by inserting calculated average values for the summer periods (delta SMOW $= -11.5$) of May–October, and snow accumulation periods (-16.9), whenever observations are missing.

A good agreement is obtained for simulations of flow after optimization of the hydrological parameters (figure 1 *a*). The simulations agree well with the observed flow peaks. A tendency to over-estimate flow during low flow situations may be caused by a possible increased storage of groundwater and flux (or both) through the deep till beneath the stream weir. However, the basic hydrological performance of BIM also seems satisfactory at Svartberget.

The ^{18}O-simulations by using optimized parameters are shown in figure 1 *b*. The main pattern of ^{18}O-variations is reflected, indicating a hydrological structure capable of simulating water residence time in the catchment. The ^{18}O-simulations might suffer, however, from the limited number of input data (precipitation ^{18}O) that restrict the ability to simulate variations during certain periods, e.g. the plateau levels in figure 1 *b*. Increased amounts of stream ^{18}O-data will probably also improve the model fit.

The observed variations in the concentrations of main chemical components such as sulphate, pH and TOC in runoff are shown in figure 1 *c*. It is apparent from the graphs that a model based on sulphate as the critical factor is not sufficient to explain the major variations in H^+-concentrations at Svartberget. A positive correlation between TOC and H^+ is evident, although sparse TOC-observations weaken any conclusions that can be made. Also the Al-dynamics seem to be governed mainly by organic components, giving low levels of inorganic Al^{3+} (Bishop & Grip 1990).

The acidity of runoff from Svartberget and Birkenes is in the same range, varying from a pH about 4.0 at peak flows to about 5.5 at Birkenes and above 6 at Svartberget during low flows, with a weighted average of 4.5. The levels of total Al are comparable, ranging from about 1 to above 35 μM l^{-1}, although Birkenes has the highest peaks and a somewhat higher average. The same applies to sulphate with an average of 110 and 135 μeq l^{-1}, respectively. Organic content is considerably higher at Svartberget reaching more than 30 mg l^{-1} during peak flows, whereas Birkenes values rarely exceed 10 mg l^{-1}.

The similarities between Birkenes and Svartberget catchments allow for comparative studies of the mechanisms determining water chemistry, focusing on strong mineral acids at Birkenes and natural organic acids at Svartberget. Although

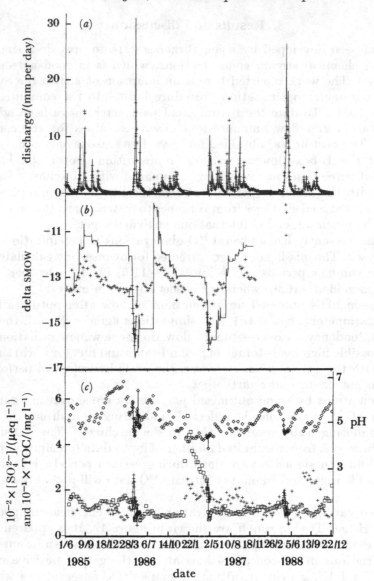

Figure 1. Svartberget catchment, 1985–88: (*a*) simulated (——) and observed (+) discharge; (*b*) simulated (——) and observed (+) stream ^{18}O; (*c*) observed stream pH (◇), SO_4^{2-} (□) and total organic carbon (TOC) (+).

organic acids seem to be correlated to acid episodes, the impact from deposited strong acids most likely influences the acidity even at Svartberget, i.e. it might be incorrect to conclude that the acidity here is all natural (Bishop *et al.* 1990). The pH-range being considered is in the buffer region of many humic and fulvic acids. Hence, the high organic content of Svartberget is likely to modify the effects of additional strong acids.

Clearly, the dynamics of Al are very much governed by organic components. At Birkenes the inorganic species dominate, whereas Svartberget has a maximum of 15 % inorganic Al. As the Al speciation is very important when considering biological effects, the interaction between Al and organic acids should be included in the model. As the flowpaths, other physiochemical variables and even biological factors may be of importance to organic matter behaviour, these factors should be considered in future model development.

4. Conclusions

1. The BIM has a hydrological sub-model that is adaptable to the Svartberget catchment, supporting its general applicability and making it useful for the study of water chemistry at Svartberget.

2. The chemical modelling of BIM is based on the dynamics of sulphate and is successful in simulating the composition of runoff from catchments dominated by strong mineral acids.

3. The current version of BIM is not able to simulate water chemistry in a catchment rich in organic acids, e.g. Svartberget.

4. To interpret data and to possibly verify the hypothesis on the key role played by organic acids in natural waters, BIM should be developed further to include the dynamics of organics, e.g. K_a-values, complex formation, etc.

5. The Svartberget catchment provides a data base suitable for such studies. Comparisons with other calibrated catchments differing in acid loadings and natural organic acid content should be undertaken to reveal interaction between organic acids and deposited mineral acids.

6. Knowledge about the development of acidification, including a possible recovery, should benefit from modelling work. Organic acid dynamics seem to be a key point to improve in model performance.

The valuable advices and comments given by Professor Hans Martin Seip and Alex Stone, Department of Chemistry, University of Oslo, are very much appreciated.

References

Bishop, K. H. & Grip, H. 1990 *Swedish environmental protection board*, report. (In the press.)

Bishop, K. H., Grip, H. & O'Neill, A. 1990 *J. Hydrol.* (In the press.)

Brakke, D. F., Henriksen, A. & Norton, S. A. 1987 *Nature, Lond.* **329**, 432–434.

Christophersen, N., Seip, H. M. & Wright, R. F. 1982 *Wat. Resour. Res.* **18**, 977–996.

Driscoll, C. T., Fuller, R. D. & Schecher, W. D. 1989 *Wat. Air Soil Pollut.* **43**, 21–40.

Hooper, R. P., Stone, A., Christophersen, N., de Grobois, E. & Seip, H. M. 1988 *Wat. Resour. Res.* **24**, 1308–1316.

Lam, D. C. L., Bobba, A. G., Bourbonniere, R. A., Howell, G. & Thompson, M. E. 1989 *Wat. Air Soil Pollut.* **46**, 277–287.

Marmorek, D., Bernard, D. P., Jones, M. L., Rattie, L. P. & Sullivan, T. J. 1987 *ESSA-report.* Vancouver, British Columbia: Environmental and Social Systems Analysts Ltd.

Rustad, S., Christophersen, N., Seip, H. M. & Dillon, P. J. 1986 *Can. J. Fish. aquat. Sci.* **43**, 625–633.

Tipping, E., Backes, C. A. & Hurley, M. A. 1988 *Wat. Res.* **22**, 321–326.

General discussion

KEVIN H. BISHOP (*Swedish Faculty of Forestry, Department of Forest Site Research, S-901 83 Umeå, Sweden*). I had intended to raise a question about soil-base saturation in the application of the MAGIC mode to the Birkenes stream. But so many serious questions have already been raised about the models presented this afternoon that I don't think the modellers need to field yet another criticism. Instead it is perhaps time to ask if the task given to the modellers by SWAP was a realistic one. According to the summary prepared by Sir John Mason and Professor Hans Martin Seip, the main objective of SWAP was to make better predictions of future trends. That goal should be realized by models, but as we have seen today, it is extremely difficult to identify and validate models of surface water acidification. The criticisms of the models should not be taken as an attack on the skills of the modellers who have responded creatively and thoughtfully to the challenges presented by the data. It would be better to think of the cogency and power of the questions raised about the modelling presentations today as a sign of the heightened awareness of the need to validate models and the possibility for searching dialogue within the research community. Both are vital to progress and both have been facilitated by SWAP. But we should not lose sight of the fact that SWAP was launched in 1984 on the assumption that five years of research would in fact lead to models reliable enough to provide an objective basis for environmental policy decisions. It must be considered fortunate that no government waited for SWAP's models and predictions before making difficult policy decisions concerning acidification. If any government had waited, it would have done so in vain. Furthermore the initial fears of many Scandinavians that SWAP would lead to unwarranted delays in the decision-making process would have been proven telling prescient. Hopefully the scientific community will have learned from the humbling experience of trying to model acidification and plans for future environmental research will be informed by a more realistic expectation of what we can hope to reliably predict about the health of complex ecosystems.

Main conclusions of the SWAP research programme

FORMULATED BY THE MANAGEMENT GROUP

1. Acidified lakes and streams without, or with impoverished fish populations, occur mainly in areas that receive high levels of acid deposition from the atmosphere and have soils derived from granite or other rocks of similar composition that are resistant to weathering and low in exchangeable elements such as calcium and magnesium. Catchments with thin soils are particularly sensitive with respect to the rate and extent of acidification.

2. Examination of the remains of diatoms and other biological material in lake sediments laid down over centuries has established that many lakes in southern Norway and Sweden and in the U.K. have undergone progressive acidification from *ca.* 1850 until very recently. The magnitude of this acidification is appreciably greater than any that has occurred in the past 10000 years and has marched in parallel with accelerated industrial development, as indicated by increases in several trace pollutants in the sediments. These changes and the extent of inferred acidification are geographically correlated with the intensity of acid deposition and with the geo-chemical status of the catchment.

3. For a given input of acid deposition, the degree of acidification of lakes and streams is largely determined by the structure and chemistry of the mineral and organic soils, and the pathways that the incoming rainwater takes through the soil. These factors determine both the nature and duration of the many chemical and biological reactions that influence the final quality of the water that emerges in the streams.

4. The evidence points convincingly to atmospheric deposition, largely of acidifying compounds of sulphur and, to a lesser extent, of nitrogen as the main cause of acidification. However, forests may enhance acidification by acting as efficient filters and collectors of acid from the atmosphere in polluted areas, and by taking up metal cations, and the acidification of some lakes may be attributed to changes in land use or agricultural practice.

5. There is evidence that, in the past decade, there has been a significant decrease in the acidity of rain and snow as a result of reduced emissions of sulphur dioxide, and that this is reflected in a small decline in the acidity and sulphate content of some lakes. However, there are signs, especially in Norwegian lakes, that the effects of reduced concentrations of sulphate are being partially offset by increases in nitrate.

6. Fish populations, especially of salmon and trout, cannot survive in lakes and streams if the pH of the water remains below a critical level of about pH 5 for long (depending on the species and age of the fish and the chemical composition of the water). The fish are killed by the action of increased acidity and of inorganic forms of aluminium leached out of the soil by the acidified water. The effects of aluminium are ameliorated by the presence of organic acids (e.g. from peat) that complex the aluminium and render it less toxic to fish, and possibly if calcium is present in sufficiently high concentration. However, in regression analyses based on a survey of

over 1000 lakes in southern Norway in which 14 variables in the regressions were studied, most of the variance in fishery status could be accounted for by pH, inorganic aluminium and altitude.

7. Fish are very vulnerable to the short, sharp episodes of high acidity and aluminium that occur in streams following heavy rains or snow melt. In these episodes much of the water flows through acid soils where it is enriched in available aluminium but spends relatively little or no time in the deeper layers where it would be neutralized.

8. From carefully controlled laboratory experiments and intensive field studies, it is now possible to relate fish survival to the concentrations of acid, aluminium and calcium in the water and to estimate the likely toxic effects of acidic episodes of differing severity, frequency and duration. The detailed mechanisms of fish death are complex. Some forms of aluminium and H^+ ions inhibit sodium uptake and increase sodium loss, so reducing body sodium content leading, eventually, to circulatory failure. The deleterious effects of inorganic aluminium can be largely counteracted if calcium is present in sufficient concentrations.

9. Acidification and release of aluminium also leads to changes in the populations of micro-organisms, lower plants and aquatic invertebrates. The effects of such changes in the ecosystem can include the availability of food for some life stages of brown trout and other fish.

10. The possibilities of the recovery of streams and lakes depend on the long-term balance between the catchment input and output of cations, such as Ca^{2+} and Mg^{2+} that are exchangeable for H^+ ions. The main input of these cations in the affected parts of Scandinavia and the UK is normally by chemical weathering; the supply by atmospheric deposition is much less than that of acidifying substances. The direct cause of acidification of lakes and streams is the excess anions of strong acids, as sulphate reduction and denitrification play only a minor role in most of the affected ecosystems. The long-term resistance of a catchment area is therefore closely related to the release of cations such as Ca^{2+}, Mg^{2+}, Na^+ and K^+. In regions covered by the last glaciation, these cations are produced mainly by weathering of minerals. Considerable progress has been made within SWAP in the determination of weathering rates as a function of mineral species, particle size, pH, production of organic ligands in the ecosystem and the history of the soil. Estimates by different methods agree in most cases within a factor of two or three. It appears that even a reduction by 60% of acid deposition would not be enough to create steady-state conditions suitable for fish in those areas that are most strongly acidified.

11. The rate at which streams and lakes will recover in response to reduced emission and deposition of acidic substances will also depend on such factors as the residence time of water in groundwater and lakes, on the release of sulphate from earlier deposition, which is retained in the soil, and on changes in the land use within the catchment. In thin soils with little storage of sulphur compounds, recovery may be quite rapid. In deeper soils containing large accumulated stores of sulphur compounds, it may take several years or even decades for this to be leached out and recovery may be much slower. Recovery or restoration may be aided by liming the catchment, but this may have undesirable effects such as increased nitrification.

12. There is evidence of increased nitrate deposition but this has been only partly reflected by the increase in its concentration in surface waters, mainly because of uptake by vegetation. As the system has limited storage capacity, an additional burden of acidification could develop over years.

General discussion

THE LORD LEWIS (*Robinson College, Cambridge, U.K.*). Mr Chairman, Director, Ladies and Gentlemen. May I say how impressed I have been listening over the past few days to the developments in the SWAP programme. I attend this conference as a person who has had no direct contact with the programme but having an interest in the results. It would be perhaps presumptuous on my part to try and judge the outcome of the work, but I would like to take this opportunity of indicating those parts that have been of direct interest and concern to me, and so perhaps give a very personal view of the programme.

Clearly the recognition by the initiating committee that there would be a strongly integrated field programme necessarily, involving the interaction of different disciplines, institutions and research groups across the member countries, was of paramount importance to the success of the programme. The further recognition that field experiments would need to be supplemented by laboratory studies of key processes, namely studies on chemical weathering of rocks and soils, speciation and toxicity of aluminium salts, the role of organic acids in surface water chemistry and the physiological effects on fish and water quality has been a significant success. The determination of the previous history of the pH profile of lake sediments from the analysis of the distribution of acid-sensitive species of diatoms was a piece of scientific detection of the highest order.

An important aspect, to a chemist, of the work discussed is the excellent survey carried out to ensure compatibility and direct comparability between the different measurements. The workshop on standardization procedures and intercalibration of instrumentation with uniform methods of data analyses is clearly essential to ensure the reliability of data sets and to recognize their limitation in the computer modelling techniques that have been developed. The old problem – basic to analytical procedures – of sampling, has clearly been of major concern. The basis of any projections in this, as in other fields of science, depends heavily upon the knowledge of the reliability of the data set from which the correlations are derived.

As with other programmes involving ecological studies, it has been observed that the significance of certain problems that had not been recognized or suspected at the beginning of the study can become of paramount importance. The Great Lakes Studies in the U.S.A. and Canada illustrated dramatically this effect in environment studies. What had initially been a recognition of a problem and study in the eutrophication of the lakes disclosed major other pollution problems involving contamination with P.C.B., pesticides and heavy metals, with a surprisingly divergent series of sources of the pollution. The appreciation of airborne pollutions and the vast areas that it can cover was a major surprise, and emphasized the potential global nature of the pollution problems. The present study once again places great emphasis on this problem.

The large distances that may be covered by airborne pollution do however complicate the reliability of detection of sources. A study being done at the University of East Anglia by Professor Liss has indicated that the presence of sulphur dioxide at specific times of the year may be associated with the production

[511]

of dimethyl sulphide from the sea via the action of microorganisms on seaweeds. Clearly this had been a potential source for years, but it has recently been claimed to produce possibly as much as 40 % of the sulphur dioxide detected in some of the southern lakes in Norway during some of the summer months. An interesting question to me as a chemist is the chemistry involved in the transformation of the dimethyl sulphide to sulphuric acid/sulphur dioxide; ozone has been attributed as one of the active components of this process, and the production of this can be associated with recent pollution sources such as the motor car.

As a chemist I have also been particularly interested in the role played by various metal ions in the biological processes described during this conference. Particular mention has been made of aluminium. The inorganic chemistry of this element is complicated; the role of organic acids, particularly the weak acids leached from many soils in complexing with aluminium, is well recognized, and this may virtually exclude the metal from any biological system that depends upon it being present as the hydration ion. In contrasting iron and aluminium, both of which exist in the trivalent state and the structural chemistries of which are very similar, it is perhaps important to recognize the great difference in the rate at which the two sets of species react. Aluminium complexes, in contrast to iron, undergo reactions very slowly, and in any biological processes rate can be a very important controlling factor. Clearly any comparisons involving iron and aluminium must recognize this difference. Both metals in high pH will also readily condense to produce polymeric aggregates, with liberation of protons and formation of hydroxy or oxo-bridges. This involves a complex equilibria in solution with variation in pH. As the rate of attainment of these polymerized or depolymerized states will clearly be different for aluminium and iron, it is possible that some effects may be kinetically rather than thermodynamically controlled. As condensation reactions are generally entropically favoured, temperature effects may also be of considerable importance.

Similarly, the comparison between calcium and aluminium in biological systems is not necessarily straightforward. Calcium shows very much less tendency for the formation of polymeric species. As a divalent ion, in contrast to the solid state, the hydrate calcium ion is smaller in size than the hydrated aluminium ion, reflecting the smaller ionic charge. The question of calcium–aluminium dependence in biological systems that has been discussed in the conference must also reflect this. In general, metals of higher charge are more effective as coordinating species to ligand groups. However, in biological systems the presence of a site for metal coordination may impose rather strict stereochemical requirements on the metal ion.

Al^{3+} has a much more limited range of stereochemical arrangements than Ca^{2+}; Al^{3+} is generally found to coordinate with octahedral, tripyramidal or tetrahedral stereochemistry, and with small variations from these stereochemistries. However, Ca^{2+} shows a vast range of both coordination environment and coordination number, reflecting the more ionic nature of the bonding. This allows the calcium ion, although potentially less thermodynamically favourable, to be able to adapt more readily to any stereochemical requirement of the active site, and the greater potential acceptor capacity of Al^{3+} may be countered by the higher coordination flexibility of the Ca^{2+} species. When we consider the vast range of elements that appears in soils, it is necessary to consider alternative processes involving other metals, as indicated by the contribution of Dr Exley on the potential implication of silicon in the chemistry of these systems.

Clearly the chemical problems that have been posed by these studies are a natural outcome of any successful study. The overall impression I have is of a well-integrated programme between the component countries that has isolated many of the initial difficulties, solved many of the problems and detected others.

I hope that this is a study that will continue under its own momentum; much that has been achieved must be continued, particularly the accumulation of assessed data for developing, testing, and the study of the biological and chemical problems that still remain. In any environmental studies it must be accepted that today's solutions may provide us with tomorrow's problems. A clear continuous assessment of the whole area covered by this conference must be carried forward, and the organizers are to be congratulated on the success to date.

I. G. LITTLEWOOD (*Institute of Hydrology, Wallingford, Oxfordshire, OX10 8BB, U.K.*). I should like to question the significance of 'piston flow' as a streamflow generation mechanism during events. According to several researchers, peak streamflows comprise mostly 'old' water 'forced out under pressure from the influx of new water'. These ideas have been employed during presentations and discussions at this symposium and receive a measure of credibility by appearing in the discussion paper. In response to a question put earlier in the week by Dr Hornung ('How old is old?') it should be noted that there is no ambiguity of definition in many published accounts of isotope hydrograph separation that claim that peak streamflows comprise mostly (more than 70%) old water; old water is liquid-phase water in (or on) the catchment before the rainfall or snowmelt (new water) that causes the runoff event to occur. If the isotope hydrograph separations are correct a displacement mechanism of streamflow generation during events must be dominant at the catchment scale.

The idea of piston flow, or displacement of pre-event water, is not new and undoubtedly the process does occur within catchments as a small-scale phenomenon. However, the variation of the mixing ratio (old:new in this case) for a point on a stream is the result of a synthesis of all hydrological processes in the catchment above that point (not just piston flow). The weight of evidence from my investigations of the hydrological and hydrochemical responses of several streams in the Llyn Brianne catchment, Wales strongly supports streamflow generation during events by predominantly non-displacement mechanisms. Much of the literature, and many practising hydrologists, refer to 'runoff' events, not 'displacement' events. May I suggest, therefore, that caution be exercised in the use of the piston flow concept (and in the use of related terms and phrases) in mechanistic descriptions and models of stream-chemistry dynamics for acidification studies.

KEVIN H. BISHOP (*Swedish Faculty of Forestry, Department of Forest Site Research, S-901 83 Umeå, Sweden*). I would like to respond to the uneasiness expressed by Dr Ian Littlewood about the implications being drawn from isotope hydrograph separation (IHS) that a large proportion of the water in a stream during an acid episode is pre-event water displaced from the hillslope. In the two decades during which the IHS has been employed, scientists have been puzzling over the large component of pre-event water that IHS keeps finding in almost every headwater stream investigated. Because those results challenged conventional wisdom about runoff generation, IHS has been exposed to considerable critical scrutiny. However, numerous investigations have been unable to disprove the general conclusion from IHS that much of the stormflow in headwater streams is pre-event water. Certainly the scientific community must continue to challenge IHS and is implications, but too much published evidence has been amassed to simply dismiss the provocative implications of IHS because of a vague sense that IHS gives results that are somehow counter-intuitive. I hope that my work, including the paper presented in session II, is making some progress towards explaining both how a large component of pre-event water contributes to stormflow and why that pre-event water can become so acid in the course of an acid episode.

KEVIN H. BISHOP. In his summary of the hydrological field studies presented in session II, Professor Walløe noted that distinctly different flow pathways had been identified at Birkenes, Allt a'Mharcaidh and Svartberget. He went on to conclude that the hydrology of the three sites was so varied that it would be difficult to make useful generalizations about hydrological processes that can help predict the response of catchments to acid deposition. Although Professor Walløe's conclusions were quite reasonable ones on the basis of the work presented by the hydrologists, myself included, there is a common thread running through all of the hydrological investigations. Clearly though, not enough emphasis was made in the presentations to draw attention to the common thread: the crucial role played by preferred flow pathways. After having consulted with researchers working at each of the field sites over the tea break, I would like to try briefly to redress the oversight by offering a few comments on the importance of preferred flow pathways in the generation of stormflow and the potential significance of preferred flow pathways for the rate at which catchments respond to changes in acid deposition.

At each field site, the bulk of stormflow was found to move in localized regions of high saturated hydraulic conductivity. At Birkenes and Svartberget these preferred flow pathways lay at the top of the soil profile and were activated by the rise in the water table in response to rainfall or snowmelt. At Allt a'Mharcaidh, the arrangement of the preferred flow pathways was far more complex. But even there, the tremendous increase in flux through the hillslope associated with acid episodes was concentrated in highly localized flow pathways that bypassed much of the soil matrix.

A major chemical implication of these preferred flow pathways is that when considering acid episodes, it would be incorrect to model the catchment as a well mixed tank reactor. The chemistry of the episodes is determined by the chemistry of the preferred flow pathways. The rapid flux of precipitation through those preferred flow pathways makes it likely that changes in precipitation chemistry would appear first along those preferred flow pathways and hence in the chemistry of episodes. This conference has done much to emphasize how critical the frequency and duration of episodes are to the ecosystem. Thus the importance of preferred flow pathways in acid episodes might even suggest that the aquatic ecosystem is more sensitive to changes in acid deposition than would be predicted by spatially lumped acidification models. Of course that hypothesis remains speculative until the nature of the chemical interaction between preferred flow pathways and the rest of the soil is elucidated. However, the rapid response to changes in acid deposition observed in the RAIN project, the Lake Gårdsjön catchment manipulations and the Scottish sulphate budgets reported at this conference all differ from the pessimistic response to changes in sulphate deposition predicted by many of the lumped models. Apparently something about those lumped models is wrong, and the treatment of hydrology that does not take account of preferred flow pathways may well be a part of that.

C. O. Tamm (*Stavgårdsgatan 11, S-161 37 Broma, Sweden*). The emphasis within SWAP has been given to two themes within the very broad concept of chemical processes in the soil affecting freshwater quality, viz. mineral weathering as a counteracting force to acid deposition, and aluminium chemistry as important for water quality and water biota. Considerable progress have been made on these two themes within SWAP but also thanks to investigations outside our programme.

The session compared favourably to most earlier discussions of weathering (i) by making the necessary distinctions between important concepts such as current against historical weathering, and weathering of primary minerals as distinguished from ion-exchange processes and formation/dissolution of secondary minerals and amorphous substances; (ii) by attempts to use more than one method on the same site. Several methods have been used, some of them recently invented. One such new method used zirconium as internal standard, another one used the differences in strontium isotope ratio in old bed-rock and in seawater to distinguish between strontium (and indirectly calcium) derived from bed-rock from cyclic strontium in runoff. In addition, catchment balances with silicium considered as a conservative element have been used, as well as mineralogical methods, improved with i.a. electronic scanning microscopy. The potential and the limitation of each method were well recognized.

The results presented show that we now have a much better, although not complete, understanding of rate-determining parameters such as texture, mineral species and organic ligands. The influence of time was particularly well demonstrated by a study of a Scottish chronosequence of soils. There was a general agreement on the rate of weathering, with differences between estimates with different methods often differing by less than a factor of two. There were good indications that the rate of the actual weathering is higher than that of the historical records in areas with high acid deposition.

Aluminium chemistry must still be considered as a difficult area, but thanks to the SWAP studies we have now methods intercalibrated between laboratories that make it possible to distinguish in a reproducible way aluminium fractions with widely different reactivities and toxicities. Better understanding of the rates of transfer between various aluminium fractions in lake and river water, as well as in soil and soil water, and the factors that control aluminium concentrations and speciation is still to be achieved.

The next two big challenges in the study of chemical processes affected by acidification of soils and waters, will be the consequences of increased nitrogen deposition and the mobilization of heavy metals from increasing acidity of soils, in addition to direct supply from a polluted atmosphere.

Author index

Andersen, D. O. *See* Vogt *et al.*
Andersen, R. *See* Bravington *et al.*
Andersen, S., Christophersen, N., Mulder, J., Seip, H. M. & Vogt, R. D. Aluminium solubility in the various soil horizons in an acidified catchment, 155; *see also* Vogt *et al.*
Anderson, H. A. *See* Ferrier *et al.*; *see also* Neal *et al.*; *see also* Walker *et al.*

Bache, B. W. *See* Townsend *et al.*
Bain, D. C., Mellor, A., Wilson, M. J. & Duthie, D. M. L. Weathering in Scottish and Norwegian catchments, 223.
Battarbee, R. W. *See* Renberg & Battarbee.
Beck, M. B., Kleissen, F. M. & Wheater, H. S. Identification of hydrological processes of surface water acidification, 477; *see also* Wheater *et al.*
Beddington, J. R. *See* Bravington *et al.*
Birks, H. J. B., Juggins, S. & Line, J. M. Lake surface-water chemistry reconstructions from palaeolimnological data, 301.
Bishop, K. H. *See* Grip & Bishop.
Bishop, K. H., Grip, H. & Piggott, E. H. Spate-specific flow pathways in an episodically acid stream, 107; *see also* Taugbøl *et al.*; *see also* Townsend *et al.*
Bjørnstad, H. E. *See* Salbu *et al.*
Blakar, I. A. *See* Seip *et al.*
Blakar, Ingaard A., Digernes, I. & Seip, H. M. Precipitation and streamwater chemistry at an alpine catchment in central Norway, 69.
Bravington, M. V., Rosenberg, A. A., Andersen, R., Muniz, I. P. & Beddington, J. R. Modelling and quantitative analysis of the impact of water quality on the dynamics of fish populations, 467.
Broberg, O. Elemental sulphur and sodium sulphate treatment of catchments in the Gårdsjön area, southwest Sweden. Effects on phosphorus, nitrogen and DOC, 193.

Christophersen, N. *See* Andersen *et al.*
Christophersen, N., Hauhs, M., Mulder, J., Seip, H. M. & Vogt, R. D. Hydrogeochemical processes in the Birkenes catchment, 97; *see also* Neal *et al.*
Christophersen, N., Neal, C., Seip, H. M. & Stone, A. Hydrochemical models for simulation of present and future short-term changes in stream chemistry: development and status, 445; *see also* Seip *et al.*; *see also* Stone-*et al.*; *see also* Vogt *et al.*
Cosby, B. J. *See* Whitehead *et al.*

Dalziel, T. R. K. *See* Morris *et al.*
Digernes, I. *See* Blakar *et al.*
Duthie, D. M. L. *See* Bain *et al.*

Eddy, F. B. *See* Potts *et al.*

Ferrier, R. C. & Harriman, R. Pristine, transitional and acidified studies in Scotland, 9; *see also* Harriman *et al.*
Ferrier, R. C., Miller, J. D., Walker, T. A. B. & Anderson, H. A. Hydrochemical changes associated with vegetation and soils, 57; *see also* Neal *et al.*; *see also* Stone *et al.*; *see also* Walker *et al.*; *see also* Wheater *et al.*
Frick, K. G. & Herrmann, J. Aluminium and pH effects on sodium-ion regulation in mayflies, 409.

Gillespie, E. *See* Harriman *et al.*
Grip, H. & Bishop, K. H. Chemical dynamics of an acid stream rich in dissolved organics, 75; *see also* Bishop *et al.*; *see also* Seip *et al.*; *see also* Taugbøl *et al.*

Harriman, R. *See* Ferrier & Harriman.

Subject index